Two-dimensional Two-product Cubic Systems, Vol. I

Albert C. J. Luo

Two-dimensional Two-product Cubic Systems, Vol. I

Different Product Structure Vector Fields

Springer

Albert C. J. Luo
Dept of Mech & Ind Eng, RmEB2064
School of Engineering
Edwardsville, IL, USA

ISBN 978-3-031-48486-5 ISBN 978-3-031-48487-2 (eBook)
https://doi.org/10.1007/978-3-031-48487-2

This Springer imprint is published by the registered company Springer Nature Switzerland AG
The registered company address is: Gewerbestrasse 11, 6330 Cham, Switzerland

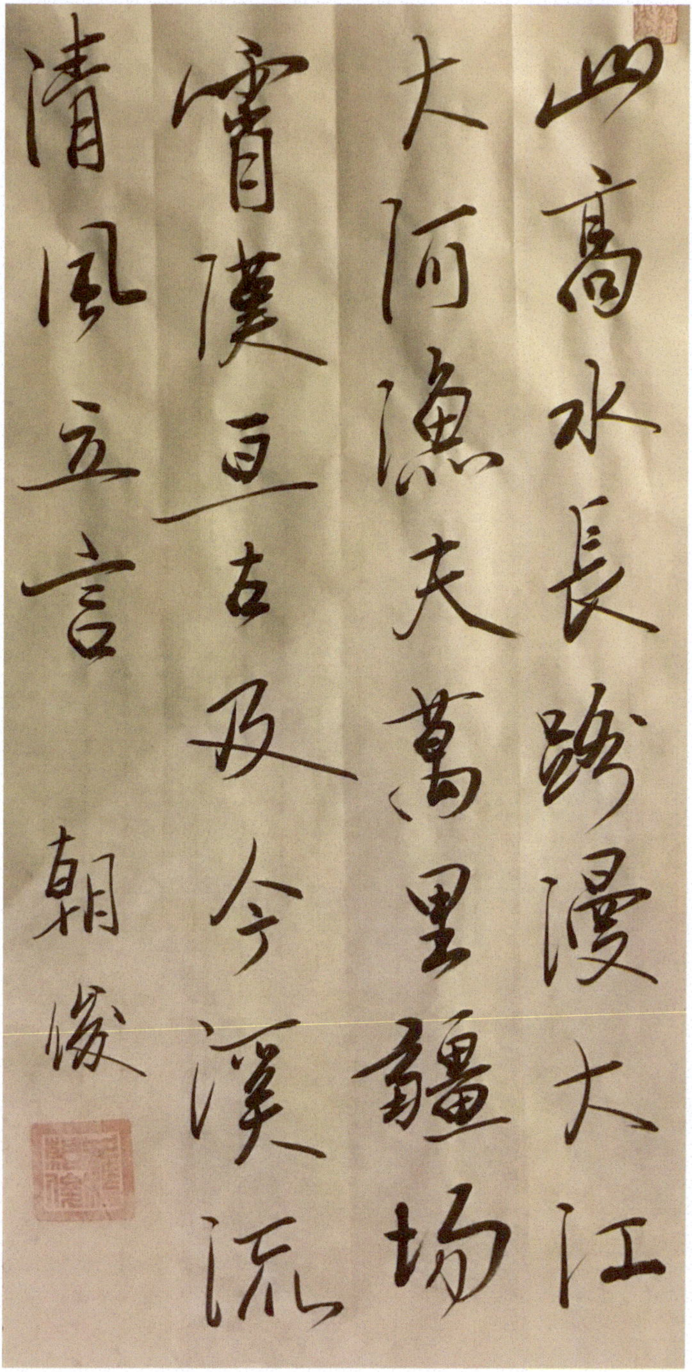

山高水长路漫，大江大河渔夫，万里疆场霄汉，亘古及今，溪流清风立言。

—朝俊【唤新年】辛丑

Preface

In this book, nonlinear dynamics and singularity of cubic dynamical systems possessing two cubic vector fields of different product structures (*i.e., self-linear and crossing-quadratic product and crossing-linear and self-quadratic product*) are presented. The equilibrium and flow singularity and bifurcations are discussed through the appearing and switching bifurcations. The appearing bifurcations are parabola-saddles and saddle-source (sink) equilibrium singularity with hyperbolic-to-hyperbolic-secant flow and inflection-source (sink) flow singularities. Hybrid networks of higher-order equilibriums and flows with simple equilibriums and hyperbolic flows are obtained. The corresponding switching bifurcations for saddle-source (sink) with hyperbolic-to-hyperbolic-secant flow, and parabola-saddles with inflection-source (sink) flows are based on parabola-source (sink), parabola-saddles, and inflection-saddles infinite-equilibriums. Hybrid networks of simple equilibriums and hyperbolic flows are presented, and the corresponding switching bifurcations given by parabola-saddles and inflection-source (sink) on the infinite-equilibriums.

In Chap. 1, nonlinear dynamics and singularity of cubic dynamical systems possessing two cubic vector fields of different product structures are discussed. The equilibrium singularity and bifurcations are presented through the appearing and switching bifurcations. Parabola-saddles, saddle-source (sink) singularities with the hyperbolic-to-hyperbolic-secant flow and inflection-source (sink) singularities in the two different product cubic systems are discussed, and the corresponding bifurcation dynamics are discussed. The corresponding switching bifurcations are discussed. Such nonlinear dynamics and singularity properties are presented through a theorem.

In Chap. 2, parabola-saddles, saddle-sources (sinks) singularities with the hyperbolic-to hyperbolic-secant flow and inflection-source (sink) singularities in the two different product cubic systems are discussed, and the corresponding bifurcation dynamics are discussed. The switching bifurcations of the parabola-saddle with hyperbolic-to-hyperbolic-secant flows and the saddle-source (sink) with inflection-source (sink) flows are presented, and the switching bifurcations of the saddle-source (sink) with hyperbolic-to-hyperbolic-secant flow and the parabola-saddle with inflection-source (sink) flow are also discussed.

In Chap. 3, inflection-source (sink) flows and parabola-saddles with hyperbolic flows and simple equilibriums are discussed, and the corresponding networks are presented for inflection flows and parabola-saddles with simple flow and equilibriums. The corresponding switching bifurcations are discussed, which include parabola-source (sink) infinite-equilibriums, inflection-source (sink) infinite-equilibriums, and two infinite-equilibriums of parabola-source (sink) and inflection-source (sink).

In Chap. 4, saddle-source (sink) flows and hyperbolic flow singularity with hyperbolic flows, positive or negative saddles, and center are discussed, and the corresponding networks are presented for saddle-source (sink) and hyperbolic-flow singularity with simple flow and equilibriums. The corresponding switching bifurcations are discussed, including up-down and down-up infinite-equilibriums, inflection-source (sink) infinite-equilibriums, and two infinite-equilibriums of inflection-source (sink) and up-down (down-up)-saddles.

In Chap. 5, equilibrium networks of saddle, center, saddle, source, and sink with hyperbolic flows are discussed, and the corresponding switching bifurcations are discussed. The inflection-source (sink) infinite-equilibriums exist in the vertical and horizontal directions. The parabola-saddles are for a hyperbolic-secant flow with a paralleled saddle and for a hyperbolic-secant flow with a paralleled center. The inflection-sources (sinks) are for a hyperbolic-secant flow with a connected source (sink) and for a hyperbolic flow with a connected saddle.

Finally, the author hopes the materials presented herein can provide a better understanding of cubic nonlinear systems in science and engineering.

Edwardsville, IL, USA Albert C. J. Luo

Contents

Chapter 1
Cubic Systems with Two Different Product Structure Vector Fields

In this chapter, nonlinear dynamics and singularity of cubic dynamical systems possessing two cubic vector fields of different product structures (*i.e., self-linear and crossing-quadratic product and crossing-linear and self-quadratic product*) are discussed. The equilibrium singularity and bifurcations are presented through the appearing and switching bifurcations. Parabola-saddles, saddle-source (sink) singularities with the hyperbolic-to-hyperbolic-secant flow and inflection-source (sink) singularities in the two different product cubic systems are discussed, and the corresponding bifurcation dynamics are discussed. The corresponding switching bifurcations are discussed. Such nonlinear dynamics and singularity properties are presented through a theorem.

1.1 Different Product Cubic Vector Fields

In this section, two different product-cubic dynamical systems will be discussed. The cubic nonlinear vector fields in such cubic systems consist of a self-linear and crossing-quadratic product, and a crossing-linear and self-quadratic product. The corresponding dynamical behaviors of the two different product cubic systems will be presented through the following theorem.

Theorem 1.1 Consider a two-dimensional, cubic dynamical system with two cubic vector fields of different product structures (i.e., self-linear and crossing-quadratic product and crossing-linear and self-quadratic product) as

$$\begin{aligned} \dot{x}_{j_2} &= a_{j_2 j_2 0}(x_{j_1} - a_{j_2 j_1 1})(x_{j_2}^2 + B_{j_2 j_2} x_{j_2} + C_{j_2 j_2}), \\ \dot{x}_{j_1} &= a_{j_1 j_1 0}(x_{j_1} - a_{j_1 j_1 1})(x_{j_2}^2 + B_{j_1 j_2} x_{j_2} + C_{j_1 j_2}) \end{aligned} \tag{1.1}$$

with

A. C. J. Luo, *Two-dimensional Two-product Cubic Systems, Vol. I*,
https://doi.org/10.1007/978-3-031-48487-2_1

$$\Delta_{j_1j_2} = B_{j_1j_2}^2 - 4C_{j_1j_2}, \Delta_{j_2j_2} = B_{j_2j_2}^2 - 4C_{j_2j_2}. \tag{1.2}$$

(i) For $\Delta_{j_1j_2} < 0$ *and* $\Delta_{j_2j_2} < 0$*, the standard form is*

$$\begin{aligned}\dot{x}_{j_2} &= a_{j_2j_20}(x_{j_1} - a_{j_2j_11})\left[(x_{j_2} - a_{j_2j_2})^2 + b_{j_2j_2}\right],\\ \dot{x}_{j_1} &= a_{j_1j_10}(x_{j_1} - a_{j_1j_11})\left[(x_{j_2} - a_{j_1j_2})^2 + b_{j_1j_2}\right]\end{aligned} \tag{1.3}$$

where

$$\begin{aligned}a_{j_1j_2} &= -\frac{1}{2}B_{j_1j_2}, b_{j_1j_2} = \frac{1}{4}(-\Delta_{j_1j_2}),\\ a_{j_2j_2} &= -\frac{1}{2}B_{j_2j_2}, b_{j_2j_2} = \frac{1}{4}(-\Delta_{j_2j_2}).\end{aligned} \tag{1.4}$$

(i_1) The first integral manifold for $a_{j_1j_11} \neq a_{j_2j_11}$ *is*

$$\begin{aligned}&\left[(x_{j_1} - a_{j_1j_11}) - (x_{j_10} - a_{j_1j_11})\right] + (a_{j_1j_11} - a_{j_2j_11})\ln\frac{|x_{j_1} - a_{j_1j_11}|}{|x_{j_10} - a_{j_1j_11}|}\\ &= \frac{a_{j_1j_10}}{a_{j_2j_20}}\left\{\left[(x_{j_2} - a_{j_2j_2}) - (x_{j_20} - a_{j_2j_2})\right]\right.\\ &+(a_{j_2j_2} - a_{j_1j_2})\ln\frac{|(x_{j_2} - a_{j_2j_2})^2 + b_{j_2j_2}|}{|(x_{j_20} - a_{j_2j_2})^2 + b_{j_2j_2}|}\\ &+\left[(a_{j_2j_2} - a_{j_1j_2})^2 + b_{j_1j_2} - b_{j_2j_2}\right]\\ &\left.\times\frac{1}{\sqrt{b_{j_2j_2}}}(\arctan\frac{x_{j_2} - a_{j_2j_2}}{\sqrt{b_{j_2j_2}}} - \arctan\frac{x_{j_20} - a_{j_2j_2}}{\sqrt{b_{j_2j_2}}})\right\}.\end{aligned} \tag{1.5}$$

(i_{1a}) The equilibrium of $x_{j_1}^* = a_{j_1j_11}$ *has the following properties:*

- For $a_{j_1j_10} > 0$ and $a_{j_2j_20}(a_{j_1j_11} - a_{j_2j_11}) > 0$,

$$(a_{j_1j_11}, \dot{x}_{j_2}) = \underbrace{(\text{SO}, \text{pF})}_{\text{source flow }(+)}. \tag{1.6}$$

The equilibrium of $x_{j_1}^* = a_{j_1j_11}$*is an* (SO,pF)-positive source flow.

- *For* $a_{j_1j_10} < 0$ *and* $a_{j_2j_20}(a_{j_1j_11} - a_{j_2j_11}) > 0$,

$$(a_{j_1j_11}, \dot{x}_{j_2}) = \underbrace{(\text{SI}, \text{pF})}_{\text{sink flow }(+)}. \tag{1.7}$$

The equilibrium of $x_{j_1}^* = a_{j_1j_11}$ *is an* (SI,pF)-positive sink flow.

- *For* $a_{j_1j_1 0} > 0$ *and* $a_{j_2j_2 0}(a_{j_1j_1 1} - a_{j_2j_1 1}) < 0$,

$$(a_{j_1j_1 1}, \dot{x}_{j_2}) = \underbrace{(\mathrm{SO}, \mathrm{nF})}_{\text{source flow }(-)}. \tag{1.8}$$

The equilibrium of $x_{j_1}^* = a_{j_1j_1 1}$ *is an* (SO,nF)-negative source flow.

- *For* $a_{j_1j_1 0} < 0$ *and* $a_{j_2j_2 0}(a_{j_1j_1 1} - a_{j_2j_1 1}) < 0$,

$$(a_{j_1j_1 1}, \dot{x}_{j_2}) = \underbrace{(\mathrm{SI}, \mathrm{nF})}_{\text{sink flow }(-)}. \tag{1.9}$$

The equilibrium *of* $x_{j_1}^* = a_{j_1j_1 1}$ *is an* (SI,nF)-negative sink flow.

(i_{1b}) The equilibrium of $x_{j_1}^* = a_{j_2j_1 1}$ *has the following properties:*

- *For* $a_{j_1j_1 0}(a_{j_2j_1 1} - a_{j_1j_1 1}) > 0$ *and* $a_{j_2j_2 0} > 0$,

$$(\dot{x}_{j_1}, a_{j_2j_1 1}) = \underbrace{(\mathrm{pF}, \mathrm{UP})}_{\text{up-parabola flow }(+)}. \tag{1.10}$$

The equilibrium of $x_{j_1}^* = a_{j_2j_1 1}$ *is a* (pF,UP)-positive up-parabola flow.

- *For* $a_{j_1j_1 0}(a_{j_2j_1 1} - a_{j_1j_1 1}) < 0$ *and* $a_{j_2j_2 0} > 0$,

$$(\dot{x}_{j_1}, a_{j_2j_1 1}) = \underbrace{(\mathrm{nF}, \mathrm{DP})}_{\text{down-parabola flow }(-)}. \tag{1.11}$$

The equilibrium of $x_{j_1}^* = a_{j_2j_1 1}$ *is an* (nF,DP)-negative down-parabola flow.

- *For* $a_{j_1j_1 0}(a_{j_2j_1 1} - a_{j_1j_1 1}) > 0$ *and* $a_{j_2j_2 0} < 0$,

$$(\dot{x}_{j_1}, a_{j_2j_1 1}) = \underbrace{(\mathrm{pF}, \mathrm{DP})}_{\text{down-parabola flow }(+)}. \tag{1.12}$$

The equilibrium of $x_{j_1}^* = a_{j_2j_1 1}$ *is a* (pF,DP)-positive down-parabola flow.

- *For* $a_{j_1j_1 0}(a_{j_2j_1 1} - a_{j_1j_1 1}) < 0$ *and* $a_{j_2j_2 0} < 0$,

$$(\dot{x}_{j_1}, a_{j_2j_1 1}) = \underbrace{(\mathrm{nF}, \mathrm{UP})}_{\text{up-parabola flow }(-)}. \tag{1.13}$$

The equilibrium of $x_{j_1}^* = a_{j_2j_1 1}$ *is an* (nF,UP)-negative up-parabola flow.

(i_2) The first integral manifold for $a_{j_1j_1 1}=a_{j_2j_1 1}$ is

$$
\begin{aligned}
\left[(x_{j_1}-a_{j_1j_1 1})-(x_{j_1 0}-a_{j_1j_1 1})\right]=\frac{a_{j_1j_1 0}}{a_{j_2j_2 0}}\Big\{&\left[(x_{j_2}-a_{j_2j_2})-(x_{j_2 0}-a_{j_2j_2})\right]\\
&+(a_{j_2j_2}-a_{j_1j_2})\ln\frac{|(x_{j_2}-a_{j_2j_2})^2+b_{j_2j_2}|}{|(x_{j_2 0}-a_{j_2j_2})^2+b_{j_2j_2}|}\\
&+\left[(a_{j_2j_2}-a_{j_1j_2})^2+b_{j_1j_2}-b_{j_2j_2}\right]\\
&\times\frac{1}{\sqrt{b_{j_2j_2}}}\Big(\arctan\frac{x_{j_2}-a_{j_2j_2}}{\sqrt{b_{j_2j_2}}}-\arctan\frac{x_{j_2 0}-a_{j_2j_2}}{\sqrt{b_{j_2j_2}}}\Big)\Big\}.
\end{aligned}
\tag{1.14}
$$

The infinite-equilibrium of $x_{j_1}^*=a_{j_1j_1 1}=a_{j_2j_1 1}$ with $\bar{x}_{j_2}\in(-\infty,\infty)$ has the following properties:

- *For $a_{j_1j_1 0}>0$ and $a_{j_2j_2 0}>0$,*

$$
(a_{j_1j_1 1},\bar{x}_{j_2})=\underbrace{(\text{SO},\text{II})}_{\text{Increasing-inflection source}}. \tag{1.15}
$$

  The *infinite-equilibrium of $x_{j_1}^*=a_{j_1j_1 1}=a_{j_2j_1 1}$ is an* (SO,II)-*increasing-inflection source.*

- *For $a_{j_1j_1 0}<0$ and $a_{j_2j_2 0}>0$,*

$$
(a_{j_1j_1 1},\bar{x}_{j_2})=\underbrace{(\text{SI},\text{DI})}_{\text{Decreasing-inflection sink}}. \tag{1.16}
$$

  The *infinite-equilibrium of $x_{j_1}^*=a_{j_1j_1 1}=a_{j_2j_1 1}$ is an* (SI,DI)-*decreasing-inflection sink.*

- *For $a_{j_1j_1 0}>0$ and $a_{j_2j_2 0}<0$,*

$$
(a_{j_1j_1 1},\bar{x}_{j_2})=\underbrace{(\text{SO},\text{DI})}_{\text{Decreasing-inflection source}}. \tag{1.17}
$$

  The *infinite-equilibrium of $x_{j_1}^*=a_{j_1j_1 1}=a_{j_2j_1 1}$ is an* (SO,DI)-*decreasing-inflection source.*

- *For $a_{j_1j_1 0}<0$ and $a_{j_2j_2 0}<0$,*

$$
(a_{j_1j_1 1},\bar{x}_{j_2})=\underbrace{(\text{SI},\text{II})}_{\text{Increasing-inflection sink}}. \tag{1.18}
$$

The infinite-equilibrium of $x_{j_1}^* = a_{j_1j_11} = a_{j_2j_11}$ *is an* (SI,II)-*increasing-inflection sink.*

(ii) For $\Delta_{j_1j_2} < 0$ *and* $\Delta_{j_2j_2} = 0$, *the standard form is*

$$\begin{aligned} \dot{x}_{j_2} &= a_{j_2j_20}(x_{j_1} - a_{j_2j_11})(x_{j_2} - a_{j_2j_21})^2, \\ \dot{x}_{j_1} &= a_{j_1j_10}(x_{j_1} - a_{j_1j_11})\left[(x_{j_2} - a_{j_1j_2})^2 + b_{j_1j_2}\right] \end{aligned} \tag{1.19}$$

where

$$\begin{aligned} a_{j_1j_2} &= -\frac{1}{2}B_{j_1j_2}, b_{j_1j_2} = \frac{1}{4}(-\Delta_{j_1j_2}), \\ a_{j_2j_21} &= -\frac{1}{2}B_{j_2j_2}. \end{aligned} \tag{1.20}$$

(ii$_1$) The first integral manifold for $a_{j_1j_11} \neq a_{j_2j_11}$ *is*

$$\begin{aligned} &\left[(x_{j_1} - a_{j_1j_11}) - (x_{j_10} - a_{j_1j_11})\right] + (a_{j_1j_11} - a_{j_2j_11})\ln\frac{|x_{j_1} - a_{j_1j_11}|}{|x_{j_10} - a_{j_1j_11}|} \\ &= \frac{a_{j_1j_10}}{a_{j_2j_20}}\left\{\left[(x_{j_2} - a_{j_2j_21}) - (x_{j_20} - a_{j_2j_21})\right]\right. \\ &\quad + 2(a_{j_2j_21} - a_{j_1j_2})\ln\frac{|x_{j_2} - a_{j_2j_21}|}{|x_{j_20} - a_{j_2j_21}|} \\ &\quad \left. - \left[(a_{j_2j_21} - a_{j_1j_2})^2 + b_{j_1j_2}\right]\left(\frac{1}{x_{j_2} - a_{j_2j_21}} - \frac{1}{x_{j_20} - a_{j_2j_21}}\right)\right\}. \end{aligned} \tag{1.21}$$

(ii$_{1a}$) The equilibrium of $(x_{j_1}^*, x_{j_2}^*) = (a_{j_1j_11}, a_{j_2j_21})$ *has the following properties:*

- *For* $a_{j_1j_10} > 0$ *and* $a_{j_2j_20}(a_{j_1j_11} - a_{j_2j_11}) > 0$,

$$(a_{j_1j_11}, a_{j_2j_21}) = \underbrace{(\text{SO}, \text{US})}_{\text{upper-saddle-source}}. \tag{1.22}$$

The equilibrium *of* $(x_{j_1}^*, x_{j_2}^*) = (a_{j_1j_11}, a_{j_2j_21})$ *is an* (SO,US)-*upper-saddle-source.*

- *For* $a_{j_1j_10} < 0$ *and* $a_{j_2j_20}(a_{j_1j_11} - a_{j_2j_11}) > 0$,

$$(a_{j_1j_11}, a_{j_2j_21}) = \underbrace{(\text{SI}, \text{US})}_{\text{upper-saddle-sink}}. \tag{1.23}$$

The equilibrium of $(x_{j_1}^*, x_{j_2}^*) = (a_{j_1j_11}, a_{j_2j_21})$ *is an* (SI,US)-*upper-saddle-sink.*

- *For* $a_{j_1j_10} > 0$ *and* $a_{j_2j_20}(a_{j_1j_11} - a_{j_2j_11}) < 0$,

$$(a_{j_1j_11}, a_{j_2j_21}) = \underbrace{(\mathrm{SO}, \mathrm{LS})}_{\text{lower-saddle-source}}. \tag{1.24}$$

The equilibrium *of* $(x^*_{j_1}, x^*_{j_2}) = (a_{j_1j_11}, a_{j_2j_21})$ *is an* (SO,LS)-*lower-saddle-source.*

- *For* $a_{j_1j_10} < 0$ *and* $a_{j_2j_20}(a_{j_1j_11} - a_{j_2j_11}) < 0$,

$$(a_{j_1j_11}, a_{j_2j_21}) = \underbrace{(\mathrm{SI}, \mathrm{LS})}_{\text{lower-saddle-source}}. \tag{1.25}$$

The equilibrium of $(x^*_{j_1}, x^*_{j_2}) = (a_{j_1j_11}, a_{j_2j_21})$ *is an* (SI,LS)-*lower-saddle-sink.*

– The saddle-source and saddle-sink are the appearing bifurcations of saddle-source and saddle-sink.

(ii$_{1b}$*) The equilibrium of* $(x^*_{j_1}, x^*_{j_2}) = (a_{j_2j_11}, a_{j_2j_21})$ *has the following properties:*

- *For* $a_{j_1j_10}(a_{j_2j_11} - a_{j_1j_11}) > 0$ *and* $a_{j_2j_20} > 0$,

$$(a_{j_2j_11}, a_{j_2j_21}) = \underbrace{(\mathrm{pF}, \mathrm{UP}:\mathrm{UP})}_{\text{hyperbolic-secant-to-hyperbolic flow }(+)}. \tag{1.26}$$

The equilibrium of $(x^*_{j_1}, x^*_{j_2}) = (a_{j_2j_11}, a_{j_2j_21})$ *is a* (pF,UP:UP)-*positive hyperbolic-secant-to-hyperbolic flow.*

- *For* $a_{j_1j_10}(a_{j_2j_11} - a_{j_1j_11}) < 0$ *and* $a_{j_2j_20} > 0$,

$$(a_{j_2j_11}, a_{j_2j_21}) = \underbrace{(\mathrm{nF}, \mathrm{DP}:\mathrm{DP})}_{\text{hyperbolic-to-hyperbolic-secant flow }(-)}. \tag{1.27}$$

The equilibrium of $(x^*_{j_1}, x^*_{j_2}) = (a_{j_1j_11}, a_{j_2j_21})$*is an* (nF,DP:DP)-*negative hyperbolic-to-hyperbolic-secant flow.*

- *For* $a_{j_1j_10}(a_{j_2j_11} - a_{j_1j_11}) > 0$ *and* $a_{j_2j_20} < 0$,

$$(a_{j_2j_11}, a_{j_2j_21}) = \underbrace{(\mathrm{pF}, \mathrm{DP}:\mathrm{DP})}_{\text{hyperbolic-to-hyperbolic-secant flow }(+)}. \tag{1.28}$$

The equilibrium of $(x^*_{j_1}, x^*_{j_2}) = (a_{j_1j_11}, a_{j_2j_21})$ *is a* (pF,DP:DP)-*positive hyperbolic-to-hyperbolic-secant flow.*

- *For* $a_{j_1j_10}(a_{j_2j_11} - a_{j_1j_11}) < 0$ *and* $a_{j_2j_20} < 0$,

$$(a_{j_2j_11}, a_{j_2j_21}) = \underbrace{(\mathrm{nF}, \mathrm{UP}:\mathrm{UP})}_{\text{hyperbolic-secant-to-hyperbolic flow }(-)}. \tag{1.29}$$

The equilibrium of $(x^*_{j_1}, x^*_{j_2}) = (a_{j_1j_1}1, a_{j_2j_2}1)$ *is an* (nF,UP:UP)-*negative hyperbolic-secant-to-hyperbolic flow.*

(ii₂) The first integral manifold for $a_{j_1j_1}1 = a_{j_2j_1}1$ *is*

$$
\begin{aligned}
&\left[(x_{j_1} - a_{j_1j_1}1) - (x_{j_10} - a_{j_1j_1}1)\right] \\
&= \frac{a_{j_1j_10}}{a_{j_2j_20}} \left\{ \left[(x_{j_2} - a_{j_2j_2}1) - (x_{j_20} - a_{j_2j_2}1)\right] \right. \\
&\quad + 2(a_{j_2j_2}1 - a_{j_1j_2}) \ln \frac{|x_{j_2} - a_{j_2j_2}1|}{|x_{j_20} - a_{j_2j_2}1|} \\
&\quad \left. - \left[(a_{j_2j_2}1 - a_{j_1j_2})^2 + b_{j_1j_2}\right] \left(\frac{1}{x_{j_2} - a_{j_2j_2}1} - \frac{1}{x_{j_20} - a_{j_2j_2}1}\right) \right\}.
\end{aligned}
\tag{1.30}
$$

(ii₂ₐ) The infinite-equilibrium of $x^*_{j_1} = a_{j_1j_1}1 = a_{j_2j_1}1$ *with* $\bar{x}_{j_2} \neq a_{j_2j_2}1$ *has the following properties:*

- *For* $a_{j_1j_10} > 0$ *and* $a_{j_2j_20}(\bar{x}_{j_2} - a_{j_2j_2}1)^2 > 0$,

$$(a_{j_1j_1}1, \bar{x}_{j_2}) = \underbrace{(\mathrm{SO}, \mathrm{II})}_{\text{increasing-inflection source}}. \tag{1.31}$$

The infinite-equilibrium of $x^*_{j_1} = a_{j_1j_1}1 = a_{j_2j_1}1$ *is an* (SO,II)-*increasing-inflection source.*

- *For* $a_{j_1j_10} < 0$ *and* $a_{j_2j_20}(\bar{x}_{j_2} - a_{j_2j_2}1)^2 > 0$,

$$(a_{j_1j_1}1, \bar{x}_{j_2}) = \underbrace{(\mathrm{SI}, \mathrm{DI})}_{\text{decreasing-inflection sink}}. \tag{1.32}$$

The infinite-equilibrium of $x^*_{j_1} = a_{j_1j_1}1 = a_{j_2j_1}1$ *is an* (SI,DI)-*decreasing inflection sink.*

- *For* $a_{j_1j_10} > 0$ *and* $a_{j_2j_20}(\bar{x}_{j_2} - a_{j_2j_2}1)^2 < 0$,

$$(a_{j_1j_1}1, \bar{x}_{j_2}) = \underbrace{(\mathrm{SO}, \mathrm{DI})}_{\text{decreasing-inflection source}}. \tag{1.33}$$

The infinite-equilibrium of $x^*_{j_1} = a_{j_1j_1}1 = a_{j_2j_1}1$ *is an* (SO,DI)-*decreasing inflection source.*

- *For* $a_{j_1j_10} < 0$ *and* $a_{j_2j_20}(\bar{x}_{j_2} - a_{j_2j_2}1)^2 < 0$,

$$(a_{j_1j_1}1, \bar{x}_{j_2}) = \underbrace{(\mathrm{SI}, \mathrm{II})}_{\text{increasing-inflection sink}}. \tag{1.34}$$

The infinite-equilibrium of $x_{j_1}^* = a_{j_1j_1 1} = a_{j_2j_1 1}$ *is an* (SI,II)-*increasing-inflection sink.*

(ii$_{2b}$) The equilibrium of $(x_{j_1}^*, x_{j_2}^*) = (a_{j_1j_1 1}, a_{j_2j_2 1})$ *has the following properties:*

- *For* $a_{j_1j_1 0} > 0$ *and* $a_{j_2j_2 0} > 0$,

$$(a_{j_1j_1 1}, a_{j_2j_2 1}) = \underbrace{(\text{SO}, \text{II} : \text{II})}_{\text{increasing-inflection source}}. \tag{1.35}$$

The equilibrium of $(x_{j_1}^*, x_{j_2}^*) = (a_{j_1j_1 1}, a_{j_2j_2 1})$ *is an* (SO,II:II)-*increasing-inflection source.*

- *For* $a_{j_1j_1 0} < 0$ *and* $a_{j_2j_2 0} > 0$,

$$(a_{j_1j_1 1}, a_{j_2j_2 1}) = \underbrace{(\text{SI}, \text{DI} : \text{DI})}_{\text{decreasing-inflection sink}}. \tag{1.36}$$

The equilibrium of $(x_{j_1}^*, x_{j_2}^*) = (a_{j_1j_1 1}, a_{j_2j_2 1})$ *is an* (SI,DI:DI)-*decreasing-inflection sink.*

- *For* $a_{j_1j_1 0} > 0$ *and* $a_{j_2j_2 0} < 0$,

$$(a_{j_1j_1 1}, a_{j_2j_2 1}) = \underbrace{(\text{SO}, \text{DI} : \text{DI})}_{\text{decreasing-inflection source}}. \tag{1.37}$$

The equilibrium of $(x_{j_1}^*, x_{j_2}^*) = (a_{j_1j_1 1}, a_{j_2j_2 1})$ *is an* (SO,DI:DI)-*decreasing-inflection source.*

- *For* $a_{j_1j_1 0} < 0$ *and* $a_{j_2j_2 0} < 0$,

$$(a_{j_1j_1 1}, a_{j_2j_2 1}) = \underbrace{(\text{SI}, \text{II} : \text{II})}_{\text{increasing-inflection sink}}. \tag{1.38}$$

The equilibrium of $(x_{j_1}^*, x_{j_2}^*) = (a_{j_1j_1 1}, a_{j_2j_2 1})$ *is an* (SI,II:II)-*increasing-inflection sink.*

(iii) For $\Delta_{j_1j_2} < 0$ *and* $\Delta_{j_2j_2} > 0$, *the standard form is*

$$\begin{aligned} \dot{x}_{j_2} &= a_{j_2j_2 0}(x_{j_1} - a_{j_2j_1 1})(x_{j_2} - a_{j_2j_2 1})(x_{j_2} - a_{j_2j_2 2}), \\ \dot{x}_{j_1} &= a_{j_1j_1 0}(x_{j_1} - a_{j_1j_1 1})\left[(x_{j_2} - a_{j_1j_2})^2 + b_{j_1j_2}\right] \end{aligned} \tag{1.39}$$

where

$$
\begin{aligned}
&a_{j_1 j_2} = -\frac{1}{2}B_{j_1 j_2}, b_{j_1 j_2} = \frac{1}{4}(-\Delta_{j_1 j_2}),\\
&b_{j_2 j_2 1}, b_{j_2 j_2 2} = -\frac{1}{2}(B_{j_2 j_2} \pm \sqrt{\Delta_{j_2 j_2}}),\\
&\{a_{j_2 j_2 1}, a_{j_2 j_2 2}\} = \mathrm{sort}\{b_{j_2 j_2 1}, b_{j_2 j_2 2}\}, a_{j_2 j_2 1} < a_{j_2 j_2 2}.
\end{aligned} \tag{1.40}
$$

(iii_1) The first integral manifold for $a_{j_1 j_1 1} \neq a_{j_2 j_1 1}$ is

$$
\begin{aligned}
&\left[(x_{j_1} - a_{j_1 j_1 1}) - (x_{j_1 0} - a_{j_1 j_1 1})\right] + (a_{j_1 j_1 1} - a_{j_2 j_1 1}) \ln \frac{|x_{j_1} - a_{j_1 j_1 1}|}{|x_{j_1 0} - a_{j_1 j_1 1}|}\\
&= \frac{a_{j_1 j_1 0}}{a_{j_2 j_2 0}} \left\{ \left[(x_{j_2} - a_{j_2 j_2 1}) - (x_{j_2 0} - a_{j_2 j_2 1})\right] \right.\\
&\quad + \sum\nolimits_{s_1 = 1}^{2} (a_{j_2 j_2 s_1} - a_{j_1 j_2}) \ln \frac{|x_{j_2} - a_{j_2 j_2 s_1}|}{|x_{j_2 0} - a_{j_2 j_2 s_1}|}\\
&\quad + \left[\prod\nolimits_{s_1 = 1}^{2} (a_{j_2 j_2 s_1} - a_{j_1 j_2}) + b_{j_1 j_2}\right]\\
&\quad \left. \times \sum\nolimits_{s_1 = 1, s_1 \neq s_2}^{2} \frac{1}{(a_{j_2 j_2 s_1} - a_{j_1 j_2 s_2})} \ln \frac{|x_{j_2} - a_{j_2 j_2 s_1}|}{|x_{j_2 0} - a_{j_2 j_2 s_1}|} \right\}.
\end{aligned} \tag{1.41}
$$

(iii_{1a}) The equilibrium of $(x^*_{j_1}, x^*_{j_2}) = (a_{j_1 j_1 1}, a_{j_2 j_2 s_1})$ (s_1, $s_2 \in \{1, 2\}$, $s_1 \neq s_2$) has the following properties:

- *For $a_{j_1 j_1 0} > 0$ and $a_{j_2 j_2 0}(a_{j_1 j_1 1} - a_{j_2 j_1 1})(a_{j_2 j_2 s_1} - a_{j_2 j_2 s_2}) > 0$,*

$$
(a_{j_1 j_1 1}, a_{j_2 j_2 s_1}) = \underbrace{(\mathrm{SO}, \mathrm{SO})}_{\text{source}}. \tag{1.42}
$$

  The equilibrium of $(x^*_{j_1}, x^*_{j_2}) = (a_{j_1 j_1 1}, a_{j_2 j_2 s_1})$ is an (SO,SO)-source.
- *For $a_{j_1 j_1 0} < 0$ and $a_{j_2 j_2 0}(a_{j_1 j_1 1} - a_{j_2 j_1 1})(a_{j_2 j_2 s_1} - a_{j_2 j_2 s_2}) > 0$,*

$$
(a_{j_1 j_1 1}, a_{j_2 j_2 s_1}) = \underbrace{(\mathrm{SI}, \mathrm{SO})}_{\text{saddle}}. \tag{1.43}
$$

  The equilibrium of $(x^*_{j_1}, x^*_{j_2}) = (a_{j_1 j_1 1}, a_{j_2 j_2 s_1})$ is an (SI,SO)-saddle.
- *For $a_{j_1 j_1 0} > 0$ and $a_{j_2 j_2 0}(a_{j_1 j_1 1} - a_{j_2 j_1 1})(a_{j_2 j_2 s_1} - a_{j_2 j_2 s_2}) < 0$,*

$$
(a_{j_1 j_1 1}, a_{j_2 j_2 s_1}) = \underbrace{(\mathrm{SO}, \mathrm{SI})}_{\text{saddle}}. \tag{1.44}
$$

  The equilibrium of $(x^*_{j_1}, x^*_{j_2}) = (a_{j_1 j_1 1}, a_{j_2 j_2 s_1})$ is an (SO,SI)-saddle.

- *For* $a_{j_1j_10}<0$ *and* $a_{j_2j_20}(a_{j_1j_11}-a_{j_2j_11})(a_{j_2j_2s_1}-a_{j_2j_2s_2})<0$,

$$(a_{j_1j_11},a_{j_2j_2s_1})=\underbrace{(\mathrm{SI},\mathrm{SI})}_{\text{sink}}. \tag{1.45}$$

The equilibrium of $(x_{j_1}^*,x_{j_2}^*)=(a_{j_1j_11},a_{j_2j_2s_1})$ *is an* (SI,SI)*-sink.*

(iii$_{1b}$*) The equilibrium of* $(x_{j_1}^*,x_{j_2}^*)=(a_{j_2j_11},a_{j_2j_2s_1})$ $(s_1,\ s_2\in\{1,2\},\ s_1\neq s_2)$ *has the following properties:*

- *For* $a_{j_1j_10}(a_{j_2j_11}-a_{j_1j_11})>0$ *and* $a_{j_2j_20}(a_{j_2j_2s_1}-a_{j_2j_2s_2})>0$,

$$(a_{j_2j_11},a_{j_2j_2s_1})=\underbrace{(\mathrm{pF},\mathrm{DP}:\mathrm{UP})}_{\text{hyperbolic flow }(+)}. \tag{1.46}$$

The equilibrium of $(x_{j_1}^*,x_{j_2}^*)=(a_{j_2j_11},a_{j_2j_2s_1})$ *is a* (pF,UP:UP)*-positive hyperbolic flow.*

- *For* $a_{j_1j_10}(a_{j_2j_11}-a_{j_1j_11})<0$ *and* $a_{j_2j_20}(a_{j_2j_2s_1}-a_{j_2j_2s_2})>0$,

$$(a_{j_2j_11},a_{j_2j_2s_1})=\underbrace{(\mathrm{nF},\mathrm{UP}:\mathrm{DP})}_{\text{hyperbolic-secant flow }(-)}. \tag{1.47}$$

The equilibrium of $(x_{j_1}^*,x_{j_2}^*)=(a_{j_1j_11},a_{j_2j_2s_1})$ *is an* (nF,UP:DP)*-positive hyperbolic secant flow.*

- *For* $a_{j_1j_10}(a_{j_2j_11}-a_{j_1j_11})>0$ *and* $a_{j_2j_20}(a_{j_2j_2s_1}-a_{j_2j_2s_2})<0$,

$$(a_{j_2j_11},a_{j_2j_2s_1})=\underbrace{(\mathrm{pF},\mathrm{UP}:\mathrm{DP})}_{\text{hyperbolic-secant flow }(+)}. \tag{1.48}$$

The equilibrium of $(x_{j_1}^*,x_{j_2}^*)=(a_{j_1j_11},a_{j_2j_2s_1})$ *is a* (pF,UP:DP)*-positive hyperbolic secant flow.*

- *For* $a_{j_1j_10}(a_{j_2j_11}-a_{j_1j_11})<0$ *and* $a_{j_2j_20}(a_{j_2j_2s_1}-a_{j_2j_2s_2})<0$,

$$(a_{j_2j_11},a_{j_2j_2s_1})=\underbrace{(\mathrm{nF},\mathrm{DP}:\mathrm{UP})}_{\text{hyperbolic flow }(-)}. \tag{1.49}$$

The equilibrium of $(x_{j_1}^*,x_{j_2}^*)=(a_{j_1j_11},a_{j_2j_2s_1})$ *is an* (nF,DP:UP)*-negative hyperbolic flow.*

(iii$_2$) The first integral manifold for $a_{j_1j_1 1}=a_{j_2j_1 1}$ is

$$\begin{aligned}
&[(x_{j_1}-a_{j_1j_1 1})-(x_{j_1 0}-a_{j_1j_1 1})]\\
&=\frac{a_{j_1j_1 0}}{a_{j_2j_2 0}}\Big\{[(x_{j_2}-a_{j_2j_2 1})-(x_{j_2 0}-a_{j_2j_2 1})]\\
&\quad+\sum\nolimits_{s_1=1}^{2}(a_{j_2j_2s_1}-a_{j_1j_2})\ln\frac{|x_{j_2}-a_{j_2j_2s_1}|}{|x_{j_2 0}-a_{j_2j_2s_1}|}\\
&\quad+\Big[\prod\nolimits_{s_1=1}^{2}(a_{j_2j_2s_1}-a_{j_1j_2})+b_{j_1j_2}\Big]\\
&\quad\times(\sum\nolimits_{s_1=1,s_1\neq s_2}^{2}\frac{1}{(a_{j_2j_2s_1}-a_{j_1j_2s_2})}\ln\frac{|x_{j_2}-a_{j_2j_2s_1}|}{|x_{j_2 0}-a_{j_2j_2s_1}|})\Big\}.
\end{aligned}\tag{1.50}$$

(iii$_{2a}$) The infinite-equilibrium of $x_{j_1}^*=a_{j_1j_1 1}=a_{j_2j_1 1}$ with $\overline{x}_{j_2}\neq a_{j_2j_2s_1},a_{j_2j_2s_2}$ has the following properties:

- *For $a_{j_1j_1 0}>0$ and $a_{j_2j_2 0}(\overline{x}_{j_2}-a_{j_2j_2s_1})(\overline{x}_{j_2}-a_{j_2j_2s_2})>0$,*

$$(a_{j_1j_1 1},\overline{x}_{j_2})=\underbrace{(\text{SO, II})}_{\text{increasing-inflection source}}. \tag{1.51}$$

  The infinite-equilibrium of $x_{j_1}^*=a_{j_1j_1 1}=a_{j_2j_1 1}$ is an (SO,II)-increasing-inflection source.

- *For $a_{j_1j_1 0}<0$ and $a_{j_2j_2 0}(\overline{x}_{j_2}-a_{j_2j_2s_1})(\overline{x}_{j_2}-a_{j_2j_2s_2})>0$,*

$$(a_{j_1j_1 1},\overline{x}_{j_2})=\underbrace{(\text{SI, DI})}_{\text{decreasing-inflection sink}}. \tag{1.52}$$

  The infinite-equilibrium of $x_{j_1}^*=a_{j_1j_1 1}=a_{j_2j_1 1}$ is an (SI,DI)-decreasing inflection sink.

- *For $a_{j_1j_1 0}>0$ and $a_{j_2j_2 0}(\overline{x}_{j_2}-a_{j_2j_2s_1})(\overline{x}_{j_2}-a_{j_2j_2s_2})<0$,*

$$(a_{j_1j_1 1},\overline{x}_{j_2})=\underbrace{(\text{SO, DI})}_{\text{decreasing-inflection source}}. \tag{1.53}$$

  The infinite-equilibrium of $x_{j_1}^*=a_{j_1j_1 1}=a_{j_2j_1 1}$ is an (SO,DI)-decreasing-inflection source.

- *For $a_{j_1j_1 0}<0$ and $a_{j_2j_2 0}(\overline{x}_{j_2}-a_{j_2j_2s_1})(\overline{x}_{j_2}-a_{j_2j_2s_2})<0$,*

$$(a_{j_1j_1 1},\overline{x}_{j_2})=\underbrace{(\text{SI, II})}_{\text{increasing-inflection sink}}. \tag{1.54}$$

  The infinite-equilibrium of $x_{j_1}^*=a_{j_1j_1 1}=a_{j_2j_1 1}$ is an (SI,II)-increasing-inflection sink.

(iii$_{2b}$) The equilibrium of $(x_{j_1}^*, x_{j_2}^*) = (a_{j_1 j_1 1}, a_{j_2 j_2 s_1})$ *(s_1, $s_2 \in \{1,2\}$; $s_1 \neq s_2$) has the following properties:*

- *For* $a_{j_1 j_1 0} > 0$ *and* $a_{j_2 j_2 0}(a_{j_2 j_2 s_1} - a_{j_2 j_2 s_2}) > 0$,

$$(a_{j_2 j_1 1}, a_{j_2 j_2 s_1}) = \underbrace{(\mathrm{UP}, \mathrm{US})}_{\text{up-parabola upper-saddle}}. \tag{1.55}$$

  The equilibrium of $(x_{j_1}^*, x_{j_2}^*) = (a_{j_1 j_1 1}, a_{j_2 j_2 s_1})$ *is a* (UP,US)-*up-parabola upper-saddle.*
- *For* $a_{j_1 j_1 0} < 0$ *and* $a_{j_2 j_2 0}(a_{j_2 j_2 s_1} - a_{j_2 j_2 s_2}) > 0$,

$$(a_{j_2 j_1 1}, a_{j_2 j_2 s_1}) = \underbrace{(\mathrm{DP}, \mathrm{US})}_{\text{down-parabola upper-saddle}}. \tag{1.56}$$

  The equilibrium of $(x_{j_1}^*, x_{j_2}^*) = (a_{j_1 j_1 1}, a_{j_2 j_2 s_1})$ *is a* (DP,US)-*down-parabola upper-saddle.*
- *For* $a_{j_1 j_1 0} > 0$ *and* $a_{j_2 j_2 0}(a_{j_2 j_2 s_1} - a_{j_2 j_2 s_2}) < 0$,

$$(a_{j_2 j_1 1}, a_{j_2 j_2 s_1}) = \underbrace{(\mathrm{DP}, \mathrm{LS})}_{\text{down-parabola lower-saddle}}. \tag{1.57}$$

  The equilibrium of $(x_{j_1}^*, x_{j_2}^*) = (a_{j_1 j_1 1}, a_{j_2 j_2 s_1})$ *is a* (DP:LS)-*down-parabola lower-saddle.*
- *For* $a_{j_1 j_1 0} < 0$ *and* $a_{j_2 j_2 0}(a_{j_2 j_2 s_1} - a_{j_2 j_2 s_2}) < 0$,

$$(a_{j_2 j_1 1}, a_{j_2 j_2 s_1}) = \underbrace{(\mathrm{UP}, \mathrm{LS})}_{\text{up-parabola lower-saddle}}. \tag{1.58}$$

  The equilibrium of $(x_{j_1}^*, x_{j_2}^*) = (a_{j_1 j_1 1}, a_{j_2 j_2 s_1})$ *is an* (UP,LS)-*up-parabola lower-saddle.*

(iv) For $\Delta_{j_1 j_2} = 0$ *and* $\Delta_{j_2 j_2} < 0$, *the standard form is*

$$\begin{aligned} \dot{x}_{j_2} &= a_{j_2 j_2 0}(x_{j_1} - a_{j_2 j_1 1})\left[(x_{j_2} - a_{j_2 j_2})^2 + b_{j_2 j_2}\right], \\ \dot{x}_{j_1} &= a_{j_1 j_1 0}(x_{j_1} - a_{j_1 j_1 1})(x_{j_2} - a_{j_1 j_2 1})^2 \end{aligned} \tag{1.59}$$

where

$$\begin{aligned} a_{j_1 j_2 1} &= -\frac{1}{2} B_{j_1 j_2}, \\ a_{j_2 j_2} &= -\frac{1}{2} B_{j_2 j_2}, b_{j_2 j_2} = \frac{1}{4}(-\Delta_{j_2 j_2}). \end{aligned} \tag{1.60}$$

(iv$_1$) The first integral manifold for $a_{j_1j_11} \neq a_{j_2j_11}$ is

$$\begin{aligned}
&[(x_{j_1} - a_{j_1j_11}) - (x_{j_10} - a_{j_1j_11})] + (a_{j_1j_11} - a_{j_2j_11}) \ln \frac{|x_{j_1} - a_{j_1j_11}|}{|x_{j_10} - a_{j_1j_11}|} \\
&= \frac{a_{j_1j_10}}{a_{j_2j_20}} \{ [(x_{j_2} - a_{j_2j_2}) - (x_{j_20} - a_{j_2j_2})] \\
&+ (a_{j_2j_2} - a_{j_1j_21}) \ln \frac{|(x_{j_2} - a_{j_2j_2})^2 + b_{j_2j_2}|}{|(x_{j_20} - a_{j_2j_2})^2 + b_{j_2j_2}|} \\
&+ [(a_{j_2j_2} - a_{j_1j_21})^2 - b_{j_2j_2}] \\
&\times \frac{1}{\sqrt{b_{j_2j_2}}} (\arctan \frac{x_{j_2} - a_{j_2j_2}}{\sqrt{b_{j_2j_2}}} - \arctan \frac{x_{j_20} - a_{j_2j_2}}{\sqrt{b_{j_2j_2}}}) \}.
\end{aligned} \tag{1.61}$$

(iv$_{1a}$) The equilibrium of $(x_{j_1}^*, x_{j_2}^*) = (a_{j_1j_11}, a_{j_1j_21})$ has the following properties:

- *For $a_{j_1j_10} > 0$ and $a_{j_2j_20}(a_{j_1j_11} - a_{j_2j_11}) > 0$,*

$$(a_{j_1j_11}, a_{j_1j_21}) = \underbrace{(_{\text{DI:II}}\text{SO, pF})}_{\text{inflection source flow } (+)}. \tag{1.62}$$

  The equilibrium of $(x_{j_1}^*, x_{j_2}^*) = (a_{j_1j_11}, a_{j_1j_21})$ is a ($_{\text{DI:II}}$SO, pF)*-positive inflection source flow.*
- *For $a_{j_1j_10} < 0$ and $a_{j_2j_20}(a_{j_1j_11} - a_{j_2j_11}) > 0$,*

$$(a_{j_1j_11}, a_{j_1j_21}) = \underbrace{(_{\text{II:DI}}\text{SI, pF})}_{\text{inflection sink flow } (+)}. \tag{1.63}$$

  The equilibrium of $(x_{j_1}^*, x_{j_2}^*) = (a_{j_1j_11}, a_{j_1j_21})$ is an ($_{\text{II:DI}}$SI, pF)*-positive inflection sink flow.*
- *For $a_{j_1j_10} > 0$ and $a_{j_2j_20}(a_{j_1j_11} - a_{j_2j_11}) < 0$,*

$$(a_{j_1j_11}, a_{j_1j_21}) = \underbrace{(_{\text{II:DI}}\text{SO, nF})}_{\text{inflection source flow } (-)}. \tag{1.64}$$

  The equilibrium of $(x_{j_1}^*, x_{j_2}^*) = (a_{j_1j_11}, a_{j_1j_21})$ is an ($_{\text{II:DI}}$SO, nF)*-negative inflection source flow.*
- *For $a_{j_1j_10} < 0$ and $a_{j_2j_20}(a_{j_1j_11} - a_{j_2j_11}) < 0$,*

$$(a_{j_1j_11}, a_{j_1j_21}) = \underbrace{(_{\text{DI:II}}\text{SI, nF})}_{\text{inflection sink flow } (-)}. \tag{1.65}$$

  The equilibrium of $(x_{j_1}^*, x_{j_2}^*) = (a_{j_1j_11}, a_{j_1j_21})$ is a ($_{\text{DI:II}}$SI, nF)*-negative inflection sink flow.*

(vi_{1b}) The equilibrium of $(x_{j_2}^*, x_{j_1}^*) = (a_{j_1j_2 1}, a_{j_2j_1 1})$ *has the following properties:*

- *For* $a_{j_1j_1 0}(a_{j_2j_1 1} - a_{j_1j_1 1}) > 0$ *and* $a_{j_2j_2 0} > 0$,

$$(a_{j_1j_2 1}, a_{j_2j_1 1}) = \underbrace{(\text{US}, \text{UP})}_{\text{up-parabola upper-saddle}}. \tag{1.66}$$

  The equilibrium of $x_{j_1}^* = a_{j_2j_1 1}$ *is a* (US,UP)*-up-parabola upper-saddle.*
- *For* $a_{j_1j_1 0}(a_{j_2j_1 1} - a_{j_1j_1 1}) < 0$ *and* $a_{j_2j_2 0} > 0$,

$$(a_{j_1j_2 1}, a_{j_2j_1 1}) = \underbrace{(\text{LS}, \text{DP})}_{\text{down-parabola lower-saddle}}. \tag{1.67}$$

  The equilibrium of $x_{j_1}^* = a_{j_2j_1 1}$ *is an* (LS,DP)*-down-parabola lower-saddle.*
- *For* $a_{j_1j_1 0}(a_{j_2j_1 1} - a_{j_1j_1 1}) > 0$ *and* $a_{j_2j_2 0} < 0$,

$$(a_{j_1j_2 1}, a_{j_2j_1 1}) = \underbrace{(\text{US}, \text{DP})}_{\text{down-parabola upper-saddle}}. \tag{1.68}$$

  The equilibrium of $x_{j_1}^* = a_{j_2j_1 1}$ *is a* (US,DP)*-down-parabola upper-saddle.*
- *For* $a_{j_1j_1 0}(a_{j_2j_1 1} - a_{j_1j_1 1}) < 0$ *and* $a_{j_2j_2 0} < 0$,

$$(a_{j_1j_2 1}, a_{j_2j_1 1}) = \underbrace{(\text{LS}, \text{UP})}_{\text{up-parabola lower-saddle}}. \tag{1.69}$$

  The equilibrium of $x_{j_1}^* = a_{j_2j_1 1}$ *is an* (LS,UP)*-up-parabola lower-saddle.*

(iv_2) The first integral manifold for $a_{j_1j_1 1} = a_{j_2j_1 1}$ *is*

$$\begin{aligned}\left[(x_{j_1} - a_{j_1j_1 1}) - (x_{j_1 0} - a_{j_1j_1 1})\right] = \frac{a_{j_1j_1 0}}{a_{j_2j_2 0}} \Big\{ &\left[(x_{j_2} - a_{j_2j_2}) - (x_{j_2 0} - a_{j_2j_2})\right] \\ &+ (a_{j_2j_2} - a_{j_1j_2 1}) \ln \frac{|(x_{j_2} - a_{j_2j_2})^2 + b_{j_2j_2}|}{|(x_{j_2 0} - a_{j_2j_2})^2 + b_{j_2j_2}|} \\ &+ \left[(a_{j_2j_2} - a_{j_1j_2 1})^2 - b_{j_2j_2}\right] \\ &\times \frac{1}{\sqrt{b_{j_2j_2}}} \Big(\arctan \frac{x_{j_2} - a_{j_2j_2}}{\sqrt{b_{j_2j_2}}} - \arctan \frac{x_{j_2 0} - a_{j_2j_2}}{\sqrt{b_{j_2j_2}}} \Big) \Big\}.\end{aligned} \tag{1.70}$$

(iv$_{2a}$) The infinite-equilibrium of $x^*_{j_1} = a_{j_1 j_1 1} = a_{j_2 j_1 1}$ *with* $\overline{x}_{j_2} \neq a_{j_1 j_2 1}$ *has the following properties:*

- *For* $a_{j_1 j_1 0}(\overline{x}_{j_2} - a_{j_1 j_2 1})^2 > 0$ *and* $a_{j_2 j_2 0} > 0$,

$$(a_{j_1 j_1 1}, \overline{x}_{j_2}) = \underbrace{(\text{SO}, \text{II})}_{\text{increasing-inflection source}}. \tag{1.71}$$

The infinite-equilibrium of $x^*_{j_1} = a_{j_1 j_1 1} = a_{j_2 j_1 1}$ *is an* (SO,II)*-increasing-inflection source.*

- *For* $a_{j_1 j_1 0}(\overline{x}_{j_2} - a_{j_1 j_2 1})^2 < 0$ *and* $a_{j_2 j_2 0} > 0$,

$$(a_{j_1 j_1 1}, \overline{x}_{j_2}) = \underbrace{(\text{SI}, \text{DI})}_{\text{decreasing-inflection sink}}. \tag{1.72}$$

The infinite-equilibrium of $x^*_{j_1} = a_{j_1 j_1 1} = a_{j_2 j_1 1}$ *is an* (SI,DI)*-decreasing-inflection sink.*

- *For* $a_{j_1 j_1 0}(\overline{x}_{j_2} - a_{j_1 j_2 1})^2 > 0$ *and* $a_{j_2 j_2 0} < 0$,

$$(a_{j_1 j_1 1}, \overline{x}_{j_2}) = \underbrace{(\text{SO}, \text{DI})}_{\text{decreasing-inflection source}}. \tag{1.73}$$

The infinite-equilibrium of $x^*_{j_1} = a_{j_1 j_1 1} = a_{j_2 j_1 1}$ *is an* (SO,DI)*-decreasing-inflection source.*

- *For* $a_{j_1 j_1 0}(\overline{x}_{j_2} - a_{j_1 j_2 1})^2 < 0$ *and* $a_{j_2 j_2 0} < 0$,

$$(a_{j_1 j_1 1}, \overline{x}_{j_2}) = \underbrace{(\text{SI}, \text{II})}_{\text{increasing-inflection sink}}. \tag{1.74}$$

The infinite-equilibrium of $x^*_{j_1} = a_{j_1 j_1 1} = a_{j_2 j_1 1}$ *is an* (SI,II)*-increasing-inflection sink.*

(iv$_{2b}$) The equilibrium of $(x^*_{j_1}, x^*_{j_2}) = (a_{j_1 j_1 1}, a_{j_1 j_2 1})$ *with* $a_{j_1 j_1 1} = a_{j_2 j_1 1}$ *has the following properties:*

- *For* $a_{j_1 j_1 0} > 0$ *and* $a_{j_2 j_2 0} > 0$,

$$(a_{j_1 j_1 1}, a_{j_1 j_2 1}) = \underbrace{(\text{SO}, \text{II} : \text{II})}_{\text{increasing-inflection source}}. \tag{1.75}$$

The equilibrium of $(x^*_{j_1}, x^*_{j_2}) = (a_{j_1 j_1 1}, a_{j_1 j_2 1})$ *is an* (SO,II:II)*-increasing-inflection source of the first kind.*

- *For* $a_{j_1j_10}<0$ *and* $a_{j_2j_20}>0$,

$$(a_{j_1j_11},a_{j_1j_21})=\underbrace{(\mathrm{SI},\mathrm{DI}:\mathrm{DI})}_{\text{decreasing-inflection sink}}. \tag{1.76}$$

The equilibrium of $(x_{j_1}^*,x_{j_2}^*)=(a_{j_1j_11},a_{j_1j_21})$ *is an* (SI,DI:DI)-*decreasing-inflection sink of the first kind.*

- *For* $a_{j_1j_10}>0$ *and* $a_{j_2j_20}<0$,

$$(a_{j_1j_11},a_{j_1j_21})=\underbrace{(\mathrm{SO},\mathrm{DI}:\mathrm{DI})}_{\text{decreasing-inflection source}}. \tag{1.77}$$

The equilibrium of $(x_{j_1}^*,x_{j_2}^*)=(a_{j_1j_11},a_{j_1j_21})$ *is an* (SO,DI:DI)-*decreasing-inflection source of the first kind.*

- *For* $a_{j_1j_10}<0$ *and* $a_{j_2j_20}<0$,

$$(a_{j_1j_11},a_{j_1j_21})=\underbrace{(\mathrm{SI},\mathrm{II}:\mathrm{II})}_{\text{increasing-inflection sink}}. \tag{1.78}$$

The equilibrium of $(x_{j_1}^*,x_{j_2}^*)=(a_{j_1j_11},a_{j_1j_21})$ *is an* (SI,II:II)-*increasing-inflection sink of the first kind.*

(v) For $\Delta_{j_1j_2}=0$ *and* $\Delta_{j_2j_2}=0$, *the standard form is*

$$\begin{aligned}\dot{x}_{j_2}&=a_{j_2j_20}(x_{j_1}-a_{j_2j_11})(x_{j_2}-a_{j_2j_21})^2,\\ \dot{x}_{j_1}&=a_{j_1j_10}(x_{j_1}-a_{j_1j_11})(x_{j_2}-a_{j_1j_21})^2\end{aligned} \tag{1.79}$$

where

$$a_{j_1j_21}=-\frac{1}{2}B_{j_1j_2},\,a_{j_2j_21}=-\frac{1}{2}B_{j_2j_2}. \tag{1.80}$$

(v₁) The first integral manifold for $a_{j_2j_21}\neq a_{j_1j_21}$ *is*

$$\begin{aligned}&\big[(x_{j_1}-a_{j_1j_11})-(x_{j_10}-a_{j_1j_11})\big]+(a_{j_1j_11}-a_{j_2j_11})\ln\frac{|x_{j_1}-a_{j_1j_11}|}{|x_{j_10}-a_{j_1j_11}|}\\ &=\frac{a_{j_1j_10}}{a_{j_2j_20}}\Big\{\big[(x_{j_2}-a_{j_2j_21})-(x_{j_20}-a_{j_2j_21})\big]\\ &\quad+2(a_{j_2j_21}-a_{j_1j_21})\ln\frac{|x_{j_2}-a_{j_2j_21}|}{|x_{j_20}-a_{j_2j_21}|}\\ &\quad-(a_{j_2j_21}-a_{j_1j_21})^2\Big(\frac{1}{x_{j_2}-a_{j_2j_21}}-\frac{1}{x_{j_20}-a_{j_2j_21}}\Big)\Big\}.\end{aligned} \tag{1.81}$$

(v_{1a}) The equilibrium of $(x^*_{j_1}, x^*_{j_2}) = (a_{j_1j_1}1, a_{j_2j_2}1)$ *with* $a_{j_2j_2 1} \neq a_{j_1j_2 1}$ *has the following properties:*

- *For* $a_{j_1j_1 0}(a_{j_2j_2 1} - a_{j_1j_2 1})^2 > 0$ *and* $a_{j_2j_2 0}(a_{j_1j_1 1} - a_{j_2j_1 1}) > 0$,

$$(a_{j_1j_1}1, a_{j_2j_2}1) = \underbrace{(\mathrm{SO}, \mathrm{US})}_{\text{upper-saddle-source}} . \tag{1.82}$$

The equilibrium of $(x^*_{j_1}, x^*_{j_2}) = (a_{j_1j_1}1, a_{j_2j_2}1)$ *is an* (SO,US)-*upper-saddle-source.*

- *For* $a_{j_1j_1 0}(a_{j_2j_2 1} - a_{j_1j_2 1})^2 < 0$ *and* $a_{j_2j_2 0}(a_{j_1j_1 1} - a_{j_2j_1 1}) > 0$,

$$(a_{j_1j_1}1, a_{j_2j_2}1) = \underbrace{(\mathrm{SI}, \mathrm{US})}_{\text{upper-saddle-sink}} . \tag{1.83}$$

The equilibrium of $(x^*_{j_1}, x^*_{j_2}) = (a_{j_1j_1}1, a_{j_2j_2}1)$ *is an* (SI,US)-*upper-saddle-sink.*

- *For* $a_{j_1j_1 0}(a_{j_2j_2 1} - a_{j_1j_2 1})^2 > 0$ *and* $a_{j_2j_2 0}(a_{j_1j_1 1} - a_{j_2j_1 1}) < 0$,

$$(a_{j_1j_1}1, a_{j_2j_2}1) = \underbrace{(\mathrm{SO}, \mathrm{LS})}_{\text{lower-saddle-source}} . \tag{1.84}$$

The equilibrium of $(x^*_{j_1}, x^*_{j_2}) = (a_{j_1j_1}1, a_{j_2j_2}1)$ *is an* (SO,LS)-*lower-saddle-source.*

- *For* $a_{j_1j_1 0}(a_{j_2j_2 1} - a_{j_1j_2 1})^2 < 0$ *and* $a_{j_2j_2 0}(a_{j_1j_1 1} - a_{j_2j_1 1}) < 0$,

$$(a_{j_1j_1}1, a_{j_2j_2}1) = \underbrace{(\mathrm{SI}, \mathrm{LS})}_{\text{lower-saddle-source}} . \tag{1.85}$$

The equilibrium of $(x^*_{j_1}, x^*_{j_2}) = (a_{j_1j_1}1, a_{j_2j_2}1)$ *is an* (SI,LS)-*lower-saddle-sink.*

– *The saddle-source and saddle-sink are the appearing bifurcations of saddle-source and saddle-sink, respectively.*

(v_{1b}) The equilibrium of $(x^*_{j_1}, x^*_{j_2}) = (a_{j_1j_1}1, a_{j_1j_2}1)$ *with* $a_{j_2j_2 1} \neq a_{j_1j_2 1}$ *has the following properties:*

- *For* $a_{j_1j_1 0} > 0$ *and* $a_{j_2j_2 0}(a_{j_1j_1 1} - a_{j_2j_1 1})(a_{j_1j_2 1} - a_{j_2j_2 1})^2 > 0$,

$$(a_{j_1j_1}1, a_{j_1j_2}1) = \underbrace{(_{\mathrm{DI:II}}\mathrm{SO}, \mathrm{pF})}_{\text{inflection-source flow }(+)} . \tag{1.86}$$

The equilibrium of $(x^*_{j_1}, x^*_{j_2}) = (a_{j_1j_1}1, a_{j_1j_2}1)$ *is a* ($_{\mathrm{DI:II}}$SO, pF)-*positive inflection-source flow.*

- *For* $a_{j_1j_10}<0$ *and* $a_{j_2j_20}(a_{j_1j_11}-a_{j_2j_11})(a_{j_1j_21}-a_{j_2j_21})^2>0$,

$$(a_{j_1j_11},a_{j_1j_21})=\underbrace{({}_{\text{II:DI}}\text{SI, pF})}_{\text{inflection sink flow }(+)}. \tag{1.87}$$

The equilibrium of $(x^*_{j_1},x^*_{j_2})=(a_{j_1j_11},a_{j_1j_21})$ *is an* (${}_{\text{II:DI}}$SI, pF)*-positive inflection-sink flow.*

- *For* $a_{j_1j_10}>0$ *and* $a_{j_2j_20}(a_{j_1j_11}-a_{j_2j_11})(a_{j_1j_21}-a_{j_2j_21})^2<0$,

$$(a_{j_1j_11},a_{j_1j_21})=\underbrace{({}_{\text{II:DI}}\text{SO, nF})}_{\text{inflection source flow }(-)}. \tag{1.88}$$

The equilibrium of $(x^*_{j_1},x^*_{j_2})=(a_{j_1j_11},a_{j_1j_21})$ *is an* (${}_{\text{II:DI}}$SO, nF)*-negative inflection-source flow.*

- *For* $a_{j_1j_10}<0$ *and* $a_{j_2j_20}(a_{j_1j_11}-a_{j_2j_11})(a_{j_1j_21}-a_{j_2j_21})^2<0$,

$$(a_{j_1j_11},a_{j_1j_21})=\underbrace{({}_{\text{DI:II}}\text{SI, nF})}_{\text{inflection sink flow }(-)}. \tag{1.89}$$

The equilibrium of $(x^*_{j_1},x^*_{j_2})=(a_{j_1j_11},a_{j_1j_21})$ *is a* (${}_{\text{DI:II}}$SI, nF)*-negative inflection-sink flow.*

(v_{1c}*) The equilibrium of* $(x^*_{j_2},x^*_{j_1})=(a_{j_1j_21},a_{j_2j_11})$ *with* $a_{j_2j_11}\neq a_{j_1j_11}$ *has the following properties:*

- *For* $a_{j_1j_10}(a_{j_2j_11}-a_{j_1j_11})>0$ *and* $a_{j_2j_20}(a_{j_1j_21}-a_{j_2j_21})^2>0$,

$$(a_{j_1j_21},a_{j_2j_11})=\underbrace{(\text{US, UP})}_{\text{up-parabola upper-saddle}}. \tag{1.90}$$

The equilibrium of $(x^*_{j_2},x^*_{j_1})=(a_{j_1j_21},a_{j_2j_11})$ *is a* (US,UP)*-up-parabola upper-saddle.*

- *For* $a_{j_1j_10}(a_{j_2j_11}-a_{j_1j_11})<0$ *and* $a_{j_2j_20}(a_{j_1j_21}-a_{j_2j_21})^2>0$,

$$(a_{j_1j_21},a_{j_2j_11})=\underbrace{(\text{LS, DP})}_{\text{down-parabola lower-saddle}}. \tag{1.91}$$

The equilibrium of $(x^*_{j_2},x^*_{j_1})=(a_{j_1j_21},a_{j_2j_11})$ *is an* (LS,DP)*-down-parabola lower-saddle.*

- *For* $a_{j_1j_1 0}(a_{j_2j_1 1} - a_{j_1j_1 1}) > 0$ *and* $a_{j_2j_2 0}(a_{j_1j_2 1} - a_{j_2j_2 1})^2 < 0$,

$$(a_{j_1j_2 1}, a_{j_2j_1 1}) = \underbrace{(\mathrm{US}, \mathrm{DP})}_{\text{down-parabola upper-saddle}}. \tag{1.92}$$

The equilibrium of $(x^*_{j_2}, x^*_{j_1}) = (a_{j_1j_2 1}, a_{j_2j_1 1})$ *is a* (US,DP)-*down-parabola upper-saddle.*

- *For* $a_{j_1j_1 0}(a_{j_2j_1 1} - a_{j_1j_1 1}) > 0$ *and* $a_{j_2j_2 0}(a_{j_1j_2 1} - a_{j_2j_2 1})^2 < 0$,

$$(a_{j_1j_2 1}, a_{j_2j_1 1}) = \underbrace{(\mathrm{LS}, \mathrm{UP})}_{\text{up-parabola lower-saddle}}. \tag{1.93}$$

The equilibrium of $(x^*_{j_2}, x^*_{j_1}) = (a_{j_1j_2 1}, a_{j_2j_1 1})$ *is an* (LS,UP)-*up-parabola lower-saddle.*

– *The parabola upper-saddle and lower-saddle are the appearing bifurcations of positive and negative saddles with clockwise and counterclockwise centers, respectively.*

(v_{1d}) *The equilibrium of* $(x^*_{j_1}, x^*_{j_2}) = (a_{j_2j_1 1}, a_{j_2j_2 1})$ *with* $a_{j_2j_2 1} \neq a_{j_1j_2 1}$ *has the following properties:*

- *For* $a_{j_1j_1 0}(a_{j_2j_1 1} - a_{j_1j_1 1})(a_{j_2j_2 1} - a_{j_1j_2 1})^2 > 0$ *and* $a_{j_2j_2 0} > 0$,

$$(a_{j_2j_1 1}, a_{j_2j_2 1}) = \underbrace{(\mathrm{pF}, \mathrm{UP}:\mathrm{UP})}_{\text{hyperbolic-secant-to-hyperbolic flow }(+)}. \tag{1.94}$$

The equilibrium of $(x^*_{j_1}, x^*_{j_2}) = (a_{j_2j_1 1}, a_{j_2j_2 1})$ *is a* (pF,UP:UP)-*positive hyperbolic-secant-to-hyperbolic flow.*

- *For* $a_{j_1j_1 0}(a_{j_2j_1 1} - a_{j_1j_1 1})(a_{j_2j_2 1} - a_{j_1j_2 1})^2 < 0$ *and* $a_{j_2j_2 0} > 0$,

$$(a_{j_2j_1 1}, a_{j_2j_2 1}) = \underbrace{(\mathrm{nF}, \mathrm{DP}:\mathrm{DP})}_{\text{hyperbolic-to-hyperbolic-secant flow }(-)}. \tag{1.95}$$

The equilibrium of $(x^*_{j_1}, x^*_{j_2}) = (a_{j_1j_1 1}, a_{j_1j_2 1})$ *is an* (nF,DP:DP)-*negative hyperbolic-to-hyperbolic-secant flow.*

- *For* $a_{j_1j_1 0}(a_{j_2j_1 1} - a_{j_1j_1 1})(a_{j_2j_2 1} - a_{j_1j_2 1})^2 > 0$ *and* $a_{j_2j_2 0} < 0$,

$$(a_{j_2j_1 1}, a_{j_2j_2 1}) = \underbrace{(\mathrm{pF}, \mathrm{DP}:\mathrm{DP})}_{\text{hyperbolic-to-hyperbolic-secant flow }(+)}. \tag{1.96}$$

The equilibrium of $(x^*_{j_1}, x^*_{j_2}) = (a_{j_1j_1 1}, a_{j_1j_2 1})$ *is a* (pF,DP:DP)-*positive hyperbolic-to-hyperbolic-secant flow.*

- *For* $a_{j_1j_10}(a_{j_2j_11}-a_{j_1j_11})(a_{j_2j_21}-a_{j_1j_21})^2<0$ *and* $a_{j_2j_20}<0$,

$$(a_{j_2j_11},a_{j_2j_21})=\underbrace{(\text{nF},\text{UP}:\text{UP})}_{\text{hyperbolic-secant-to-hyperbolic flow }(-)}. \tag{1.97}$$

The equilibrium of $(x^*_{j_1},x^*_{j_2})=(a_{j_1j_11},a_{j_1j_21})$ *is an* (nF,UP:UP)-*hyperbolic-to-hyperbolic-secant flow.*

(v_2*) The first integral manifold for* $a_{j_2j_21}=a_{j_1j_21}$ and $a_{j_1j_11}\neq a_{j_2j_11}$ *is*

$$\begin{aligned}&\left[(x_{j_1}-a_{j_1j_11})-(x_{j_10}-a_{j_1j_11})\right]+(a_{j_1j_11}-a_{j_2j_11})\ln\frac{|x_{j_1}-a_{j_1j_11}|}{|x_{j_10}-a_{j_1j_11}|}\\&=\frac{a_{j_1j_10}}{a_{j_2j_20}}\left[(x_{j_2}-a_{j_2j_21})-(x_{j_20}-a_{j_2j_21})\right].\end{aligned} \tag{1.98}$$

(v_{2a}*) The infinite-equilibrium of* $x^*_{j_2}=a_{j_1j_21}=a_{j_2j_21}$ *with* $\bar{x}_{j_1}\neq a_{j_1j_11},a_{j_2j_11}$ *for* $a_{j_1j_11}\neq a_{j_2j_11}$ *has the following properties:*

- *For* $a_{j_1j_10}(\bar{x}_{j_1}-a_{j_1j_11})>0$ *and* $a_{j_2j_20}(\bar{x}_{j_1}-a_{j_2j_11})>0$,

$$(\bar{x}_{j_1},a_{j_2j_21})=\underbrace{(\text{II},\text{US})}_{\text{increasing-inflection upper-saddle}}. \tag{1.99}$$

The infinite-equilibrium of $x^*_{j_2}=a_{j_1j_21}=a_{j_2j_21}$ *is an* (II,US)-*increasing-inflection upper-saddle.*

- *For* $a_{j_1j_10}(\bar{x}_{j_1}-a_{j_1j_11})<0$ *and* $a_{j_2j_20}(\bar{x}_{j_1}-a_{j_2j_11})>0$,

$$(\bar{x}_{j_1},a_{j_2j_21})=\underbrace{(\text{DI},\text{US})}_{\text{decreasing-inflection upper-saddle}}. \tag{1.100}$$

The infinite-equilibrium of $x^*_{j_2}=a_{j_1j_21}=a_{j_2j_21}$ *is a* (DI,US)-*decreasing-inflection upper-saddle.*

- *For* $a_{j_1j_10}(\bar{x}_{j_1}-a_{j_1j_11})>0$ *and* $a_{j_2j_20}(\bar{x}_{j_1}-a_{j_2j_11})<0$,

$$(\bar{x}_{j_1},a_{j_2j_21})=\underbrace{(\text{DI},\text{LS})}_{\text{decreasing-inflection lower-saddle}}. \tag{1.101}$$

The infinite-equilibrium of $x^*_{j_2}=a_{j_1j_21}=a_{j_2j_21}$ *is a* (DI,LS)-*decreasing-inflection lower-saddle.*

- *For* $a_{j_1j_10}(\overline{x}_{j_1}-a_{j_1j_11})<0$ *and* $a_{j_2j_20}(\overline{x}_{j_1}-a_{j_2j_11})<0$,

$$(\overline{x}_{j_1},a_{j_2j_21})=\underbrace{(\text{II},\text{LS})}_{\text{increasing-inflection lower-saddle}}. \tag{1.102}$$

The infinite-equilibrium of $x^*_{j_2}=a_{j_1j_21}=a_{j_2j_21}$ *is an (II,LS)-increasing-inflection lower-saddle.*

(v_{2b}*) The equilibrium of* $(x^*_{j_1},x^*_{j_2})=(a_{j_1j_11},a_{j_2j_21})$ *with* $a_{j_2j_21}=a_{j_1j_21}$ *has the following properties:*

- *For* $a_{j_1j_10}>0$ *and* $a_{j_2j_20}(a_{j_1j_11}-a_{j_2j_11})>0$,

$$(a_{j_1j_11},a_{j_2j_21})=\underbrace{(\text{DI}:\text{II},\text{US})}_{\text{inflection upper-saddle}}. \tag{1.103}$$

The equilibrium of $(x^*_{j_1},x^*_{j_2})=(a_{j_1j_11},a_{j_2j_21})$ *is a* (DI:II,US)*-inflection upper-saddle.*

- *For* $a_{j_1j_10}<0$ *and* $a_{j_2j_20}(a_{j_1j_11}-a_{j_2j_11})>0$,

$$(a_{j_1j_11},a_{j_2j_21})=\underbrace{(\text{II}:\text{DI},\text{US})}_{\text{inflection upper-saddle}}. \tag{1.104}$$

The equilibrium of $(x^*_{j_1},x^*_{j_2})=(a_{j_1j_11},a_{j_2j_21})$ *is an* (II:DI,US)*-inflection upper-saddle.*

- *For* $a_{j_1j_10}>0$ *and* $a_{j_2j_20}(a_{j_1j_11}-a_{j_2j_11})<0$,

$$(a_{j_1j_11},a_{j_2j_21})=\underbrace{(\text{II}:\text{DI},\text{LS})}_{\text{inflection upper-saddle}}. \tag{1.105}$$

The equilibrium of $(x^*_{j_1},x^*_{j_2})=(a_{j_1j_11},a_{j_2j_21})$ *is an* (II:DI,LS)*-inflection lower-saddle.*

- *For* $a_{j_1j_10}<0$ *and* $a_{j_2j_20}(a_{j_1j_11}-a_{j_2j_11})<0$,

$$(a_{j_1j_11},a_{j_2j_21})=\underbrace{(\text{DI}:\text{II},\text{US})}_{\text{inflection lower-saddle}}. \tag{1.106}$$

The equilibrium of $(x^*_{j_1},x^*_{j_2})=(a_{j_1j_11},a_{j_2j_21})$ *is a* (DI:II,LS)*-inflection lower-saddle.*

(v_{2c}) The equilibrium of $(x_{j_2}^*, x_{j_1}^*) = (a_{j_1j_2 1}, a_{j_2j_1 1})$ with $a_{j_2j_2 1} = a_{j_1j_2 1}$ has the following properties:

- *For $a_{j_1j_1 0}(a_{j_2j_1 1} - a_{j_2j_1 1}) > 0$ and $a_{j_2j_2 0}(a_{j_1j_2 1} - a_{j_2j_2 1})^2 > 0$,*

$$(a_{j_1j_2 1}, a_{j_2j_2 1}) = \underbrace{(\mathrm{US}, \mathrm{UP})}_{\text{up-parabola upper-saddle}} . \tag{1.107}$$

  The equilibrium of $(x_{j_2}^*, x_{j_1}^*) = (a_{j_1j_2 1}, a_{j_2j_1 1})$ is a (US,UP)-*up-parabola upper-saddle.*
- *For $a_{j_1j_1 0}(a_{j_2j_1 1} - a_{j_2j_1 1}) < 0$ and $a_{j_2j_2 0}(a_{j_1j_2 1} - a_{j_2j_2 1})^2 > 0$,*

$$(a_{j_1j_2 1}, a_{j_2j_1 1}) = \underbrace{(\mathrm{LS}, \mathrm{DP})}_{\text{down-parabola lower-saddle}} . \tag{1.108}$$

  The equilibrium of $(x_{j_2}^*, x_{j_1}^*) = (a_{j_1j_2 1}, a_{j_2j_1 1})$ is an (LS,DP)-*down-parabola lower-saddle.*
- *For $a_{j_1j_1 0}(a_{j_2j_1 1} - a_{j_2j_1 1}) > 0$ and $a_{j_2j_2 0}(a_{j_1j_2 1} - a_{j_2j_2 1})^2 < 0$,*

$$(a_{j_1j_2 1}, a_{j_2j_1 1}) = \underbrace{(\mathrm{US}, \mathrm{DP})}_{\text{down-parabola upper-saddle}} . \tag{1.109}$$

  The equilibrium of $(x_{j_2}^*, x_{j_1}^*) = (a_{j_1j_2 1}, a_{j_2j_1 1})$ is a (US,DP)-*down-parabola upper-saddle.*
- *For $a_{j_1j_1 0}(a_{j_2j_1 1} - a_{j_2j_1 1}) < 0$ and $a_{j_2j_2 0}(a_{j_1j_2 1} - a_{j_2j_2 1})^2 < 0$,*

$$(a_{j_1j_2 1}, a_{j_2j_1 1}) = \underbrace{(\mathrm{LS}, \mathrm{UP})}_{\text{up-parabola lower-saddle}} . \tag{1.110}$$

  The equilibrium of $(x_{j_2}^*, x_{j_1}^*) = (a_{j_1j_2 1}, a_{j_2j_1 1})$ is an (LS,UP)-*up-parabola lower-saddle.*

(v_3) The first integral manifold for $a_{j_1j_1 1} = a_{j_2j_1 1}$ and *$a_{j_2j_2 1} \neq a_{j_1j_2 1}$ is*

$$\begin{aligned}
\left[(x_{j_1} - a_{j_1j_1 1}) - (x_{j_1 0} - a_{j_1j_1 1})\right] = \frac{a_{j_1j_1 0}}{a_{j_2j_2 0}} \Big\{ & \left[(x_{j_2} - a_{j_2j_2 1}) - (x_{j_2 0} - a_{j_2j_2 1})\right] \\
& + 2(a_{j_2j_2 1} - a_{j_1j_2 1}) \ln \frac{|x_{j_2} - a_{j_2j_2 1}|}{|x_{j_2 0} - a_{j_2j_2 1}|} \\
& - (a_{j_2j_2 1} - a_{j_1j_2 1})^2 \Big(\frac{1}{x_{j_2} - a_{j_2j_2 1}} - \frac{1}{x_{j_2 0} - a_{j_2j_2 1}}\Big) \Big\}.
\end{aligned} \tag{1.111}$$

(v_{3a}) *The infinite-equilibrium of* $x_{j_1}^* = a_{j_1j_1 1} = a_{j_2j_1 1}$ *with* $\overline{x}_{j_2} \neq a_{j_1j_2 1}, a_{j_2j_2 1}$ *for* $a_{j_2j_2 1} \neq a_{j_1j_2 1}$ *has the following properties:*

- *For* $a_{j_1j_1 0}(\overline{x}_{j_2} - a_{j_1j_2 1})^2 > 0$ *and* $a_{j_2j_2 0}(\overline{x}_{j_2} - a_{j_2j_2 1})^2 > 0$,

$$(a_{j_1j_1 1}, \overline{x}_{j_2}) = \underbrace{(\text{SO}, \text{II})}_{\text{increasing-inflection source}}. \tag{1.112}$$

The infinite-equilibrium of $x_{j_1}^* = a_{j_1j_1 1} = a_{j_2j_1 1}$ *is an* (SO,II)*-increasing-inflection source.*

- *For* $a_{j_1j_1 0}(\overline{x}_{j_2} - a_{j_1j_2 1})^2 < 0$ *and* $a_{j_2j_2 0}(\overline{x}_{j_2} - a_{j_2j_2 1})^2 > 0$,

$$(a_{j_1j_1 1}, \overline{x}_{j_2}) = \underbrace{(\text{SI}, \text{DI})}_{\text{increasing-inflection sink}}. \tag{1.113}$$

The infinite-equilibrium of $x_{j_1}^* = a_{j_1j_1 1} = a_{j_2j_1 1}$ *is an* (SI,DI)*-decreasing-inflection sink.*

- *For* $a_{j_1j_1 0}(\overline{x}_{j_2} - a_{j_1j_2 1})^2 > 0$ *and* $a_{j_2j_2 0}(\overline{x}_{j_2} - a_{j_2j_2 1})^2 < 0$,

$$(a_{j_1j_1 1}, \overline{x}_{j_2}) = \underbrace{(\text{SO}, \text{DI})}_{\text{increasing-inflection source}}. \tag{1.114}$$

The infinite-equilibrium of $x_{j_1}^* = a_{j_1j_1 1} = a_{j_2j_1 1}$ *is an* (SO,DI)*-decreasing-inflection source.*

- *For* $a_{j_1j_1 0}(\overline{x}_{j_2} - a_{j_1j_2 1})^2 < 0$ *and* $a_{j_2j_2 0}(\overline{x}_{j_2} - a_{j_2j_2 1})^2 < 0$,

$$(a_{j_1j_1 1}, \overline{x}_{j_2}) = \underbrace{(\text{SI}, \text{II})}_{\text{increasing-inflection sink}}. \tag{1.115}$$

The infinite-equilibrium of $x_{j_1}^* = a_{j_1j_1 1} = a_{j_2j_1 1}$ *is an* (SI,II)*-increasing-inflection sink.*

(v_{3b}) *The equilibrium of* $(x_{j_1}^*, x_{j_2}^*) = (a_{j_1j_1 1}, a_{j_1j_2 1})$ *with* $a_{j_1j_1 1} = a_{j_2j_1 1}$ *has the following properties:*

- *For* $a_{j_1j_1 0} > 0$ *and* $a_{j_2j_2 0}(a_{j_1j_2 1} - a_{j_2j_2 1})^2 > 0$,

$$(a_{j_1j_1 1}, a_{j_1j_2 1}) = \underbrace{(\text{SO}, \text{II} : \text{II})}_{\text{increasing-inflection source}}. \tag{1.116}$$

The equilibrium of $(x^*_{j_1}, x^*_{j_2}) = (a_{j_1j_1}1, a_{j_1j_2}1)$ *is an* (SO,II:II)*-increasing-inflection source of the first kind.*

- *For* $a_{j_1j_10} < 0$ *and* $a_{j_2j_20}(a_{j_1j_2}1 - a_{j_2j_2}1)^2 > 0$,

$$(a_{j_1j_1}1, a_{j_1j_2}1) = \underbrace{(\mathrm{SI}, \mathrm{DI} : \mathrm{DI})}_{\text{decreasing-inflection sink}} . \tag{1.117}$$

The equilibrium of $(x^*_{j_1}, x^*_{j_2}) = (a_{j_1j_1}1, a_{j_1j_2}1)$ *is an* (SI,DI:DI)*-decreasing-inflection sink of the first kind.*

- *For* $a_{j_1j_10} > 0$ *and* $a_{j_2j_20}(a_{j_1j_2}1 - a_{j_2j_2}1)^2 < 0$,

$$(a_{j_1j_1}1, a_{j_1j_2}1) = \underbrace{(\mathrm{SO}, \mathrm{DI} : \mathrm{DI})}_{\text{decreasing-inflection source}} . \tag{1.118}$$

The equilibrium of $(x^*_{j_1}, x^*_{j_2}) = (a_{j_1j_1}1, a_{j_1j_2}1)$ *is an* (SO,DI:DI)*-decreasing-inflection source of the first kind.*

- *For* $a_{j_1j_10} < 0$ *and* $a_{j_2j_20}(a_{j_1j_2}1 - a_{j_2j_2}1)^2 < 0$,

$$(a_{j_1j_1}1, a_{j_1j_2}1) = \underbrace{(\mathrm{SI}, \mathrm{II} : \mathrm{II})}_{\text{increasing-inflection sink}} . \tag{1.119}$$

The equilibrium of $(x^*_{j_1}, x^*_{j_2}) = (a_{j_1j_1}1, a_{j_1j_2}1)$ *is an* (SI,II:II)*-increasing-inflection sink of the first kind.*

(v_{3c}*) The equilibrium of* $(x^*_{j_2}, x^*_{j_1}) = (a_{j_1j_1}1, a_{j_2j_2}1)$ *with* $a_{j_1j_1}1 = a_{j_2j_1}1$ *has the following properties:*

- *For* $a_{j_1j_10}(a_{j_2j_2}1 - a_{j_1j_2}1)^2 > 0$ *and* $a_{j_2j_20} > 0$,

$$(a_{j_1j_1}1, a_{j_2j_2}1) = \underbrace{(\mathrm{SO}, \mathrm{II} : \mathrm{II})}_{\text{increasing-inflection source}} . \tag{1.120}$$

The equilibrium of $(x^*_{j_2}, x^*_{j_1}) = (a_{j_1j_1}1, a_{j_2j_2}1)$ *is an* (SO,II:II)*-increasing-inflection source of the second kind.*

- *For* $a_{j_1j_10}(a_{j_2j_2}1 - a_{j_1j_2}1)^2 < 0$ *and* $a_{j_2j_20} > 0$,

$$(a_{j_1j_1}1, a_{j_2j_2}1) = \underbrace{(\mathrm{SI}, \mathrm{DI} : \mathrm{DI})}_{\text{decreasing-inflection sink}} . \tag{1.121}$$

The equilibrium of $(x_{j_2}^*, x_{j_1}^*) = (a_{j_1j_1 1}, a_{j_2j_2 1})$ *is an (SI,DI:DI)-decreasing-inflection sink of the second kind.*

- *For* $a_{j_1j_1 0}(a_{j_2j_2 1} - a_{j_1j_2 1})^2 > 0$ *and* $a_{j_2j_2 0} < 0$,

$$(a_{j_1j_1 1}, a_{j_2j_2 1}) = \underbrace{(\mathrm{SO}, \mathrm{DI}:\mathrm{DI})}_{\text{decreasing-inflection source}}. \tag{1.122}$$

The equilibrium of $(x_{j_2}^*, x_{j_1}^*) = (a_{j_1j_1 1}, a_{j_2j_2 1})$ *is an* (SO,DI:DI)*-decreasing-inflection source of the second kind.*

- *For* $a_{j_1j_1 0}(a_{j_2j_2 1} - a_{j_1j_2 1})^2 < 0$ *and* $a_{j_2j_2 0} < 0$,

$$(a_{j_1j_1 1}, a_{j_2j_2 1}) = \underbrace{(\mathrm{SI}, \mathrm{II}:\mathrm{II})}_{\text{increasing-inflection sink}}. \tag{1.123}$$

The equilibrium of $(x_{j_2}^*, x_{j_1}^*) = (a_{j_1j_1 1}, a_{j_2j_2 1})$ *is an* (SI,II:II)*-increasing-inflection sink of the second kind.*

(v_4) The first integral manifold for $a_{j_1j_1 1} = a_{j_2j_1 1}$ and $a_{j_2j_2 1} = a_{j_1j_2 1}$ *is*

$$\left[(x_{j_1} - a_{j_1j_1 1}) - (x_{j_1 0} - a_{j_1j_1 1})\right] = \frac{a_{j_1j_1 0}}{a_{j_2j_2 0}}\left[(x_{j_2} - a_{j_2j_2 1}) - (x_{j_2 0} - a_{j_2j_2 1})\right]. \tag{1.124}$$

(v_{4a}) The infinite-equilibrium of $x_{j_1}^* = a_{j_1j_1 1} = a_{j_2j_1 1}$ *with* $\bar{x}_{j_2} \neq a_{j_1j_2 1}, a_{j_2j_2 1}$ *for* $a_{j_1j_1 1} = a_{j_2j_1 1}$ *and* $a_{j_2j_2 1} = a_{j_1j_2 1}$ *has the following properties:*

- *For* $a_{j_1j_1 0}(\bar{x}_{j_2} - a_{j_1j_2 1})^2 > 0$ *and* $a_{j_2j_2 0}(\bar{x}_{j_2} - a_{j_2j_2 1})^2 > 0$,

$$(a_{j_1j_1 1}, \bar{x}_{j_2}) = \underbrace{(\mathrm{SO}, \mathrm{II})}_{\text{increasing-inflection source}}. \tag{1.125}$$

The infinite-equilibrium of $x_{j_1}^* = a_{j_1j_1 1} = a_{j_2j_1 1}$ *is an* (SO,II)*-increasing-inflection source.*

- *For* $a_{j_1j_1 0}(\bar{x}_{j_2} - a_{j_1j_2 1})^2 < 0$ *and* $a_{j_2j_2 0}(\bar{x}_{j_2} - a_{j_2j_2 1})^2 > 0$,

$$(a_{j_1j_1 1}, \bar{x}_{j_2}) = \underbrace{(\mathrm{SI}, \mathrm{DI})}_{\text{increasing-inflection sink}}. \tag{1.126}$$

The infinite-equilibrium of $x_{j_1}^* = a_{j_1j_1 1} = a_{j_2j_1 1}$ *is an* (SI,DI)*-decreasing-inflection sink.*

- *For* $a_{j_1j_10}(\overline{x}_{j_2} - a_{j_1j_21})^2 > 0$ *and* $a_{j_2j_20}(\overline{x}_{j_2} - a_{j_2j_21})^2 < 0$,

$$(a_{j_1j_11}, \overline{x}_{j_2}) = \underbrace{(\text{SO}, \text{DI})}_{\text{increasing-inflection source}}. \tag{1.127}$$

The infinite-equilibrium of $x^*_{j_1} = a_{j_1j_11} = a_{j_2j_11}$ *is an* (SO,DI)-*decreasing-inflection source.*

- *For* $a_{j_1j_10}(\overline{x}_{j_2} - a_{j_1j_21})^2 < 0$ *and* $a_{j_2j_20}(\overline{x}_{j_2} - a_{j_2j_21})^2 < 0$,

$$(a_{j_1j_11}, \overline{x}_{j_2}) = \underbrace{(\text{SI}, \text{II})}_{\text{increasing-inflection sink}}. \tag{1.128}$$

The infinite-equilibrium of $x^*_{j_1} = a_{j_1j_11} = a_{j_2j_11}$ *is an* (SI,II)-*increasing-inflection sink.*

(v_{4b}) *The infinite-equilibrium of* $x^*_{j_2} = a_{j_2j_21} = a_{j_1j_21}$ *with* $\overline{x}_{j_1} \neq a_{j_1j_11}, a_{j_2j_11}$ *for* $a_{j_1j_11} = a_{j_2j_11}$ *and* $a_{j_2j_21} = a_{j_1j_21}$ *has the following properties:*

- *For* $a_{j_1j_10}(\overline{x}_{j_1} - a_{j_1j_11}) > 0$ *and* $a_{j_2j_20}(\overline{x}_{j_1} - a_{j_2j_11}) > 0$,

$$(\overline{x}_{j_1}, a_{j_2j_21}) = \underbrace{(\text{II}, \text{US})}_{\text{increasing-inflection upper-saddle}}. \tag{1.129}$$

The infinite-equilibrium of $x^*_{j_2} = a_{j_2j_21} = a_{j_1j_21}$ *is an* (II,US)-*increasing-inflection upper-saddle.*

- *For* $a_{j_1j_10}(\overline{x}_{j_1} - a_{j_1j_11}) < 0$ *and* $a_{j_2j_20}(\overline{x}_{j_1} - a_{j_2j_11}) > 0$,

$$(\overline{x}_{j_1}, a_{j_2j_21}) = \underbrace{(\text{DI}, \text{US})}_{\text{decreasing-inflection upper-saddle}}. \tag{1.130}$$

The infinite-equilibrium of $x^*_{j_2} = a_{j_2j_21} = a_{j_1j_21}$ *is a* (DI,US)-*decreasing-inflection upper-saddle.*

- *For* $a_{j_1j_10}(\overline{x}_{j_1} - a_{j_1j_11}) > 0$ *and* $a_{j_2j_20}(\overline{x}_{j_1} - a_{j_2j_11}) < 0$,

$$(\overline{x}_{j_1}, a_{j_2j_21}) = \underbrace{(\text{DI}, \text{LS})}_{\text{decreasing-inflection lower-saddle}}. \tag{1.131}$$

The infinite-equilibrium of $x^*_{j_2} = a_{j_2j_21} = a_{j_1j_21}$ *is a* (DI,US)-*decreasing-inflection lower-saddle.*

- *For* $a_{j_1j_10}(\overline{x}_{j_1}-a_{j_1j_11})<0$ *and* $a_{j_2j_20}(\overline{x}_{j_1}-a_{j_2j_11})<0$,

$$(\overline{x}_{j_1},a_{j_2j_21})=\underbrace{(\mathrm{II},\mathrm{LS})}_{\text{increasing-inflection lower-saddle}}. \tag{1.132}$$

The infinite-equilibrium of $x^*_{j_2}=a_{j_2j_21}=a_{j_1j_21}$ *is an* (II,LS)-*increasing-inflection lower-saddle.*

(v_{4c}) *The equilibrium of* $(x^*_{j_1},x^*_{j_2})=(a_{j_2j_11},a_{j_2j_21})=(a_{j_1j_11},a_{j_1j_21})$ *with* $a_{j_1j_11}=a_{j_2j_11}$ *and* $a_{j_2j_21}=a_{j_1j_21}$ *has the following properties:*

- *For* $a_{j_1j_10}>0$ *and* $a_{j_2j_20}>0$,

$$(a_{j_2j_11},a_{j_2j_21})=(a_{j_1j_11},a_{j_1j_21})=\underbrace{({}_{\mathrm{II:II}}\mathrm{SO},{}_{\mathrm{LS:US}}\mathrm{II})}_{\text{increasing-inflection source}}. \tag{1.133}$$

The equilibrium of $(x^*_{j_1},x^*_{j_2})=(a_{j_2j_11},a_{j_2j_21})=(a_{j_1j_11},a_{j_1j_21})$ *is an* $({}_{\mathrm{II:II}}\mathrm{SO},{}_{\mathrm{LS:US}}\mathrm{II})$-*increasing-inflection source.*

- *For* $a_{j_1j_10}<0$ *and* $a_{j_2j_20}>0$,

$$(a_{j_2j_11},a_{j_2j_21})=(a_{j_1j_11},a_{j_1j_21})=\underbrace{({}_{\mathrm{DI:DI}}\mathrm{SI},{}_{\mathrm{LS:US}}\mathrm{DI})}_{\text{decreasing-inflection sink}}. \tag{1.134}$$

The equilibrium of $(x^*_{j_1},x^*_{j_2})=(a_{j_2j_11},a_{j_2j_21})=(a_{j_1j_11},a_{j_1j_21})$ *is a* $({}_{\mathrm{DI:DI}}\mathrm{SI},{}_{\mathrm{LS:US}}\mathrm{DI})$-*decreasing-inflection sink.*

- *For* $a_{j_1j_10}>0$ *and* $a_{j_2j_20}<0$,

$$(a_{j_2j_11},a_{j_2j_21})=(a_{j_1j_11},a_{j_1j_21})=\underbrace{({}_{\mathrm{DI:DI}}\mathrm{SO},{}_{\mathrm{US:LS}}\mathrm{DI})}_{\text{decreasing-inflection source}}. \tag{1.135}$$

The equilibrium of $(x^*_{j_1},x^*_{j_2})=(a_{j_2j_11},a_{j_2j_21})=(a_{j_1j_11},a_{j_1j_21})$ *is a* $({}_{\mathrm{DI:DI}}\mathrm{SO},{}_{\mathrm{US:LS}}\mathrm{DI})$-*decreasing-inflection source.*

- *For* $a_{j_1j_10}<0$ *and* $a_{j_2j_20}<0$,

$$(a_{j_2j_11},a_{j_2j_21})=(a_{j_1j_11},a_{j_1j_21})=\underbrace{({}_{\mathrm{II:II}}\mathrm{SI},{}_{\mathrm{US:LS}}\mathrm{II})}_{\text{increasing-inflection sink}}. \tag{1.136}$$

The equilibrium of $(x^*_{j_1},x^*_{j_2})=(a_{j_2j_11},a_{j_2j_21})=(a_{j_1j_11},a_{j_1j_21})$ *is an* $({}_{\mathrm{II:II}}\mathrm{SI},{}_{\mathrm{US:LS}}\mathrm{II})$-*increasing-inflection sink.*

(vi) For $\Delta_{j_1j_2}=0$ and $\Delta_{j_2j_2}>0$, the standard form is

$$\begin{aligned}\dot{x}_{j_2}&=a_{j_2j_20}(x_{j_1}-a_{j_2j_11})(x_{j_2}-a_{j_2j_21})(x_{j_2}-a_{j_2j_22}),\\ \dot{x}_{j_1}&=a_{j_1j_10}(x_{j_1}-a_{j_1j_11})(x_{j_2}-a_{j_1j_21})^2\end{aligned} \tag{1.137}$$

where

$$\begin{aligned}a_{j_1j_21}&=-\frac{1}{2}B_{j_1j_2},\\ b_{j_2j_21},b_{j_2j_22}&=-\frac{1}{2}(B_{j_2j_2}\pm\sqrt{\Delta_{j_2j_2}}),\\ \{a_{j_2j_21},a_{j_2j_22}\}&=\text{sort}\{b_{j_2j_21},b_{j_2j_22}\},a_{j_2j_21}<a_{j_2j_22}.\end{aligned} \tag{1.138}$$

(vi_1) The first integral manifold ($a_{j_2j_2s_1}\neq a_{j_1j_21}$, $a_{j_1j_11}\neq a_{j_2j_11}$) is

$$\begin{aligned}&\left[(x_{j_1}-a_{j_1j_11})-(x_{j_10}-a_{j_1j_11})\right]+(a_{j_1j_11}-a_{j_2j_11})\ln\frac{|x_{j_1}-a_{j_1j_11}|}{|x_{j_10}-a_{j_1j_11}|}\\ &=\frac{a_{j_1j_10}}{a_{j_2j_20}}\{\left[(x_{j_2}-a_{j_2j_21})-(x_{j_20}-a_{j_2j_21})\right]\\ &\quad+\sum\nolimits_{s_1=1}^{2}\frac{(a_{j_2j_2s_1}-a_{j_1j_21})^2}{(a_{j_2j_2s_1}-a_{j_2j_2s_2})}\ln\frac{|x_{j_2}-a_{j_2j_2s_1}|}{|x_{j_20}-a_{j_2j_2s_1}|}\}.\end{aligned} \tag{1.139}$$

(vi_{1a}) The equilibrium of $(x_{j_1}^*,x_{j_2}^*)=(a_{j_1j_11},a_{j_2j_2s_1})$ (s_1, $s_2\in\{1,2\}$, $s_1\neq s_2$) has the following properties:

- *For $a_{j_1j_10}(a_{j_2j_2s_1}-a_{j_1j_21})^2>0$ and $a_{j_2j_20}(a_{j_1j_11}-a_{j_2j_11})(a_{j_2j_2s_1}-a_{j_2j_2s_2})>0$,*

$$(a_{j_1j_11},a_{j_2j_2s_1})=\underbrace{(\text{SO},\text{SO})}_{\text{source}}. \tag{1.140}$$

  The equilibrium of $(x_{j_1}^*,x_{j_2}^*)=(a_{j_1j_11},a_{j_2j_2s_1})$ is an (SO,SO)-source.
- *For $a_{j_1j_10}(a_{j_2j_2s_1}-a_{j_1j_21})^2<0$ and $a_{j_2j_20}(a_{j_1j_11}-a_{j_2j_11})(a_{j_2j_2s_1}-a_{j_2j_2s_2})>0$,*

$$(a_{j_1j_11},a_{j_2j_2s_1})=\underbrace{(\text{SI},\text{SO})}_{\text{saddle}}. \tag{1.141}$$

  The equilibrium of $(x_{j_1}^*,x_{j_2}^*)=(a_{j_1j_11},a_{j_2j_2s_1})$ is an (SI,SO)-saddle.
- *For $a_{j_1j_10}(a_{j_2j_2s_1}-a_{j_1j_21})^2>0$ and $a_{j_2j_20}(a_{j_1j_11}-a_{j_2j_11})(a_{j_2j_2s_1}-a_{j_2j_2s_2})<0$,*

$$(a_{j_1j_11},a_{j_2j_2s_1})=\underbrace{(\text{SO},\text{SI})}_{\text{saddle}}. \tag{1.142}$$

  The equilibrium of $(x_{j_1}^*,x_{j_2}^*)=(a_{j_1j_11},a_{j_2j_2s_1})$ is an (SO,SI)-saddle.

- *For* $a_{j_1j_10}(a_{j_2j_2s_1} - a_{j_1j_21})^2 < 0$ *and* $a_{j_2j_20}(a_{j_1j_11} - a_{j_2j_11})(a_{j_2j_2s_1} - a_{j_2j_2s_2}) < 0$,

$$(a_{j_1j_11}, a_{j_2j_2s_1}) = \underbrace{(\mathrm{SI}, \mathrm{SI})}_{\text{sink}}. \tag{1.143}$$

The equilibrium of $(x^*_{j_1}, x^*_{j_2}) = (a_{j_1j_11}, a_{j_2j_2s_1})$ *is an* (SI,SI)-*sink.*

(vi$_{1b}$) The equilibrium of $(x^*_{j_1}, x^*_{j_2}) = (a_{j_1j_11}, a_{j_1j_21})$ *with* $a_{j_1j_21} \neq a_{j_2j_21}, a_{j_2j_22}$ *has the following properties:*

- *For* $a_{j_1j_10} > 0$ *and* $a_{j_2j_20}(a_{j_1j_11} - a_{j_2j_11})\prod_{s_1=1}^{2}(a_{j_1j_21} - a_{j_2j_2s_1}) > 0$,

$$(a_{j_1j_11}, a_{j_1j_21}) = \underbrace{({}_{\mathrm{DI:II}}\mathrm{SO}, \mathrm{pF})}_{\text{inflection-source flow }(+)}. \tag{1.144}$$

The equilibrium of $(x^*_{j_1}, x^*_{j_2}) = (a_{j_1j_11}, a_{j_1j_21})$ *is a* (${}_{\mathrm{DI:II}}$SO, pF)-*positive inflection-source flow.*

- *For* $a_{j_1j_10} < 0$ *and* $a_{j_2j_20}(a_{j_1j_11} - a_{j_2j_11})\prod_{s_1=1}^{2}(a_{j_1j_21} - a_{j_2j_2s_1}) > 0$,

$$(a_{j_1j_11}, a_{j_1j_21}) = \underbrace{({}_{\mathrm{II:DI}}\mathrm{SI}, \mathrm{pF})}_{\text{inflection-sink flow }(+)}. \tag{1.145}$$

The equilibrium of $(x^*_{j_1}, x^*_{j_2}) = (a_{j_1j_11}, a_{j_1j_21})$ *is an* (${}_{\mathrm{II:DI}}$SI, pF)-*positive inflection-sink flow.*

- *For* $a_{j_1j_10} > 0$ *and* $a_{j_2j_20}(a_{j_1j_11} - a_{j_2j_11})\prod_{s_1=1}^{2}(a_{j_1j_21} - a_{j_2j_2s_1}) < 0$,

$$(a_{j_1j_11}, a_{j_1j_21}) = \underbrace{({}_{\mathrm{II:DI}}\mathrm{SO}, \mathrm{nF})}_{\text{inflection-source flow }(-)}. \tag{1.146}$$

The equilibrium of $(x^*_{j_1}, x^*_{j_2}) = (a_{j_1j_11}, a_{j_1j_21})$ *is an* (${}_{\mathrm{II:DI}}$SO, nF)-*negative inflection-sink flow.*

- *For* $a_{j_1j_10} < 0$ *and* $a_{j_2j_20}(a_{j_1j_11} - a_{j_2j_11})\prod_{s_1=1}^{2}(a_{j_1j_21} - a_{j_2j_2s_1}) < 0$,

$$(a_{j_1j_11}, a_{j_1j_21}) = \underbrace{({}_{\mathrm{DI:II}}\mathrm{SI}, \mathrm{nF})}_{\text{inflection-sink flow }(-)}. \tag{1.147}$$

The equilibrium of $(x^*_{j_1}, x^*_{j_2}) = (a_{j_1j_11}, a_{j_1j_21})$ *is a* (${}_{\mathrm{DI:II}}$SI, nF)-*negative inflection-sink flow.*

(vi_{1c}) The equilibrium of $(x^*_{j_2}, x^*_{j_1}) = (a_{j_1j_2 1}, a_{j_2j_1 1})$ *has the following properties:*

- *For* $a_{j_1j_1 0}(a_{j_2j_1 1} - a_{j_1j_1 1}) > 0$ *and* $a_{j_2j_2 0}\prod_{s_1=1}^{2}(a_{j_1j_2 1} - a_{j_2j_2 s_1}) > 0$,

$$(a_{j_1j_2 1}, a_{j_2j_1 1}) = \underbrace{(\text{US}, \text{UP})}_{\text{up-parabola upper-saddle}}. \tag{1.148}$$

The equilibrium of $(x^*_{j_2}, x^*_{j_1}) = (a_{j_1j_2 1}, a_{j_2j_1 1})$ *is a* (US,UP)*-up-parabola upper-saddle.*

- *For* $a_{j_1j_1 0}(a_{j_2j_1 1} - a_{j_1j_1 1}) < 0$ *and* $a_{j_2j_2 0}\prod_{s_1=1}^{2}(a_{j_1j_2 1} - a_{j_2j_2 s_1}) > 0$,

$$(a_{j_1j_2 1}, a_{j_2j_1 1}) = \underbrace{(\text{LS}, \text{DP})}_{\text{down-parabola lower-saddle}}. \tag{1.149}$$

The equilibrium of $(x^*_{j_2}, x^*_{j_1}) = (a_{j_1j_2 1}, a_{j_2j_1 1})$ *is an* (LS,DP)*-down-parabola lower-saddle.*

- *For* $a_{j_1j_1 0}(a_{j_2j_1 1} - a_{j_1j_1 1}) > 0$ *and* $a_{j_2j_2 0}\prod_{s_1=1}^{2}(a_{j_1j_2 1} - a_{j_2j_2 s_1}) < 0$,

$$(a_{j_1j_2 1}, a_{j_2j_1 1}) = \underbrace{(\text{US}, \text{DP})}_{\text{down-parabola upper-saddle}}. \tag{1.150}$$

The equilibrium of $(x^*_{j_2}, x^*_{j_1}) = (a_{j_1j_2 1}, a_{j_2j_1 1})$ *is a* (US,DP)*-down-parabola upper-saddle.*

- *For* $a_{j_1j_1 0}(a_{j_2j_1 1} - a_{j_1j_1 1}) < 0$ *and* $a_{j_2j_2 0}\prod_{s_1=1}^{2}(a_{j_1j_2 1} - a_{j_2j_2 s_1}) < 0$,

$$(a_{j_1j_2 1}, a_{j_2j_1 1}) = \underbrace{(\text{LS}, \text{UP})}_{\text{upper-parabola lower-saddle}}. \tag{1.151}$$

The equilibrium of $(x^*_{j_2}, x^*_{j_1}) = (a_{j_1j_2 1}, a_{j_2j_1 1})$ *is an* (LS,UP)*-up-parabola lower-saddle.*

(vi_{1d}) The equilibrium of $(x^*_{j_1}, x^*_{j_2}) = (a_{j_2j_1 1}, a_{j_2j_2 s_1})$ $(s_1,\ s_2 \in \{1,2\})$ *with* $a_{j_1j_2 1} \neq a_{j_2j_2 1}, a_{j_2j_2 2}$ *has the following properties:*

- *For* $a_{j_1j_1 0}(a_{j_2j_1 1} - a_{j_1j_1 1})(a_{j_2j_2 s_1} - a_{j_1j_2 1})^2 > 0$ *and* $a_{j_2j_2 0}(a_{j_2j_2 s_1} - a_{j_2j_2 s_2}) > 0$,

$$(a_{j_2j_1 1}, a_{j_2j_2 s_1}) = \underbrace{(\text{pF}, \text{DP} : \text{UP})}_{\text{hyperbolic flow }(+)}. \tag{1.152}$$

The equilibrium of $(x^*_{j_1}, x^*_{j_2}) = (a_{j_2j_1 1}, a_{j_2j_2 s_1})$ *is a* (pF, DP:UP)*-positive hyperbolic flow.*

- *For* $a_{j_1j_10}(a_{j_2j_11}-a_{j_1j_11})(a_{j_2j_2s_1}-a_{j_1j_21})^2<0$ *and* $a_{j_2j_20}(a_{j_2j_2s_1}-a_{j_2j_2s_2})>0$,

$$(a_{j_2j_11},a_{j_2j_2s_1})=\underbrace{(\text{nF},\text{UP}:\text{DP})}_{\text{hyperbolic-secant flow }(-)}. \tag{1.153}$$

The equilibrium of $(x^*_{j_1},x^*_{j_2})=(a_{j_2j_11},a_{j_2j_2s_1})$ *is an* (nF, UP:DP)*-negative hyperbolic-secant flow.*

- *For* $a_{j_1j_10}(a_{j_2j_11}-a_{j_1j_11})(a_{j_2j_2s_1}-a_{j_1j_21})^2>0$ *and* $a_{j_2j_20}(a_{j_2j_2s_1}-a_{j_2j_2s_2})<0$,

$$(a_{j_2j_11},a_{j_2j_2s_1})=\underbrace{(\text{pF},\text{UP}:\text{DP})}_{\text{hyperbolic-secant flow }(+)}. \tag{1.154}$$

The equilibrium of $(x^*_{j_1},x^*_{j_2})=(a_{j_2j_11},a_{j_2j_2s_1})$ *is a* (pF, UP:DP)*-negative hyperbolic-secant flow.*

- *For* $a_{j_1j_10}(a_{j_2j_11}-a_{j_1j_11})(a_{j_2j_2s_1}-a_{j_1j_21})^2<0$ *and* $a_{j_2j_20}(a_{j_2j_2s_1}-a_{j_2j_2s_2})<0$,

$$(a_{j_2j_11},a_{j_2j_2s_1})=\underbrace{(\text{nF},\text{DP}:\text{UP})}_{\text{hyperbolic flow }(-)}. \tag{1.155}$$

The equilibrium of $(x^*_{j_1},x^*_{j_2})=(a_{j_2j_11},a_{j_2j_2s_1})$ *is an* (nF, DP:UP)*-negative hyperbolic flow.*

(vi₂) The first integral manifold ($a_{j_2j_2s_1}=a_{j_1j_21}$ and $a_{j_1j_11}\neq a_{j_2j_11}$) *is*

$$\begin{aligned}&\left[(x_{j_1}-a_{j_1j_11})-(x_{j_10}-a_{j_1j_11})\right]+(a_{j_1j_11}-a_{j_2j_11})\ln\frac{|x_{j_1}-a_{j_1j_11}|}{|x_{j_10}-a_{j_1j_11}|}\\&=\frac{a_{j_1j_10}}{a_{j_2j_20}}\Big\{\left[(x_{j_2}-a_{j_2j_2s_2})-(x_{j_20}-a_{j_2j_2s_2})\right]\\&\quad+(a_{j_2j_2s_2}-a_{j_1j_21})\ln\frac{|x_{j_2}-a_{j_2j_2s_2}|}{|x_{j_20}-a_{j_2j_2s_2}|}\Big\}.\end{aligned} \tag{1.156}$$

(vi_{2a}) The infinite-equilibrium of $x^*_{j_2}=a_{j_1j_21}=a_{j_2j_2s_1}$ *with* $\overline{x}_{j_1}\neq a_{j_1j_11}$ *has the following properties:*

- *For* $a_{j_1j_10}(\overline{x}_{j_1}-a_{j_1j_11})>0$ *and* $a_{j_2j_20}(\overline{x}_{j_1}-a_{j_2j_11})(a_{j_2j_2s_1}-a_{j_2j_2s_2})>0$,

$$(\overline{x}_{j_1},a_{j_2j_2s_1})=\underbrace{(\text{UP},\text{SO})}_{\text{up-parabola-source}}. \tag{1.157}$$

The infinite-equilibrium of $x^*_{j_2}=a_{j_1j_21}=a_{j_2j_2s_1}$ *is a* (UP,SO)*-up-parabola-source.*

- *For* $a_{j_1j_10}(\bar{x}_{j_1}-a_{j_1j_11})<0$ *and* $a_{j_2j_20}(\bar{x}_{j_1}-a_{j_2j_11})(a_{j_2j_2s_1}-a_{j_2j_2s_2})>0$,

$$(\bar{x}_{j_1},a_{j_2j_2s_1})=\underbrace{(\mathrm{DP},\mathrm{SO})}_{\text{down-parabola-source}}. \tag{1.158}$$

  The infinite-equilibrium of $x_{j_2}^*=a_{j_1j_21}=a_{j_2j_2s_1}$ *is a* (DP,SO)-*down-parabola-source.*
- *For* $a_{j_1j_10}(\bar{x}_{j_1}-a_{j_1j_11})>0$ *and* $a_{j_2j_20}(\bar{x}_{j_1}-a_{j_2j_11})(a_{j_2j_2s_1}-a_{j_2j_2s_2})<0$,

$$(\bar{x}_{j_1},a_{j_2j_2s_1})=\underbrace{(\mathrm{DP},\mathrm{SI})}_{\text{down-parabola-sink}}. \tag{1.159}$$

  The infinite-equilibrium of $x_{j_2}^*=a_{j_1j_21}=a_{j_2j_2s_1}$ *is a* (DP,SI)-*down-parabola-sink.*
- *For* $a_{j_1j_10}(\bar{x}_{j_1}-a_{j_1j_11})<0$ *and* $a_{j_2j_20}(\bar{x}_{j_1}-a_{j_2j_11})(a_{j_2j_2s_1}-a_{j_2j_2s_2})<0$,

$$(\bar{x}_{j_1},a_{j_2j_2s_1})=\underbrace{(\mathrm{UP},\mathrm{SI})}_{\text{up-parabola-sink}}. \tag{1.160}$$

  The infinite-equilibrium of $x_{j_2}^*=a_{j_1j_21}=a_{j_2j_2s_1}$ *is a* (UP,SI)-*up-parabola-sink.*

(vi$_{2b}$) The equilibrium of $(x_{j_1}^*,x_{j_2}^*)=(a_{j_1j_11},a_{j_1j_21})$ *with* $a_{j_2j_2s_1}=a_{j_1j_21}$ *has the following properties:*

- *For* $a_{j_1j_10}>0$ *and* $a_{j_2j_20}(a_{j_1j_11}-a_{j_2j_11})(a_{j_2j_2s_1}-a_{j_2j_2s_2})>0$,

$$(a_{j_1j_11},a_{j_1j_21})\equiv(a_{j_1j_11},a_{j_2j_2s_1})=\underbrace{(\mathrm{DP}:\mathrm{UP},\mathrm{SO})}_{\text{hyperbolic source}}. \tag{1.161}$$

  The equilibrium of $(x_{j_1}^*,x_{j_2}^*)=(a_{j_1j_11},a_{j_1j_21})$ *is a* (DP:UP,SO)-*hyperbolic source.*
- *For* $a_{j_1j_10}<0$ *and* $a_{j_2j_20}(a_{j_1j_11}-a_{j_2j_11})(a_{j_2j_2s_1}-a_{j_2j_2s_2})>0$,

$$(a_{j_1j_11},a_{j_1j_21})\equiv(a_{j_1j_11},a_{j_2j_2s_1})=\underbrace{(\mathrm{UP}:\mathrm{DP},\mathrm{SO})}_{\text{hyperbolic-secant source}}. \tag{1.162}$$

  The equilibrium of $(x_{j_1}^*,x_{j_2}^*)=(a_{j_1j_11},a_{j_1j_21})$ *is a* (UP:DP,SO)-*hyperbolic-secant source.*
- *For* $a_{j_1j_10}>0$ *and* $a_{j_2j_20}(a_{j_1j_11}-a_{j_2j_11})(a_{j_2j_2s_1}-a_{j_2j_2s_2})<0$,

$$(a_{j_1j_11},a_{j_1j_21})\equiv(a_{j_1j_11},a_{j_2j_2s_1})=\underbrace{(\mathrm{UP}:\mathrm{DP},\mathrm{SI})}_{\text{hyperbolic-secant sink}}. \tag{1.163}$$

  The equilibrium of $(x_{j_1}^*,x_{j_2}^*)=(a_{j_1j_11},a_{j_1j_21})$ *is a* (UP:DP,SI)-*hyperbolic-secant sink.*

- *For* $a_{j_1j_10}<0$ *and* $a_{j_2j_20}(a_{j_1j_11}-a_{j_2j_11})(a_{j_2j_2s_1}-a_{j_2j_2s_2})<0$,

$$(a_{j_1j_11},a_{j_1j_21})\equiv(a_{j_1j_11},a_{j_2j_2s_1})=\underbrace{(\text{DP}:\text{UP},\text{SI})}_{\text{hyperbolic sink}}. \tag{1.164}$$

The equilibrium of $(x^*_{j_1},x^*_{j_2})=(a_{j_1j_11},a_{j_1j_21})$ *is a* (DP:UP,SI)-*hyperbolic sink.*

(vi_{2c}) *The equilibrium of* $(x^*_{j_2},x^*_{j_1})=(a_{j_1j_21},a_{j_2j_11})$*with* $a_{j_2j_2s_1}=a_{j_1j_21}$ *has the following properties:*

- *For* $a_{j_1j_10}(a_{j_2j_11}-a_{j_1j_11})>0$ *and* $a_{j_2j_20}(a_{j_2j_2s_1}-a_{j_2j_2s_2})>0$,

$$(a_{j_1j_21},a_{j_2j_11})=\underbrace{(\text{DP}_+:\text{UP}_+,{}_{\text{DP}}\text{SI}:{}_{\text{UP}}\text{SO})}_{\text{hyperbolic sink-to-source}}. \tag{1.165}$$

The equilibrium of $(x^*_{j_2},x^*_{j_1})=(a_{j_1j_21},a_{j_2j_11})$ *is a* (DP$_+$:UP$_+$,$_{\text{DP}}$SI:$_{\text{UP}}$SO)-*hyperbolic sink-to-source.*

- *For* $a_{j_1j_10}(a_{j_2j_11}-a_{j_1j_11})<0$ *and* $a_{j_2j_20}(a_{j_2j_2s_1}-a_{j_2j_2s_2})>0$,

$$(a_{j_1j_21},a_{j_2j_11})=\underbrace{(\text{UP}_-:\text{DP}_-,{}_{\text{UP}}\text{SI}:{}_{\text{DP}}\text{SO})}_{\text{circular sink-to-source}}. \tag{1.166}$$

The equilibrium of $(x^*_{j_2},x^*_{j_1})=(a_{j_1j_21},a_{j_2j_11})$ *is a* (UP$_-$:DP$_-$,$_{\text{UP}}$SI:$_{\text{DP}}$SO)-*circular sink-to-source.*

- *For* $a_{j_1j_10}(a_{j_2j_11}-a_{j_1j_11})>0$ *and* $a_{j_2j_20}(a_{j_2j_2s_1}-a_{j_2j_2s_2})<0$,

$$(a_{j_1j_21},a_{j_2j_11})=\underbrace{(\text{UP}_+:\text{DP}_+,{}_{\text{UP}}\text{SO}:{}_{\text{DP}}\text{SI})}_{\text{circular soure-to-sink}}. \tag{1.167}$$

The equilibrium of $(x^*_{j_2},x^*_{j_1})=(a_{j_1j_21},a_{j_2j_11})$ *is a* (UP$_+$:DP$_+$,$_{\text{UP}}$SO:$_{\text{DP}}$SI)-*circular source-to-sink.*

- *For* $a_{j_1j_10}(a_{j_2j_11}-a_{j_1j_11})<0$ *and* $a_{j_2j_20}(a_{j_2j_2s_1}-a_{j_2j_2s_2})<0$,

$$(a_{j_1j_21},a_{j_2j_11})=\underbrace{(\text{DP}_-:\text{UP}_-,{}_{\text{DP}}\text{SO}:{}_{\text{UP}}\text{SI})}_{\text{hyperbolic source-to-sink}}. \tag{1.168}$$

The equilibrium of $(x^*_{j_2},x^*_{j_1})=(a_{j_1j_21},a_{j_2j_11})$ *is a* (DP$_-$:UP$_-$,$_{\text{DP}}$SO:$_{\text{UP}}$SI)-*hyperbolic source-to-sink.*

(vi_3) The first integral manifold ($a_{j_1j_11}=a_{j_2j_11}$ and $a_{j_2j_2s_1}\neq a_{j_1j_21}$) is

$$[(x_{j_1}-a_{j_1j_11})-(x_{j_10}-a_{j_1j_11})] = \frac{a_{j_1j_10}}{a_{j_2j_20}}\Big\{[(x_{j_2}-a_{j_2j_21})-(x_{j_20}-a_{j_2j_21})] + \sum\nolimits_{s_1=1}^{2}\frac{(a_{j_2j_2s_1}-a_{j_1j_21})^2}{(a_{j_2j_2s_1}-a_{j_2j_2s_2})}\ln\frac{|x_{j_2}-a_{j_2j_2s_1}|}{|x_{j_20}-a_{j_2j_2s_1}|}\Big\}. \tag{1.169}$$

(vi_{3a}) The infinite-equilibrium of $x^*_{j_1}=a_{j_1j_11}=a_{j_2j_11}$ with $\bar{x}_{j_2}\neq a_{j_2j_2s_1}, a_{j_1j_21}$ has the following properties:

- *For $a_{j_1j_10}(\bar{x}_{j_2}-a_{j_1j_21})^2>0$ and $a_{j_2j_20}\prod_{s_1=1}^{2}(\bar{x}_{j_2}-a_{j_2j_2s_1})>0$,*

$$(a_{j_1j_11},\bar{x}_{j_2}) = \underbrace{(\text{SO, II})}_{\text{increasing-inflection source}}. \tag{1.170}$$

The infinite-equilibrium of $x^*_{j_1}=a_{j_1j_11}=a_{j_2j_11}$ is an (SO,II)*-increasing-inflection source.*

- *For $a_{j_1j_10}(\bar{x}_{j_2}-a_{j_1j_21})^2<0$ and $a_{j_2j_20}\prod_{s_1=1}^{2}(\bar{x}_{j_2}-a_{j_2j_2s_1})>0$,*

$$(a_{j_1j_11},\bar{x}_{j_2}) = \underbrace{(\text{SI, DI})}_{\text{decreasing-inflection sink}}. \tag{1.171}$$

The infinite-equilibrium of $x^*_{j_1}=a_{j_1j_11}=a_{j_2j_11}$ is an (SI,DI)*-decreasing-inflection sink.*

- *For $a_{j_1j_10}(\bar{x}_{j_2}-a_{j_1j_21})^2>0$ and $a_{j_2j_20}\prod_{s_1=1}^{2}(\bar{x}_{j_2}-a_{j_2j_2s_1})<0$,*

$$(a_{j_1j_11},\bar{x}_{j_2}) = \underbrace{(\text{SO, DI})}_{\text{decreasing-inflection source}}. \tag{1.172}$$

The infinite-equilibrium of $x^*_{j_2}=a_{j_1j_21}=a_{j_2j_2s_1}$ is an (SO,DI)*-decreasing-inflection source.*

- *For $a_{j_1j_10}(\bar{x}_{j_2}-a_{j_1j_21})^2<0$ and $a_{j_2j_20}\prod_{s_1=1}^{2}(\bar{x}_{j_2}-a_{j_2j_2s_1})<0$,*

$$(a_{j_1j_11},\bar{x}_{j_2}) = \underbrace{(\text{SI, II})}_{\text{increasing-inflection sink}}. \tag{1.173}$$

The infinite-equilibrium of $x^*_{j_1}=a_{j_1j_11}=a_{j_2j_11}$ is an (SI,II)*-increasing-inflection sink.*

(vi_{3b}) The equilibrium of $(x^*_{j_1}, x^*_{j_2}) = (a_{j_1j_11}, a_{j_1j_21})$ *with* $a_{j_1j_11} = a_{j_2j_11}$ *has the following properties:*

- *For* $a_{j_1j_10} > 0$ *and* $a_{j_2j_20}\prod_{s_1=1}^{2}(a_{j_1j_21} - a_{j_2j_2s_1}) > 0$,

$$(a_{j_1j_11}, a_{j_1j_21}) = \underbrace{(\text{SO}, \text{II} : \text{II})}_{\text{increasing-inflection source}}. \tag{1.174}$$

  The equilibrium of $(x^*_{j_1}, x^*_{j_2}) = (a_{j_1j_11}, a_{j_1j_21})$ *is an* (SO,II:II)*-increasing-inflection source of the first kind.*

- *For* $a_{j_1j_10} < 0$ *and* $a_{j_2j_20}\prod_{s_1=1}^{2}(a_{j_1j_21} - a_{j_2j_2s_1}) > 0$,

$$(a_{j_1j_11}, a_{j_1j_21}) = \underbrace{(\text{SI}, \text{DI} : \text{DI})}_{\text{decreasing-inflection sink}}. \tag{1.175}$$

  The equilibrium of $(x^*_{j_1}, x^*_{j_2}) = (a_{j_1j_11}, a_{j_1j_21})$ *is an* (SI,DI:DI)*-decreasing-inflection sink of the first kind.*

- *For* $a_{j_1j_10} > 0$ *and* $a_{j_2j_20}\prod_{s_1=1}^{2}(a_{j_1j_21} - a_{j_2j_2s_1}) < 0$,

$$(a_{j_1j_11}, a_{j_1j_21}) = \underbrace{(\text{SO}, \text{DI} : \text{DI})}_{\text{decreasing-inflection source}}. \tag{1.176}$$

  The equilibrium of $(x^*_{j_1}, x^*_{j_2}) = (a_{j_1j_11}, a_{j_1j_21})$ *is an* (SO,DI:DI)*-decreasing-inflection source of the first kind.*

- *For* $a_{j_1j_10} < 0$ *and* $a_{j_2j_20}\prod_{s_1=1}^{2}(a_{j_1j_21} - a_{j_2j_2s_1}) < 0$,

$$(a_{j_1j_11}, a_{j_1j_21}) = \underbrace{(\text{SI}, \text{II} : \text{II})}_{\text{increasing-inflection sink}}. \tag{1.177}$$

  The equilibrium of $(x^*_{j_1}, x^*_{j_2}) = (a_{j_1j_11}, a_{j_1j_21})$ *is an* (SI,II:II)*-increasing-inflection sink of the first kind.*

(vi_{3c}) The equilibrium of $(x^*_{j_1}, x^*_{j_2}) = (a_{j_1j_11}, a_{j_2j_2s_1})$ *with* $a_{j_2j_11} = a_{j_1j_11}$ *has the following properties:*

- *For* $a_{j_1j_10}(a_{j_2j_2s_1} - a_{j_1j_11})^2 > 0$ *and* $a_{j_2j_20}(a_{j_2j_2s_1} - a_{j_2j_2s_2}) > 0$,

$$(a_{j_1j_11}, a_{j_2j_2s_1}) = \underbrace{(\text{SO}, \text{DI} : \text{II})}_{\text{inflection source}}. \tag{1.178}$$

  The equilibrium of $(x^*_{j_1}, x^*_{j_2}) = (a_{j_1j_11}, a_{j_2j_2s_1})$ *is an* (SO,DI:II)*-inflection source.*

- *For* $a_{j_1j_1 0}(a_{j_2j_2s_1}-a_{j_1j_2 1})^2<0$ *and* $a_{j_2j_2 0}(a_{j_2j_2s_1}-a_{j_2j_2s_2})>0$,

$$(a_{j_1j_1 1},a_{j_2j_2s_1})=\underbrace{(\mathrm{SI,II:DI})}_{\text{inflection sink}}. \tag{1.179}$$

The equilibrium of $(x^*_{j_1},x^*_{j_2})=(a_{j_1j_1 1},a_{j_2j_2s_1})$ *is an* (SI,II:DI)-*inflection sink.*

- *For* $a_{j_1j_1 0}(a_{j_2j_2s_1}-a_{j_1j_2 1})^2>0$ *and* $a_{j_2j_2 0}(a_{j_2j_2s_1}-a_{j_2j_2s_2})<0$,

$$(a_{j_1j_1 1},a_{j_2j_2s_1})=\underbrace{(\mathrm{SO,II:DI})}_{\text{inflection source}}. \tag{1.180}$$

The equilibrium of $(x^*_{j_1},x^*_{j_2})=(a_{j_1j_1 1},a_{j_2j_2s_1})$ *is an* (SO,II:DI)-*inflection source.*

- *For* $a_{j_1j_1 0}(a_{j_2j_2s_1}-a_{j_1j_2 1})^2<0$ *and* $a_{j_2j_2 0}(a_{j_2j_2s_1}-a_{j_2j_2s_2})<0$,

$$(a_{j_1j_1 1},a_{j_2j_2s_1})=\underbrace{(\mathrm{SI,DI:II})}_{\text{inflection sink}}. \tag{1.181}$$

The equilibrium of $(x^*_{j_1},x^*_{j_2})=(a_{j_1j_1 1},a_{j_2j_2s_1})$ *is an* (SI,DI:II)-*inflection sink.*

(vi$_4$) The first integral manifold $(a_{j_1j_1 1}=a_{j_2j_1 1}\text{and}a_{j_2j_2s_1}=a_{j_1j_2 1})$ *is*

$$\begin{aligned}\left[(x_{j_1}-a_{j_1j_1 1})-(x_{j_1 0}-a_{j_1j_1 1})\right] &= \frac{a_{j_1j_1 0}}{a_{j_2j_2 0}}\Bigg\{\left[(x_{j_2}-a_{j_2j_2 1})-(x_{j_2 0}-a_{j_2j_2 1})\right]\\ &\quad +(a_{j_2j_2s_2}-a_{j_1j_2 1})\ln\frac{|x_{j_2}-a_{j_2j_2s_2}|}{|x_{j_2 0}-a_{j_2j_2s_2}|}\Bigg\}.\end{aligned} \tag{1.182}$$

The equilibrium of $(x^*_{j_1},x^*_{j_2})=(a_{j_1j_1 1},a_{j_1j_2 1})$ *with* $a_{j_1j_1 1}=a_{j_2j_1 1}$ and $a_{j_1j_2 1}=a_{j_2j_2s_1}$ *has the following properties:*

- *For* $a_{j_1j_1 0}>0$ *and* $a_{j_2j_2 0}(a_{j_2j_2s_1}-a_{j_2j_2s_2})>0$,

$$(a_{j_1j_1 1},a_{j_1j_2 1})=\underbrace{({}_{\mathrm{DI:II}}\mathrm{SO},\ {}_{\mathrm{SI:SO}}\mathrm{UP})}_{\text{up-parabola-source}}. \tag{1.183}$$

The equilibrium of $(x^*_{j_1},x^*_{j_2})=(a_{j_1j_1 1},a_{j_1j_2 1})$ *is a* $({}_{\mathrm{DI:II}}\mathrm{SO},\ {}_{\mathrm{SI:SO}}\mathrm{UP})$-up-parabola-source.

- *For* $a_{j_1j_1 0}<0$ *and* $a_{j_2j_2 0}(a_{j_2j_2s_1}-a_{j_2j_2s_2})>0$,

$$(a_{j_1j_1 1},a_{j_1j_2 1})=\underbrace{({}_{\mathrm{II:DI}}\mathrm{SI},\ {}_{\mathrm{SO:SI}}\mathrm{DP})}_{\text{down-parabola-sink}}. \tag{1.184}$$

The equilibrium of $(x^*_{j_1},x^*_{j_2})=(a_{j_1j_1 1},a_{j_1j_2 1})$ *is an* $({}_{\mathrm{II:DI}}\mathrm{SI},\ {}_{\mathrm{SO:SI}}\mathrm{DP})$-down-parabola-sink.

- *For* $a_{j_1j_10} > 0$ *and* $a_{j_2j_20}(a_{j_2j_2s_1} - a_{j_2j_2s_2}) < 0$,

$$(a_{j_1j_11}, a_{j_1j_21}) = \underbrace{({}_{\text{II:DI}}\text{SO}, {}_{\text{SI:SO}}\text{DP})}_{\text{down-parabola-source}}. \tag{1.185}$$

The equilibrium of $(x^*_{j_1}, x^*_{j_2}) = (a_{j_1j_11}, a_{j_1j_21})$ *is an* (${}_{\text{II:DI}}$SO, ${}_{\text{SI:SO}}$DP)-down-parabola-source.

- *For* $a_{j_1j_10} < 0$ *and* $a_{j_2j_20}(a_{j_2j_2s_1} - a_{j_2j_2s_2}) < 0$,

$$(a_{j_1j_11}, a_{j_1j_21}) = \underbrace{({}_{\text{DI:II}}\text{SI}, {}_{\text{SI:SO}}\text{UP})}_{\text{up-parabola-sink}}. \tag{1.186}$$

The equilibrium of $(x^*_{j_1}, x^*_{j_2}) = (a_{j_1j_11}, a_{j_1j_21})$ *is a* (${}_{\text{DI:II}}$SI, ${}_{\text{SO:SI}}$UP)-up-parabola-sink.

(vii) For $\Delta_{j_1j_2} > 0$ and $\Delta_{j_2j_2} < 0$, the standard form is

$$\begin{aligned} \dot{x}_{j_2} &= a_{j_2j_20}(x_{j_1} - a_{j_2j_11})\left[(x_{j_2} - a_{j_2j_2})^2 + b_{j_2j_2}\right], \\ \dot{x}_{j_1} &= a_{j_1j_10}(x_{j_1} - a_{j_1j_11})(x_{j_2} - a_{j_1j_21})(x_{j_2} - a_{j_1j_22}) \end{aligned} \tag{1.187}$$

where

$$\begin{aligned} & b_{j_1j_21}, b_{j_1j_22} = -\frac{1}{2}(B_{j_1j_2} \pm \sqrt{\Delta_{j_1j_2}}), \\ & \{a_{j_1j_21}, a_{j_1j_22}\} = \text{sort}\{b_{j_1j_21}, b_{j_1j_22}\}, a_{j_1j_21} < a_{j_1j_22}; \\ & a_{j_2j_2} = -\frac{1}{2}B_{j_2j_2}, b_{j_2j_2} = \frac{1}{4}(-\Delta_{j_2j_2}). \end{aligned} \tag{1.188}$$

(vii$_1$) The first integral manifold for $a_{j_1j_11} \neq a_{j_2j_11}$ *is*

$$\begin{aligned} & \left[(x_{j_1} - a_{j_1j_11}) - (x_{j_10} - a_{j_1j_11})\right] + (a_{j_1j_11} - a_{j_2j_11}) \ln \frac{|x_{j_1} - a_{j_1j_11}|}{|x_{j_10} - a_{j_1j_11}|} \\ & \quad = \frac{a_{j_1j_10}}{a_{j_2j_20}} \Big\{ \left[(x_{j_2} - a_{j_2j_2}) - (x_{j_20} - a_{j_2j_2})\right] \\ & \qquad + \sum\nolimits_{s_1=1}^{2} (a_{j_2j_2} - a_{j_1j_2s_1}) \ln \frac{|(x_{j_2} - a_{j_2j_2})^2 + b_{j_2j_2}|}{|(x_{j_20} - a_{j_2j_2})^2 + b_{j_2j_2}|} \\ & \qquad + \left[\prod\nolimits_{s_1=1}^{2} (a_{j_2j_2} - a_{j_1j_2s_1}) - b_{j_2j_2}\right] \\ & \qquad \times \frac{1}{\sqrt{b_{j_2j_2}}} \left[\arctan \frac{x_{j_2} - a_{j_2j_2}}{\sqrt{b_{j_2j_2}}} - \arctan \frac{x_{j_20} - a_{j_2j_2}}{\sqrt{b_{j_2j_2}}}\right] \Big\}. \end{aligned} \tag{1.189}$$

(vii$_{1a}$) The equilibrium of $(x^*_{j_1}, x^*_{j_2}) = (a_{j_1 j_1 1}, a_{j_1 j_2 s_1})$ *(s_1, $s_2 \in \{1,2\}$, $s_1 \neq s_2$) has the following properties:*

- *For* $a_{j_1 j_1 0}(a_{j_1 j_2 s_1} - a_{j_1 j_2 s_2}) > 0$ *and* $a_{j_2 j_2 0}(a_{j_1 j_1 1} - a_{j_2 j_1 1}) > 0$,

$$(a_{j_1 j_1 1}, a_{j_1 j_2 s_1}) = \underbrace{(\mathrm{DP:UP,pF})}_{\text{hyperbolic flow } (+)}. \tag{1.190}$$

The equilibrium of $(x^*_{j_1}, x^*_{j_2}) = (a_{j_1 j_1 1}, a_{j_1 j_2 s_1})$ *is a* (DP:UP,pF)*-positive hyperbolic flow.*

- *For* $a_{j_1 j_1 0}(a_{j_1 j_2 s_1} - a_{j_1 j_2 s_2}) < 0$ *and* $a_{j_2 j_2 0}(a_{j_1 j_1 1} - a_{j_2 j_1 1}) > 0$,

$$(a_{j_1 j_1 1}, a_{j_1 j_2 s_1}) = \underbrace{(\mathrm{UP:DP,pF})}_{\text{hyperbolic-secant flow } (+)}. \tag{1.191}$$

The equilibrium of $(x^*_{j_1}, x^*_{j_2}) = (a_{j_1 j_1 1}, a_{j_1 j_2 s_1})$ *is a (UP:DP,pF)-positive hyperbolic-secant flow.*

- *For* $a_{j_1 j_1 0}(a_{j_1 j_2 s_1} - a_{j_1 j_2 s_2}) > 0$ *and* $a_{j_2 j_2 0}(a_{j_1 j_1 1} - a_{j_2 j_1 1}) < 0$,

$$(a_{j_1 j_1 1}, a_{j_1 j_2 s_1}) = \underbrace{(\mathrm{UP:DP,nF})}_{\text{hyperbolic-secant flow } (-)}. \tag{1.192}$$

The equilibrium of $(x^*_{j_1}, x^*_{j_2}) = (a_{j_1 j_1 1}, a_{j_1 j_2 s_1})$ *is a* (UP:DP,nF)*-negative hyperbolic-secant flow.*

- *For* $a_{j_1 j_1 0}(a_{j_1 j_2 s_1} - a_{j_1 j_2 s_2}) < 0$ *and* $a_{j_2 j_2 0}(a_{j_1 j_1 1} - a_{j_2 j_1 1}) < 0$,

$$(a_{j_1 j_1 1}, a_{j_1 j_2 s_1}) = \underbrace{(\mathrm{DP:UP,nF})}_{\text{hyperbolic flow } (-)}. \tag{1.193}$$

The equilibrium of $(x^*_{j_1}, x^*_{j_2}) = (a_{j_1 j_1 1}, a_{j_1 j_2 s_1})$ *is a* (DP:UP,nF)*-negative hyperbolic flow.*

(vii$_{1b}$) The equilibrium of $(x^*_{j_2}, x^*_{j_1}) = (a_{j_1 j_2 s_1}, a_{j_2 j_1 1})$ *(s_1, $s_2 \in \{1,2\}$, $s_1 \neq s_2$) has the following properties:*

- *For* $a_{j_1 j_1 0}(a_{j_2 j_1 1} - a_{j_1 j_1 1})(a_{j_1 j_2 s_1} - a_{j_1 j_2 s_2}) > 0$ *and* $a_{j_2 j_2 0} > 0$,

$$(a_{j_1 j_2 s_1}, a_{j_2 j_1 1}) = \underbrace{(\mathrm{UP}_+, \mathrm{UP}_+)}_{\text{positive saddle}}. \tag{1.194}$$

The equilibrium of $(x^*_{j_2}, x^*_{j_1}) = (a_{j_1 j_2 s_1}, a_{j_2 j_1 1})$ *is a* (UP_+,UP_+)*-positive saddle.*

- *For* $a_{j_1j_1 0}(a_{j_2j_1 1} - a_{j_1j_1 1})(a_{j_1j_2 s_1} - a_{j_1j_2 s_2}) < 0$ *and* $a_{j_2j_2 0} > 0$,

$$(a_{j_1j_2 s_1}, a_{j_2j_1 1}) = \underbrace{(\mathrm{DP}_{+}, \mathrm{DP}_{-})}_{\text{CCW center}}. \tag{1.195}$$

The equilibrium of $(x_{j_2}^*, x_{j_1}^*) = (a_{j_1j_2 s_1}, a_{j_2j_1 1})$ *is a* (DP_+,DP_-)*-counterclockwise center.*

- *For* $a_{j_1j_1 0}(a_{j_2j_1 1} - a_{j_1j_1 1})(a_{j_1j_2 s_1} - a_{j_1j_2 s_2}) > 0$ *and* $a_{j_2j_2 0} < 0$,

$$(a_{j_1j_2 s_1}, a_{j_2j_1 1}) = \underbrace{(\mathrm{DP}_{-}, \mathrm{DP}_{+})}_{\text{CW center}}. \tag{1.196}$$

The equilibrium of $(x_{j_2}^*, x_{j_1}^*) = (a_{j_1j_2 s_1}, a_{j_2j_1 1})$ *is a* (DP_-,DP_+)*-clockwise center.*

- *For* $a_{j_1j_1 0}(a_{j_2j_1 1} - a_{j_1j_1 1})(a_{j_1j_2 s_1} - a_{j_1j_2 s_2}) < 0$ *and* $a_{j_2j_2 0} < 0$,

$$(a_{j_1j_2 s_1}, a_{j_2j_1 1}) = \underbrace{(\mathrm{UP}_{-}, \mathrm{UP}_{-})}_{\text{negative saddle}}. \tag{1.197}$$

The equilibrium of $(x_{j_2}^*, x_{j_1}^*) = (a_{j_1j_2 s_1}, a_{j_2j_1 1})$ *is a* (UP_-,UP_-)*-negative saddle.*

(vii$_2$*) The first integral manifold for* $a_{j_1j_1 1} = a_{j_2j_1 1}$ *is*

$$\begin{aligned}
&[(x_{j_1} - a_{j_1j_1 1}) - (x_{j_1 0} - a_{j_1j_1 1})] \\
&= \frac{a_{j_1j_1 0}}{a_{j_2j_2 0}}\Big\{[(x_{j_2} - a_{j_2j_2}) - (x_{j_2 0} - a_{j_2j_2})] \\
&\quad + \sum\nolimits_{s_1=1}^{2}(a_{j_2j_2} - a_{j_1j_2 s_1}) \ln \frac{|(x_{j_2} - a_{j_2j_2})^2 + b_{j_2j_2}|}{|(x_{j_2 0} - a_{j_2j_2})^2 + b_{j_2j_2}|} \\
&\quad + \Big[\prod\nolimits_{s_1=1}^{2}(a_{j_2j_2} - a_{j_1j_2 s_1}) - b_{j_2j_2}\Big] \\
&\quad \times \frac{1}{\sqrt{b_{j_2j_2}}}\Big(\arctan \frac{x_{j_2} - a_{j_2j_2}}{\sqrt{b_{j_2j_2}}} - \arctan \frac{x_{j_2 0} - a_{j_2j_2}}{\sqrt{b_{j_2j_2}}}\Big)\Big\}.
\end{aligned} \tag{1.198}$$

(vii$_{2a}$*) The infinite-equilibrium of* $x_{j_1}^* = a_{j_1j_1 1} = a_{j_2j_1 1}$ *with* $\bar{x}_{j_2} \neq a_{j_1j_2 s_1}$ *(*s_1, $s_2 \in \{1, 2\}$, $s_1 \neq s_2$*) for* $a_{j_1j_1 1} = a_{j_2j_1 1}$ *has the following properties:*

- *For* $a_{j_1j_1 0}(\bar{x}_{j_2} - a_{j_1j_2 s_1})(\bar{x}_{j_2} - a_{j_1j_2 s_2}) > 0$ *and* $a_{j_2j_2 0} > 0$,

$$(a_{j_1j_1 1}, \bar{x}_{j_2}) = \underbrace{(\mathrm{SO}, \mathrm{II})}_{\text{increasing-inflection source}}. \tag{1.199}$$

The infinite-equilibrium of $x_{j_1}^* = a_{j_1j_1 1} = a_{j_2j_1 1}$ *is an* (SO,II)*-increasing-inflection source.*

- *For* $a_{j_1j_10}(\overline{x}_{j_2} - a_{j_1j_2s_1})(\overline{x}_{j_2} - a_{j_1j_2s_2}) < 0$ *and* $a_{j_2j_20} > 0$,

$$(a_{j_1j_11}, \overline{x}_{j_2}) = \underbrace{(\text{SI}, \text{DI})}_{\text{decreasing-inflection sink}}. \tag{1.200}$$

The infinite-equilibrium of $x^*_{j_1} = a_{j_1j_11} = a_{j_2j_11}$ *is an* (SI,DI)-*decreasing-inflection sink.*

- *For* $a_{j_1j_10}(\overline{x}_{j_2} - a_{j_1j_2s_1})(\overline{x}_{j_2} - a_{j_1j_2s_2}) > 0$ *and* $a_{j_2j_20} < 0$,

$$(a_{j_1j_11}, \overline{x}_{j_2}) = \underbrace{(\text{SO}, \text{DI})}_{\text{decreasing-inflection source}}. \tag{1.201}$$

The infinite-equilibrium of $x^*_{j_1} = a_{j_1j_11} = a_{j_2j_11}$ *is an* (SO,DI)-*decreasing-inflection source.*

- *For* $a_{j_1j_10}(\overline{x}_{j_2} - a_{j_1j_2s_1})(\overline{x}_{j_2} - a_{j_1j_2s_2}) < 0$ *and* $a_{j_2j_20} < 0$,

$$(a_{j_1j_11}, \overline{x}_{j_2}) = \underbrace{(\text{SI}, \text{II})}_{\text{increasing-inflection sink}}. \tag{1.202}$$

The infinite-equilibrium of $x^*_{j_1} = a_{j_1j_11} = a_{j_2j_11}$ *is an* (SI,II)-*increasing-inflection sink.*

(vii$_{2b}$) The equilibrium of $(x^*_{j_2}, x^*_{j_1}) = (a_{j_1j_2s_1}, a_{j_2j_11})$ $(s_1, s_2 \in \{1,2\}, s_1 \neq s_2)$ *with* $a_{j_1j_11} = a_{j_2j_11}$ *has the following properties:*

- *For* $a_{j_1j_10}(a_{j_1j_2s_1} - a_{j_1j_2s_2}) > 0$ *and* $a_{j_2j_20} > 0$,

$$(a_{j_1j_2s_1}, a_{j_2j_11}) = \underbrace{({}_{\text{DI:II}}\text{UP}, {}_{\text{SI:SO}}\text{US})}_{\text{up-parabola upper-saddle}}. \tag{1.203}$$

The equilibrium of $(x^*_{j_2}, x^*_{j_1}) = (a_{j_1j_2s_1}, a_{j_2j_11})$ *is a* (${}_{\text{DI:II}}$UP,${}_{\text{SI:SO}}$US)-*up-parabola upper-saddle.*

- *For* $a_{j_1j_10}(a_{j_1j_2s_1} - a_{j_1j_2s_2}) < 0$ *and* $a_{j_2j_20} > 0$,

$$(a_{j_1j_2s_1}, a_{j_2j_11}) = \underbrace{({}_{\text{II:DI}}\text{DP}, {}_{\text{SO:SI}}\text{LS})}_{\text{down-parabola lower-saddle}}. \tag{1.204}$$

The equilibrium of $(x^*_{j_2}, x^*_{j_1}) = (a_{j_1j_2s_1}, a_{j_2j_11})$ *is an* (${}_{\text{II:DI}}$DP,${}_{\text{SO:SI}}$LS)-*down-parabola lower-saddle.*

- *For* $a_{j_1j_10}(a_{j_1j_2s_1}-a_{j_1j_2s_2})>0$ *and* $a_{j_2j_20}<0$,

$$(a_{j_1j_2s_1},a_{j_2j_11})=\underbrace{({}_{\text{II:DI}}\text{DP},{}_{\text{SI:SO}}\text{US})}_{\text{down-parabola lower-saddle}}. \tag{1.205}$$

The equilibrium of $(x^*_{j_1},x^*_{j_2})=(a_{j_1j_11},a_{j_1j_2s_1})$ *is an* $({}_{\text{II:DI}}\text{DP},{}_{\text{SI:SO}}\text{US})$*-down-parabola upper-saddle.*

- *For* $a_{j_1j_10}(a_{j_1j_2s_1}-a_{j_1j_2s_2})<0$ *and* $a_{j_2j_20}<0$,

$$(a_{j_1j_2s_1},a_{j_2j_11})=\underbrace{({}_{\text{DI:II}}\text{UP},{}_{\text{SO:SI}}\text{LS})}_{\text{up-parabola lower-saddle}}. \tag{1.206}$$

The equilibrium of $(x^*_{j_1},x^*_{j_2})=(a_{j_1j_11},a_{j_1j_2s_1})$ *is a* $({}_{\text{DI:II}}\text{UP},{}_{\text{SO:SI}}\text{LS})$*-up-parabola lower-saddle.*

(viii) For $\Delta_{j_1j_2}>0$ and $\Delta_{j_2j_2}=0$, the standard form is

$$\begin{aligned}\dot{x}_{j_2}&=a_{j_2j_20}(x_{j_1}-a_{j_2j_11})(x_{j_2}-a_{j_2j_21})^2,\\ \dot{x}_{j_1}&=a_{j_1j_10}(x_{j_1}-a_{j_1j_11})(x_{j_2}-a_{j_1j_21})(x_{j_2}-a_{j_1j_22})\end{aligned} \tag{1.207}$$

where

$$\begin{aligned}b_{j_1j_21},b_{j_1j_22}&=-\frac{1}{2}(B_{j_1j_2}\pm\sqrt{\Delta_{j_1j_2}}),\\ \{a_{j_1j_21},a_{j_1j_22}\}&=\text{sort}\{b_{j_1j_21},b_{j_1j_22}\},a_{j_1j_21}<a_{j_1j_22};\\ a_{j_2j_21}&=-\frac{1}{2}B_{j_2j_2}.\end{aligned} \tag{1.208}$$

(viii$_1$) The first integral manifold for $a_{j_2j_21}\neq a_{j_1j_21},a_{j_1j_22}$ and $a_{j_1j_11}\neq a_{j_2j_11}$ *is*

$$\begin{aligned}&[(x_{j_1}-a_{j_1j_11})-(x_{j_10}-a_{j_1j_11})]+(a_{j_1j_11}-a_{j_2j_11})\ln\frac{|x_{j_1}-a_{j_1j_11}|}{|x_{j_10}-a_{j_1j_11}|}\\ &=\frac{a_{j_1j_10}}{a_{j_2j_20}}\Big\{[(x_{j_2}-a_{j_2j_21})-(x_{j_20}-a_{j_2j_21})]\\ &\quad+\sum\nolimits_{s_1=1}^{2}(a_{j_2j_21}-a_{j_1j_2s_1})\ln\frac{|x_{j_2}-a_{j_2j_21}|}{|x_{j_20}-a_{j_2j_21}|}\\ &\quad-\prod\nolimits_{s_1=1}^{2}(a_{j_2j_21}-a_{j_1j_2s_1})(\frac{1}{x_{j_2}-a_{j_2j_21}}-\frac{1}{x_{j_20}-a_{j_2j_21}})\Big\}.\end{aligned} \tag{1.209}$$

(viii$_{1a}$) The equilibrium of $(x^*_{j_1}, x^*_{j_2}) = (a_{j_1 j_1 1}, a_{j_2 j_2 1})$ *has the following properties:*

- *For* $a_{j_1 j_1 0}\prod_{s_1=1}^{2}(a_{j_2 j_2 1} - a_{j_1 j_2 s_1}) > 0$ *and* $a_{j_2 j_2 0}(a_{j_1 j_1 1} - a_{j_2 j_1 1}) > 0$,

$$(a_{j_1 j_1 1}, a_{j_2 j_2 1}) = \underbrace{(\mathrm{SO}, \mathrm{US})}_{\text{upper-saddle-source}} . \tag{1.210}$$

The equilibrium of $(x^*_{j_1}, x^*_{j_2}) = (a_{j_1 j_1 1}, a_{j_2 j_2 1})$ *is an* (SO,US)-*upper-saddle-source.*

- *For* $a_{j_1 j_1 0}\prod_{s_1=1}^{2}(a_{j_2 j_2 1} - a_{j_1 j_2 s_1}) < 0$ *and* $a_{j_2 j_2 0}(a_{j_1 j_1 1} - a_{j_2 j_1 1}) > 0$,

$$(a_{j_1 j_1 1}, a_{j_2 j_2 1}) = \underbrace{(\mathrm{SI}, \mathrm{US})}_{\text{upper-saddle-sink}} . \tag{1.211}$$

The equilibrium of $(x^*_{j_1}, x^*_{j_2}) = (a_{j_1 j_1 1}, a_{j_2 j_2 1})$ *is an* (SI,US)-*upper-saddle-sink.*

- For $a_{j_1 j_1 0}\prod_{s_1=1}^{2}(a_{j_2 j_2 1} - a_{j_1 j_2 s_1}) > 0$ and $a_{j_2 j_2 0}(a_{j_1 j_1 1} - a_{j_2 j_1 1}) < 0$,

$$(a_{j_1 j_1 1}, a_{j_2 j_2 1}) = \underbrace{(\mathrm{SO}, \mathrm{LS})}_{\text{lower-saddle-source}} . \tag{1.212}$$

The equilibrium of $(x^*_{j_1}, x^*_{j_2}) = (a_{j_1 j_1 1}, a_{j_2 j_2 1})$ *is an* (SO,LS)-*lower-saddle-source.*

- *For* $a_{j_1 j_1 0}\prod_{s_1=1}^{2}(a_{j_2 j_2 1} - a_{j_1 j_2 s_1}) < 0$ *and* $a_{j_2 j_2 0}(a_{j_1 j_1 1} - a_{j_2 j_1 1}) < 0$,

$$(a_{j_1 j_1 1}, a_{j_2 j_2 1}) = \underbrace{(\mathrm{SI}, \mathrm{LS})}_{\text{lower-saddle-sink}} . \tag{1.213}$$

The equilibrium of $(x^*_{j_1}, x^*_{j_2}) = (a_{j_1 j_1 1}, a_{j_2 j_2 1})$ *is an* (SI,LS)-*lower-saddle-sink.*

– *The saddle-source and saddle-sink are the appearing bifurcations of saddle-source and saddle-sink.*

(viii$_{1b}$) The equilibrium of $(x^*_{j_1}, x^*_{j_2}) = (a_{j_1 j_1 1}, a_{j_1 j_2 s_1})$ *with* $a_{j_1 j_2 s_1} \neq a_{j_2 j_2 1} (s_1, s_2 \in \{1,2\}, s_1 \neq s_2)$ *has the following properties:*

- *For* $a_{j_1 j_1 0}(a_{j_1 j_2 s_1} - a_{j_1 j_2 s_2}) > 0$ *and* $a_{j_2 j_2 0}(a_{j_1 j_1 1} - a_{j_2 j_1 1})(a_{j_1 j_2 s_1} - a_{j_2 j_2 1})^2 > 0$,

$$(a_{j_1 j_1 1}, a_{j_1 j_2 s_1}) = \underbrace{(\mathrm{DP} : \mathrm{UP}, \mathrm{pF})}_{\text{hyperbolic flow }(+)} . \tag{1.214}$$

The equilibrium of $(x^*_{j_1}, x^*_{j_2}) = (a_{j_1 j_1 1}, a_{j_1 j_2 s_1})$ *is a* (DP:UP, pF)-*positive hyperbolic flow.*

- *For* $a_{j_1j_10}(a_{j_1j_2s_1}-a_{j_1j_2s_2})<0$ *and* $a_{j_2j_20}(a_{j_1j_11}-a_{j_2j_11})(a_{j_1j_2s_1}-a_{j_2j_21})^2>0$,

$$(a_{j_1j_11},a_{j_1j_2s_1})=\underbrace{(\mathrm{UP}:\mathrm{DP},\mathrm{pF})}_{\text{hyperbolic-secant flow }(+)}. \tag{1.215}$$

The equilibrium of $(x^*_{j_1},x^*_{j_2})=(a_{j_1j_11},a_{j_1j_2s_1})$ *is a* (UP:DP, pF)*-positive hyperbolic-secant flow.*

- For $a_{j_1j_10}(a_{j_1j_2s_1}-a_{j_1j_2s_2})>0$ *and* $a_{j_2j_20}(a_{j_1j_11}-a_{j_2j_11})(a_{j_1j_2s_1}-a_{j_2j_21})^2<0$,

$$(a_{j_1j_11},a_{j_1j_2s_1})=\underbrace{(\mathrm{UP}:\mathrm{DP},\mathrm{nF})}_{\text{hyperbolic-secant flow }(-)}. \tag{1.216}$$

The equilibrium of $(x^*_{j_1},x^*_{j_2})=(a_{j_1j_11},a_{j_1j_2s_1})$ *is a* (UP:DP, nF)*-negative hyperbolic-secant flow.*

- *For* $a_{j_1j_10}(a_{j_1j_2s_1}-a_{j_1j_2s_2})<0$ *and* $a_{j_2j_20}(a_{j_1j_11}-a_{j_2j_11})(a_{j_1j_2s_1}-a_{j_2j_21})^2<0$,

$$(a_{j_1j_11},a_{j_1j_2s_1})=\underbrace{(\mathrm{DP}:\mathrm{UP},\mathrm{nF})}_{\text{hyperbolic flow }(-)}. \tag{1.217}$$

The equilibrium of $(x^*_{j_1},x^*_{j_2})=(a_{j_1j_11},a_{j_1j_2s_1})$ *is a* (DP:UP, nF)*-negative hyperbolic flow.*

(viii$_{1c}$) The equilibrium of $(x^*_{j_2},x^*_{j_1})=(a_{j_1j_2s_1},a_{j_2j_11})$ $(s_1,\ s_2\in\{1,2\},\ s_1\neq s_2)$ *has the following properties:*

- *For* $a_{j_1j_10}(a_{j_2j_11}-a_{j_1j_11})(a_{j_1j_2s_1}-a_{j_1j_2s_2})>0$ *and* $a_{j_2j_20}(a_{j_1j_2s_1}-a_{j_2j_21})^2>0$,

$$(a_{j_1j_2s_1},a_{j_2j_11})=\underbrace{(\mathrm{UP}_+,\mathrm{UP}_+)}_{\text{positive saddle}}. \tag{1.218}$$

The equilibrium of $(x^*_{j_2},x^*_{j_1})=(a_{j_1j_2s_1},a_{j_2j_11})$ *is a* (UP$_+$,UP$_+$)*-positive saddle.*

- *For* $a_{j_1j_10}(a_{j_2j_11}-a_{j_1j_11})(a_{j_1j_2s_1}-a_{j_1j_2s_2})<0$ *and* $a_{j_2j_20}(a_{j_1j_2s_1}-a_{j_2j_21})^2>0$,

$$(a_{j_1j_2s_1},a_{j_2j_11})=\underbrace{(\mathrm{DP}_+,\mathrm{DP}_-)}_{\text{CCW center}}. \tag{1.219}$$

The equilibrium of $(x^*_{j_2},x^*_{j_1})=(a_{j_1j_2s_1},a_{j_2j_11})$ *is a* (DP$_+$,DP$_-$)*-counterclockwise center.*

- *For* $a_{j_1j_10}(a_{j_2j_11}-a_{j_1j_11})(a_{j_1j_2s_1}-a_{j_1j_2s_2})>0$ *and* $a_{j_2j_20}(a_{j_1j_2s_1}-a_{j_2j_21})^2<0$,

$$(a_{j_1j_2s_1},a_{j_2j_11})=\underbrace{(\mathrm{DP}_-,\mathrm{DP}_+)}_{\text{CW center}}. \tag{1.220}$$

The equilibrium of $(x^*_{j_2},x^*_{j_1})=(a_{j_1j_2s_1},a_{j_2j_11})$ *is a* (DP$_-$,DP$_+$)*-clockwise center.*

- *For* $a_{j_1j_10}(a_{j_2j_11} - a_{j_1j_11})(a_{j_1j_2s_1} - a_{j_1j_2s_2}) < 0$ *and* $a_{j_2j_20}(a_{j_1j_2s_1} - a_{j_2j_21})^2 < 0$,

$$(a_{j_1j_2s_1}, a_{j_2j_11}) = \underbrace{(\mathrm{UP}_-, \mathrm{UP}_-)}_{\text{negative saddle}}. \tag{1.221}$$

The equilibrium of $(x^*_{j_2}, x^*_{j_1}) = (a_{j_1j_2s_1}, a_{j_2j_11})$ *is a* (UP$_-$,UP$_-$)*-negative saddle.*

(viii$_{1d}$) The equilibrium of $(x^*_{j_1}, x^*_{j_2}) = (a_{j_2j_11}, a_{j_2j_21})$ *with* $a_{j_2j_21} \neq a_{j_1j_21}$ *has the following properties:*

- *For* $a_{j_1j_10}(a_{j_2j_11} - a_{j_1j_11})\prod_{s_1=1}^{2}(a_{j_2j_21} - a_{j_1j_2s_1}) > 0$ *and* $a_{j_2j_20} > 0$,

$$(a_{j_2j_11}, a_{j_2j_21}) = \underbrace{(\mathrm{pF}, \mathrm{UP}:\mathrm{UP})}_{\text{hyperbolic-secant-to-hyperbolic flow } (+)}. \tag{1.222}$$

The equilibrium of $(x^*_{j_1}, x^*_{j_2}) = (a_{j_2j_11}, a_{j_2j_21})$ *is a* (pF, UP:UP)*-positive hyperbolic-secant-to-hyperbolic flow.*

- *For* $a_{j_1j_10}(a_{j_2j_11} - a_{j_1j_11})\prod_{s_1=1}^{2}(a_{j_2j_21} - a_{j_1j_2s_1}) < 0$ *and* $a_{j_2j_20} > 0$,

$$(a_{j_2j_11}, a_{j_2j_21}) = \underbrace{(\mathrm{nF}, \mathrm{DP}:\mathrm{DP})}_{\text{hyperbolic-to-hyperbolic-secant flow } (-)}. \tag{1.223}$$

The equilibrium of $(x^*_{j_1}, x^*_{j_2}) = (a_{j_2j_11}, a_{j_2j_21})$ *is an* (nF, DP:DP)*-negative hyperbolic-to-hyperbolic-secant flow.*

- *For* $a_{j_1j_10}(a_{j_2j_11} - a_{j_1j_11})\prod_{s_1=1}^{2}(a_{j_2j_21} - a_{j_1j_2s_1}) > 0$ *and* $a_{j_2j_20} < 0$,

$$(a_{j_2j_11}, a_{j_2j_21}) = \underbrace{(\mathrm{pF}, \mathrm{DP}:\mathrm{DP})}_{\text{hyperbolic-to-hyperbolic-secant flow } (+)}. \tag{1.224}$$

The equilibrium of $(x^*_{j_1}, x^*_{j_2}) = (a_{j_2j_11}, a_{j_2j_21})$ *is a* (pF, DP:DP)*-positive hyperbolic-to-hyperbolic-secant flow.*

- *For* $a_{j_1j_10}(a_{j_2j_11} - a_{j_1j_11})\prod_{s_1=1}^{2}(a_{j_2j_21} - a_{j_1j_2s_1}) < 0$ *and* $a_{j_2j_20} < 0$,

$$(a_{j_2j_11}, a_{j_2j_21}) = \underbrace{(\mathrm{nF}, \mathrm{UP}:\mathrm{UP})}_{\text{hyperbolic-secant-to-hyperbolic flow } (-)}. \tag{1.225}$$

The equilibrium of $(x^*_{j_1}, x^*_{j_2}) = (a_{j_2j_11}, a_{j_2j_21})$ *is an* (nF, UP:UP)*-hyperbolic-secant-to-hyperbolic flow.*

(viii$_2$) The first integral manifold for $a_{j_2j_21}=a_{j_1j_2s_1}$ $(s_1,\ s_2\in\{1,2\},\ s_1\neq s_2)$ *and* $a_{j_1j_11}\neq a_{j_2j_11}$ *is*

$$
\begin{aligned}
&\left[(x_{j_1}-a_{j_1j_11})-(x_{j_10}-a_{j_1j_11})\right]+(a_{j_1j_11}-a_{j_2j_11})\ln\frac{|x_{j_1}-a_{j_1j_11}|}{|x_{j_10}-a_{j_1j_11}|}\\
&\quad=\frac{a_{j_1j_10}}{a_{j_2j_20}}\left\{\left[(x_{j_2}-a_{j_2j_21})-(x_{j_20}-a_{j_2j_21})\right]\right.\\
&\qquad\left.+(a_{j_2j_21}-a_{j_1j_2s_2})\ln\frac{|x_{j_2}-a_{j_2j_21}|}{|x_{j_20}-a_{j_2j_21}|}\right\}.
\end{aligned}
\tag{1.226}
$$

(viii$_{2a}$) The infinite-equilibrium of $x^*_{j_2}=a_{j_1j_2s_1}=a_{j_2j_21}$ *with* $\overline{x}_{j_1}\neq a_{j_1j_11}$ *has the following properties:*

- *For* $a_{j_1j_10}(\overline{x}_{j_1}-a_{j_1j_11})(a_{j_1j_2s_1}-a_{j_1j_2s_2})>0$ *and* $a_{j_2j_20}(\overline{x}_{j_1}-a_{j_2j_11})>0$,

$$
(\overline{x}_{j_1},a_{j_1j_21})=\underbrace{(\mathrm{DU},\mathrm{US})}_{\text{down-up upper-saddle}}. \tag{1.227}
$$

  The infinite-equilibrium of $x^*_{j_2}=a_{j_1j_2s_1}=a_{j_2j_21}$ *is a* (DU,US)*-down-up upper-saddle.*

- *For* $a_{j_1j_10}(\overline{x}_{j_1}-a_{j_1j_11})(a_{j_1j_2s_1}-a_{j_1j_2s_2})<0$ *and* $a_{j_2j_20}(\overline{x}_{j_1}-a_{j_2j_11})>0$,

$$
(\overline{x}_{j_1},a_{j_1j_21})=\underbrace{(\mathrm{UD},\mathrm{US})}_{\text{up-down upper-saddle}}. \tag{1.228}
$$

  The infinite-equilibrium of $x^*_{j_2}=a_{j_1j_2s_1}=a_{j_2j_21}$ *is a* (UD,US)*-up-down upper-saddle.*

- *For* $a_{j_1j_10}(\overline{x}_{j_1}-a_{j_1j_11})(a_{j_1j_2s_1}-a_{j_1j_2s_2})>0$ *and* $a_{j_2j_20}(\overline{x}_{j_1}-a_{j_2j_11})<0$,

$$
(\overline{x}_{j_1},a_{j_1j_21})=\underbrace{(\mathrm{UD},\mathrm{LS})}_{\text{up-down lower-saddle}}. \tag{1.229}
$$

  The infinite-equilibrium of $x^*_{j_2}=a_{j_1j_2s_1}=a_{j_2j_21}$ *is a* (UD,LS)*-up-down lower-saddle.*

- For $a_{j_1j_10}(\overline{x}_{j_1}-a_{j_1j_11})(a_{j_1j_2s_1}-a_{j_1j_2s_2})<0$ *and* $a_{j_2j_20}(\overline{x}_{j_1}-a_{j_2j_11})<0$,

$$
(\overline{x}_{j_1},a_{j_1j_21})=\underbrace{(\mathrm{DU},\mathrm{LS})}_{\text{down-up lower-saddle}}. \tag{1.230}
$$

  The infinite-equilibrium of $x^*_{j_2}=a_{j_1j_2s_1}=a_{j_2j_21}$ *is a* (DU,LS)*-down-up lower-saddle.*

(viii$_{2b}$) The equilibrium of $(x^*_{j_1}, x^*_{j_2}) = (a_{j_1 j_1 1}, a_{j_2 j_2 1})$ *with* $a_{j_2 j_2 1} = a_{j_1 j_2 s_1}$ *has the following properties:*

- *For* $a_{j_1 j_1 0}(a_{j_1 j_2 s_1} - a_{j_1 j_2 s_2}) > 0$ *and* $a_{j_2 j_2 0}(a_{j_1 j_1 1} - a_{j_2 j_1 1}) > 0$,

$$(a_{j_1 j_1 1}, a_{j_2 j_2 1}) = \underbrace{(\mathrm{DU} : \mathrm{UD}, \mathrm{US})}_{\text{down-up-to-up-down upper-saddle}}. \quad (1.231)$$

  The equilibrium of $(x^*_{j_1}, x^*_{j_2}) = (a_{j_1 j_1 1}, a_{j_2 j_2 1})$ *is a* (DU:UD,US)-*down-up-to-up-down upper-saddle.*

- *For* $a_{j_1 j_1 0}(a_{j_1 j_2 s_1} - a_{j_1 j_2 s_2}) < 0$ *and* $a_{j_2 j_2 0}(a_{j_1 j_1 1} - a_{j_2 j_1 1}) > 0$,

$$(a_{j_1 j_1 1}, a_{j_2 j_2 1}) = \underbrace{(\mathrm{UD} : \mathrm{DU}, \mathrm{US})}_{\text{up-down-to-down-up upper-saddle}}. \quad (1.232)$$

  The equilibrium of $(x^*_{j_1}, x^*_{j_2}) = (a_{j_1 j_1 1}, a_{j_2 j_2 1})$ *is a* (UD:DU,US)-*up-down-to-down-up upper-saddle.*

- *For* $a_{j_1 j_1 0}(a_{j_1 j_2 s_1} - a_{j_1 j_2 s_2}) > 0$ *and* $a_{j_2 j_2 0}(a_{j_1 j_1 1} - a_{j_2 j_1 1}) < 0$,

$$(a_{j_1 j_1 1}, a_{j_2 j_2 1}) = \underbrace{(\mathrm{UD} : \mathrm{DU}, \mathrm{LS})}_{\text{up-down-to-down-up lower-saddle}}. \quad (1.233)$$

  The equilibrium of $(x^*_{j_1}, x^*_{j_2}) = (a_{j_1 j_1 1}, a_{j_2 j_2 1})$ *is a* (UD:DU,LS)-*up-down-to-down-up lower-saddle.*

- For $a_{j_1 j_1 0}(a_{j_1 j_2 s_1} - a_{j_1 j_2 s_2}) < 0$ *and* $a_{j_2 j_2 0}(a_{j_1 j_1 1} - a_{j_2 j_1 1}) < 0$,

$$(a_{j_1 j_1 1}, a_{j_2 j_2 1}) = \underbrace{(\mathrm{DU} : \mathrm{UD}, \mathrm{LS})}_{\text{down-up-to-up-down lower-saddle}}. \quad (1.234)$$

  The equilibrium of $(x^*_{j_1}, x^*_{j_2}) = (a_{j_1 j_1 1}, a_{j_2 j_2 1})$ *is a* (DU:UD,LS)-*down-up-to-up-down lower-saddle.*

(viii$_{2c}$) The equilibrium of $(x^*_{j_2}, x^*_{j_1}) = (a_{j_1 j_2 s_1}, a_{j_2 j_1 1})$ *with* $a_{j_2 j_2 1} = a_{j_1 j_2 s_1}$ *has the following properties:*

- *For* $a_{j_1 j_1 0}(a_{j_2 j_1 1} - a_{j_1 j_1 1})(a_{j_1 j_2 s_1} - a_{j_1 j_2 s_2}) > 0$ *and* $a_{j_2 j_2 0} > 0$,

$$(a_{j_1 j_2 s_1}, a_{j_2 j_1 1}) = \underbrace{({}_{\mathrm{UD}}\mathrm{LS} : {}_{\mathrm{DU}}\mathrm{US}, \mathrm{DP}_- : \mathrm{UP}_+)}_{\text{hyperbolic lower-to-upper saddle}}. \quad (1.235)$$

  The equilibrium of $(x^*_{j_2}, x^*_{j_1}) = (a_{j_1 j_2 s_1}, a_{j_2 j_1 1})$ *is a* (${}_{\mathrm{UD}}$LS:${}_{\mathrm{DU}}$US, DP$_-$:UP$_+$) *hyperbolic lower-to-upper saddle.*

- For $a_{j_1j_1 0}(a_{j_2j_1 1}-a_{j_1j_1 1})(a_{j_1j_2s_1}-a_{j_1j_2s_2})<0$ and $a_{j_2j_2 0}>0$,

$$(a_{j_1j_2s_1}, a_{j_2j_1 1}) = \underbrace{({}_{\mathrm{DU}}\mathrm{US} : {}_{\mathrm{UD}}\mathrm{LS}, \mathrm{UP}_- : \mathrm{DP}_+)}_{\text{hyperbolic-secant upper-to-lower saddle}}. \tag{1.236}$$

The equilibrium of $(x^*_{j_2}, x^*_{j_1}) = (a_{j_1j_2s_1}, a_{j_2j_1 1})$ *is a* (${}_{\mathrm{DU}}$US:${}_{\mathrm{UD}}$LS, UP$_-$:DP$_+$)-*hyperbolic-secant upper-to-lower saddle.*
- For $a_{j_1j_1 0}(a_{j_2j_1 1}-a_{j_1j_1 1})(a_{j_1j_2s_1}-a_{j_1j_2s_2})>0$ and $a_{j_2j_2 0}<0$,

$$(a_{j_1j_2s_1}, a_{j_2j_1 1}) = \underbrace{({}_{\mathrm{DU}}\mathrm{LS} : {}_{\mathrm{UD}}\mathrm{US}, \mathrm{UP}_+ : \mathrm{DP}_-)}_{\text{hyperbolic-secant lower-to-upper saddle}}. \tag{1.237}$$

The equilibrium of $(x^*_{j_2}, x^*_{j_1}) = (a_{j_1j_2s_1}, a_{j_2j_1 1})$ *is a* (${}_{\mathrm{DU}}$LS:${}_{\mathrm{UD}}$US, UP$_+$:DP$_-$)-*hyperbolic-secant lower-to-upper saddle.*
- *For* $a_{j_1j_1 0}(a_{j_2j_1 1}-a_{j_1j_1 1})(a_{j_1j_2s_1}-a_{j_1j_2s_2})<0$ *and* $a_{j_2j_2 0}<0$,

$$(a_{j_1j_2s_1}, a_{j_2j_1 1}) = \underbrace{({}_{\mathrm{UD}}\mathrm{US} : {}_{\mathrm{DU}}\mathrm{LS}, \mathrm{DP}_+ : \mathrm{UP}_-)}_{\text{hypebolic upper-to-lower saddle}}. \tag{1.238}$$

The equilibrium of $(x^*_{j_2}, x^*_{j_1}) = (a_{j_1j_2s_1}, a_{j_2j_1 1})$ *is a* (${}_{\mathrm{UD}}$US:${}_{\mathrm{DU}}$LS, DP$_+$:UP$_-$)-*hyperbolic upper-to-lower saddle.*

(viii$_3$) The first integral manifold for $a_{j_2j_2 1} \neq a_{j_1j_2s_1}$ $(s_1,\ s_2 \in \{1,2\},\ s_1 \neq s_2)$ *and* $a_{j_1j_1 1} = a_{j_2j_1 1}$ *is*

$$\begin{aligned}
&\left[(x_{j_1}-a_{j_1j_1 1})-(x_{j_1 0}-a_{j_1j_1 1})\right]\\
&\quad = \frac{a_{j_1j_1 0}}{a_{j_2j_2 0}}\Big\{\left[(x_{j_2}-a_{j_2j_2 1})-(x_{j_2 0}-a_{j_2j_2 1})\right]\\
&\qquad + \sum\nolimits_{s_1=1}^{2}(a_{j_2j_2 1}-a_{j_1j_2s_1})\ln\frac{|x_{j_2}-a_{j_2j_2 1}|}{|x_{j_2 0}-a_{j_2j_2 1}|}\\
&\qquad - \prod\nolimits_{s_1=1}^{2}(a_{j_2j_2 1}-a_{j_1j_2s_1})\Big(\frac{1}{x_{j_2}-a_{j_2j_2 1}}-\frac{1}{x_{j_2 0}-a_{j_2j_2 1}}\Big)\Big\}.
\end{aligned} \tag{1.239}$$

(viii$_{3a}$) The infinite-equilibrium of $x^*_{j_1} = a_{j_1j_1 1} = a_{j_2j_1 1}$ *with* $\overline{x}_{j_2} \neq a_{j_2j_2 1}, a_{j_1j_2s_1}$ $(a_{j_1j_2s_1} \neq a_{j_2j_2 1})$ *has the following properties:*

- *For* $a_{j_1j_1 0}(\overline{x}_{j_2}-a_{j_1j_2s_1})(\overline{x}_{j_2}-a_{j_1j_2s_2})>0$ *and* $a_{j_2j_2 0}(\overline{x}_{j_2}-a_{j_2j_2 1})^2>0$,

$$(a_{j_1j_1 1}, \overline{x}_{j_2}) = \underbrace{(\mathrm{SO}, \mathrm{II})}_{\text{increasing-inflection source}}. \tag{1.240}$$

The infinite-equilibrium of $x^*_{j_1} = a_{j_1j_1 1} = a_{j_2j_1 1}$ *is an* (SO,II)-*increasing-inflection source.*

- *For* $a_{j_1j_10}(\overline{x}_{j_2}-a_{j_1j_2s_1})(\overline{x}_{j_2}-a_{j_1j_2s_2})<0$ *and* $a_{j_2j_20}(\overline{x}_{j_2}-a_{j_2j_21})^2>0$,

$$(a_{j_1j_11},\overline{x}_{j_2})=\underbrace{(\mathrm{SI},\mathrm{DI})}_{\text{decreasing-inflection sink}}. \tag{1.241}$$

The infinite-equilibrium of $x^*_{j_1}=a_{j_1j_11}=a_{j_2j_11}$ *is an* (SI,DI)*-decreasing-inflection sink.*

- *For* $a_{j_1j_10}(\overline{x}_{j_2}-a_{j_1j_2s_1})(\overline{x}_{j_2}-a_{j_1j_2s_2})>0$ *and* $a_{j_2j_20}(\overline{x}_{j_2}-a_{j_2j_21})^2<0$,

$$(a_{j_1j_11},\overline{x}_{j_2})=\underbrace{(\mathrm{SO},\mathrm{DI})}_{\text{decreasing-inflection source}}. \tag{1.242}$$

The infinite-equilibrium of $x^*_{j_1}=a_{j_1j_11}=a_{j_2j_11}$ *is an* (SO,DI)*-decreasing-inflection source.*

- For $a_{j_1j_10}(\overline{x}_{j_2}-a_{j_1j_2s_1})(\overline{x}_{j_2}-a_{j_1j_2s_2})<0$ and $a_{j_2j_20}(\overline{x}_{j_2}-a_{j_2j_21})^2<0$,

$$(a_{j_1j_11},\overline{x}_{j_2})=\underbrace{(\mathrm{SI},\mathrm{II})}_{\text{increasing-inflection sink}}. \tag{1.243}$$

The infinite-equilibrium of $x^*_{j_1}=a_{j_1j_11}=a_{j_2j_11}$ *is an* (SI,II)*-increasing-inflection sink.*

(viii$_{3b}$) The equilibrium of $(x^*_{j_2},x^*_{j_1})=(a_{j_1j_2s_1},a_{j_2j_11})$ *with* $a_{j_1j_11}=a_{j_2j_11}$ *has the following properties:*

- *For* $a_{j_1j_10}(a_{j_1j_2s_1}-a_{j_1j_2s_2})>0$ *and* $a_{j_2j_20}(a_{j_1j_2s_1}-a_{j_2j_21})^2>0$,

$$(a_{j_1j_2s_1},a_{j_2j_11})=\underbrace{({}_{\mathrm{DI:II}}\mathrm{UP},{}_{\mathrm{SI:SO}}\mathrm{US})}_{\text{up-parabola upper-saddle}}. \tag{1.244}$$

The equilibrium of $(x^*_{j_2},x^*_{j_1})=(a_{j_1j_2s_1},a_{j_2j_11})$ *is a* $({}_{\mathrm{DI:II}}\mathrm{UP},{}_{\mathrm{SI:SO}}\mathrm{US})$*-up-parabola upper-saddle.*

- *For* $a_{j_1j_10}(a_{j_1j_2s_1}-a_{j_1j_2s_2})<0$ *and* $a_{j_2j_20}(a_{j_1j_2s_1}-a_{j_2j_21})^2>0$,

$$(a_{j_1j_2s_1},a_{j_2j_11})=\underbrace{({}_{\mathrm{II:DI}}\mathrm{DP},{}_{\mathrm{SO:SI}}\mathrm{LS})}_{\text{down-parabola lower-saddle}}. \tag{1.245}$$

The equilibrium of $(x^*_{j_2},x^*_{j_1})=(a_{j_1j_2s_1},a_{j_2j_11})$ *is an* $({}_{\mathrm{II:DI}}\mathrm{DP},{}_{\mathrm{SO:SI}}\mathrm{LS})$*-down-parabola lower-saddle.*

- *For* $a_{j_1j_10}(a_{j_1j_2s_1}-a_{j_1j_2s_2})>0$ *and* $a_{j_2j_20}(a_{j_1j_2s_1}-a_{j_2j_21})^2<0$,

$$(a_{j_1j_2s_1},a_{j_2j_11})=\underbrace{({}_{\text{II:DI}}\text{DP},{}_{\text{SI:SO}}\text{US})}_{\text{down-parabola upper-saddle}}. \tag{1.246}$$

The equilibrium of $(x^*_{j_2},x^*_{j_1})=(a_{j_1j_2s_1},a_{j_2j_11})$ *is an* $({}_{\text{II:DI}}\text{DP},{}_{\text{SI:SO}}\text{US})$*-down-parabola upper-saddle.*

- *For* $a_{j_1j_10}(a_{j_1j_2s_1}-a_{j_1j_2s_2})<0$ *and* $a_{j_2j_20}(a_{j_1j_2s_1}-a_{j_2j_21})^2<0$,

$$(a_{j_1j_2s_1},a_{j_2j_11})=\underbrace{({}_{\text{DI:II}}\text{UP},{}_{\text{SO:SI}}\text{LS})}_{\text{up-parabola lower-saddle}}. \tag{1.247}$$

The equilibrium of $(x^*_{j_2},x^*_{j_1})=(a_{j_1j_2s_1},a_{j_2j_11})$ *is a* $({}_{\text{DI:II}}\text{UP},{}_{\text{SO:SI}}\text{LS})$*-up-parabola lower-saddle.*

(viii$_{3c}$) The equilibrium of $(x^*_{j_1},x^*_{j_2})=(a_{j_1j_11},a_{j_2j_21})$ *with* $a_{j_1j_11}=a_{j_2j_11}$ *has the following properties:*

- *For* $a_{j_1j_10}(a_{j_2j_21}-a_{j_1j_2s_1})(a_{j_2j_21}-a_{j_1j_2s_2})>0$ *and* $a_{j_2j_20}>0$,

$$(a_{j_1j_11},a_{j_2j_21})=\underbrace{(\text{SO},\text{II}:\text{II})}_{\text{increasing-inflection source}}. \tag{1.248}$$

The equilibrium of $(x^*_{j_1},x^*_{j_2})=(a_{j_1j_11},a_{j_2j_21})$ *is an* (SO,II:II)*-increasing-inflection source of the second kind.*

- *For* $a_{j_1j_10}(a_{j_2j_21}-a_{j_1j_2s_1})(a_{j_2j_21}-a_{j_1j_2s_2})<0$ *and* $a_{j_2j_20}>0$,

$$(a_{j_1j_11},a_{j_2j_21})=\underbrace{(\text{SI},\text{DI}:\text{DI})}_{\text{dereasing-inflection sink}}. \tag{1.249}$$

The equilibrium of $(x^*_{j_1},x^*_{j_2})=(a_{j_1j_11},a_{j_2j_21})$ *is an* (SI,DI:DI)*-deceasing-inflection sink of the second kind.*

- *For* $a_{j_1j_10}(a_{j_2j_21}-a_{j_1j_2s_1})(a_{j_2j_21}-a_{j_1j_2s_2})>0$ *and* $a_{j_2j_20}<0$,

$$(a_{j_1j_11},a_{j_2j_21})=\underbrace{(\text{SO},\text{DI}:\text{DI})}_{\text{dereasing-inflection source}}. \tag{1.250}$$

The equilibrium of $(x^*_{j_1},x^*_{j_2})=(a_{j_1j_11},a_{j_2j_21})$ *is an* (SO,DI:DI)*-decreasing-inflection source of the second kind.*

- *For* $a_{j_1j_10}(a_{j_2j_21}-a_{j_1j_2s_1})(a_{j_2j_21}-a_{j_1j_2s_2})<0$ *and* $a_{j_2j_20}<0$,

$$(a_{j_1j_11},a_{j_2j_21}) = \underbrace{(\mathrm{SI},\mathrm{II}:\mathrm{II})}_{\text{increasing-inflection sink}}. \tag{1.251}$$

The equilibrium of $(x^*_{j_1},x^*_{j_2})=(a_{j_1j_11},a_{j_2j_21})$ *is an* (SI,II:II)-*increasing-inflection sink of the second kind.*

(viii$_4$*) The first integral manifold for* $a_{j_2j_21}=a_{j_1j_2s_1}$ $(s_1,\ s_2\in\{1,2\},\ s_1\neq s_2)$ *and* $a_{j_1j_11}=a_{j_2j_11}$ *is*

$$\begin{aligned}\left[(x_{j_1}-a_{j_1j_11})-(x_{j_10}-a_{j_1j_11})\right] = \frac{a_{j_1j_10}}{a_{j_2j_20}}\Big\{&\left[(x_{j_2}-a_{j_2j_21})-(x_{j_20}-a_{j_2j_21})\right]\\ &+(a_{j_2j_21}-a_{j_1j_2s_2})\ln\frac{|x_{j_2}-a_{j_2j_21}|}{|x_{j_20}-a_{j_2j_21}|}\Big\}.\end{aligned} \tag{1.252}$$

(viii$_{4a}$*) The equilibrium of* $(x^*_{j_1},x^*_{j_2})=(a_{j_1j_11},a_{j_1j_2s_1})$ *with* $a_{j_1j_11}=a_{j_2j_11}$ and $a_{j_1j_2s_1}=a_{j_2j_21}$ *has the following properties:*

- *For* $a_{j_1j_10}(a_{j_1j_2s_1}-a_{j_1j_2s_2})>0$ *and* $a_{j_2j_20}(a_{j_1j_2s_1}-a_{j_2j_21})^2>0$,

$$(a_{j_1j_11},a_{j_1j_2s_1}) = \underbrace{({}_{\mathrm{LS:US}}\mathrm{DU},{}_{\mathrm{DI}}\mathrm{SI}:{}_{\mathrm{II}}\mathrm{SO})}_{\text{down-up inflection sink-to-source}}. \tag{1.253}$$

The equilibrium of $(x^*_{j_1},x^*_{j_2})=(a_{j_1j_11},a_{j_1j_2s_1})$ *is an* ($_{\mathrm{LS:US}}$DU, $_{\mathrm{DI}}$SI:$_{\mathrm{II}}$SO)-down-up inflection sink-to-source.

- *For* $a_{j_1j_10}(a_{j_1j_2s_1}-a_{j_1j_2s_2})<0$ *and* $a_{j_2j_20}(a_{j_1j_2s_1}-a_{j_2j_21})^2>0$,

$$(a_{j_1j_11},a_{j_1j_2s_1}) = \underbrace{({}_{\mathrm{LS:US}}\mathrm{UD},{}_{\mathrm{II}}\mathrm{SO}:{}_{\mathrm{DI}}\mathrm{SI})}_{\text{up-down inflection source-to-sink}}. \tag{1.254}$$

The equilibrium of $(x^*_{j_1},x^*_{j_2})=(a_{j_1j_11},a_{j_1j_2s_1})$ *is an* ($_{\mathrm{LS:US}}$UD, $_{\mathrm{II}}$SO:$_{\mathrm{DI}}$SI)-up-down inflection source-to-sink.

- *For* $a_{j_1j_10}(a_{j_1j_2s_1}-a_{j_1j_2s_2})>0$ *and* $a_{j_2j_20}(a_{j_1j_2s_1}-a_{j_2j_21})^2<0$,

$$(a_{j_1j_11},a_{j_1j_2s_1}) = \underbrace{({}_{\mathrm{US:LS}}\mathrm{UD},{}_{\mathrm{II}}\mathrm{SI}:{}_{\mathrm{DI}}\mathrm{SO})}_{\text{up-down inflection sink-to-source}}. \tag{1.255}$$

The equilibrium of $(x^*_{j_1},x^*_{j_2})=(a_{j_1j_11},a_{j_1j_2s_1})$ *is a* ($_{\mathrm{US:LS}}$UD, $_{\mathrm{II}}$SI:$_{\mathrm{DI}}$SO)-up-down inflection sink-to-source.

- *For* $a_{j_1j_10}(a_{j_1j_2s_1} - a_{j_1j_2s_2}) < 0$ *and* $a_{j_2j_20}(a_{j_1j_2s_1} - a_{j_2j_21})^2 < 0$,

$$(a_{j_1j_11}, a_{j_1j_2s_1}) = \underbrace{({}_{\text{US:LS}}\text{DU}, {}_{\text{DI}}\text{SO}:{}_{\text{II}}\text{SI})}_{\text{down-up inflection source-to-sink}}. \tag{1.256}$$

The equilibrium of $(x^*_{j_1}, x^*_{j_2}) = (a_{j_1j_11}, a_{j_1j_2s_1})$ *is a* (${}_{\text{US:LS}}$DU, ${}_{\text{DI}}$SI:${}_{\text{II}}$SO)-down-up inflection sink-to-source.

(viii$_{4d}$) The equilibrium of $(x^*_{j_2}, x^*_{j_1}) = (a_{j_1j_2s_2}, a_{j_2j_11})$ *with* $a_{j_1j_11} = a_{j_2j_11}$ *and* $a_{j_1j_2s_1} = a_{j_2j_21}$ *has the following properties:*

- *For* $a_{j_1j_10}(a_{j_1j_2s_2} - a_{j_1j_2s_1}) > 0$ *and* $a_{j_2j_20}(a_{j_1j_2s_2} - a_{j_2j_21})^2 > 0$,

$$(a_{j_1j_2s_2}, a_{j_2j_11}) = \underbrace{({}_{\text{DI:II}}\text{UP}, {}_{\text{SI:SO}}\text{US})}_{\text{up-parabola upper-saddle}}. \tag{1.257}$$

The equilibrium of $(x^*_{j_2}, x^*_{j_1}) = (a_{j_1j_2s_2}, a_{j_2j_11})$ *is a* (${}_{\text{DI:II}}$UP,${}_{\text{SI:SO}}$US)*-up-parabola upper-saddle.*

- *For* $a_{j_1j_10}(a_{j_1j_2s_2} - a_{j_1j_2s_1}) < 0$ *and* $a_{j_2j_20}(a_{j_1j_2s_2} - a_{j_2j_21})^2 > 0$,

$$(a_{j_1j_2s_2}, a_{j_2j_11}) = \underbrace{({}_{\text{II:DI}}\text{DP}, {}_{\text{SO:SI}}\text{LS})}_{\text{down-parabola lower-saddle}}. \tag{1.258}$$

The equilibrium of $(x^*_{j_2}, x^*_{j_1}) = (a_{j_1j_2s_2}, a_{j_2j_11})$ *is an* (${}_{\text{II:DI}}$DP,${}_{\text{SO:SI}}$LS)*-down-parabola lower-saddle.*

- *For* $a_{j_1j_10}(a_{j_1j_2s_2} - a_{j_1j_2s_1}) > 0$ *and* $a_{j_2j_20}(a_{j_1j_2s_2} - a_{j_2j_21})^2 < 0$,

$$(a_{j_1j_2s_2}, a_{j_2j_11}) = \underbrace{({}_{\text{II:DI}}\text{DP}, {}_{\text{SI:SO}}\text{US})}_{\text{down-parabola upper-saddle}}. \tag{1.259}$$

The equilibrium of $(x^*_{j_2}, x^*_{j_1}) = (a_{j_1j_2s_2}, a_{j_2j_11})$ *is an* (${}_{\text{II:DI}}$DP,${}_{\text{SI:SO}}$US)*-down-parabola upper-saddle.*

- *For* $a_{j_1j_10}(a_{j_1j_2s_2} - a_{j_1j_2s_1}) < 0$ *and* $a_{j_2j_20}(a_{j_1j_2s_2} - a_{j_2j_21})^2 < 0$,

$$(a_{j_1j_2s_2}, a_{j_2j_11}) = \underbrace{({}_{\text{DI:II}}\text{UP}, {}_{\text{SO:SI}}\text{LS})}_{\text{up-parabola lower-saddle}}. \tag{1.260}$$

The equilibrium of $(x^*_{j_2}, x^*_{j_1}) = (a_{j_1j_2s_2}, a_{j_2j_11})$ *is a* (${}_{\text{DI:II}}$UP,${}_{\text{SO:SI}}$LS)*-up-parabola lower-saddle.*

(ix) For $\Delta_{j_1j_2} > 0$ and $\Delta_{j_2j_2} > 0$, the standard form is

$$\begin{aligned}\dot{x}_{j_2} &= a_{j_2j_20}(x_{j_1} - a_{j_2j_11})(x_{j_2} - a_{j_2j_21})(x_{j_2} - a_{j_2j_22}),\\ \dot{x}_{j_1} &= a_{j_1j_10}(x_{j_1} - a_{j_1j_11})(x_{j_2} - a_{j_1j_21})(x_{j_2} - a_{j_1j_22})\end{aligned} \tag{1.261}$$

where

$$\begin{aligned}&b_{j_1j_21}, b_{j_1j_22} = -\frac{1}{2}(B_{j_1j_2} \pm \sqrt{\Delta_{j_1j_2}}),\\ &\{a_{j_1j_21}, a_{j_1j_22}\} = \text{sort}\{b_{j_1j_21}, b_{j_1j_22}\}, a_{j_1j_21} < a_{j_1j_22};\\ &b_{j_2j_21}, b_{j_2j_22} = -\frac{1}{2}(B_{j_2j_2} \pm \sqrt{\Delta_{j_2j_2}}),\\ &\{a_{j_2j_21}, a_{j_2j_22}\} = \text{sort}\{b_{j_2j_21}, b_{j_2j_22}\}, a_{j_2j_21} < a_{j_2j_22}.\end{aligned} \tag{1.262}$$

(ix_1) The first integral manifold ($a_{j_2j_2s_1} \neq a_{j_1j_21}$) is

$$\begin{aligned}&\left[(x_{j_1} - a_{j_1j_11}) - (x_{j_10} - a_{j_1j_11})\right] + (a_{j_1j_11} - a_{j_2j_11}) \ln \frac{|x_{j_1} - a_{j_1j_11}|}{|x_{j_10} - a_{j_1j_11}|}\\ &= \frac{a_{j_1j_10}}{a_{j_2j_20}}\left[(x_{j_2} - x_{j_20}) + \sum\nolimits_{l_1=1}^{2} \frac{\prod_{s_1=1}^{2}(a_{j_2j_2l_1} - a_{j_1j_2s_1})}{(a_{j_2j_2l_1} - a_{j_2j_2l_2})} \ln \frac{|x_{j_2} - a_{j_2j_2l_1}|}{|x_{j_20} - a_{j_2j_2l_1}|}\right].\end{aligned} \tag{1.263}$$

(ix_{1a}) The equilibrium of $(x^*_{j_1}, x^*_{j_2}) = (a_{j_1j_11}, a_{j_2j_2s_1})$ ($s_1, s_2 \in \{1, 2\}$, $s_1 \neq s_2$) has the following properties:

- *For* $a_{j_1j_10}\prod_{l_1=1}^{2}(a_{j_2j_2s_1} - a_{j_1j_2l_1}) > 0$ *and* $a_{j_2j_20}(a_{j_1j_11} - a_{j_2j_11})(a_{j_2j_2s_1} - a_{j_2j_2s_2}) > 0$,

$$(a_{j_1j_11}, a_{j_2j_2s_1}) = \underbrace{(\text{SO}, \text{SO})}_{\text{source}}. \tag{1.264}$$

The equilibrium of $(x^*_{j_1}, x^*_{j_2}) = (a_{j_1j_11}, a_{j_2j_2s_1})$ is an (SO,SO)-source.

- *For* $a_{j_1j_10}\prod_{l_1=1}^{2}(a_{j_2j_2s_1} - a_{j_1j_2l_1}) < 0$ *and* $a_{j_2j_20}(a_{j_1j_11} - a_{j_2j_11})(a_{j_2j_2s_1} - a_{j_2j_2s_2}) > 0$,

$$(a_{j_1j_11}, a_{j_2j_2s_1}) = \underbrace{(\text{SI}, \text{SO})}_{\text{saddle}}. \tag{1.265}$$

The equilibrium of $(x^*_{j_1}, x^*_{j_2}) = (a_{j_1j_11}, a_{j_2j_2s_1})$ is an (SI,SO)-saddle.

- *For* $a_{j_1j_10}\prod_{l_1=1}^{2}(a_{j_2j_2s_1} - a_{j_1j_2l_1}) > 0$ *and* $a_{j_2j_20}(a_{j_1j_11} - a_{j_2j_11})(a_{j_2j_2s_1} - a_{j_2j_2s_2}) < 0$,

$$(a_{j_1j_11}, a_{j_2j_2s_1}) = \underbrace{(\text{SO}, \text{SI})}_{\text{saddle}}. \tag{1.266}$$

The equilibrium of $(x^*_{j_1}, x^*_{j_2}) = (a_{j_1j_11}, a_{j_2j_2s_1})$ is an (SO,SI)-saddle.

- *For* $a_{j_1j_10}\prod_{l_1=1}^{2}(a_{j_2j_2s_1}-a_{j_1j_2l_1})<0$ *and* $a_{j_2j_20}(a_{j_1j_11}-a_{j_2j_11})(a_{j_2j_2s_1}-a_{j_2j_2s_2})<0$,

$$(a_{j_1j_11},a_{j_2j_2s_1})=\underbrace{(\text{SI},\text{SI})}_{\text{sink}}. \tag{1.267}$$

The equilibrium of $(x_{j_1}^*,x_{j_2}^*)=(a_{j_1j_11},a_{j_2j_2s_1})$ *is an* (SI,SI)*-sink.*

(ix$_{1b}$*) The equilibrium of* $(x_{j_1}^*,x_{j_2}^*)=(a_{j_1j_11},a_{j_1j_2l_1})$ $(l_1,\ l_2\in\{1,2\},\ l_1\neq l_2)$ *with* $a_{j_1j_2l_1}\neq a_{j_2j_21},a_{j_2j_22}$ *has the following properties:*

- *For* $a_{j_1j_10}(a_{j_1j_2l_1}-a_{j_1j_2l_2})>0$ *and* $a_{j_2j_20}(a_{j_1j_11}-a_{j_2j_11})\prod_{s_1=1}^{2}(a_{j_1j_2l_1}-a_{j_2j_2s_1})>0$,

$$(a_{j_1j_11},a_{j_1j_2l_1})=\underbrace{(\text{DP}:\text{UP},\text{pF})}_{\text{hyperbolic flow }(+)}. \tag{1.268}$$

The equilibrium of $(x_{j_1}^*,x_{j_2}^*)=(a_{j_1j_11},a_{j_1j_2l_1})$ *is a* (DP:UP,pF)*-positive hyperbolic flow.*

- *For* $a_{j_1j_10}(a_{j_1j_2l_1}-a_{j_1j_2l_2})<0$ *and* $a_{j_2j_20}(a_{j_1j_11}-a_{j_2j_11})\prod_{s_1=1}^{2}(a_{j_1j_2l_1}-a_{j_2j_2s_1})>0$,

$$(a_{j_1j_11},a_{j_1j_2l_1})=\underbrace{(\text{UP}:\text{DP},\text{pF})}_{\text{hyperbolic-secant flow }(+)}. \tag{1.269}$$

The equilibrium of $(x_{j_1}^*,x_{j_2}^*)=(a_{j_1j_11},a_{j_1j_2l_1})$ *is a* (UP:DP,pF)*-positive hyperbolic-secant flow.*

- For $a_{j_1j_10}(a_{j_1j_2l_1}-a_{j_1j_2l_2})>0$ and $a_{j_2j_20}(a_{j_1j_11}-a_{j_2j_11})\prod_{s_1=1}^{2}(a_{j_1j_2l_1}-a_{j_2j_2s_1})<0$,

$$(a_{j_1j_11},a_{j_1j_2l_1})=\underbrace{(\text{UP}:\text{DP},\text{nF})}_{\text{hyperbolic-secant flow }(-)}. \tag{1.270}$$

The equilibrium of $(x_{j_1}^*,x_{j_2}^*)=(a_{j_1j_11},a_{j_1j_2l_1})$ *is a* (UP:DP,nF)*-negative hyperbolic-secant flow.*

- *For* $a_{j_1j_10}(a_{j_1j_2l_1}-a_{j_1j_2l_2})<0$ *and* $a_{j_2j_20}(a_{j_1j_11}-a_{j_2j_11})\prod_{s_1=1}^{2}(a_{j_1j_2l_1}-a_{j_2j_2s_1})<0$,

$$(a_{j_1j_11},a_{j_1j_2l_1})=\underbrace{(\text{DP}:\text{UP},\text{nF})}_{\text{hyperbolic flow }(-)}. \tag{1.271}$$

The equilibrium of $(x_{j_1}^*,x_{j_2}^*)=(a_{j_1j_11},a_{j_1j_2l_1})$ *is a* (DP:UP,nF)*-negative hyperbolic flow.*

(ix_{1c}) The equilibrium of $(x_{j_2}^*, x_{j_1}^*) = (a_{j_1 j_2 l_1}, a_{j_2 j_1 1})$ *(l_1, $l_2 \in \{1,2\}$, $l_1 \neq l_2$) has the following properties:*

- *For* $a_{j_1 j_1 0}(a_{j_2 j_1 1} - a_{j_1 j_1 1})(a_{j_1 j_2 l_1} - a_{j_1 j_2 l_2}) > 0$ *and* $a_{j_2 j_2 0}\prod_{s_1=1}^{2}(a_{j_1 j_2 l_1} - a_{j_2 j_2 s_1}) > 0$,

$$(a_{j_1 j_2 l_1}, a_{j_2 j_1 1}) = \underbrace{(\mathrm{UP}_+, \mathrm{UP}_+)}_{\text{positive saddle}}. \tag{1.272}$$

The equilibrium of $(x_{j_2}^*, x_{j_1}^*) = (a_{j_1 j_2 l_1}, a_{j_2 j_1 1})$ *is a* $(\mathrm{UP}_+,\mathrm{UP}_+)$*-positive saddle.*

- *For* $a_{j_1 j_1 0}(a_{j_2 j_1 1} - a_{j_1 j_1 1})(a_{j_1 j_2 l_1} - a_{j_1 j_2 l_2}) < 0$ *and* $a_{j_2 j_2 0}\prod_{s_1=1}^{2}(a_{j_1 j_2 l_1} - a_{j_2 j_2 s_1}) > 0$,

$$(a_{j_1 j_2 l_1}, a_{j_2 j_1 1}) = \underbrace{(\mathrm{DP}_+, \mathrm{DP}_-)}_{\text{CCW center}}. \tag{1.273}$$

The equilibrium of $(x_{j_2}^*, x_{j_1}^*) = (a_{j_1 j_2 l_1}, a_{j_2 j_1 1})$ *is a* $(\mathrm{DP}_+,\mathrm{DP}_-)$*-counterclockwise center.*

- *For* $a_{j_1 j_1 0}(a_{j_2 j_1 1} - a_{j_1 j_1 1})(a_{j_1 j_2 l_1} - a_{j_1 j_2 l_2}) > 0$ *and* $a_{j_2 j_2 0}\prod_{s_1=1}^{2}(a_{j_1 j_2 l_1} - a_{j_2 j_2 s_1}) < 0$,

$$(a_{j_1 j_2 l_1}, a_{j_2 j_1 1}) = \underbrace{(\mathrm{DP}_-, \mathrm{DP}_+)}_{\text{CW center}}. \tag{1.274}$$

The equilibrium of $(x_{j_2}^*, x_{j_1}^*) = (a_{j_1 j_2 l_1}, a_{j_2 j_1 1})$ *is a* $(\mathrm{DP}_-,\mathrm{DP}_+)$ *clockwise center.*

- *For* $a_{j_1 j_1 0}(a_{j_2 j_1 1} - a_{j_1 j_1 1})(a_{j_1 j_2 l_1} - a_{j_1 j_2 l_2}) < 0$ *and* $a_{j_2 j_2 0}\prod_{s_1=1}^{2}(a_{j_1 j_2 l_1} - a_{j_2 j_2 s_1}) < 0$,

$$(a_{j_1 j_2 l_1}, a_{j_2 j_1 1}) = \underbrace{(\mathrm{UP}_-, \mathrm{UP}_-)}_{\text{negative saddle}}. \tag{1.275}$$

The equilibrium of $(x_{j_1}^*, x_{j_2}^*) = (a_{j_1 j_2 l_1}, a_{j_2 j_1 1})$ *is a* $(\mathrm{UP}_-,\mathrm{UP}_-)$*-negative saddle.*

(ix_{1d}) The equilibrium of $(x_{j_1}^*, x_{j_2}^*) = (a_{j_2 j_1 1}, a_{j_2 j_2 s_1})$ *(l_1, $l_2 \in \{1,2\}$, $l_1 \neq l_2$) with* $a_{j_1 j_2 l_1} \neq a_{j_2 j_2 1}, a_{j_2 j_2 2}$ *has the following properties:*

- *For* $a_{j_1 j_1 0}(a_{j_2 j_1 1} - a_{j_1 j_1 1})\prod_{l_1=1}^{2}(a_{j_2 j_2 s_1} - a_{j_1 j_2 l_1}) > 0$ *and* $a_{j_2 j_2 0}(a_{j_2 j_2 s_1} - a_{j_2 j_2 s_2}) > 0$,

$$(a_{j_2 j_1 1}, a_{j_2 j_2 s_1}) = \underbrace{(\mathrm{pF}, \mathrm{DP}:\mathrm{UP})}_{\text{hyperbolic flow }(+)}. \tag{1.276}$$

The equilibrium of $(x_{j_1}^*, x_{j_2}^*) = (a_{j_2 j_1 1}, a_{j_2 j_2 s_1})$ *is a* (pF, DP:UP)*-positive hyperbolic flow.*

- *For* $a_{j_1j_10}(a_{j_2j_11}-a_{j_1j_11})\prod_{l_1=1}^{2}(a_{j_2j_2s_1}-a_{j_1j_2l_1})<0$ *and* $a_{j_2j_20}(a_{j_2j_2s_1}-a_{j_2j_2s_2})>0$,

$$(a_{j_2j_11},a_{j_2j_2s_1})=\underbrace{(\text{nF},\text{DP}:\text{UP})}_{\text{hyperbolic-secant flow }(-)}. \tag{1.277}$$

The equilibrium of $(x_{j_1}^*,x_{j_2}^*)=(a_{j_2j_11},a_{j_2j_2s_1})$ *is an* (nF,UP:DP)-*negative hyperbolic-secant flow.*

- *For* $a_{j_1j_10}(a_{j_2j_11}-a_{j_1j_11})\prod_{l_1=1}^{2}(a_{j_2j_2s_1}-a_{j_1j_2l_1})>0$ *and* $a_{j_2j_20}(a_{j_2j_2s_1}-a_{j_2j_2s_2})<0$,

$$(a_{j_2j_11},a_{j_2j_2s_1})=\underbrace{(\text{nF},\text{UP}:\text{DP})}_{\text{hyperbolic-secant flow }(+)}. \tag{1.278}$$

The equilibrium of $(x_{j_1}^*,x_{j_2}^*)=(a_{j_2j_11},a_{j_2j_2s_1})$ *is an* (pF,UP:DP)-*negative hyperbolic-secant flow.*

- *For* $a_{j_1j_10}(a_{j_2j_11}-a_{j_1j_11})\prod_{l_1=1}^{2}(a_{j_2j_2s_1}-a_{j_1j_2l_1})<0$ *and* $a_{j_2j_20}(a_{j_2j_2s_1}-a_{j_2j_2s_2})<0$,

$$(a_{j_2j_11},a_{j_2j_2s_1})=\underbrace{(\text{nF},\text{DP}:\text{UP})}_{\text{hyperbolic flow }(-)}. \tag{1.279}$$

The equilibrium of $(x_{j_1}^*,x_{j_2}^*)=(a_{j_2j_11},a_{j_2j_2s_1})$ *is an* (nF,DP:UP)-*negative hyperbolic flow.*

(ix₂) The first integral manifold ($a_{j_2j_2s_1}=a_{j_1j_2l_1}$ and $a_{j_2j_2s_2}\neq a_{j_1j_2l_2}$, $s_1,s_2,l_1,l_2\in\{1,2\}$; $s_1\neq s_2$, $l_1\neq l_2$*) is*

$$\begin{aligned}&\left[(x_{j_1}-a_{j_1j_11})-(x_{j_10}-a_{j_1j_11})\right]+(a_{j_1j_11}-a_{j_2j_11})\ln\frac{|x_{j_1}-a_{j_1j_11}|}{|x_{j_10}-a_{j_1j_11}|}\\&=\frac{a_{j_1j_10}}{a_{j_2j_20}}\Big\{\left[(x_{j_2}-a_{j_2j_2s_2})-(x_{j_20}-a_{j_2j_2s_2})\right]\\&\quad+(a_{j_2j_2s_2}-a_{j_1j_2l_2})\ln\frac{|x_{j_2}-a_{j_2j_2s_2}|}{|x_{j_20}-a_{j_2j_2s_2}|}\Big\},\end{aligned} \tag{1.280}$$

and the first integral manifold ($a_{j_2j_21}=a_{j_1j_22}$ and $a_{j_2j_22}=a_{j_1j_22}$*) is*

$$(x_{j_1}-x_{j_10})+(a_{j_1j_11}-a_{j_2j_11})\ln\frac{|x_{j_1}-a_{j_1j_11}|}{|x_{j_10}-a_{j_1j_11}|}=\frac{a_{j_1j_10}}{a_{j_2j_20}}(x_{j_2}-x_{j_20}). \tag{1.281}$$

(ix_{2a}) The infinite-equilibrium of $x_{j_2}^* = a_{j_1j_2l_1} = a_{j_2j_2s_1}$ with $\overline{x}_{j_1} \neq a_{j_1j_1 1}$ has the following properties:

- *For* $a_{j_1j_10}(\overline{x}_{j_1} - a_{j_1j_11})(a_{j_1j_2l_1} - a_{j_1j_2l_2}) > 0$ *and* $a_{j_2j_20}(\overline{x}_{j_1} - a_{j_2j_11})(a_{j_2j_2s_1} - a_{j_2j_2s_2}) > 0$,

$$(\overline{x}_{j_1}, a_{j_2j_2s_1}) = \underbrace{(\text{II}, \text{SO})}_{\text{increasing-inflection source}}. \tag{1.282}$$

The infinite-equilibrium of $x_{j_2}^* = a_{j_1j_2l_1} = a_{j_2j_2s_1}$ *is an* (II,SO)*-increasing-inflection source.*

- *For* $a_{j_1j_10}(\overline{x}_{j_1} - a_{j_1j_11})(a_{j_1j_2l_1} - a_{j_1j_2l_2}) < 0$ *and* $a_{j_2j_20}(\overline{x}_{j_1} - a_{j_2j_11})(a_{j_2j_2s_1} - a_{j_2j_2s_2}) > 0$,

$$(\overline{x}_{j_1}, a_{j_2j_2s_1}) = \underbrace{(\text{DI}, \text{SO})}_{\text{decreasing-inflection source}}. \tag{1.283}$$

The infinite-equilibrium of $x_{j_2}^* = a_{j_1j_2l_1} = a_{j_2j_2s_1}$ *is a* (DI,SO)*-decreasing-inflection source.*

- *For* $a_{j_1j_10}(\overline{x}_{j_1} - a_{j_1j_11})(a_{j_1j_2l_1} - a_{j_1j_2l_2}) > 0$ *and* $a_{j_2j_20}(\overline{x}_{j_1} - a_{j_2j_11})(a_{j_2j_2s_1} - a_{j_2j_2s_2}) < 0$,

$$(\overline{x}_{j_1}, a_{j_2j_2s_1}) = \underbrace{(\text{DI}, \text{SI})}_{\text{decreasing-inflection sink}}. \tag{1.284}$$

The infinite-equilibrium of $x_{j_2}^* = a_{j_1j_2l_1} = a_{j_2j_2s_1}$ *is a* (DI,SI)*-decreasing-inflection sink.*

- *For* $a_{j_1j_10}(\overline{x}_{j_1} - a_{j_1j_11})(a_{j_1j_2l_1} - a_{j_1j_2l_2}) < 0$ *and* $a_{j_2j_20}(\overline{x}_{j_1} - a_{j_2j_11})(a_{j_2j_2s_1} - a_{j_2j_2s_2}) < 0$,

$$(\overline{x}_{j_1}, a_{j_2j_2s_1}) = \underbrace{(\text{II}, \text{SI})}_{\text{increasing-inflection sink}}. \tag{1.285}$$

The infinite-equilibrium of $x_{j_2}^* = a_{j_1j_2l_1} = a_{j_2j_2s_1}$ *is an* (II,SI)*-increasing-inflection sink.*

(ix_{2b}) The equilibrium of $(x_{j_1}^*, x_{j_2}^*) = (a_{j_1j_11}, a_{j_2j_2s_1})$ with $a_{j_2j_2s_1} = a_{j_1j_2l_1}$ has the following properties:

- *For* $a_{j_1j_10}(a_{j_1j_2l_1} - a_{j_1j_2l_2}) > 0$ *and* $a_{j_2j_20}(a_{j_1j_11} - a_{j_2j_11})(a_{j_2j_2s_1} - a_{j_2j_2s_2}) > 0$,

$$(a_{j_1j_11}, a_{j_2j_2s_1}) = \underbrace{(\text{DI} : \text{II}, \text{SO})}_{\text{inflection source}}. \tag{1.286}$$

The equilibrium of $(x_{j_1}^*, x_{j_2}^*) = (a_{j_1j_11}, a_{j_2j_2s_1})$ *is a* (DI:II,SO)*-inflection-source.*

- *For* $a_{j_1j_10}(a_{j_1j_2l_1}-a_{j_1j_2l_2})<0$ *and* $a_{j_2j_20}(a_{j_1j_11}-a_{j_2j_11})(a_{j_2j_2s_1}-a_{j_2j_2s_2})>0$,

$$(a_{j_1j_11},a_{j_2j_2s_1})=\underbrace{(\text{II}:\text{DI},\text{SO})}_{\text{inflection source}}. \tag{1.287}$$

The equilibrium of $(x_{j_1}^*,x_{j_2}^*)=(a_{j_1j_11},a_{j_2j_2s_1})$ *is an* (II:DI,SO)-*inflection-source.*

- *For* $a_{j_1j_10}(a_{j_1j_2l_1}-a_{j_1j_2l_2})>0$ *and* $a_{j_2j_20}(a_{j_1j_11}-a_{j_2j_11})(a_{j_2j_2s_1}-a_{j_2j_2s_2})<0$,

$$(a_{j_1j_11},a_{j_2j_2s_1})=\underbrace{(\text{II}:\text{DI},\text{SI})}_{\text{inflection sink}}. \tag{1.288}$$

The equilibrium of $(x_{j_1}^*,x_{j_2}^*)=(a_{j_1j_11},a_{j_2j_2s_1})$ *is an* (II:DI,SI)-*inflection-sink.*

- *For* $a_{j_1j_10}(a_{j_1j_2l_1}-a_{j_1j_2l_2})<0$ *and* $a_{j_2j_20}(a_{j_1j_11}-a_{j_2j_11})(a_{j_2j_2s_1}-a_{j_2j_2s_2})<0$,

$$(a_{j_1j_11},a_{j_2j_2s_1})=\underbrace{(\text{DI}:\text{II},\text{SI})}_{\text{inflection sink}}. \tag{1.289}$$

The equilibrium of $(x_{j_1}^*,x_{j_2}^*)=(a_{j_1j_11},a_{j_2j_2s_1})$ *is a* (DI:II,SI)-*inflection-sink.*

(ix$_{2c}$) The equilibrium of $(x_{j_2}^*,x_{j_1}^*)=(a_{j_1j_2l_1},a_{j_2j_11})$ *with* $a_{j_2j_2s_1}=a_{j_1j_2l_1}$ *has the following properties:*

- *For* $a_{j_1j_10}(a_{j_2j_11}-a_{j_1j_11})(a_{j_1j_2l_1}-a_{j_1j_2l_2})>0$ *and* $a_{j_2j_20}(a_{j_2j_2s_1}-a_{j_2j_2s_2})>0$,

$$(a_{j_1j_2l_1},a_{j_2j_11})=\underbrace{({}_{\text{SI:SO}}\text{US},\ {}_{\text{DI:II}}\text{UP})}_{\text{up-parabola upper-saddle}}. \tag{1.290}$$

The equilibrium of $(x_{j_2}^*,x_{j_1}^*)=(a_{j_1j_2l_1},a_{j_2j_11})$ *is an* (${}_{\text{SI:SO}}$US, ${}_{\text{DI:II}}$UP)-*up-parabola upper-saddle.*

- *For* $a_{j_1j_10}(a_{j_2j_11}-a_{j_1j_11})(a_{j_1j_2l_1}-a_{j_1j_2l_2})<0$ *and* $a_{j_2j_20}(a_{j_2j_2s_1}-a_{j_2j_2s_2})>0$,

$$(a_{j_1j_2l_1},a_{j_2j_11})=\underbrace{({}_{\text{SI:SO}}\text{US},\ {}_{\text{II:DI}}\text{DP})}_{\text{down-parabola upper-saddle}}. \tag{1.291}$$

The equilibrium of $(x_{j_2}^*,x_{j_1}^*)=(a_{j_1j_2l_1},a_{j_2j_11})$ *is an* (${}_{\text{SI:SO}}$US, ${}_{\text{II:DI}}$DP)-*down-parabola upper-saddle.*

- *For* $a_{j_1j_10}(a_{j_2j_11}-a_{j_1j_11})(a_{j_1j_2l_1}-a_{j_1j_2l_2})>0$ *and* $a_{j_2j_20}(a_{j_2j_2s_1}-a_{j_2j_2s_2})<0$,

$$(a_{j_1j_2l_1},a_{j_2j_11})=\underbrace{({}_{\text{SO:SI}}\text{LS},\ {}_{\text{II:DI}}\text{DP})}_{\text{down-parabola lower-saddle}}. \tag{1.292}$$

The equilibrium of $(x_{j_2}^*,x_{j_1}^*)=(a_{j_1j_2l_1},a_{j_2j_11})$ *is an* (${}_{\text{SO:SI}}$LS, ${}_{\text{II:DI}}$DP)-*down-parabola lower-saddle.*

- For $a_{j_1j_10}(a_{j_2j_11}-a_{j_1j_11})(a_{j_1j_2l_1}-a_{j_1j_2l_2})<0$ *and* $a_{j_2j_20}(a_{j_2j_2s_1}-a_{j_2j_2s_2})<0$,

$$(a_{j_1j_2l_1},a_{j_2j_11})=\underbrace{({}_{\text{SO:SI}}\text{LS},\ {}_{\text{DI:II}}\text{UP})}_{\text{up-parabola lower-saddle}}. \tag{1.293}$$

The equilibrium of $(x_{j_2}^*,x_{j_1}^*)=(a_{j_1j_2l_1},a_{j_2j_11})$ *is an* $({}_{\text{SO:SI}}\text{LS},\ {}_{\text{DI:II}}\text{UP})$-*up-parabola lower-saddle.*

(ix₃) The first integral manifold for $a_{j_2j_2s_1}\neq a_{j_1j_2l_1}$ $(s_1,s_2,l_1,l_2\in\{1,2\},s_1\neq s_2,l_1\neq l_2)$ *and* $a_{j_1j_11}=a_{j_2j_11}$ *is*

$$\begin{aligned}&\left[(x_{j_1}-a_{j_1j_11})-(x_{j_10}-a_{j_1j_11})\right]\\&\quad=\frac{a_{j_1j_10}}{a_{j_2j_20}}\left[(x_{j_2}-x_{j_20})+\sum\nolimits_{s_1=1}^{2}\frac{\prod_{l_1=1}^{2}(a_{j_2j_2s_1}-a_{j_1j_2l_1})}{(a_{j_2j_2s_1}-a_{j_2j_2s_2})}\ln\frac{|x_{j_2}-a_{j_2j_2s_1}|}{|x_{j_20}-a_{j_2j_2s_1}|}\right].\end{aligned} \tag{1.294}$$

(ix₃ₐ) The infinite-equilibrium of $x_{j_1}^*=a_{j_1j_11}=a_{j_2j_11}$ *with* $\bar{x}_{j_2}\neq a_{j_2j_2s_1},a_{j_1j_2l_1}$ $(a_{j_1j_2s_1},a_{j_1j_2s_2}\neq a_{j_2j_2l_1},a_{j_2j_2l_2})$ *has the following properties:*

- *For* $a_{j_1j_10}\prod_{l_1=1}^{2}(\bar{x}_{j_2}-a_{j_1j_2l_1})>0$ *and* $a_{j_2j_20}\prod_{s_1=1}^{2}(\bar{x}_{j_2}-a_{j_2j_2s_1})>0$,

$$(a_{j_1j_11},\bar{x}_{j_2})=\underbrace{(\text{SO},\text{II})}_{\text{increasing-inflection source}}. \tag{1.295}$$

The infinite-equilibrium of $x_{j_1}^*=a_{j_1j_11}=a_{j_2j_11}$ *is an* (SO,II)-*increasing-inflection source.*

- *For* $a_{j_1j_10}\prod_{l_1=1}^{2}(\bar{x}_{j_2}-a_{j_1j_2l_1})<0$ *and* $a_{j_2j_20}\prod_{s_1=1}^{2}(\bar{x}_{j_2}-a_{j_2j_2s_1})>0$,

$$(a_{j_1j_11},\bar{x}_{j_2})=\underbrace{(\text{SI},\text{DI})}_{\text{decreasing-inflection sink}}. \tag{1.296}$$

The infinite-equilibrium of $x_{j_1}^*=a_{j_1j_11}=a_{j_2j_11}$ *is an* (SI,DI)-*decreasing-inflection sink.*

- *For* $a_{j_1j_10}\prod_{l_1=1}^{2}(\bar{x}_{j_2}-a_{j_1j_2l_1})>0$ *and* $a_{j_2j_20}\prod_{s_1=1}^{2}(\bar{x}_{j_2}-a_{j_2j_2s_1})<0$,

$$(a_{j_1j_11},\bar{x}_{j_2})=\underbrace{(\text{SO},\text{DI})}_{\text{decreasing-inflection source}}. \tag{1.297}$$

The infinite-equilibrium of $x_{j_1}^*=a_{j_1j_11}=a_{j_2j_11}$ *is an* (SO,DI)-*decreasing-inflection source.*

- *For* $a_{j_1j_10}\prod_{l_1=1}^{2}(\bar{x}_{j_2}-a_{j_1j_2l_1})<0$ *and* $a_{j_2j_20}\prod_{s_1=1}^{2}(\bar{x}_{j_2}-a_{j_2j_2s_1})<0$,

$$(a_{j_1j_1 1},\bar{x}_{j_2})=\underbrace{(\text{SI, II})}_{\text{increasing-inflection sink}}. \tag{1.298}$$

The infinite-equilibrium of $x_{j_1}^*=a_{j_1j_1 1}=a_{j_2j_1 1}$ *is an* (SI,II)*-increasing-inflection sink.*

(ix$_{3b}$) The equilibrium of $(x_{j_2}^*,x_{j_1}^*)=(a_{j_1j_2l_1},a_{j_2j_1 1})$ *with* $a_{j_1j_1 1}=a_{j_2j_1 1}$ *has the following properties:*

- *For* $a_{j_1j_10}(a_{j_1j_2l_1}-a_{j_1j_2l_2})>0$ *and* $a_{j_2j_20}\prod_{s_1=1}^{2}(a_{j_1j_2l_1}-a_{j_2j_2s_1})>0$,

$$(a_{j_1j_2l_1},a_{j_2j_1 1})=\underbrace{({}_{\text{DI:II}}\text{UP},\ {}_{\text{SI:SO}}\text{US})}_{\text{up-parabola upper-saddle}}. \tag{1.299}$$

The equilibrium of $(x_{j_2}^*,x_{j_1}^*)=(a_{j_1j_2l_1},a_{j_2j_1 1})$ *is a* (${}_{\text{DI:II}}$UP, ${}_{\text{SI:SO}}$US)*-up-parabola upper-saddle.*

- *For* $a_{j_1j_10}(a_{j_1j_2l_1}-a_{j_1j_2l_2})<0$ *and* $a_{j_2j_20}\prod_{s_1=1}^{2}(a_{j_1j_2l_1}-a_{j_2j_2s_1})>0$,

$$(a_{j_1j_2l_1},a_{j_2j_1 1})=\underbrace{({}_{\text{II:DI}}\text{DP},\ {}_{\text{SO:SI}}\text{LS})}_{\text{down-parabola lower-saddle}}. \tag{1.300}$$

The equilibrium of $(x_{j_2}^*,x_{j_1}^*)=(a_{j_1j_2l_1},a_{j_2j_1 1})$ *is an* (${}_{\text{II:DI}}$DP, ${}_{\text{SO:SI}}$LS)*-down-parabola lower-saddle.*

- *For* $a_{j_1j_10}(a_{j_1j_2l_1}-a_{j_1j_2l_2})>0$ *and* $a_{j_2j_20}\prod_{s_1=1}^{2}(a_{j_1j_2l_1}-a_{j_2j_2s_1})<0$,

$$(a_{j_1j_2l_1},a_{j_2j_1 1})=\underbrace{({}_{\text{II:DI}}\text{DP},\ {}_{\text{SI:SO}}\text{US})}_{\text{down-parabola upper-saddle}}. \tag{1.301}$$

The equilibrium of $(x_{j_2}^*,x_{j_1}^*)=(a_{j_1j_2l_1},a_{j_2j_1 1})$ *is an* (${}_{\text{II:DI}}$DP, ${}_{\text{SI:SO}}$US)*-down-parabola upper-saddle.*

- *For* $a_{j_1j_10}(a_{j_1j_2l_1}-a_{j_1j_2l_2})<0$ *and* $a_{j_2j_20}\prod_{s_1=1}^{2}(a_{j_1j_2l_1}-a_{j_2j_2s_1})<0$,

$$(a_{j_1j_2l_1},a_{j_2j_1 1})=\underbrace{({}_{\text{DI:II}}\text{UP},\ {}_{\text{SI:SO}}\text{LS})}_{\text{up-parabola lower-saddle}}. \tag{1.302}$$

The equilibrium of $(x_{j_2}^*,x_{j_1}^*)=(a_{j_1j_2l_1},a_{j_2j_1 1})$ *is a* (${}_{\text{DI:II}}$UP, ${}_{\text{SO:SI}}$LS)*-up-parabola lower-saddle.*

(ix_{3c}) The equilibrium of $(x_{j_1}^*, x_{j_2}^*) = (a_{j_1j_1 1}, a_{j_2j_2s_1})$ *with* $a_{j_1j_1 1} = a_{j_2j_1 1}$ *has the following properties:*

- *For* $a_{j_1j_1 0}\prod_{l_1=1}^{2}(a_{j_2j_2s_1} - a_{j_1j_2l_1}) > 0$ *and* $a_{j_2j_2 0}(a_{j_2j_2s_1} - a_{j_2j_2s_2}) > 0$,

$$(a_{j_1j_1 1}, a_{j_2j_2s_1}) = \underbrace{(\mathrm{SO}, \mathrm{DI} : \mathrm{II})}_{\text{inflection source}}. \tag{1.303}$$

The equilibrium of $(x_{j_1}^*, x_{j_2}^*) = (a_{j_1j_1 1}, a_{j_2j_2s_1})$ *is an* (SO,DI:II)*-inflection source.*

- *For* $a_{j_1j_1 0}\prod_{l_1=1}^{2}(a_{j_2j_2s_1} - a_{j_1j_2l_1}) > 0$ *and* $a_{j_2j_2 0}(a_{j_2j_2s_1} - a_{j_2j_2s_2}) > 0$,

$$(a_{j_1j_1 1}, a_{j_2j_2s_1}) = \underbrace{(\mathrm{SI}, \mathrm{II} : \mathrm{DI})}_{\text{inflection sink}}. \tag{1.304}$$

The equilibrium of $(x_{j_1}^*, x_{j_2}^*) = (a_{j_1j_1 1}, a_{j_2j_2s_1})$ *is an* (SI,II:DI)*-inflection sink.*

- *For* $a_{j_1j_1 0}\prod_{l_1=1}^{2}(a_{j_2j_2s_1} - a_{j_1j_2l_1}) > 0$ *and* $a_{j_2j_2 0}(a_{j_2j_2s_1} - a_{j_2j_2s_2}) < 0$,

$$(a_{j_1j_1 1}, a_{j_2j_2s_1}) = \underbrace{(\mathrm{SO}, \mathrm{II} : \mathrm{DI})}_{\text{inflection source}}. \tag{1.305}$$

The equilibrium of $(x_{j_1}^*, x_{j_2}^*) = (a_{j_1j_1 1}, a_{j_2j_2s_1})$ *is an (SO,II:DI)-inflection source.*

- *For* $a_{j_1j_1 0}\prod_{l_1=1}^{2}(a_{j_2j_2s_1} - a_{j_1j_2l_1}) < 0$ *and* $a_{j_2j_2 0}(a_{j_2j_2s_1} - a_{j_2j_2s_2}) < 0$,

$$(a_{j_1j_1 1}, a_{j_2j_2s_1}) = \underbrace{(\mathrm{SI}, \mathrm{DI} : \mathrm{II})}_{\text{inflection sink}}. \tag{1.306}$$

The equilibrium of $(x_{j_1}^*, x_{j_2}^*) = (a_{j_1j_1 1}, a_{j_2j_2s_1})$ *is an* (SI,DI:II)*-inflection sink.*

(ix_4) The first integral manifold for $a_{j_2j_2s_1} = a_{j_1j_2l_1}$ $(s_1, s_2, l_1, l_2 \in \{1,2\}, s_1 \neq s_2, l_1 \neq l_2)$ *and* $a_{j_1j_1 1} = a_{j_2j_1 1}$ *is*

$$\begin{aligned}&\left[(x_{j_1} - a_{j_1j_1 1}) - (x_{j_1 0} - a_{j_1j_1 1})\right]\\&= \frac{a_{j_1j_1 0}}{a_{j_2j_2 0}}\left[(x_{j_2} - x_{j_2 0}) + (a_{j_2j_2s_2} - a_{j_1j_2l_2})\ln\frac{|x_{j_2} - a_{j_2j_2l_2}|}{|x_{j_2 0} - a_{j_2j_2l_2}|}\right].\end{aligned} \tag{1.307}$$

The first integral manifold for $a_{j_2j_2 1} = a_{j_1j_2 1}$ and $a_{j_2j_2 2} = a_{j_1j_2 2}$ with $a_{j_1j_1 1} = a_{j_2j_1 1}$ is

$$(x_{j_1} - x_{j_1 0}) = \frac{a_{j_1j_1 0}}{a_{j_2j_2 0}}(x_{j_2} - x_{j_2 0}). \tag{1.308}$$

The equilibrium of $(x_{j_1}^*, x_{j_2}^*) = (a_{j_1j_11}, a_{j_2j_2s_1})$ *with* $a_{j_1j_11} = a_{j_2j_11}$ and $a_{j_2j_2s_1} = a_{j_1j_2l_1}$ $(s_1, s_2, l_1, l_2 \in \{1,2\}; s_1 \neq s_2, l_1 \neq l_2)$ *has the following properties:*

- *For* $a_{j_1j_10}(a_{j_1j_2l_1} - a_{j_1j_2l_2}) > 0$ *and* $a_{j_2j_20}(a_{j_2j_2s_1} - a_{j_2j_2s_2}) > 0$,

$$(a_{j_1j_11}, a_{j_2j_2s_1}) = \underbrace{({}_{\mathrm{SI:SO}}\mathrm{II}, {}_{\mathrm{SI:SO}}\mathrm{II})}_{\text{increasing-inflection saddle}}. \tag{1.309}$$

  The equilibrium of $(x_{j_1}^*, x_{j_2}^*) = (a_{j_1j_11}, a_{j_2j_2s_1})$ *is an* $({}_{\mathrm{SI:SO}}\mathrm{II}, {}_{\mathrm{SI:SO}}\mathrm{II})$*-increasing-inflection saddle.*

- *For* $a_{j_1j_10}(a_{j_1j_2s_1} - a_{j_1j_2s_2}) < 0$ *and* $a_{j_2j_20}(a_{j_2j_2s_1} - a_{j_2j_2s_2}) > 0$,

$$(a_{j_1j_11}, a_{j_2j_2s_1}) = \underbrace{({}_{\mathrm{SI:SO}}\mathrm{DI}, {}_{\mathrm{SO:SI}}\mathrm{DI})}_{\text{decreasing-inflection saddle}}. \tag{1.310}$$

  The equilibrium of $(x_{j_1}^*, x_{j_2}^*) = (a_{j_1j_11}, a_{j_2j_2s_1})$ *is an* $({}_{\mathrm{SI:SO}}\mathrm{DI}, {}_{\mathrm{SO:SI}}\mathrm{DI})$*-decreasing-inflection saddle.*

- *For* $a_{j_1j_10}(a_{j_1j_2s_1} - a_{j_1j_2s_2}) > 0$ *and* $a_{j_2j_20}(a_{j_2j_2s_1} - a_{j_2j_2s_2}) < 0$,

$$(a_{j_1j_11}, a_{j_2j_2s_1}) = \underbrace{({}_{\mathrm{SO:SI}}\mathrm{DI}, {}_{\mathrm{SI:SO}}\mathrm{DI})}_{\text{decreasing-inflection saddle}}. \tag{1.311}$$

  The equilibrium of $(x_{j_1}^*, x_{j_2}^*) = (a_{j_1j_11}, a_{j_2j_2s_1})$ *is an* $({}_{\mathrm{SO:SI}}\mathrm{DI}, {}_{\mathrm{SI:SO}}\mathrm{DI})$*-decreasing-inflection saddle.*

- *For* $a_{j_1j_10}(a_{j_1j_2s_1} - a_{j_1j_2s_2}) < 0$ *and* $a_{j_2j_20}(a_{j_2j_2s_1} - a_{j_2j_2s_2}) < 0$,

$$(a_{j_1j_11}, a_{j_2j_2s_1}) = \underbrace{({}_{\mathrm{SO:SI}}\mathrm{II}, {}_{\mathrm{SO:SI}}\mathrm{II})}_{\text{increasing-inflection saddle}}. \tag{1.312}$$

  The equilibrium of $(x_{j_1}^*, x_{j_2}^*) = (a_{j_1j_11}, a_{j_2j_2s_1})$ *is an* $({}_{\mathrm{SO:SI}}\mathrm{II}, {}_{\mathrm{SO:SI}}\mathrm{II})$*-increasing-inflection saddle.*

1.2 Proof of Theorem

Consider a two-dimensional, nonlinear dynamical system as

$$\begin{aligned} \dot{x}_{j_2} &= a_{j_2j_20}(x_{j_1} - a_{j_2j_11})(x_{j_2}^2 + B_{j_2j_2}x_{j_2} + C_{j_2j_2}), \\ \dot{x}_{j_1} &= a_{j_1j_10}(x_{j_1} - a_{j_1j_11})(x_{j_2}^2 + B_{j_1j_2}x_{j_2} + C_{j_1j_2}) \end{aligned}$$

with

$$\Delta_{j_2j_2} = B^2_{j_2j_2} - 4C_{j_2j_2},$$
$$\Delta_{j_1j_2} = B^2_{j_1j_2} - 4C_{j_1j_2}.$$

(i) For $\Delta_{j_1j_2} < 0$ and $\Delta_{j_2j_2} < 0$, the standard form is

$$\dot{x}_{j_2} = a_{j_2j_20}(x_{j_1} - a_{j_2j_11})\left[(x_{j_2} - a_{j_2j_2})^2 + b_{j_2j_2}\right],$$
$$\dot{x}_{j_1} = a_{j_1j_10}(x_{j_1} - a_{j_1j_11})\left[(x_{j_2} - a_{j_1j_2})^2 + b_{j_1j_2}\right]$$

where

$$a_{j_2j_2} = -\frac{1}{2}B_{j_2j_2}, b_{j_2j_2} = \frac{1}{4}(-\Delta_{j_1j_2});$$
$$a_{j_1j_2} = -\frac{1}{2}B_{j_1j_2}, b_{j_1j_2} = \frac{1}{4}(-\Delta_{j_1j_2}).$$

In phase space,

$$\frac{dx_{j_1}}{dx_{j_2}} = \frac{a_{j_1j_10}}{a_{j_2j_20}}\frac{(x_{j_1} - a_{j_1j_11})\left[(x_{j_2} - a_{j_1j_2})^2 + b_{j_1j_2}\right]}{(x_{j_1} - a_{j_2j_11})\left[(x_{j_2} - a_{j_2j_2})^2 + b_{j_2j_2}\right]},$$

and the deformation of the foregoing equation is

$$\left[1 + \frac{(a_{j_1j_11} - a_{j_2j_11})}{(x_{j_1} - a_{j_1j_11})}\right]dx_{j_1} = \frac{a_{j_1j_10}}{a_{j_2j_20}}\left[1 + \frac{2(a_{j_2j_2} - a_{j_1j_2})(x_{j_2} - a_{j_2j_2})}{(x_{j_2} - a_{j_2j_2})^2 + b_{j_2j_2}}\right.$$
$$\left. + \frac{(a_{j_2j_2} - a_{j_1j_2})^2 + b_{j_1j_2} - b_{j_2j_2}}{(x_{j_2} - a_{j_2j_2})^2 + b_{j_2j_2}}\right]dx_{j_2}.$$

With an initial condition of (x_{j_10}, x_{j_20}) at $t = t_0$, the integration of the above equation gives

$$\left[(x_{j_1} - a_{j_1j_11}) - (x_{j_10} - a_{j_1j_11})\right] + (a_{j_1j_11} - a_{j_2j_11})\ln\frac{|x_{j_1} - a_{j_1j_11}|}{|x_{j_10} - a_{j_1j_11}|}$$
$$= \frac{a_{j_1j_10}}{a_{j_2j_20}}\left\{\left[(x_{j_2} - a_{j_2j_2}) - (x_{j_20} - a_{j_2j_2})\right]\right.$$
$$+ (a_{j_2j_2} - a_{j_1j_2})\ln\frac{|(x_{j_2} - a_{j_2j_2})^2 + b_{j_2j_2}|}{|(x_{j_20} - a_{j_2j_2})^2 + b_{j_2j_2}|}$$
$$+ \left[(a_{j_2j_2} - a_{j_1j_2})^2 + b_{j_2j_1} - b_{j_2j_2}\right]$$
$$\left. \times \frac{1}{\sqrt{b_{j_2j_2}}}\left(\arctan\frac{x_{j_2} - a_{j_2j_2}}{\sqrt{b_{j_2j_2}}} - \arctan\frac{x_{j_20} - a_{j_2j_2}}{\sqrt{b_{j_2j_2}}}\right)\right\}.$$

(i_1) At $x_{j_1}^* = a_{j_1 j_1 1}$, the variational equation is

$$\Delta\dot{x}_{j_1} = a_{j_1 j_1 0}\Delta x_{j_1}\left[(\overline{x}_{j_2} - a_{j_1 j_2})^2 + b_{j_1 j_2}\right].$$

Because of

$$(\overline{x}_{j_2} - a_{j_1 j_2})^2 + b_{j_1 j_2} > 0,$$

the flows at $x_{j_1}^* = a_{j_1 j_1 1}$ in the x_{j_1}-direction are source and sink for $a_{j_1 j_1 0} > 0$ and $a_{j_1 j_1 0} < 0$, respectively. For $a_{j_1 j_1 1} \neq a_{j_2 j_1 1}$, let

$$\dot{x}_{j_2} = a_{j_2 j_2 0}(a_{j_1 j_1 1} - a_{j_2 j_1 1})\left[(\overline{x}_{j_2} - a_{j_2 j_2})^2 + b_{j_2 j_2}\right],$$

and

$$(\overline{x}_{j_2} - a_{j_2 j_2})^2 + b_{j_2 j_2} > 0.$$

The flows at $x_{j_1}^* = a_{j_1 j_1 1}$ in the x_{j_2}-direction are positive and negative for $a_{j_2 j_2 0} > 0$ and $a_{j_2 j_2 0} < 0$, respectively.

Thus, the equilibrium of $x_{j_1}^* = a_{j_1 j_1 1}$ has the following properties as in Eqs. (1.6)–(1.9):

- For $a_{j_1 j_1 0} > 0$ and $a_{j_2 j_2 0}(a_{j_1 j_1 1} - a_{j_2 j_1 1}) > 0$,

$$(a_{j_1 j_1 1}, \dot{x}_{j_2}) = \underbrace{(\mathrm{SO}, \mathrm{pF})}_{\text{source flow } (+)}.$$

  The equilibrium of $x_{j_1}^* = a_{j_1 j_1 1}$ is an (SO,pF)-positive source flow.
- For $a_{j_1 j_1 0} < 0$ and $a_{j_2 j_2 0}(a_{j_1 j_1 1} - a_{j_2 j_1 1}) > 0$,

$$(a_{j_1 j_1 1}, \dot{x}_{j_2}) = \underbrace{(\mathrm{SI}, \mathrm{pF})}_{\text{sink flow } (+)}.$$

  The equilibrium of $x_{j_1}^* = a_{j_1 j_1 1}$ is an (SI,pF)-positive sink flow.
- For $a_{j_1 j_1 0} > 0$ and $a_{j_2 j_2 0}(a_{j_1 j_1 1} - a_{j_2 j_1 1}) < 0$,

$$(a_{j_1 j_1 1}, \dot{x}_{j_2}) = \underbrace{(\mathrm{SO}, \mathrm{nF})}_{\text{source flow } (-)}.$$

  The equilibrium of $x_{j_1}^* = a_{j_1 j_1 1}$ is an (SO,nF)-negative source flow.

- For $a_{j_1j_10} < 0$ and $a_{j_2j_20}(a_{j_1j_11} - a_{j_2j_11}) < 0$,

$$(a_{j_1j_11}, \dot{x}_{j_2}) = \underbrace{(\mathrm{SI}, \mathrm{nF})}_{\text{sink flow }(-)}.$$

The equilibrium of $x_{j_1}^* = a_{j_1j_11}$ is an (SI,pF)-negative sink flow.

($\mathrm{i}_{1\mathrm{b}}$) At $x_{j_1}^* = a_{j_2j_11}$, in phase space, for $a_{j_1j_11} \neq a_{j_2j_11}$

$$\frac{dx_{j_2}}{dx_{j_1}}\bigg|_{x_{j_1}^* = a_{j_2j_11}} = \frac{a_{j_2j_20}}{a_{j_1j_10}} \frac{(x_{j_1} - a_{j_2j_11})\left[(\overline{x}_{j_2} - a_{j_2j_2})^2 + b_{j_2j_2}\right]}{(x_{j_1} - a_{j_1j_11})\left[(\overline{x}_{j_2} - a_{j_1j_2})^2 + b_{j_1j_2}\right]}\bigg|_{x_{j_1}^* = a_{j_2j_11}} = 0.$$

If

$$\frac{d^2x_{j_2}}{dx_{j_1}^2}\bigg|_{x_{j_1}^* = a_{j_2j_11}} = \frac{a_{j_2j_20}}{a_{j_1j_10}} \frac{(\overline{x}_{j_2} - a_{j_2j_2})^2 + b_{j_2j_2}}{(a_{j_2j_11} - a_{j_1j_11})\left[(\overline{x}_{j_2} - a_{j_1j_2})^2 + b_{j_1j_2}\right]} > 0,$$

there is an up-parabola flow. If

$$\frac{d^2x_{j_2}}{dx_{j_1}^2}\bigg|_{x_{j_1}^* = a_{j_2j_11}} = \frac{a_{j_2j_20}}{a_{j_1j_10}} \frac{(\overline{x}_{j_2} - a_{j_2j_2})^2 + b_{j_2j_2}}{(a_{j_2j_11} - a_{j_1j_11})\left[(\overline{x}_{j_2} - a_{j_1j_2})^2 + b_{j_1j_2}\right]} < 0,$$

there is a down-parabola flow. Let

$$\dot{x}_{j_1} = a_{j_1j_10}(a_{j_2j_11} - a_{j_1j_11})\left[(\overline{x}_{j_2} - a_{j_1j_2})^2 + b_{j_1j_2}\right],$$

and

$$(\overline{x}_{j_2} - a_{j_1j_2})^2 + b_{j_1j_2} > 0.$$

The parabola flows at $x_{j_1}^* = a_{j_2j_11}$ in the x_{j_2}-direction are positive and negative for $a_{j_1j_10}(a_{j_2j_11} - a_{j_1j_11}) > 0$ and $a_{j_1j_10}(a_{j_2j_11} - a_{j_1j_11}) < 0$, respectively. Similarly,

$$(\overline{x}_{j_2} - a_{j_2j_2})^2 + b_{j_2j_2} > 0.$$

Thus, the equilibrium of $x_{j_1}^* = a_{j_2j_11}$ has the following properties as in Eqs. (1.10)–(1.13):

- For $a_{j_1j_1 0}(a_{j_2j_1 1} - a_{j_1j_1 1}) > 0$ and $a_{j_2j_2 0} > 0$,

$$(\dot{x}_{j_1}, a_{j_2j_1 1}) = \underbrace{(\text{pF}, \text{UP})}_{\text{up-parabola flow } (+)}.$$

The equilibrium of $x_{j_1}^* = a_{j_2j_1 1}$ is a (pF,UP)-positive up-parabola flow.

- For $a_{j_1j_1 0}(a_{j_2j_1 1} - a_{j_1j_1 1}) < 0$ and $a_{j_2j_2 0} > 0$,

$$(\dot{x}_{j_1}, a_{j_2j_1 1}) = \underbrace{(\text{nF}, \text{DP})}_{\text{down-parabola flow } (-)}.$$

The equilibrium of $x_{j_1}^* = a_{j_2j_1 1}$ is an (nF,DP)-negative down-parabola flow.

- For $a_{j_1j_1 0}(a_{j_2j_1 1} - a_{j_1j_1 1}) > 0$ and $a_{j_2j_2 0} < 0$,

$$(\dot{x}_{j_1}, a_{j_2j_1 1}) = \underbrace{(\text{pF}, \text{DP})}_{\text{down-parabola flow } (+)}.$$

The equilibrium of $x_{j_1}^* = a_{j_2j_1 1}$ is a (pF,DP)-positive down-parabola flow.

- For $a_{j_1j_1 0}(a_{j_2j_1 1} - a_{j_1j_1 1}) < 0$ and $a_{j_2j_2 0} < 0$,

$$(\dot{x}_{j_1}, a_{j_2j_1 1}) = \underbrace{(\text{nF}, \text{UP})}_{\text{up-parabola flow } (-)}.$$

The equilibrium of $x_{j_1}^* = a_{j_2j_1 1}$ is an (nF,UP)-negative up-parabola flow.

(i_2) For $a_{j_1j_1 1} = a_{j_2j_1 1}$, in phase space,

$$\frac{dx_{j_1}}{dx_{j_2}} = \frac{a_{j_1j_1 0}}{a_{j_2j_2 0}} \frac{(x_{j_2} - a_{j_1j_2})^2 + b_{j_1j_2}}{(x_{j_2} - a_{j_2j_2})^2 + b_{j_2j_2}},$$

and the deformation of the foregoing equation is

$$dx_{j_1} = \frac{a_{j_1j_1 0}}{a_{j_2j_2 0}} \left[1 + \frac{2(a_{j_2j_2} - a_{j_1j_2})(x_{j_2} - a_{j_2j_2})}{(x_{j_2} - a_{j_2j_2})^2 + b_{j_2j_2}} \right.$$
$$\left. + \frac{(a_{j_2j_2} - a_{j_1j_2})^2 + b_{j_1j_2} - b_{j_2j_2}}{(x_{j_2} - a_{j_2j_2})^2 + b_{j_2j_2}} \right] dx_{j_2}.$$

With an initial condition of $(x_{j_1 0}, x_{j_2 0})$ at $t = t_0$, the integration of the above equation gives the first integral manifold for $a_{j_1j_1 1} = a_{j_2j_1 1}$ as

$$
\begin{aligned}
&\left[(x_{j_1} - a_{j_1 j_1 1}) - (x_{j_1 0} - a_{j_1 j_1 1})\right] \\
&\quad = \frac{a_{j_1 j_1 0}}{a_{j_2 j_2 0}} \Big\{ \left[(x_{j_2} - a_{j_2 j_2}) - (x_{j_2 0} - a_{j_2 j_2})\right] \\
&\qquad + (a_{j_2 j_2} - a_{j_1 j_2}) \ln \frac{|(x_{j_2} - a_{j_2 j_2})^2 + b_{j_2 j_2}|}{|(x_{j_2 0} - a_{j_2 j_2})^2 + b_{j_2 j_2}|} \\
&\qquad + \left[(a_{j_2 j_2} - a_{j_1 j_2})^2 + b_{j_1 j_2} - b_{j_2 j_2}\right] \\
&\qquad \times \frac{1}{\sqrt{b_{j_2 j_2}}} \Big(\arctan \frac{x_{j_2} - a_{j_2 j_2}}{\sqrt{b_{j_2 j_2}}} - \arctan \frac{x_{j_2 0} - a_{j_2 j_2}}{\sqrt{b_{j_2 j_2}}} \Big) \Big\}.
\end{aligned}
$$

In phase space, at $x_{j_1}^* = a_{j_1 j_1 1} = a_{j_2 j_1 1}$, if

$$
\frac{dx_{j_2}}{dx_{j_1}}\Big|_{x_{j_1}^* = a_{j_1 j_1 1}} = \frac{a_{j_2 j_2 0}}{a_{j_1 j_1 0}} \frac{(x_{j_2} - a_{j_2 j_2})^2 + b_{j_2 j_2}}{(x_{j_2} - a_{j_1 j_2})^2 + b_{j_1 j_2}}\Big|_{x_{j_1}^* = a_{j_1 j_1 1}} > 0,
$$

there is an increasing-inflection flow, and if

$$
\frac{dx_{j_2}}{dx_{j_1}}\Big|_{x_{j_1}^* = a_{j_1 j_1 1}} = \frac{a_{j_2 j_2 0}}{a_{j_1 j_1 0}} \frac{(x_{j_2} - a_{j_2 j_2})^2 + b_{j_2 j_2}}{(x_{j_2} - a_{j_1 j_2})^2 + b_{j_1 j_2}}\Big|_{x_{j_1}^* = a_{j_1 j_1 1}} < 0,
$$

there is a decreasing-inflection flow. The variational equation at $x_{j_1}^* = a_{j_1 j_1 1}$ is

$$
\Delta \dot{x}_{j_1} = a_{j_1 j_1 0} \left[(\bar{x}_{j_2} - a_{j_1 j_2})^2 + b_{j_1 j_2}\right] \Delta x_{j_1}
$$

and

$$
(\bar{x}_{j_2} - a_{j_1 j_2})^2 + b_{j_1 j_2} > 0.
$$

Thus, the infinite-equilibrium of $x_{j_1}^* = a_{j_1 j_1 1} = a_{j_2 j_1 1}$ with $\bar{x}_{j_2} \in (-\infty, \infty)$ has the following properties as in Eqs. (1.15)–(1.18):

- For $a_{j_1 j_1 0} > 0$ and $a_{j_2 j_2 0} > 0$,

$$
(a_{j_1 j_1 1}, \bar{x}_{j_2}) = \underbrace{(\text{SO}, \text{II})}_{\text{increasing-infelction source}}.
$$

  The infinite-equilibrium of $x_{j_1}^* = a_{j_1 j_1 1} = a_{j_2 j_1 1}$ is an (SO,II)-increasing-inflection source.

- For $a_{j_1j_10} < 0$ and $a_{j_2j_20} > 0$,

$$(a_{j_1j_11}, \bar{x}_{j_2}) = \underbrace{(\text{SI, DI})}_{\text{decreasing-infelction sink}}.$$

The infinite-equilibrium of $x^*_{j_1} = a_{j_1j_11} = a_{j_2j_11}$ is an (SI,DI)-decreasing-inflection sink.

- For $a_{j_1j_10} > 0$ and $a_{j_2j_20} < 0$,

$$(a_{j_1j_11}, \bar{x}_{j_2}) = \underbrace{(\text{SO, DI})}_{\text{decreasing-infelction source}}.$$

The infinite-equilibrium of $x^*_{j_1} = a_{j_1j_11} = a_{j_2j_11}$ is an (SO,DI)-decreasing-inflection source.

- For $a_{j_1j_10} < 0$ and $a_{j_2j_20} < 0$,

$$(a_{j_1j_11}, \bar{x}_{j_2}) = \underbrace{(\text{SI, II})}_{\text{increasing-infelction sink}}.$$

The infinite-equilibrium of $x^*_{j_1} = a_{j_1j_11} = a_{j_2j_11}$ is an (SI,II)-increasing-inflection sink.

(ii) For $\Delta_{j_1j_2} < 0$ and $\Delta_{j_2j_2} = 0$, the standard form is

$$\begin{aligned}
\dot{x}_{j_2} &= a_{j_2j_20}(x_{j_1} - a_{j_2j_11})(x_{j_2} - a_{j_2j_21})^2, \\
\dot{x}_{j_1} &= a_{j_1j_10}(x_{j_1} - a_{j_1j_11})\left[(x_{j_2} - a_{j_1j_2})^2 + b_{j_1j_2}\right]
\end{aligned}$$

where

$$\begin{aligned}
a_{j_1j_2} &= -\frac{1}{2}B_{j_1j_2}, b_{j_1j_2} = \frac{1}{4}(-\Delta_{j_1j_2}), \\
a_{j_2j_21} &= -\frac{1}{2}B_{j_2j_2}.
\end{aligned}$$

(ii$_1$) In phase space,

$$\frac{dx_{j_1}}{dx_{j_2}} = \frac{a_{j_1j_10}}{a_{j_2j_20}} \frac{(x_{j_1} - a_{j_1j_11})\left[(x_{j_2} - a_{j_1j_2})^2 + b_{j_1j_2}\right]}{(x_{j_1} - a_{j_2j_11})(x_{j_2} - a_{j_2j_21})^2},$$

and the deformation of the foregoing equation is

$$\left[1+\frac{(a_{j_1j_1 1}-a_{j_2j_1 1})}{(x_{j_1}-a_{j_1j_1 1})}\right]dx_{j_1}=\frac{a_{j_1j_1 0}}{a_{j_2j_2 0}}\left[1+\frac{2(a_{j_2j_2 1}-a_{j_1j_2})(x_{j_2}-a_{j_2j_2 1})}{(x_{j_2}-a_{j_2j_2 1})^2}\right.$$
$$\left.+\frac{(a_{j_2j_2 1}-a_{j_1j_2})^2+b_{j_1j_2}}{(x_{j_2}-a_{j_2j_2 1})^2}\right]dx_{j_2}.$$

With an initial condition of $(x_{j_1 0}, x_{j_2 0})$ at $t=t_0$, the integration of the above equation gives

$$\left[(x_{j_1}-a_{j_1j_1 1})-(x_{j_1 0}-a_{j_1j_1 1})\right]+(a_{j_1j_1 1}-a_{j_2j_1 1})\ln\frac{|x_{j_1}-a_{j_1j_1 1}|}{|x_{j_1 0}-a_{j_1j_1 1}|}$$
$$=\frac{a_{j_1j_1 0}}{a_{j_2j_2 0}}\left\{\left[(x_{j_2}-a_{j_2j_2 1})-(x_{j_2 0}-a_{j_2j_2 1})\right]\right.$$
$$+2(a_{j_2j_2 1}-a_{j_1j_2})\ln\frac{|x_{j_2}-a_{j_2j_2 1}|}{|x_{j_2 0}-a_{j_2j_2 1}|}$$
$$\left.-\left[(a_{j_2j_2 1}-a_{j_1j_2})^2+b_{j_2j_1}\right]\left(\frac{1}{x_{j_2}-a_{j_2j_2 1}}-\frac{1}{x_{j_2 0}-a_{j_2j_2 1}}\right)\right\}.$$

(ii$_{1a}$) At $(x^*_{j_1}, x^*_{j_2})=(a_{j_1j_1 1}, a_{j_2j_2 1})$, the variational equations are

$$\Delta\dot{x}_{j_2}=a_{j_2j_2 0}(a_{j_1j_1 1}-a_{j_2j_1 1})(\Delta x_{j_2})^2,$$
$$\Delta\dot{x}_{j_1}=a_{j_1j_1 0}\Delta x_{j_1}\left[(a_{j_2j_2 1}-a_{j_1j_2})^2+b_{j_1j_2}\right].$$

Because of

$$\left[(a_{j_2j_2 1}-a_{j_1j_2})^2+b_{j_1j_2}\right]>0,$$

the flows at $(x^*_{j_1}, x^*_{j_2})=(a_{j_1j_1 1}, a_{j_2j_2 1})$ in the x_{j_1} -direction are source and sink for $a_{j_1j_1 0}>0$ and $a_{j_1j_1 0}<0$, respectively. The flows at $(x^*_{j_1}, x^*_{j_2})=(a_{j_1j_1 1}, a_{j_2j_2 1})$ in the x_{j_2} -direction are upper-saddle and lower-saddle for $a_{j_2j_2 0}(a_{j_1j_1 1}-a_{j_2j_1 1})>0$ and $a_{j_2j_2 0}(a_{j_1j_1 1}-a_{j_2j_1 1})<0$, respectively.

Therefore, the equilibrium of $(x^*_{j_1}, x^*_{j_2})=(a_{j_1j_1 1}, a_{j_2j_2 1})$ has the following properties as in Eqs. (1.22)–(1.25):

- For $a_{j_1j_1 0}>0$ and $a_{j_2j_2 0}(a_{j_1j_1 1}-a_{j_2j_1 1})>0$,

$$(a_{j_1j_1 1}, a_{j_2j_2 1})=\underbrace{(\text{SO}, \text{US})}_{\text{upper-saddle-source}}.$$

The equilibrium of $(x^*_{j_1}, x^*_{j_2})=(a_{j_1j_1 1}, a_{j_2j_2 1})$ is an (SO,US)-upper-saddle-source.

- For $a_{j_1j_1 0} < 0$ and $a_{j_2j_2 0}(a_{j_1j_1 1} - a_{j_2j_1 1}) > 0$,

$$(a_{j_1j_1 1}, a_{j_2j_2 1}) = \underbrace{(\mathrm{SI}, \mathrm{US})}_{\text{upper-saddle-sink}}.$$

  The equilibrium of $(x_{j_1}^*, x_{j_2}^*) = (a_{j_1j_1 1}, a_{j_2j_2 1})$ is an (SI,US)-upper-saddle-sink.
- For $a_{j_1j_1 0} > 0$ and $a_{j_2j_2 0}(a_{j_1j_1 1} - a_{j_2j_1 1}) < 0$,

$$(a_{j_1j_1 1}, a_{j_2j_2 1}) = \underbrace{(\mathrm{SO}, \mathrm{LS})}_{\text{lower-saddle-source}}.$$

  The equilibrium of $(x_{j_1}^*, x_{j_2}^*) = (a_{j_1j_1 1}, a_{j_2j_2 1})$ is an (SO,LS)-lower-saddle-source.
- For $a_{j_1j_1 0} < 0$ and $a_{j_2j_2 0}(a_{j_1j_1 1} - a_{j_2j_1 1}) < 0$,

$$(a_{j_1j_1 1}, a_{j_2j_2 1}) = \underbrace{(\mathrm{SI}, \mathrm{LS})}_{\text{lower-saddle-source}}.$$

  The equilibrium of $(x_{j_1}^*, x_{j_2}^*) = (a_{j_1j_1 1}, a_{j_2j_2 1})$ is an (SI,LS)-lower-saddle-sink.

– The saddle-source and saddle-sink are the appearing bifurcations of saddle-source and saddle-sink.

(ii$_{1b}$) At $x_{j_1}^* = a_{j_2j_1 1}$, in phase space, for $a_{j_1j_1 1} \neq a_{j_2j_1 1}$,

$$\frac{dx_{j_2}}{dx_{j_1}}\bigg|_{x_{j_1}^* = a_{j_2j_1 1}} = \frac{a_{j_2j_2 0}}{a_{j_1j_1 0}} \frac{(x_{j_1} - a_{j_2j_1 1})(\overline{x}_{j_2} - a_{j_2j_2 1})^2}{(x_{j_1} - a_{j_1j_1 1})\left[(\overline{x}_{j_2} - a_{j_1j_2})^2 + b_{j_1j_2}\right]}\bigg|_{x_{j_1}^* = a_{j_2j_1 1}} = 0.$$

If

$$\frac{d^2x_{j_2}}{dx_{j_1}^2}\bigg|_{x_{j_1}^* = a_{j_2j_1 1}} = \frac{a_{j_2j_2 0}}{a_{j_1j_1 0}} \frac{(\overline{x}_{j_2} - a_{j_2j_2 1})^2}{(a_{j_2j_1 1} - a_{j_1j_1 1})\left[(\overline{x}_{j_2} - a_{j_1j_2})^2 + b_{j_1j_2}\right]} > 0,$$

there is an up-parabola flow. If

$$\frac{d^2x_{j_2}}{dx_{j_1}^2}\bigg|_{x_{j_1}^* = a_{j_2j_1 1}} = \frac{a_{j_2j_2 0}}{a_{j_1j_1 0}} \frac{(\overline{x}_{j_2} - a_{j_2j_2 1})^2}{(a_{j_2j_1 1} - a_{j_1j_1 1})\left[(\overline{x}_{j_2} - a_{j_1j_2})^2 + b_{j_1j_2}\right]} < 0,$$

there is a down-parabola flow. Let

$$\dot{x}_{j_1} = a_{j_1j_1 0}(a_{j_2j_1 1} - a_{j_1j_1 1})\left[(\overline{x}_{j_2} - a_{j_1j_2})^2 + b_{j_1j_2}\right],$$

and

.

$$(\overline{x}_{j_2} - a_{j_1j_2})^2 + b_{j_1j_2} > 0.$$

The parabola flows at $x^*_{j_1} = a_{j_2j_1 1}$ in the x_{j_2}-direction are positive and negative for $a_{j_1j_1 0}(a_{j_2j_1 1} - a_{j_1j_1 1}) > 0$ and $a_{j_1j_1 0}(a_{j_2j_1 1} - a_{j_1j_1 1}) < 0$, respectively. For $\overline{x}_{j_2} \neq a_{j_2j_2 1}$,

$$(\overline{x}_{j_2} - a_{j_2j_2 1})^2 > 0.$$

Therefore, the equilibrium of $(x^*_{j_1}, x^*_{j_2}) = (a_{j_2j_1 1}, a_{j_2j_2 1})$ has the following properties as in Eqs. (1.26)–(1.29):

- For $a_{j_1j_1 0}(a_{j_2j_1 1} - a_{j_1j_1 1}) > 0$ and $a_{j_2j_2 0} > 0$,

  $$(a_{j_2j_1 1}, a_{j_2j_2 1}) = \underbrace{(\text{pF}, \text{UP} : \text{UP})}_{\text{hyperbolic-secant-to-hyperbolic flow } (+)}.$$

  The equilibrium of $(x^*_{j_1}, x^*_{j_2}) = (a_{j_2j_1 1}, a_{j_2j_2 1})$ is a (pF,UP:UP)-positive hyperbolic-secant-to-hyperbolic flow.
- For $a_{j_1j_1 0}(a_{j_2j_1 1} - a_{j_1j_1 1}) < 0$ and $a_{j_2j_2 0} > 0$,

  $$(a_{j_2j_1 1}, a_{j_2j_2 1}) = \underbrace{(\text{nF}, \text{DP} : \text{DP})}_{\text{hyperbolic-to-hyperbolic-secant flow } (-)}.$$

  The equilibrium of $(x^*_{j_1}, x^*_{j_2}) = (a_{j_1j_1 1}, a_{j_2j_2 1})$ is an (nF,DP:DP)-negative hyperbolic-to-hyperbolic-secant flow.
- For $a_{j_1j_1 0}(a_{j_2j_1 1} - a_{j_1j_1 1}) > 0$ and $a_{j_2j_2 0} < 0$,

  $$(a_{j_2j_1 1}, a_{j_2j_2 1}) = \underbrace{(\text{pF}, \text{DP} : \text{DP})}_{\text{hyperbolic-to-hyperbolic-secant flow } (+)}.$$

  The equilibrium of $(x^*_{j_1}, x^*_{j_2}) = (a_{j_1j_1 1}, a_{j_2j_2 1})$ is a (pF,DP:DP)-positive hyperbolic-to-hyperbolic-secant flow.
- For $a_{j_1j_1 0}(a_{j_2j_1 1} - a_{j_1j_1 1}) < 0$ and $a_{j_2j_2 0} < 0$,

  $$(a_{j_2j_1 1}, a_{j_2j_2 1}) = \underbrace{(\text{nF}, \text{UP} : \text{UP})}_{\text{hyperbolic-secant-to-hyperbolic flow } (-)}.$$

  The equilibrium of $(x^*_{j_1}, x^*_{j_2}) = (a_{j_1j_1 1}, a_{j_2j_2 1})$ is an (nF,UP:UP)-negative hyperbolic-secant-to-hyperbolic flow.

(ii_2) In phase space, for $a_{j_1j_11} = a_{j_2j_11}$,

$$\frac{dx_{j_1}}{dx_{j_2}} = \frac{a_{j_1j_10}}{a_{j_2j_20}} \frac{(x_{j_2} - a_{j_1j_2})^2 + b_{j_1j_2}}{(x_{j_2} - a_{j_2j_21})^2},$$

and the deformation of the foregoing equation is

$$dx_{j_1} = \frac{a_{j_1j_10}}{a_{j_2j_20}} \Big[1 + \frac{2(a_{j_2j_21} - a_{j_1j_2})(x_{j_2} - a_{j_2j_21})}{(x_{j_2} - a_{j_2j_21})^2} + \frac{(a_{j_2j_21} - a_{j_1j_2})^2 + b_{j_1j_2}}{(x_{j_2} - a_{j_2j_21})^2}\Big] dx_{j_2}.$$

With an initial condition of (x_{j_10}, x_{j_20}) at $t = t_0$, the integration of the above equation gives the first integral manifold for $a_{j_1j_11} = a_{j_2j_11}$ as

$$\begin{aligned}
&\big[(x_{j_1} - a_{j_1j_11}) - (x_{j_10} - a_{j_1j_11})\big] \\
&= \frac{a_{j_1j_10}}{a_{j_2j_20}} \Big\{ \big[(x_{j_2} - a_{j_2j_21}) - (x_{j_20} - a_{j_2j_21})\big] \\
&+2(a_{j_2j_21} - a_{j_1j_2}) \ln \frac{|x_{j_2} - a_{j_2j_21}|}{|x_{j_20} - a_{j_2j_21}|} \\
&- \big[(a_{j_2j_21} - a_{j_1j_2})^2 + b_{j_1j_2}\big] \Big(\frac{1}{x_{j_2} - a_{j_2j_21}} - \frac{1}{x_{j_20} - a_{j_2j_21}}\Big) \Big\}.
\end{aligned}$$

(ii_{2a}) In phase space, at $x^*_{j_1} = a_{j_1j_11} = a_{j_2j_11}$, if

$$\frac{dx_{j_2}}{dx_{j_1}}\Big|_{x^*_{j_1} = a_{j_1j_11}} = \frac{a_{j_2j_20}}{a_{j_1j_10}} \frac{(\bar{x}_{j_2} - a_{j_2j_21})^2}{(\bar{x}_{j_2} - a_{j_1j_2})^2 + b_{j_1j_2}} > 0,$$

there is an increasing-inflection flow, and if

$$\frac{dx_{j_2}}{dx_{j_1}}\Big|_{x^*_{j_1} = a_{j_1j_11}} = \frac{a_{j_2j_20}}{a_{j_1j_10}} \frac{(\bar{x}_{j_2} - a_{j_2j_21})^2}{(\bar{x}_{j_2} - a_{j_1j_2})^2 + b_{j_1j_2}} < 0,$$

there is a decreasing-inflection flow. The variational equation is

$$\Delta \dot{x}_{j_1} = a_{j_1j_10} \big[(\bar{x}_{j_2} - a_{j_1j_2})^2 + b_{j_1j_2}\big] \Delta x_{j_1}$$

and

$$(\bar{x}_{j_2} - a_{j_1j_2})^2 + b_{j_1j_2} > 0.$$

The *inflection* flows are sink and source for $a_{j_1j_10} > 0$ and $a_{j_1j_10} < 0$, respectively.

Therefore, the infinite-equilibrium of $x_{j_1}^* = a_{j_1j_11} = a_{j_2j_11}$ with $\bar{x}_{j_2} \neq a_{j_2j_21}$ has the following properties as in Eqs. (1.31)–(1.34):

- For $a_{j_1j_10} > 0$ and $a_{j_2j_20}(\bar{x}_{j_2} - a_{j_2j_21})^2 > 0$,

$$(a_{j_1j_11}, \bar{x}_{j_2}) = \underbrace{(\text{SO, II})}_{\text{increasing-infelction source}}.$$

  The infinite-equilibrium of $x_{j_1}^* = a_{j_1j_11} = a_{j_2j_11}$ is an (SO,II)-increasing-inflection source.
- For $a_{j_1j_10} < 0$ and $a_{j_2j_20}(\bar{x}_{j_2} - a_{j_2j_21})^2 > 0$,

$$(a_{j_1j_11}, \bar{x}_{j_2}) = \underbrace{(\text{SI, DI})}_{\text{decreasing-infelction sink}}.$$

  The infinite-equilibrium of $x_{j_1}^* = a_{j_1j_11} = a_{j_2j_11}$ is an (SI,DI)-decreasing-inflection sink.
- For $a_{j_1j_10} > 0$ and $a_{j_2j_20}(\bar{x}_{j_2} - a_{j_2j_21})^2 < 0$,

$$(a_{j_1j_11}, \bar{x}_{j_2}) = \underbrace{(\text{SO, DI})}_{\text{decreasing-infelction source}}.$$

  The infinite-equilibrium of $x_{j_1}^* = a_{j_1j_11} = a_{j_2j_11}$ is an (SO,DI)-decreasing-inflection source.
- For $a_{j_1j_10} < 0$ and $a_{j_2j_20}(\bar{x}_{j_2} - a_{j_2j_21})^2 < 0$,

$$(a_{j_1j_11}, \bar{x}_{j_2}) = \underbrace{(\text{SI, II})}_{\text{increasing-infelction sink}}.$$

  The infinite-equilibrium of $x_{j_1}^* = a_{j_1j_11} = a_{j_2j_11}$ is an (SI,II)-increasing-inflection sink.

($\text{ii}_{2\text{b}}$) From case ($\text{ii}_{2\text{a}}$), for $\bar{x}_{j_2} = a_{j_2j_21}$, thus, the equilibrium of $(x_{j_1}^*, x_{j_2}^*) = (a_{j_1j_11}, a_{j_2j_21})$ has the following properties as in Eqs. (1.35)–(1.38):

- For $a_{j_1j_10} > 0$ and $a_{j_2j_20} > 0$,

$$(a_{j_1j_11}, a_{j_2j_21}) = \underbrace{(\text{SO, II : II})}_{\text{increasing-inflection source}}.$$

  The equilibrium of $(x_{j_1}^*, x_{j_2}^*) = (a_{j_1j_11}, a_{j_2j_21})$ is an (SO,II:II)-increasing-inflection source.

- For $a_{j_1j_10} < 0$ and $a_{j_2j_20} > 0$,

$$(a_{j_1j_11}, a_{j_2j_21}) = \underbrace{(\text{SI}, \text{DI} : \text{DI})}_{\text{decreasing-inflection sink}}.$$

The equilibrium of $(x^*_{j_1}, x^*_{j_2}) = (a_{j_1j_11}, a_{j_2j_21})$ is an (SI,DI:DI)-decreasing-inflection sink.

- For $a_{j_1j_10} > 0$ and $a_{j_2j_20} < 0$,

$$(a_{j_1j_11}, a_{j_2j_21}) = \underbrace{(\text{SO}, \text{DI} : \text{DI})}_{\text{decreasing-inflection source}}.$$

The equilibrium of $(x^*_{j_1}, x^*_{j_2}) = (a_{j_1j_11}, a_{j_2j_21})$ is an (SO,DI:DI)-decreasing-inflection source.

- For $a_{j_1j_10} < 0$ and $a_{j_2j_20} < 0$,

$$(a_{j_1j_11}, a_{j_2j_21}) = \underbrace{(\text{SI}, \text{II} : \text{II})}_{\text{increasing-inflection sink}}.$$

The equilibrium of $(x^*_{j_1}, x^*_{j_2}) = (a_{j_1j_11}, a_{j_2j_21})$ is an (SI,II:II)-increasing-inflection sink.

(iii) For $\Delta_{j_1j_2} < 0$ and $\Delta_{j_2j_2} > 0$, the standard form is

$$\begin{aligned}
\dot{x}_{j_2} &= a_{j_2j_20}(x_{j_1} - a_{j_2j_11})(x_{j_2} - a_{j_2j_21})(x_{j_2} - a_{j_2j_22}),\\
\dot{x}_{j_1} &= a_{j_1j_10}(x_{j_1} - a_{j_1j_11})\left[(x_{j_2} - a_{j_1j_2})^2 + b_{j_1j_2}\right]
\end{aligned}$$

where

$$\begin{aligned}
&a_{j_1j_2} = -\frac{1}{2}B_{j_1j_2}, b_{j_1j_2} = \frac{1}{4}(-\Delta_{j_1j_2}),\\
&b_{j_2j_21}, b_{j_2j_22} = -\frac{1}{2}(B_{j_2j_2} \pm \sqrt{\Delta_{j_2j_2}}),\\
&\{a_{j_2j_21}, a_{j_2j_22}\} = \text{sort}\{b_{j_2j_21}, b_{j_2j_22}\}, a_{j_2j_21} < a_{j_2j_22}.
\end{aligned}$$

(iii_1) In phase space,

$$\frac{dx_{j_1}}{dx_{j_2}} = \frac{a_{j_1j_10}}{a_{j_2j_20}} \frac{(x_{j_1} - a_{j_1j_11})\left[(x_{j_2} - a_{j_1j_2})^2 + b_{j_1j_2}\right]}{(x_{j_1} - a_{j_2j_11})(x_{j_2} - a_{j_2j_21})(x_{j_2} - a_{j_2j_22})},$$

and the deformation of the foregoing equation is

$$\left[1+\frac{(a_{j_1j_1 1}-a_{j_2j_1 1})}{(x_{j_1}-a_{j_1j_1 1})}\right]dx_{j_1}=\frac{a_{j_1j_1 0}}{a_{j_2j_2 0}}\left[1+\frac{(a_{j_2j_2 1}-a_{j_1j_2})}{(x_{j_2}-a_{j_2j_2 1})}+\frac{(a_{j_2j_2 2}-a_{j_1j_2})}{(x_{j_2}-a_{j_2j_2 2})}\right.$$
$$\left.+\frac{(a_{j_2j_2 1}-a_{j_1j_2})(a_{j_2j_2 2}-a_{j_1j_2})+b_{j_1j_2}}{(x_{j_2}-a_{j_2j_2 1})(x_{j_2}-a_{j_2j_2 2})}\right]dx_{j_2}.$$

With an initial condition of $(x_{j_1 0},x_{j_2 0})$ at $t=t_0$, the integration of the above equation gives

$$\begin{aligned}
&[(x_{j_1}-a_{j_1j_1 1})-(x_{j_1 0}-a_{j_1j_1 1})]+\frac{a_{j_1j_1 1}-a_{j_2j_1 1}}{x_{j_1}-a_{j_1j_1 1}}\ln\frac{|x_{j_1}-a_{j_1j_1 1}|}{|x_{j_1 0}-a_{j_1j_1 1}|}\\
&=\frac{a_{j_1j_1 0}}{a_{j_2j_2 0}}\Big\{[(x_{j_2}-a_{j_2j_2 1})-(x_{j_2 0}-a_{j_2j_2 1})]\\
&\quad+\sum\nolimits_{s_1=1}^{2}(a_{j_2j_2 s_1}-a_{j_1j_2})\ln\frac{|x_{j_2}-a_{j_2j_2 s_1}|}{|x_{j_2 0}-a_{j_2j_2 s_1}|}\\
&\quad+\Big[\prod\nolimits_{s_1=1}^{2}(a_{j_2j_2 s_1}-a_{j_1j_2})+b_{j_2j_1}\Big]\\
&\quad\times\Big(\sum\nolimits_{s_1=1,s_1\neq s_2}^{2}\frac{1}{a_{j_2j_2 s_1}-a_{j_1j_2 s_2}}\ln\frac{|x_{j_2}-a_{j_2j_2 s_1}|}{|x_{j_2 0}-a_{j_2j_2 s_1}|}\Big)\Big\}.
\end{aligned}$$

(iii_{1a}) At $(x^*_{j_1},x^*_{j_2})=(a_{j_1j_1 1},a_{j_2j_2 s_1})$, the variational equations are

$$\begin{aligned}
\Delta\dot{x}_{j_2}&=a_{j_2j_2 0}(a_{j_1j_1 1}-a_{j_2j_1 1})(a_{j_2j_2 s_1}-a_{j_2j_2 s_2})\Delta x_{j_2},\\
\Delta\dot{x}_{j_1}&=a_{j_1j_1 0}\left[(a_{j_2j_2 s_1}-a_{j_1j_2})^2+b_{j_1j_2}\right]\Delta x_{j_1}.
\end{aligned}$$

Because of

$$\left[(a_{j_2j_2 s_1}-a_{j_1j_2})^2+b_{j_1j_2}\right]>0,$$

the flows at $(x^*_{j_1},x^*_{j_2})=(a_{j_1j_1 1},a_{j_2j_2 s_1})$ in the x_{j_1}-direction are source and sink for $a_{j_1j_1 0}>0$ and $a_{j_1j_1 0}<0$, respectively. The flows at $(x^*_{j_1},x^*_{j_2})=(a_{j_1j_1 1},a_{j_2j_2 s_1})$ in the x_{j_2}-direction are source and sink for $a_{j_2j_2 0}(a_{j_1j_1 1}-a_{j_2j_1 1})(a_{j_2j_2 s_1}-a_{j_2j_2 s_2})>0$ and $a_{j_2j_2 0}(a_{j_1j_1 1}-a_{j_2j_1 1})(a_{j_2j_2 s_1}-a_{j_2j_2 s_2})<0$, respectively.

Therefore, the equilibrium of $(x^*_{j_1},x^*_{j_2})=(a_{j_1j_1 1},a_{j_2j_2 s_1})$ ($s_1,s_2\in\{1,2\}$, $s_1\neq s_2$) has the following properties as in Eqs. (1.42)–(1.45):

- For $a_{j_1j_1 0}>0$ and $a_{j_2j_2 0}(a_{j_1j_1 1}-a_{j_2j_1 1})(a_{j_2j_2 s_1}-a_{j_2j_2 s_2})>0$,

$$(a_{j_1j_1 1},a_{j_2j_2 s_1})=\underbrace{(\mathrm{SO},\mathrm{SO})}_{\text{source}}.$$

The equilibrium of $(x^*_{j_1},x^*_{j_2})=(a_{j_1j_1 1},a_{j_2j_2 s_1})$ is an (SO,SO)-source.

- For $a_{j_1j_10}<0$ and $a_{j_2j_20}(a_{j_1j_11}-a_{j_2j_11})(a_{j_2j_2s_1}-a_{j_2j_2s_2})>0$,

$$(a_{j_1j_11},a_{j_2j_2s_1})=\underbrace{(\text{SI, SO})}_{\text{saddle}}.$$

The equilibrium of $(x_{j_1}^*,x_{j_2}^*)=(a_{j_1j_11},a_{j_2j_2s_1})$ is an (SI,SO)-saddle.

- For $a_{j_1j_10}>0$ and $a_{j_2j_20}(a_{j_1j_11}-a_{j_2j_11})(a_{j_2j_2s_1}-a_{j_2j_2s_2})<0$,

$$(a_{j_1j_11},a_{j_2j_2s_1})=\underbrace{(\text{SO, SI})}_{\text{saddle}}.$$

The equilibrium of $(x_{j_1}^*,x_{j_2}^*)=(a_{j_1j_11},a_{j_2j_2s_1})$ is an (SO,SI)-saddle.

- For $a_{j_1j_10}<0$ and $a_{j_2j_20}(a_{j_1j_11}-a_{j_2j_11})(a_{j_2j_2s_1}-a_{j_2j_2s_2})<0$,

$$(a_{j_1j_11},a_{j_2j_2s_1})=\underbrace{(\text{SI, SI})}_{\text{sink}}.$$

The equilibrium of $(x_{j_1}^*,x_{j_2}^*)=(a_{j_1j_11},a_{j_2j_2s_1})$ is an (SI,SI)-sink.

(iii$_{1b}$) At $x_{j_1}^*=a_{j_2j_11}$, in phase space, for $a_{j_1j_11}\neq a_{j_2j_11}$,

$$\left.\frac{dx_{j_2}}{dx_{j_1}}\right|_{x_{j_1}^*=a_{j_2j_11}}=\frac{a_{j_2j_20}}{a_{j_1j_10}}\left.\frac{(x_{j_1}-a_{j_2j_11})(\overline{x}_{j_2}-a_{j_2j_21})(\overline{x}_{j_2}-a_{j_2j_22})}{(x_{j_1}-a_{j_1j_11})\left[(\overline{x}_{j_2}-a_{j_1j_2})^2+b_{j_1j_2}\right]}\right|_{x_{j_1}^*=a_{j_2j_11}}=0.$$

If

$$\left.\frac{d^2x_{j_2}}{dx_{j_1}^2}\right|_{x_{j_1}^*=a_{j_2j_11}}=\frac{a_{j_2j_20}}{a_{j_1j_10}}\frac{(\overline{x}_{j_2}-a_{j_2j_21})(\overline{x}_{j_2}-a_{j_2j_22})}{(a_{j_2j_11}-a_{j_1j_11})\left[(\overline{x}_{j_2}-a_{j_1j_2})^2+b_{j_1j_2}\right]}>0,$$

there is an up-parabola flow. If

$$\left.\frac{d^2x_{j_2}}{dx_{j_1}^2}\right|_{x_{j_1}^*=a_{j_2j_11}}=\frac{a_{j_2j_20}}{a_{j_1j_10}}\frac{(\overline{x}_{j_2}-a_{j_2j_21})(\overline{x}_{j_2}-a_{j_2j_22})}{(a_{j_2j_11}-a_{j_1j_11})\left[(\overline{x}_{j_2}-a_{j_1j_2})^2+b_{j_1j_2}\right]}<0,$$

there is a down-parabola flow. Let

$$\dot{x}_{j_1}=a_{j_1j_10}(a_{j_2j_11}-a_{j_1j_11})\left[(\overline{x}_{j_2}-a_{j_1j_2})^2+b_{j_1j_2}\right],$$

and

$$(\overline{x}_{j_2}-a_{j_1j_2})^2+b_{j_1j_2}>0.$$

The parabola flows at $x_{j_1}^* = a_{j_2 j_1 1}$ in the x_{j_2}-direction are positive and negative for $a_{j_1 j_1 0}(a_{j_2 j_1 1} - a_{j_1 j_1 1}) > 0$ and $a_{j_1 j_1 0}(a_{j_2 j_1 1} - a_{j_1 j_1 1}) < 0$, respectively.

Therefore, the equilibrium of $(x_{j_1}^*, x_{j_2}^*) = (a_{j_2 j_1 1}, a_{j_2 j_2 s_1})$ $(s_1, s_2 \in \{1,2\}, s_1 \neq s_2)$ has the following properties as in Eqs. (1.46)–(1.49):

- For $a_{j_1 j_1 0}(a_{j_2 j_1 1} - a_{j_1 j_1 1}) > 0$ and $a_{j_2 j_2 0}(a_{j_2 j_2 s_1} - a_{j_2 j_2 s_2}) > 0$,

$$(a_{j_2 j_1 1}, a_{j_2 j_2 s_1}) = \underbrace{(\text{pF}, \text{DP} : \text{UP})}_{\text{hyperbolic flow } (+)}.$$

The equilibrium of $(x_{j_1}^*, x_{j_2}^*) = (a_{j_2 j_1 1}, a_{j_2 j_2 s_1})$ is a (pF,UP:UP)-positive hyperbolic flow.

- For $a_{j_1 j_1 0}(a_{j_2 j_1 1} - a_{j_1 j_1 1}) < 0$ and $a_{j_2 j_2 0}(a_{j_2 j_2 s_1} - a_{j_2 j_2 s_2}) > 0$,

$$(a_{j_2 j_1 1}, a_{j_2 j_2 s_1}) = \underbrace{(\text{nF}, \text{UP} : \text{DP})}_{\text{hyperbolic-secant flow } (-)}.$$

The equilibrium of $(x_{j_1}^*, x_{j_2}^*) = (a_{j_1 j_1 1}, a_{j_2 j_2 s_1})$ is an (nF,UP:DP)-negative hyperbolic-secant flow.

- For $a_{j_1 j_1 0}(a_{j_2 j_1 1} - a_{j_1 j_1 1}) > 0$ and $a_{j_2 j_2 0}(a_{j_2 j_2 s_1} - a_{j_2 j_2 s_2}) < 0$,

$$(a_{j_2 j_1 1}, a_{j_2 j_2 s_1}) = \underbrace{(\text{pF}, \text{UP} : \text{DP})}_{\text{hyperbolic-secant flow } (+)}.$$

The equilibrium of $(x_{j_1}^*, x_{j_2}^*) = (a_{j_1 j_1 1}, a_{j_2 j_2 s_1})$ is a (pF,UP:DP)-positive hyperbolic-secant flow.

- For $a_{j_1 j_1 0}(a_{j_2 j_1 1} - a_{j_1 j_1 1}) < 0$ and $a_{j_2 j_2 0}(a_{j_2 j_2 s_1} - a_{j_2 j_2 s_2}) < 0$,

$$(a_{j_2 j_1 1}, a_{j_2 j_2 s_1}) = \underbrace{(\text{nF}, \text{DP} : \text{UP})}_{\text{hyperbolic flow } (-)}.$$

The equilibrium of $(x_{j_1}^*, x_{j_2}^*) = (a_{j_1 j_1 1}, a_{j_2 j_2 s_1})$ is an (nF,DP:UP)-negative hyperbolic flow.

(iii$_2$) For $a_{j_1 j_1 1} = a_{j_2 j_1 1}$, in phase space,

$$\frac{dx_{j_1}}{dx_{j_2}} = \frac{a_{j_1 j_1 0}}{a_{j_2 j_2 0}} \frac{\left[(x_{j_2} - a_{j_1 j_2})^2 + b_{j_1 j_2}\right]}{(x_{j_2} - a_{j_2 j_2 1})(x_{j_2} - a_{j_2 j_2 2})},$$

and the deformation of the foregoing equation is

$$
\begin{aligned}
dx_{j_1} = \frac{a_{j_1j_10}}{a_{j_2j_20}} & \left[1 + \frac{(a_{j_2j_21} - a_{j_1j_2})}{(x_{j_2} - a_{j_2j_21})} + \frac{(a_{j_2j_22} - a_{j_1j_2})}{(x_{j_2} - a_{j_2j_22})} \right. \\
& \left. + \frac{(a_{j_2j_21} - a_{j_1j_2})(a_{j_2j_22} - a_{j_1j_2}) + b_{j_1j_2}}{(x_{j_2} - a_{j_2j_21})(x_{j_2} - a_{j_2j_22})} \right] dx_{j_2}.
\end{aligned}
$$

With an initial condition of (x_{j_10}, x_{j_20}) at $t = t_0$, the integration of the above equation gives the first integral manifold for $a_{j_1j_11} = a_{j_2j_11}$ as

$$
\begin{aligned}
& [(x_{j_1} - a_{j_1j_11}) - (x_{j_10} - a_{j_1j_11})] \\
& = \frac{a_{j_1j_10}}{a_{j_2j_20}} \left\{ [(x_{j_2} - a_{j_2j_21}) - (x_{j_20} - a_{j_2j_21})] \right. \\
& + \sum\nolimits_{s_1=1}^{2} (a_{j_2j_2s_1} - a_{j_1j_2}) \ln \frac{|x_{j_2} - a_{j_2j_2s_1}|}{|x_{j_20} - a_{j_2j_2s_1}|} \\
& + \left[\prod\nolimits_{s_1=1}^{2} (a_{j_2j_2s_1} - a_{j_1j_2}) + b_{j_1j_2} \right] \\
& \left. \times \left[\sum\nolimits_{s_1=1, s_1 \neq s_2}^{2} \frac{1}{(a_{j_2j_2s_1} - a_{j_1j_2s_2})} \ln \frac{|x_{j_2} - a_{j_2j_2s_1}|}{|x_{j_20} - a_{j_2j_2s_1}|} \right] \right\}.
\end{aligned}
$$

(iii$_{2a}$) In phase space, at $x^*_{j_1} = a_{j_1j_11} = a_{j_2j_11}$, if

$$
\frac{dx_{j_2}}{dx_{j_1}}\bigg|_{x^*_{j_1} = a_{j_1j_11}} = \frac{a_{j_2j_20}}{a_{j_1j_10}} \frac{(\bar{x}_{j_2} - a_{j_2j_21})(\bar{x}_{j_2} - a_{j_2j_22})}{(\bar{x}_{j_2} - a_{j_1j_2})^2 + b_{j_1j_2}} > 0,
$$

there is an increasing-inflection flow, and if

$$
\frac{dx_{j_2}}{dx_{j_1}}\bigg|_{x^*_{j_1} = a_{j_1j_11}} = \frac{a_{j_2j_20}}{a_{j_1j_10}} \frac{(\bar{x}_{j_2} - a_{j_2j_21})(\bar{x}_{j_2} - a_{j_2j_22})}{(\bar{x}_{j_2} - a_{j_1j_2})^2 + b_{j_1j_2}} < 0,
$$

there is a decreasing-inflection flow. The variational equation is

$$
\Delta \dot{x}_{j_1} = a_{j_1j_10} \left[(\bar{x}_{j_2} - a_{j_1j_2})^2 + b_{j_1j_2} \right] \Delta x_{j_1}
$$

and

$$
(\bar{x}_{j_2} - a_{j_1j_2})^2 + b_{j_1j_2} > 0.
$$

The inflection flows are sink and source for $a_{j_1j_10} > 0$ and $a_{j_1j_10} < 0$, respectively.

Therefore, the infinite-equilibrium of $x^*_{j_1} = a_{j_1j_11} = a_{j_2j_11}$ has the following properties as in Eqs. (1.51)–(1.54):

- For $a_{j_1j_10} > 0$ and $a_{j_2j_20}(\overline{x}_{j_2} - a_{j_2j_2s_1})(\overline{x}_{j_2} - a_{j_2j_2s_2}) > 0$,

$$(a_{j_1j_11}, \overline{x}_{j_2}) = \underbrace{(\text{SO, II})}_{\text{increasing-infelction source}}.$$

The infinite-equilibrium of $x^*_{j_1} = a_{j_1j_11} = a_{j_2j_11}$ is an (SO,II)-increasing-inflection source.

- For $a_{j_1j_10} < 0$ and $a_{j_2j_20}(\overline{x}_{j_2} - a_{j_2j_2s_1})(\overline{x}_{j_2} - a_{j_2j_2s_2}) > 0$,

$$(a_{j_1j_11}, \overline{x}_{j_2}) = \underbrace{(\text{SI, DI})}_{\text{decreasing-infelction sink}}.$$

The infinite-equilibrium of $x^*_{j_1} = a_{j_1j_11} = a_{j_2j_11}$ is an (SI,DI)-decreasing-inflection sink.

- For $a_{j_1j_10} > 0$ and $a_{j_2j_20}(\overline{x}_{j_2} - a_{j_2j_2s_1})(\overline{x}_{j_2} - a_{j_2j_2s_2}) < 0$,

$$(a_{j_1j_11}, \overline{x}_{j_2}) = \underbrace{(\text{SO, DI})}_{\text{decreasing-infelction source}}.$$

The infinite-equilibrium of $x^*_{j_1} = a_{j_1j_11} = a_{j_2j_11}$ is an (SO,DI)-decreasing-inflection source.

- For $a_{j_1j_10} < 0$ and $a_{j_2j_20}(\overline{x}_{j_2} - a_{j_2j_2s_1})(\overline{x}_{j_2} - a_{j_2j_2s_2}) < 0$,

$$(a_{j_1j_11}, \overline{x}_{j_2}) = \underbrace{(\text{SI, II})}_{\text{increasing-infelction sink}}.$$

The infinite-equilibrium of $x^*_{j_1} = a_{j_1j_11} = a_{j_2j_11}$ is an (SI,II)-increasing-inflection sink.

(iii$_{2b}$) From case (iii$_{2a}$), for $\overline{x}_{j_2} = a_{j_2j_2s_1}$, therefore, the equilibrium of $(x^*_{j_1}, x^*_{j_2}) = (a_{j_1j_11}, a_{j_2j_2s_1})$ (s_1, $s_2 \in \{1, 2\}$; $s_1 \neq s_2$) has the following properties as in Eqs. (1.55)–(1.58):

- For $a_{j_1j_10} > 0$ and $a_{j_2j_20}(a_{j_2j_2s_1} - a_{j_2j_2s_2}) > 0$,

$$(a_{j_2j_11}, a_{j_2j_2s_1}) = \underbrace{(\text{UP, US})}_{\text{up-parabola upper-saddle}}.$$

The equilibrium of $(x^*_{j_1}, x^*_{j_2}) = (a_{j_1j_11}, a_{j_2j_2s_1})$ is a (UP,US)-up-parabola upper-saddle.

- For $a_{j_1j_10}<0$ and $a_{j_2j_20}(a_{j_2j_2s_1}-a_{j_2j_2s_2})>0$,

$$(a_{j_2j_11},a_{j_2j_2s_1})=\underbrace{(\mathrm{DP},\mathrm{US})}_{\text{down-parabola upper-saddle}}.$$

The equilibrium of $(x_{j_1}^*,x_{j_2}^*)=(a_{j_1j_11},a_{j_2j_2s_1})$ is a (DP,US)-down-parabola upper-saddle.

- For $a_{j_1j_10}>0$ and $a_{j_2j_20}(a_{j_2j_2s_1}-a_{j_2j_2s_2})<0$,

$$(a_{j_2j_11},a_{j_2j_2s_1})=\underbrace{(\mathrm{DP},\mathrm{LS})}_{\text{down-parabola lower-saddle}}.$$

The equilibrium of $(x_{j_1}^*,x_{j_2}^*)=(a_{j_1j_11},a_{j_2j_2s_1})$ is a (DP:LS)-down-parabola lower-saddle.

- For $a_{j_1j_10}<0$ and $a_{j_2j_20}(a_{j_2j_2s_1}-a_{j_2j_2s_2})<0$,

$$(a_{j_2j_11},a_{j_2j_2s_1})=\underbrace{(\mathrm{UP},\mathrm{LS})}_{\text{up-parabola lower-saddle}}.$$

The equilibrium of $(x_{j_1}^*,x_{j_2}^*)=(a_{j_1j_11},a_{j_2j_2s_1})$ is a (UP,LS)-up-parabola lower-saddle.

(iv) For $\Delta_{j_1j_2}=0$ and $\Delta_{j_2j_2}<0$, the standard form is

$$\begin{aligned}\dot{x}_{j_2}&=a_{j_2j_20}(x_{j_1}-a_{j_2j_11})\left[(x_{j_2}-a_{j_2j_2})^2+b_{j_2j_2}\right],\\ \dot{x}_{j_1}&=a_{j_1j_10}(x_{j_1}-a_{j_1j_11})(x_{j_2}-a_{j_1j_21})^2\end{aligned}$$

where

$$\begin{aligned}a_{j_1j_21}&=-\frac{1}{2}B_{j_1j_2},\\ a_{j_2j_2}&=-\frac{1}{2}B_{j_2j_2},\, b_{j_2j_2}=\frac{1}{4}(-\Delta_{j_2j_2}).\end{aligned}$$

In phase space,

$$\frac{dx_{j_1}}{dx_{j_2}}=\frac{a_{j_1j_10}}{a_{j_2j_20}}\frac{(x_{j_1}-a_{j_1j_11})(x_{j_2}-a_{j_1j_21})^2}{(x_{j_1}-a_{j_2j_11})\left[(x_{j_2}-a_{j_2j_2})^2+b_{j_2j_2}\right]},$$

and the deformation of the foregoing equation is

$$\left[1+\frac{(a_{j_1j_1}1-a_{j_2j_1}1)}{(x_{j_1}-a_{j_1j_1}1)}\right]dx_{j_1}=\frac{a_{j_1j_1}0}{a_{j_2j_2}0}\left[1+\frac{2(a_{j_2j_2}-a_{j_1j_2}1)(x_{j_2}-a_{j_2j_2})}{(x_{j_2}-a_{j_2j_2})^2+b_{j_2j_2}}\right.$$
$$\left.+\frac{(a_{j_2j_2}-a_{j_1j_2}1)^2-b_{j_2j_2}}{(x_{j_2}-a_{j_2j_2})^2+b_{j_2j_2}}\right]dx_{j_2}.$$

With an initial condition of $(x_{j_1 0}, x_{j_2 0})$ at $t = t_0$, the integration of the above equation gives

$$\begin{aligned}&\left[(x_{j_1}-a_{j_1j_1}1)-(x_{j_1 0}-a_{j_1j_1}1)\right]+(a_{j_1j_1}1-a_{j_2j_1}1)\ln\frac{|x_{j_1}-a_{j_1j_1}1|}{|x_{j_1 0}-a_{j_1j_1}1|}\\&=\frac{a_{j_1j_1}0}{a_{j_2j_2}0}\left\{\left[(x_{j_2}-a_{j_2j_2})-(x_{j_2 0}-a_{j_2j_2})\right]\right.\\&+(a_{j_2j_2}-a_{j_1j_2}1)\ln\frac{|(x_{j_2}-a_{j_2j_2})^2+b_{j_2j_2}|}{|(x_{j_2 0}-a_{j_2j_2})^2+b_{j_2j_2}|}\\&\left.+\left[(a_{j_2j_2}-a_{j_1j_2}1)^2-b_{j_2j_2}\right]\frac{1}{\sqrt{b_{j_2j_2}}}\left(\arctan\frac{x_{j_2}-a_{j_2j_2}}{\sqrt{b_{j_2j_2}}}-\arctan\frac{x_{j_2 0}-a_{j_2j_2}}{\sqrt{b_{j_2j_2}}}\right)\right\}.\end{aligned}$$

(iv$_1$) Consider two cases (I) $x^*_{j_2}=a_{j_1j_2}1$ and (II) $x^*_{j_1}=a_{j_1j_1}1$.
(I) In phase space, at $x^*_{j_2}=a_{j_1j_2}1$ with $\overline{x}_{j_1}\neq a_{j_1j_1}1$,

$$\begin{aligned}&\left.\frac{dx_{j_1}}{dx_{j_2}}\right|_{x^*_{j_2}=a_{j_1j_2}1}=\frac{a_{j_1j_1}0}{a_{j_2j_2}0}\left.\frac{(\overline{x}_{j_1}-a_{j_1j_1}1)(x_{j_2}-a_{j_1j_2}1)^2}{(\overline{x}_{j_1}-a_{j_2j_1}1)\left[(x_{j_2}-a_{j_2j_2})^2+b_{j_2j_2}\right]}\right|_{x^*_{j_2}=a_{j_1j_2}1}=0,\\&\left.\frac{d^2x_{j_1}}{dx^2_{j_2}}\right|_{x^*_{j_2}=a_{j_1j_2}1}=0.\end{aligned}$$

If

$$\left.\frac{d^3x_{j_1}}{dx^3_{j_2}}\right|_{x^*_{j_2}=a_{j_1j_2}1}=\frac{a_{j_1j_1}0}{a_{j_2j_2}0}\frac{2(\overline{x}_{j_1}-a_{j_1j_1}1)}{(\overline{x}_{j_1}-a_{j_2j_1}1)\left[(a_{j_1j_2}1-a_{j_2j_2})^2+b_{j_2j_2}\right]}>0,$$

there is a third-order increasing-inflection flow at $x^*_{j_2}=a_{j_1j_2}1$ with $\overline{x}_{j_1}\neq a_{j_1j_1}1$. If

$$\left.\frac{d^3x_{j_1}}{dx^3_{j_2}}\right|_{x^*_{j_2}=a_{j_1j_2}1}=\frac{a_{j_1j_1}0}{a_{j_2j_2}0}\frac{2(\overline{x}_{j_1}-a_{j_1j_1}1)}{(\overline{x}_{j_1}-a_{j_2j_1}1)\left[(a_{j_1j_2}1-a_{j_2j_2})^2+b_{j_2j_2}\right]}<0,$$

there is a third-order decreasing-inflection flow at $x^*_{j_2}=a_{j_1j_2}1$ with $\overline{x}_{j_1}\neq a_{j_1j_1}1$. Let

$$\dot{x}_{j_2}=a_{j_2j_2}0(\overline{x}_{j_1}-a_{j_2j_1}1)\left[(a_{j_1j_2}1-a_{j_2j_2})^2+b_{j_2j_2}\right],$$

with

$$(a_{j_1j_2 1}-a_{j_2j_2})^2+b_{j_2j_2}>0.$$

The third-order inflection flows at $x^*_{j_2}=a_{j_1j_2 1}$ with $\overline{x}_{j_1}\neq a_{j_1j_1 1}$ are positive and negative for $a_{j_2j_2 0}(\overline{x}_{j_1}-a_{j_2j_1 1})>0$ and $a_{j_2j_2 0}(\overline{x}_{j_1}-a_{j_2j_1 1})<0$, respectively.

Thus, the equilibrium of $x^*_{j_2}=a_{j_1j_2 1}$ with $\overline{x}_{j_1}\neq a_{j_1j_1 1}$ has the following properties:

- For $a_{j_1j_1 0}(\overline{x}_{j_1}-a_{j_1j_1 1})>0$ and $a_{j_2j_2 0}(\overline{x}_{j_1}-a_{j_2j_1 1})>0$,

$$(a_{j_1j_2 1},\dot{x}_{j_2})=\underbrace{(\mathrm{II},\mathrm{pF})}_{\text{increasing-inflection flow }(+)}.$$

- For $a_{j_1j_1 0}(\overline{x}_{j_1}-a_{j_1j_1 1})<0$ and $a_{j_2j_2 0}(\overline{x}_{j_1}-a_{j_2j_1 1})>0$,

$$(a_{j_1j_2 1},\dot{x}_{j_2})=\underbrace{(\mathrm{DI},\mathrm{pF})}_{\text{decreasing-inflection flow }(+)}.$$

- For $a_{j_1j_1 0}(\overline{x}_{j_1}-a_{j_1j_1 1})>0$ and $a_{j_2j_2 0}(\overline{x}_{j_1}-a_{j_2j_1 1})<0$,

$$(a_{j_1j_2 1},\dot{x}_{j_2})=\underbrace{(\mathrm{DI},\mathrm{nF})}_{\text{decreasing-inflection flow }(-)}.$$

- For $a_{j_1j_1 0}(\overline{x}_{j_1}-a_{j_1j_1 1})<0$ and $a_{j_2j_2 0}(\overline{x}_{j_1}-a_{j_2j_1 1})<0$,

$$(a_{j_1j_2 1},\dot{x}_{j_2})=\underbrace{(\mathrm{II},\mathrm{nF})}_{\text{increasing-inflection flow }(-)}.$$

(II) The variational equation at $x^*_{j_1}=a_{j_1j_1 1}$ with $\overline{x}_{j_2}\neq a_{j_1j_2 1}$ is

$$\Delta\dot{x}_{j_1}=a_{j_1j_1 0}(\overline{x}_{j_2}-a_{j_1j_2 1})^2\Delta x_{j_1}$$

and

$$\dot{x}_{j_2}=a_{j_2j_2 0}(a_{j_1j_1 1}-a_{j_2j_1 1})\left[(\overline{x}_{j_2}-a_{j_2j_2})^2+b_{j_2j_2}\right].$$

Thus, the equilibrium of $x^*_{j_2}=a_{j_1j_2 1}$ with $\overline{x}_{j_1}\neq a_{j_1j_1 1}$ has the following properties:

- For $a_{j_1j_10}(\overline{x}_{j_2} - a_{j_1j_21})^2 > 0$ and $a_{j_2j_20}(a_{j_1j_11} - a_{j_2j_11}) > 0$,

$$(a_{j_1j_11}, \dot{x}_{j_2}) = \underbrace{(\text{SO}, \text{pF})}_{\text{source flow }(+)}.$$

- For $a_{j_1j_10}(\overline{x}_{j_2} - a_{j_1j_21})^2 < 0$ and $a_{j_2j_20}(a_{j_1j_11} - a_{j_2j_11}) > 0$,

$$(a_{j_1j_11}, \dot{x}_{j_2}) = \underbrace{(\text{SI}, \text{pF})}_{\text{sink flow }(+)}.$$

- For $a_{j_1j_10}(\overline{x}_{j_2} - a_{j_1j_21})^2 > 0$ and $a_{j_2j_20}(a_{j_1j_11} - a_{j_2j_11}) < 0$,

$$(a_{j_1j_11}, \dot{x}_{j_2}) = \underbrace{(\text{SO}, \text{nF})}_{\text{source flow }(-)}.$$

- For $a_{j_1j_10}(\overline{x}_{j_2} - a_{j_1j_21})^2 < 0$ and $a_{j_2j_20}(a_{j_1j_11} - a_{j_2j_11}) < 0$,

$$(a_{j_1j_11}, \dot{x}_{j_2}) = \underbrace{(\text{SI}, \text{nF})}_{\text{sink flow }(-)}.$$

Therefore, the equilibrium of $(x^*_{j_1}, x^*_{j_2}) = (a_{j_1j_11}, a_{j_1j_21})$ has the following properties as in Eqs. (1.62)–(1.65):

- For $a_{j_1j_10} > 0$ and $a_{j_2j_20}(a_{j_1j_11} - a_{j_2j_11}) > 0$,

$$(a_{j_1j_11}, a_{j_1j_21}) = \underbrace{({}_{\text{DI:II}}\text{SO}, \text{pF})}_{\text{inflection-source flow }(+)}.$$

  The equilibrium of $(x^*_{j_1}, x^*_{j_2}) = (a_{j_1j_11}, a_{j_1j_21})$ is a $({}_{\text{DI:II}}\text{SO}, \text{pF})$-positive inflection-source flow.

- For $a_{j_1j_10} < 0$ and $a_{j_2j_20}(a_{j_1j_11} - a_{j_2j_11}) > 0$,

$$(a_{j_1j_11}, a_{j_1j_21}) = \underbrace{({}_{\text{II:DI}}\text{SI}, \text{pF})}_{\text{inflection-sink flow }(+)}.$$

  The equilibrium of $(x^*_{j_1}, x^*_{j_2}) = (a_{j_1j_11}, a_{j_1j_21})$ is an $({}_{\text{II:DI}}\text{SI}, \text{pF})$-positive inflection-sink flow.

- For $a_{j_1j_10} > 0$ and $a_{j_2j_20}(a_{j_1j_11} - a_{j_2j_11}) < 0$,

$$(a_{j_1j_11}, a_{j_1j_21}) = \underbrace{(\text{II:DI}\text{SO, nF})}_{\text{inflection-source flow }(-)}.$$

The equilibrium of $(x_{j_1}^*, x_{j_2}^*) = (a_{j_1j_11}, a_{j_1j_21})$ is an (II:DISO, nF)-negative inflection-source flow.

- For $a_{j_1j_10} < 0$ and $a_{j_2j_20}(a_{j_1j_11} - a_{j_2j_11}) < 0$,

$$(a_{j_1j_11}, a_{j_1j_21}) = \underbrace{(\text{DI:II}\text{SI, nF})}_{\text{inflection-sink flow }(-)}.$$

The equilibrium of $(x_{j_1}^*, x_{j_2}^*) = (a_{j_1j_11}, a_{j_1j_21})$ is a (DI:IISI, nF)-negative inflection-sink flow.

(iv$_{1b}$) For $x_{j_1}^* = a_{j_2j_11}$, in phase space, for $a_{j_2j_11} \neq a_{j_1j_11}$,

$$\frac{dx_{j_2}}{dx_{j_1}}\bigg|_{x_{j_1}^* = a_{j_2j_11}} = \frac{a_{j_2j_20}}{a_{j_1j_10}} \frac{(x_{j_1} - a_{j_2j_11})\left[(x_{j_2} - a_{j_2j_2})^2 + b_{j_2j_2}\right]}{(x_{j_1} - a_{j_1j_11})(x_{j_2} - a_{j_1j_21})^2}\bigg|_{x_{j_1}^* = a_{j_2j_11}} = 0,$$

If

$$\frac{d^2x_{j_2}}{dx_{j_1}^2}\bigg|_{x_{j_1}^* = a_{j_2j_11}} = \frac{a_{j_2j_20}}{a_{j_1j_10}} \frac{(\overline{x}_{j_2} - a_{j_2j_2})^2 + b_{j_2j_2}}{(a_{j_2j_11} - a_{j_1j_11})(\overline{x}_{j_2} - a_{j_1j_21})^2} > 0,$$

there is an up-parabola flow. If

$$\frac{d^2x_{j_2}}{dx_{j_1}^2}\bigg|_{x_{j_1}^* = a_{j_2j_11}} = \frac{a_{j_2j_20}}{a_{j_1j_10}} \frac{(\overline{x}_{j_2} - a_{j_2j_2})^2 + b_{j_2j_2}}{(a_{j_2j_11} - a_{j_1j_11})(\overline{x}_{j_2} - a_{j_1j_21})^2} < 0,$$

there is a down-parabola flow. Let

$$\dot{x}_{j_1} = a_{j_1j_10}(a_{j_2j_11} - a_{j_1j_11})(\overline{x}_{j_2} - a_{j_1j_21})^2,$$

and for $\overline{x}_{j_2} \neq a_{j_1j_21}$

$$(\overline{x}_{j_2} - a_{j_1j_21})^2 > 0.$$

The parabola flows at $x_{j_1}^* = a_{j_2 j_1 1}$ in the x_{j_2}-direction are positive and negative for $a_{j_1 j_1 0}(a_{j_2 j_1 1} - a_{j_1 j_1 1}) > 0$ and $a_{j_1 j_1 0}(a_{j_2 j_1 1} - a_{j_1 j_1 1}) < 0$, respectively. Thus,

- For $a_{j_1 j_1 0}(a_{j_2 j_1 1} - a_{j_1 j_1 1})(\bar{x}_{j_2} - a_{j_1 j_2 1})^2 > 0$ and $a_{j_2 j_2 0} > 0$,

$$(\dot{x}_{j_1}, a_{j_2 j_1 1}) = \underbrace{(\text{pF, UP})}_{\text{up-parabola flow } (+)}.$$

- For $a_{j_1 j_1 0}(a_{j_2 j_1 1} - a_{j_1 j_1 1})(\bar{x}_{j_2} - a_{j_1 j_2 1})^2 < 0$ and $a_{j_2 j_2 0} > 0$,

$$(\dot{x}_{j_1}, a_{j_2 j_1 1}) = \underbrace{(\text{nF, DP})}_{\text{down-parabola flow } (-)}.$$

- For $a_{j_1 j_1 0}(a_{j_2 j_1 1} - a_{j_1 j_1 1})(\bar{x}_{j_2} - a_{j_1 j_2 1})^2 > 0$ and $a_{j_2 j_2 0} < 0$,

$$(\dot{x}_{j_1}, a_{j_2 j_1 1}) = \underbrace{(\text{pF, DP})}_{\text{down-parabola flow } (+)}.$$

- For $a_{j_1 j_1 0}(a_{j_2 j_1 1} - a_{j_1 j_1 1})(\bar{x}_{j_2} - a_{j_1 j_2 1})^2 < 0$ and $a_{j_2 j_2 0} < 0$,

$$(\dot{x}_{j_1}, a_{j_2 j_1 1}) = \underbrace{(\text{nF, UP})}_{\text{up-parabola flow } (-)}.$$

From the above analysis with case (I) of ($\text{iv}_{1\text{a}}$), therefore, the equilibrium of $(x_{j_2}^*, x_{j_1}^*) = (a_{j_1 j_2 1}, a_{j_2 j_1 1})$ has the following properties as in Eqs. (1.66)–(1.69):

- For $a_{j_1 j_1 0}(a_{j_2 j_1 1} - a_{j_1 j_1 1}) > 0$ and $a_{j_2 j_2 0} > 0$,

$$(a_{j_1 j_2 1}, a_{j_2 j_1 1}) = \underbrace{(\text{US, UP})}_{\text{up-parabola upper-saddle}}.$$

  The equilibrium of $x_{j_1}^* = a_{j_2 j_1 1}$ is a (US,UP)-up-parabola upper-saddle.
- For $a_{j_1 j_1 0}(a_{j_2 j_1 1} - a_{j_1 j_1 1}) < 0$ and $a_{j_2 j_2 0} > 0$,

$$(a_{j_1 j_2 1}, a_{j_2 j_1 1}) = \underbrace{(\text{LS, DP})}_{\text{down-parabola lower-saddle}}.$$

  The equilibrium of $x_{j_1}^* = a_{j_2 j_1 1}$ is an (LS,DP)-down-parabola lower-saddle.

- For $a_{j_1j_10}(a_{j_2j_11}-a_{j_1j_11})>0$ and $a_{j_2j_20}<0$,

$$(a_{j_1j_21},a_{j_2j_11})=\underbrace{(\mathrm{US},\mathrm{DP})}_{\text{down-parabola upper-saddle}}.$$

  The equilibrium of $x_{j_1}^*=a_{j_2j_11}$ is a (US,DP)-down-parabola upper-saddle.
- For $a_{j_1j_10}(a_{j_2j_11}-a_{j_1j_11})<0$ and $a_{j_2j_20}<0$,

$$(a_{j_1j_21},a_{j_2j_11})=\underbrace{(\mathrm{LS},\mathrm{UP})}_{\text{up-parabola lower-saddle}}.$$

  The equilibrium of $x_{j_1}^*=a_{j_2j_11}$ is an (LS,UP)-up-parabola lower-saddle.

(iv_2) For $a_{j_1j_11}=a_{j_2j_11}$, in phase space,

$$\frac{dx_{j_1}}{dx_{j_2}}=\frac{a_{j_1j_10}}{a_{j_2j_20}}\frac{(x_{j_2}-a_{j_1j_21})^2}{(x_{j_2}-a_{j_2j_2})^2+b_{j_2j_2}},$$

and the deformation of the foregoing equation is

$$\begin{aligned}dx_{j_1}=\frac{a_{j_1j_10}}{a_{j_2j_20}}&\Big[1+\frac{2(a_{j_2j_2}-a_{j_1j_21})(x_{j_2}-a_{j_2j_2})}{(x_{j_2}-a_{j_2j_2})^2+b_{j_2j_2}}\\&+\frac{(a_{j_2j_2}-a_{j_1j_21})^2-b_{j_2j_2}}{(x_{j_2}-a_{j_2j_2})^2+b_{j_2j_2}}\Big]dx_{j_2}.\end{aligned}$$

With an initial condition of (x_{j_10},x_{j_20}) at $t=t_0$, the integration of the above equation gives the first integral manifold for $a_{j_1j_11}=a_{j_2j_11}$ as

$$\begin{aligned}\big[(x_{j_1}-a_{j_1j_11})-(x_{j_10}-a_{j_1j_11})\big]=\frac{a_{j_1j_10}}{a_{j_2j_20}}&\Big\{\big[(x_{j_2}-a_{j_2j_2})-(x_{j_20}-a_{j_2j_2})\big]\\&+(a_{j_2j_2}-a_{j_1j_21})\ln\frac{|(x_{j_2}-a_{j_2j_2})^2+b_{j_2j_2}|}{|(x_{j_20}-a_{j_2j_2})^2+b_{j_2j_2}|}\\&+\big[(a_{j_2j_2}-a_{j_1j_21})^2-b_{j_2j_2}\big]\\&\times\frac{1}{\sqrt{b_{j_2j_2}}}\Big(\arctan\frac{x_{j_2}-a_{j_2j_2}}{\sqrt{b_{j_2j_2}}}-\arctan\frac{x_{j_20}-a_{j_2j_2}}{\sqrt{b_{j_2j_2}}}\Big)\Big\}.\end{aligned}$$

(iv_{2a}) For $x_{j_1}^*=a_{j_1j_11}=a_{j_2j_11}$ with $\bar{x}_{j_2}\neq a_{j_1j_21}$, if

$$\frac{dx_{j_2}}{dx_{j_1}}\Big|_{x_{j_1}^*=a_{j_1j_11}}=\frac{a_{j_2j_20}}{a_{j_1j_10}}\frac{(\bar{x}_{j_2}-a_{j_2j_2})^2+b_{j_2j_2}}{(\bar{x}_{j_2}-a_{j_1j_21})^2}\Big|_{x_{j_1}^*=a_{j_1j_11}}>0,$$

there is an increasing-inflection flow, and if

$$\frac{dx_{j_2}}{dx_{j_1}}\bigg|_{x^*_{j_1}=a_{j_1j_1 1}} = \frac{a_{j_2j_2 0}}{a_{j_1j_1 0}}\frac{(\overline{x}_{j_2}-a_{j_2j_2})^2+b_{j_2j_2}}{(\overline{x}_{j_2}-a_{j_1j_2 1})^2}\bigg|_{x^*_{j_1}=a_{j_1j_1 1}} < 0,$$

there is a decreasing-inflection flow. The variational equation at $x^*_{j_1}=a_{j_1j_1 1}$ is

$$\Delta\dot{x}_{j_1} = a_{j_1j_1 0}(\overline{x}_{j_2}-a_{j_1j_2 1})^2\Delta x_{j_1}$$

and for $\overline{x}_{j_2}\neq a_{j_1j_2 1}$

$$(\overline{x}_{j_2}-a_{j_1j_2 1})^2 > 0.$$

The inflection flows for $a_{j_1j_1 0}(\overline{x}_{j_2}-a_{j_1j_2 1})^2>0$ and $a_{j_1j_1 0}(\overline{x}_{j_2}-a_{j_1j_2 1})^2<0$ are source and sink, respectively.

Therefore, the infinite-equilibrium of $x^*_{j_1}=a_{j_1j_1 1}=a_{j_2j_1 1}$ with $\overline{x}_{j_2}\neq a_{j_1j_2 1}$ has the following properties as in Eqs. (1.71)–(1.74):

- For $a_{j_1j_1 0}(\overline{x}_{j_2}-a_{j_1j_2 1})^2>0$ and $a_{j_2j_2 0}>0$,

$$(a_{j_1j_1 1},\overline{x}_{j_2}) = \underbrace{(\text{SO, II})}_{\text{increasing-inflection source}}.$$

  The infinite-equilibrium of $x^*_{j_1}=a_{j_1j_1 1}=a_{j_2j_1 1}$ is an (SO,II)-increasing-inflection source.
- For $a_{j_1j_1 0}(\overline{x}_{j_2}-a_{j_1j_2 1})^2<0$ and $a_{j_2j_2 0}>0$,

$$(a_{j_1j_1 1},\overline{x}_{j_2}) = \underbrace{(\text{SI, DI})}_{\text{decreasing-inflection sink}}.$$

  The infinite-equilibrium of $x^*_{j_1}=a_{j_1j_1 1}=a_{j_2j_1 1}$ is an (SI,DI)-decreasing-inflection sink.
- For $a_{j_1j_1 0}(\overline{x}_{j_2}-a_{j_1j_2 1})^2>0$ and $a_{j_2j_2 0}<0$,

$$(a_{j_1j_1 1},\overline{x}_{j_2}) = \underbrace{(\text{SO, DI})}_{\text{decreasing-inflection source}}.$$

  The infinite-equilibrium of $x^*_{j_1}=a_{j_1j_1 1}=a_{j_2j_1 1}$ is an (SO,DI)-decreasing-inflection source.
- For $a_{j_1j_1 0}(\overline{x}_{j_2}-a_{j_1j_2 1})^2<0$ and $a_{j_2j_2 0}<0$,

$$(a_{j_1j_1 1},\overline{x}_{j_2}) = \underbrace{(\text{SI, II})}_{\text{increasing-inflection sink}}.$$

  The infinite-equilibrium of $x^*_{j_1}=a_{j_1j_1 1}=a_{j_2j_1 1}$ is an (SI,II)-increasing-inflection sink.

(iv_{2b}) From case (iv_{2a}), for $\bar{x}_{j_2} = a_{j_1 j_2 1}$, therefore, the equilibrium of $(x^*_{j_1}, x^*_{j_2}) = (a_{j_1 j_1 1}, a_{j_1 j_2 1})$ with $a_{j_1 j_1 1} = a_{j_2 j_1 1}$ has the following properties as in Eqs. (1.75)–(1.78):

- For $a_{j_1 j_1 0} > 0$ and $a_{j_2 j_2 0} > 0$,

$$(a_{j_1 j_1 1}, a_{j_1 j_2 1}) = \underbrace{(\mathrm{SO}, \mathrm{II} : \mathrm{II})}_{\text{increasing-inflection source}}.$$

  The equilibrium of $(x^*_{j_1}, x^*_{j_2}) = (a_{j_1 j_1 1}, a_{j_1 j_2 1})$ is an (SO,II:II)-increasing-inflection source of the first kind.
- For $a_{j_1 j_1 0} < 0$ and $a_{j_2 j_2 0} > 0$,

$$(a_{j_1 j_1 1}, a_{j_1 j_2 1}) = \underbrace{(\mathrm{SI}, \mathrm{DI} : \mathrm{DI})}_{\text{decreasing-inflection sink}}.$$

  The equilibrium of $(x^*_{j_1}, x^*_{j_2}) = (a_{j_1 j_1 1}, a_{j_1 j_2 1})$ is an (SI,DI:DI)-decreasing-inflection sink of the first kind.
- For $a_{j_1 j_1 0} > 0$ and $a_{j_2 j_2 0} < 0$,

$$(a_{j_1 j_1 1}, a_{j_1 j_2 1}) = \underbrace{(\mathrm{SO}, \mathrm{DI} : \mathrm{DI})}_{\text{decreasing-inflection source}}.$$

  The equilibrium of $(x^*_{j_1}, x^*_{j_2}) = (a_{j_1 j_1 1}, a_{j_1 j_2 1})$ is an (SO,DI:DI)-decreasing-inflection source of the first kind.
- For $a_{j_1 j_1 0} < 0$ and $a_{j_2 j_2 0} < 0$,

$$(a_{j_1 j_1 1}, a_{j_1 j_2 1}) = \underbrace{(\mathrm{SI}, \mathrm{II} : \mathrm{II})}_{\text{increasing-inflection sink}}.$$

  The equilibrium of $(x^*_{j_1}, x^*_{j_2}) = (a_{j_1 j_1 1}, a_{j_1 j_2 1})$ is an (SI,II:II)-increasing-inflection sink of the first kind.

(v) For $\Delta_{j_1 j_2} = 0$ and $\Delta_{j_2 j_2} = 0$, the standard form is

$$\begin{aligned}
\dot{x}_{j_2} &= a_{j_2 j_2 0}(x_{j_1} - a_{j_2 j_1 1})(x_{j_2} - a_{j_2 j_2 1})^2, \\
\dot{x}_{j_1} &= a_{j_1 j_1 0}(x_{j_1} - a_{j_1 j_1 1})(x_{j_2} - a_{j_1 j_2 1})^2
\end{aligned}$$

where

$$a_{j_1 j_2 1} = -\frac{1}{2} B_{j_1 j_2}, a_{j_2 j_2 1} = -\frac{1}{2} B_{j_2 j_2}.$$

(v_1) For $a_{j_2j_21} \neq a_{j_1j_21}$, in phase space,

$$\frac{dx_{j_1}}{dx_{j_2}} = \frac{a_{j_1j_10}}{a_{j_2j_20}} \frac{(x_{j_1} - a_{j_1j_11})(x_{j_2} - a_{j_1j_21})^2}{(x_{j_1} - a_{j_2j_11})(x_{j_2} - a_{j_2j_21})^2},$$

and the deformation of the foregoing equation is

$$\left[1 + \frac{(a_{j_1j_11} - a_{j_2j_11})}{(x_{j_1} - a_{j_1j_11})}\right] dx_{j_1} = \frac{a_{j_1j_10}}{a_{j_2j_20}} \left[1 + \frac{2(a_{j_2j_21} - a_{j_1j_21})}{(x_{j_2} - a_{j_2j_21})} + \frac{(a_{j_2j_21} - a_{j_1j_21})^2}{(x_{j_2} - a_{j_2j_21})^2}\right] dx_{j_2}.$$

With an initial condition of (x_{j_10}, x_{j_20}) at $t = t_0$, the integration of the above equation gives

$$\begin{aligned}
&\left[(x_{j_1} - a_{j_1j_11}) - (x_{j_10} - a_{j_1j_11})\right] + (a_{j_1j_11} - a_{j_2j_11}) \ln \frac{|x_{j_1} - a_{j_1j_11}|}{|x_{j_10} - a_{j_1j_11}|} \\
&= \frac{a_{j_1j_10}}{a_{j_2j_20}} \Big\{ \left[(x_{j_2} - a_{j_2j_21}) - (x_{j_20} - a_{j_2j_21})\right] \\
&\quad + 2(a_{j_2j_21} - a_{j_1j_21}) \ln \frac{|x_{j_2} - a_{j_2j_21}|}{|x_{j_20} - a_{j_2j_21}|} \\
&\quad - (a_{j_2j_21} - a_{j_1j_21})^2 \Big(\frac{1}{x_{j_2} - a_{j_2j_21}} - \frac{1}{x_{j_20} - a_{j_2j_21}}\Big) \Big\}.
\end{aligned}$$

($\mathrm{v}_{1\mathrm{a}}$) For $(x_{j_1}^*, x_{j_2}^*) = (a_{j_1j_11}, a_{j_2j_21})$, the variational equations are

$$\begin{aligned}
\Delta\dot{x}_{j_2} &= a_{j_2j_20}(a_{j_1j_11} - a_{j_2j_11})(\Delta x_{j_2})^2, \\
\Delta\dot{x}_{j_1} &= a_{j_1j_10}(a_{j_2j_21} - a_{j_1j_21})^2 \Delta x_{j_1}.
\end{aligned}$$

The flows at $(x_{j_1}^*, x_{j_2}^*) = (a_{j_1j_11}, a_{j_2j_21})$ in the x_{j_1}-direction are source and sink for $a_{j_1j_10} > 0$ and $a_{j_1j_10} < 0$, respectively. The flows at $(x_{j_1}^*, x_{j_2}^*) = (a_{j_1j_11}, a_{j_2j_21})$ in the x_{j_2}-direction are upper-saddle and lower-saddle for $a_{j_2j_20}(a_{j_1j_11} - a_{j_2j_11}) > 0$ and $a_{j_2j_20}(a_{j_1j_11} - a_{j_2j_11}) < 0$, respectively.

Therefore, the equilibrium of $(x_{j_1}^*, x_{j_2}^*) = (a_{j_1j_11}, a_{j_2j_21})$ with $a_{j_2j_21} \neq a_{j_1j_21}$ has the following properties as in Eqs. (1.82)–(1.85):

- For $a_{j_1j_10}(a_{j_2j_21} - a_{j_1j_21})^2 > 0$ and $a_{j_2j_20}(a_{j_1j_11} - a_{j_2j_11}) > 0$,

$$(a_{j_1j_11}, a_{j_2j_21}) = \underbrace{(\mathrm{SO}, \mathrm{US})}_{\text{upper-saddle-source}}.$$

The equilibrium of $(x_{j_1}^*, x_{j_2}^*) = (a_{j_1j_11}, a_{j_2j_21})$ is an (SO,US)-upper-saddle-source.

- For $a_{j_1j_10}(a_{j_2j_21}-a_{j_1j_21})^2<0$ and $a_{j_2j_20}(a_{j_1j_11}-a_{j_2j_11})>0$,

$$(a_{j_1j_11},a_{j_2j_21})=\underbrace{(\text{SI, US})}_{\text{upper-saddle-sink}}.$$

The equilibrium of $(x^*_{j_1},x^*_{j_2})=(a_{j_1j_11},a_{j_2j_21})$ is an (SI,US)-upper-saddle-sink.

- For $a_{j_1j_10}(a_{j_2j_21}-a_{j_1j_21})^2>0$ and $a_{j_2j_20}(a_{j_1j_11}-a_{j_2j_11})<0$,

$$(a_{j_1j_11},a_{j_2j_21})=\underbrace{(\text{SO, LS})}_{\text{lower-saddle-source}}.$$

The equilibrium of $(x^*_{j_1},x^*_{j_2})=(a_{j_1j_11},a_{j_2j_21})$ is an (SO,LS)-lower-saddle-source.

- For $a_{j_1j_10}(a_{j_2j_21}-a_{j_1j_21})^2<0$ and $a_{j_2j_20}(a_{j_1j_11}-a_{j_2j_11})<0$,

$$(a_{j_1j_11},a_{j_2j_21})=\underbrace{(\text{SI, LS})}_{\text{lower-saddle-source}}.$$

The equilibrium of $(x^*_{j_1},x^*_{j_2})=(a_{j_1j_11},a_{j_2j_21})$ is an (SI,LS)-lower-saddle-sink.

– The saddle-source and saddle-sink are the appearing bifurcations of saddle-source and saddle-sink, respectively.

($\text{v}_{1\text{b}}$) In phase space, at $x^*_{j_2}=a_{j_1j_21}$ with $\overline{x}_{j_1}\neq a_{j_1j_11}$,

$$\frac{dx_{j_1}}{dx_{j_2}}\bigg|_{x^*_{j_2}=a_{j_1j_21}}=\frac{a_{j_1j_10}\,(\overline{x}_{j_1}-a_{j_1j_11})(x_{j_2}-a_{j_1j_21})^2}{a_{j_2j_20}\,(\overline{x}_{j_1}-a_{j_2j_11})(x_{j_2}-a_{j_2j_21})^2}\bigg|_{x^*_{j_2}=a_{j_1j_21}}=0,$$
$$\frac{d^2x_{j_1}}{dx^2_{j_2}}\bigg|_{x^*_{j_2}=a_{j_1j_21}}=0.$$

If

$$\frac{d^3x_{j_1}}{dx^3_{j_2}}\bigg|_{x^*_{j_2}=a_{j_1j_21}}=\frac{a_{j_1j_10}}{a_{j_2j_20}}\frac{2(\overline{x}_{j_1}-a_{j_1j_11})}{(\overline{x}_{j_1}-a_{j_2j_11})(a_{j_1j_21}-a_{j_2j_21})^2}>0,$$

there is an increasing-inflection flow at $x^*_{j_2}=a_{j_1j_21}$ with $\overline{x}_{j_1}\neq a_{j_1j_11}$. If

$$\frac{d^3x_{j_1}}{dx^3_{j_2}}\bigg|_{x^*_{j_2}=a_{j_1j_21}}=\frac{a_{j_1j_10}}{a_{j_2j_20}}\frac{2(\overline{x}_{j_1}-a_{j_1j_11})}{(\overline{x}_{j_1}-a_{j_2j_11})(a_{j_1j_21}-a_{j_2j_21})^2}<0,$$

there is a decreasing-inflection flow at $x^*_{j_2}=a_{j_1j_21}$ with $\overline{x}_{j_1}\neq a_{j_1j_11}$. Let

$$\dot{x}_{j_2}=a_{j_2j_20}(\overline{x}_{j_1}-a_{j_2j_11})(a_{j_1j_21}-a_{j_2j_21})^2\neq 0.$$

The inflection flows at $x^*_{j_2}=a_{j_1j_21}$ with $\overline{x}_{j_1}\neq a_{j_1j_11}$ are positive and negative for $a_{j_2j_20}>0$ and $a_{j_2j_20}<0$, respectively.

Thus, the equilibrium of $x_{j_2}^* = a_{j_1j_2 1}$ with $\overline{x}_{j_1} \neq a_{j_1j_1 1}$ has the following properties:

- For $a_{j_1j_1 0}(\overline{x}_{j_1} - a_{j_1j_1 1}) > 0$ and $a_{j_2j_2 0}(\overline{x}_{j_1} - a_{j_2j_1 1})(a_{j_1j_2 1} - a_{j_2j_2 1})^2 > 0$,

$$(a_{j_1j_2 1}, \dot{x}_{j_2}) = \underbrace{(\mathrm{II}, \mathrm{pF})}_{\text{increasing-inflection flow } (+)}.$$

- For $a_{j_1j_1 0}(\overline{x}_{j_1} - a_{j_1j_1 1}) < 0$ and $a_{j_2j_2 0}(\overline{x}_{j_1} - a_{j_2j_1 1})(a_{j_1j_2 1} - a_{j_2j_2 1})^2 > 0$,

$$(a_{j_1j_2 1}, \dot{x}_{j_2}) = \underbrace{(\mathrm{DI}, \mathrm{pF})}_{\text{decreasing-inflection flow } (+)}.$$

- For $a_{j_1j_1 0}(\overline{x}_{j_1} - a_{j_1j_1 1}) > 0$ and $a_{j_2j_2 0}(\overline{x}_{j_1} - a_{j_2j_1 1})(a_{j_1j_2 1} - a_{j_2j_2 1})^2 < 0$,

$$(a_{j_1j_2 1}, \dot{x}_{j_2}) = \underbrace{(\mathrm{DI}, \mathrm{nF})}_{\text{decreasing-inflection flow } (-)}.$$

- For $a_{j_1j_1 0}(\overline{x}_{j_1} - a_{j_1j_1 1}) < 0$ and $a_{j_2j_2 0}(\overline{x}_{j_1} - a_{j_2j_1 1})(a_{j_1j_2 1} - a_{j_2j_2 1})^2 < 0$,

$$(a_{j_1j_2 1}, \dot{x}_{j_2}) = \underbrace{(\mathrm{II}, \mathrm{nF})}_{\text{increasing-inflection flow } (-)}.$$

The variational equation at $x_{j_1}^* = a_{j_1j_1 1}$ with $\overline{x}_{j_2} \neq a_{j_1j_2 1}$ is

$$\Delta\dot{x}_{j_1} = a_{j_1j_1 0}(\overline{x}_{j_2} - a_{j_1j_2 1})^2 \Delta x_{j_1}$$

and

$$\dot{x}_{j_2} = a_{j_2j_2 0}(a_{j_1j_1 1} - a_{j_2j_1 1})(\overline{x}_{j_2} - a_{j_2j_2 1})^2.$$

Thus, the equilibrium of $x_{j_2}^* = a_{j_1j_2 1}$ with $\overline{x}_{j_1} \neq a_{j_1j_1 1}$ has the following properties:

- For $a_{j_1j_1 0}(\overline{x}_{j_2} - a_{j_1j_2 1})^2 > 0$ and $a_{j_2j_2 0}(a_{j_1j_1 1} - a_{j_2j_1 1})(\overline{x}_{j_2} - a_{j_2j_2 1})^2 > 0$,

$$(a_{j_1j_1 1}, \dot{x}_{j_2}) = \underbrace{(\mathrm{SO}, \mathrm{pF})}_{\text{source flow } (+)}.$$

- For $a_{j_1j_1 0}(\overline{x}_{j_2} - a_{j_1j_2 1})^2 < 0$ and $a_{j_2j_2 0}(a_{j_1j_1 1} - a_{j_2j_1 1})(\overline{x}_{j_2} - a_{j_2j_2 1})^2 > 0$,

$$(a_{j_1j_1 1}, \dot{x}_{j_2}) = \underbrace{(\text{SI, pF})}_{\text{sink flow }(+)}.$$

- For $a_{j_1j_1 0}(\bar{x}_{j_2} - a_{j_1j_2 1})^2 > 0$ and $a_{j_2j_2 0}(a_{j_1j_1 1} - a_{j_2j_1 1})(\bar{x}_{j_2} - a_{j_2j_2 1})^2 < 0$,

$$(a_{j_1j_1 1}, \dot{x}_{j_2}) = \underbrace{(\text{SO, nF})}_{\text{source flow }(-)}.$$

- For $a_{j_1j_1 0}(\bar{x}_{j_2} - a_{j_1j_2 1})^2 < 0$ and $a_{j_2j_2 0}(a_{j_1j_1 1} - a_{j_2j_1 1})(\bar{x}_{j_2} - a_{j_2j_2 1})^2 < 0$,

$$(a_{j_1j_1 1}, \dot{x}_{j_2}) = \underbrace{(\text{SI, nF})}_{\text{sink flow }(-)}.$$

From the foregoing analysis, the equilibrium of $(x^*_{j_1}, x^*_{j_2}) = (a_{j_1j_1 1}, a_{j_1j_2 1})$ with $a_{j_2j_2 1} \neq a_{j_1j_2 1}$ has the following properties as in Eqs. (1.86)–(1.89):

- For $a_{j_1j_1 0} > 0$ and $a_{j_2j_2 0}(a_{j_1j_1 1} - a_{j_2j_1 1})(a_{j_1j_2 1} - a_{j_2j_2 1})^2 > 0$,

$$(a_{j_1j_1 1}, a_{j_1j_2 1}) = \underbrace{({}_{\text{DI:II}}\text{SO, pF})}_{\text{inflection-source flow }(+)}.$$

The equilibrium of $(x^*_{j_1}, x^*_{j_2}) = (a_{j_1j_1 1}, a_{j_1j_2 1})$ is a (${}_{\text{DI:II}}$SO, pF)-positive inflection-source flow.

- For $a_{j_1j_1 0} < 0$ and $a_{j_2j_2 0}(a_{j_1j_1 1} - a_{j_2j_1 1})(a_{j_1j_2 1} - a_{j_2j_2 1})^2 > 0$,

$$(a_{j_1j_1 1}, a_{j_1j_2 1}) = \underbrace{({}_{\text{II:DI}}\text{SI, pF})}_{\text{inflection-source flow }(+)}.$$

The equilibrium of $(x^*_{j_1}, x^*_{j_2}) = (a_{j_1j_1 1}, a_{j_1j_2 1})$ is an (${}_{\text{II:DI}}$SI, pF)-positive inflection-sink flow.

- For $a_{j_1j_1 0} > 0$ and $a_{j_2j_2 0}(a_{j_1j_1 1} - a_{j_2j_1 1})(a_{j_1j_2 1} - a_{j_2j_2 1})^2 < 0$,

$$(a_{j_1j_1 1}, a_{j_1j_2 1}) = \underbrace{({}_{\text{II:DI}}\text{SO, nF})}_{\text{inflection-source flow }(-)}.$$

The equilibrium of $(x^*_{j_1}, x^*_{j_2}) = (a_{j_1j_1 1}, a_{j_1j_2 1})$ is an (${}_{\text{II:DI}}$SI, nF)-negative inflection-source flow.

- For $a_{j_1j_1 0} < 0$ and $a_{j_2j_2 0}(a_{j_1j_1 1} - a_{j_2j_1 1})(a_{j_1j_2 1} - a_{j_2j_2 1})^2 < 0$,

$$(a_{j_1j_11}, a_{j_1j_21}) = \underbrace{(_{\text{DI:II}}\text{SI}, \text{nF})}_{\text{inflection sink flow } (-)}.$$

The equilibrium of $(x_{j_1}^*, x_{j_2}^*) = (a_{j_1j_11}, a_{j_1j_21})$ is a $(_{\text{DI:II}}\text{SI}, \text{nF})$-negative inflection-sink flow.

(v_{1c}) For $x_{j_1}^* = a_{j_2j_11}$, in phase space, for $a_{j_2j_11} \neq a_{j_1j_11}$,

$$\frac{dx_{j_2}}{dx_{j_1}}\bigg|_{x_{j_1}^* = a_{j_2j_11}} = \frac{a_{j_2j_20}}{a_{j_1j_10}} \frac{(x_{j_1} - a_{j_2j_11})(x_{j_2} - a_{j_2j_21})^2}{(x_{j_1} - a_{j_1j_11})(x_{j_2} - a_{j_1j_21})^2}\bigg|_{x_{j_1}^* = a_{j_2j_11}} = 0,$$

If

$$\frac{d^2x_{j_2}}{dx_{j_1}^2}\bigg|_{x_{j_1}^* = a_{j_2j_11}} = \frac{a_{j_2j_20}}{a_{j_1j_10}} \frac{(\overline{x}_{j_2} - a_{j_2j_21})^2}{(a_{j_2j_11} - a_{j_1j_11})(\overline{x}_{j_2} - a_{j_1j_21})^2} > 0,$$

there is an up-parabola flow. If

$$\frac{d^2x_{j_2}}{dx_{j_1}^2}\bigg|_{x_{j_1}^* = a_{j_2j_11}} = \frac{a_{j_2j_20}}{a_{j_1j_10}} \frac{(\overline{x}_{j_2} - a_{j_2j_21})^2}{(a_{j_2j_11} - a_{j_1j_11})(\overline{x}_{j_2} - a_{j_1j_21})^2} < 0,$$

there is a down-parabola flow. Let

$$\dot{x}_{j_1} = a_{j_1j_10}(a_{j_2j_11} - a_{j_1j_11})(\overline{x}_{j_2} - a_{j_1j_21})^2,$$

and for $\overline{x}_{j_2} \neq a_{j_1j_21}$,

$$(\overline{x}_{j_2} - a_{j_1j_21})^2 > 0.$$

The parabola flows at $x_{j_1}^* = a_{j_2j_11}$ in the x_{j_2}-direction are positive and negative for $a_{j_1j_10}(a_{j_2j_11} - a_{j_1j_11}) > 0$ and $a_{j_1j_10}(a_{j_2j_11} - a_{j_1j_11}) < 0$, respectively. Thus,

- For $a_{j_1j_10}(a_{j_2j_11} - a_{j_1j_11})(\overline{x}_{j_2} - a_{j_1j_21})^2 > 0$ and $a_{j_2j_20}(\overline{x}_{j_2} - a_{j_2j_21})^2 > 0$,

$$(\dot{x}_{j_1}, a_{j_2j_11}) = \underbrace{(\text{pF}, \text{UP})}_{\text{up-parabola flow } (+)}.$$

- For $a_{j_1j_10}(a_{j_2j_11} - a_{j_1j_11})(\overline{x}_{j_2} - a_{j_1j_21})^2 < 0$ and $a_{j_2j_20}(\overline{x}_{j_2} - a_{j_2j_21})^2 > 0$,

$$(\dot{x}_{j_1}, a_{j_2j_11}) = \underbrace{(\text{nF}, \text{DP})}_{\text{down-parabola flow } (-)}.$$

- For $a_{j_1j_1 0}(a_{j_2j_1 1} - a_{j_1j_1 1})(\overline{x}_{j_2} - a_{j_1j_2 1})^2 > 0$ and $a_{j_2j_2 0}(\overline{x}_{j_2} - a_{j_2j_2 1})^2 < 0$,

$$(\dot{x}_{j_1}, a_{j_2j_1 1}) = \underbrace{(\mathrm{pF}, \mathrm{DP})}_{\text{down-parabola flow } (+)}.$$

- For $a_{j_1j_1 0}(a_{j_2j_1 1} - a_{j_1j_1 1})(\overline{x}_{j_2} - a_{j_1j_2 1})^2 < 0$ and $a_{j_2j_2 0}(\overline{x}_{j_2} - a_{j_2j_2 1})^2 < 0$,

$$(\dot{x}_{j_1}, a_{j_2j_1 1}) = \underbrace{(\mathrm{nF}, \mathrm{UP})}_{\text{up-parabola flow } (-)}.$$

Therefore, for $\overline{x}_{j_2} = a_{j_1j_2 1}$, the equilibrium of $(x^*_{j_2}, x^*_{j_1}) = (a_{j_1j_2 1}, a_{j_2j_1 1})$ with $a_{j_2j_1 1} \neq a_{j_1j_1 1}$ has the following properties as in Eqs. (1.90)–(1.93):

- For $a_{j_1j_1 0}(a_{j_2j_1 1} - a_{j_1j_1 1}) > 0$ and $a_{j_2j_2 0}(a_{j_1j_2 1} - a_{j_2j_2 1})^2 > 0$,

$$(a_{j_1j_2 1}, a_{j_2j_1 1}) = \underbrace{(\mathrm{US}, \mathrm{UP})}_{\text{up-parabola upper-saddle}}.$$

  The equilibrium of $(x^*_{j_2}, x^*_{j_1}) = (a_{j_1j_2 1}, a_{j_2j_1 1})$ is a (US,UP)-up-parabola upper-saddle.
- For $a_{j_1j_1 0}(a_{j_2j_1 1} - a_{j_1j_1 1}) < 0$ and $a_{j_2j_2 0}(a_{j_1j_2 1} - a_{j_2j_2 1})^2 > 0$,

$$(a_{j_1j_2 1}, a_{j_2j_1 1}) = \underbrace{(\mathrm{LS}, \mathrm{DP})}_{\text{down-parabola lower-saddle}}.$$

  The equilibrium of $(x^*_{j_2}, x^*_{j_1}) = (a_{j_1j_2 1}, a_{j_2j_1 1})$ is an (LS,DP)-down-parabola lower-saddle.
- For $a_{j_1j_1 0}(a_{j_2j_1 1} - a_{j_1j_1 1}) > 0$ and $a_{j_2j_2 0}(a_{j_1j_2 1} - a_{j_2j_2 1})^2 < 0$,

$$(a_{j_1j_2 1}, a_{j_2j_1 1}) = \underbrace{(\mathrm{US}, \mathrm{DP})}_{\text{down-parabola upper-saddle}}.$$

  The equilibrium of $(x^*_{j_2}, x^*_{j_1}) = (a_{j_1j_2 1}, a_{j_2j_1 1})$ is a (US,DP)-down-parabola upper-saddle.
- For $a_{j_1j_1 0}(a_{j_2j_1 1} - a_{j_1j_1 1}) > 0$ and $a_{j_2j_2 0}(a_{j_1j_2 1} - a_{j_2j_2 1})^2 < 0$,

$$(a_{j_1j_2 1}, a_{j_2j_1 1}) = \underbrace{(\mathrm{LS}, \mathrm{UP})}_{\text{up-parabola lower-saddle}}.$$

  The equilibrium of $(x^*_{j_2}, x^*_{j_1}) = (a_{j_1j_2 1}, a_{j_2j_1 1})$ is an (LS,UP)-up-parabola lower-saddle.

– The parabola upper-saddle and lower-saddle are the appearing bifurcations of positive and negative saddles with clockwise and counterclockwise centers, respectively.

($\mathrm{v}_{1\mathrm{d}}$) From the proof of ($\mathrm{v}_{1\mathrm{c}}$), for $\bar{x}_{j_2} = a_{j_2 j_2 1}$, the equilibrium of $(x^*_{j_1}, x^*_{j_2}) = (a_{j_2 j_1 1}, a_{j_2 j_2 1})$ with $a_{j_2 j_2 1} \neq a_{j_1 j_2 1}$ has the following properties as in Eqs. (1.94)–(1.97):

- For $a_{j_1 j_1 0}(a_{j_2 j_1 1} - a_{j_1 j_1 1})(a_{j_2 j_2 1} - a_{j_1 j_2 1})^2 > 0$ and $a_{j_2 j_2 0} > 0$,

$$(a_{j_2 j_1 1}, a_{j_2 j_2 1}) = \underbrace{(\mathrm{pF}, \mathrm{UP} : \mathrm{UP})}_{\text{hyperbolic-secant-to-hyperbolic flow } (+)}.$$

The equilibrium of $(x^*_{j_1}, x^*_{j_2}) = (a_{j_2 j_1 1}, a_{j_2 j_2 1})$ is a (pF, UP:UP)-positive hyperbolic-secant-to-hyperbolic flow.

- For $a_{j_1 j_1 0}(a_{j_2 j_1 1} - a_{j_1 j_1 1})(a_{j_2 j_2 1} - a_{j_1 j_2 1})^2 < 0$ and $a_{j_2 j_2 0} > 0$,

$$(a_{j_2 j_1 1}, a_{j_2 j_2 1}) = \underbrace{(\mathrm{nF}, \mathrm{DP} : \mathrm{DP})}_{\text{hyperbolic-to-hyperbolic-secant flow } (-)}.$$

The equilibrium of $(x^*_{j_1}, x^*_{j_2}) = (a_{j_1 j_1 1}, a_{j_1 j_2 1})$ is an (nF,DP:DP)-negative hyperbolic-to-hyperbolic-secant flow.

- For $a_{j_1 j_1 0}(a_{j_2 j_1 1} - a_{j_1 j_1 1})(a_{j_2 j_2 1} - a_{j_1 j_2 1})^2 > 0$ and $a_{j_2 j_2 0} < 0$,

$$(a_{j_2 j_1 1}, a_{j_2 j_2 1}) = \underbrace{(\mathrm{pF}, \mathrm{DP} : \mathrm{DP})}_{\text{hyperbolic-to-hyperbolic-secant flow } (+)}.$$

The equilibrium of $(x^*_{j_1}, x^*_{j_2}) = (a_{j_1 j_1 1}, a_{j_1 j_2 1})$ is a (pF, DP:DP)-positive hyperbolic-to-hyperbolic-secant flow.

- For $a_{j_1 j_1 0}(a_{j_2 j_1 1} - a_{j_1 j_1 1})(a_{j_2 j_2 1} - a_{j_1 j_2 1})^2 < 0$ and $a_{j_2 j_2 0} < 0$,

$$(a_{j_2 j_1 1}, a_{j_2 j_2 1}) = \underbrace{(\mathrm{nF}, \mathrm{UP} : \mathrm{UP})}_{\text{hyperbolic-secant-to-hyperbolic flow } (-)}.$$

The equilibrium of $(x^*_{j_1}, x^*_{j_2}) = (a_{j_1 j_1 1}, a_{j_1 j_2 1})$ is an (nF, UP:UP)-negative hyperbolic-secant-to-hyperbolic flow.

(v_2) For $a_{j_2 j_2 1} = a_{j_1 j_2 1}$, we have

$$\frac{dx_{j_1}}{dx_{j_2}} = \frac{a_{j_1 j_1 0}}{a_{j_2 j_2 0}} \frac{(x_{j_1} - a_{j_1 j_1 1})}{(x_{j_1} - a_{j_2 j_1 1})},$$

and the deformation of the foregoing equation is

$$\left[1+\frac{(a_{j_1j_1}1-a_{j_2j_1}1)}{(x_{j_1}-a_{j_1j_1}1)}\right]dx_{j_1}=\frac{a_{j_1j_1}0}{a_{j_2j_2}0}dx_{j_2}.$$

With an initial condition of $(x_{j_1 0}, x_{j_2 0})$ at $t = t_0$, the integration of the above equation gives the first integral manifold for $a_{j_2j_2}1 = a_{j_1j_2}1$ as

$$\begin{aligned}&\left[(x_{j_1}-a_{j_1j_1}1)-(x_{j_1 0}-a_{j_1j_1}1)\right]+(a_{j_1j_1}1-a_{j_2j_1}1)\ln\frac{|x_{j_1}-a_{j_1j_1}1|}{|x_{j_1 0}-a_{j_1j_1}1|}\\&=\frac{a_{j_1j_1}0}{a_{j_2j_2}0}\left[(x_{j_2}-a_{j_2j_2}1)-(x_{j_1 0}-a_{j_2j_2}1)\right].\end{aligned}$$

(v_{2a}) In phase space, at $x^*_{j_2} = a_{j_1j_2}1 = a_{j_2j_2}1$ with $\bar{x}_{j_1} \neq a_{j_1j_1}1, a_{j_2j_1}1$, if

$$\left.\frac{dx_{j_1}}{dx_{j_2}}\right|_{x^*_{j_2}=a_{j_1j_2}1}=\frac{a_{j_1j_1}0}{a_{j_2j_2}0}\frac{(\bar{x}_{j_1}-a_{j_1j_1}1)}{(\bar{x}_{j_1}-a_{j_2j_1}1)}>0,$$

there is an increasing-inflection flow at $x^*_{j_2} = a_{j_1j_2}1$ with $\bar{x}_{j_1} \neq a_{j_1j_1}1, a_{j_2j_1}1$. If

$$\left.\frac{dx_{j_1}}{dx_{j_2}}\right|_{x^*_{j_2}=a_{j_1j_2}1}=\frac{a_{j_1j_1}0}{a_{j_2j_2}0}\frac{(\bar{x}_{j_1}-a_{j_1j_1}1)}{(\bar{x}_{j_1}-a_{j_2j_1}1)}>0,$$

there is a decreasing-inflection flow at $x^*_{j_2} = a_{j_1j_2}1$ with $\bar{x}_{j_1} \neq a_{j_1j_1}1, a_{j_2j_1}1$. The variational equation is

$$\Delta\dot{x}_{j_2}=a_{j_2j_2}0(\bar{x}_{j_1}-a_{j_2j_1}1)(\Delta x_{j_2})^2.$$

The inflection flows at $x^*_{j_2} = a_{j_1j_2}1$ with $\bar{x}_{j_1} \neq a_{j_1j_1}1$ are upper-saddle and lower-saddle for $a_{j_2j_2}0(\bar{x}_{j_1} - a_{j_2j_1}1) > 0$ and $a_{j_2j_2}0(\bar{x}_{j_1} - a_{j_2j_1}1) < 0$, respectively.

Therefore, the infinite-equilibrium of $x^*_{j_2} = a_{j_1j_2}1 = a_{j_2j_2}1$ with $\bar{x}_{j_1} \neq a_{j_1j_1}1, a_{j_2j_1}1$ has the following properties as in Eqs. (1.99)–(1.102):

- For $a_{j_1j_1}0(\bar{x}_{j_1} - a_{j_1j_1}1) > 0$ and $a_{j_2j_2}0(\bar{x}_{j_1} - a_{j_2j_1}1) > 0$,

$$(\bar{x}_{j_1}, a_{j_1j_2}1)=\underbrace{(\mathrm{II},\mathrm{US})}_{\text{increasing-inflection upper-saddle}}.$$

The infinite-equilibrium of $x^*_{j_2} = a_{j_1j_2}1 = a_{j_2j_2}1$ is an (II,US)-increasing-inflection upper-saddle.

- For $a_{j_1j_1 0}(\overline{x}_{j_1} - a_{j_1j_1 1}) < 0$ and $a_{j_2j_2 0}(\overline{x}_{j_1} - a_{j_2j_1 1}) > 0$,

$$(\overline{x}_{j_1}, a_{j_1j_2 1}) = \underbrace{(\mathrm{DI}, \mathrm{US})}_{\text{decreasing-inflection upper-saddle}}.$$

The equilibrium of $x^*_{j_2} = a_{j_1j_2 1} = a_{j_2j_2 1}$ is a (DI,US)-decreasing-inflection upper-saddle.

- For $a_{j_1j_1 0}(\overline{x}_{j_1} - a_{j_1j_1 1}) > 0$ and $a_{j_2j_2 0}(\overline{x}_{j_1} - a_{j_2j_1 1}) < 0$,

$$(\overline{x}_{j_1}, a_{j_1j_2 1}) = \underbrace{(\mathrm{DI}, \mathrm{LS})}_{\text{decreasing-inflection lower-saddle}}.$$

The equilibrium of $x^*_{j_2} = a_{j_1j_2 1} = a_{j_2j_2 1}$ is a (DI,LS)-decreasing-inflection lower-saddle.

- For $a_{j_1j_1 0}(\overline{x}_{j_1} - a_{j_1j_1 1}) < 0$ and $a_{j_2j_2 0}(\overline{x}_{j_1} - a_{j_2j_1 1}) < 0$,

$$(\overline{x}_{j_1}, a_{j_1j_2 1}) = \underbrace{(\mathrm{II}, \mathrm{LS})}_{\text{increasing-inflection lower-saddle}}.$$

The equilibrium of $x^*_{j_2} = a_{j_1j_2 1} = a_{j_2j_2 1}$ is an (II,LS)-increasing-inflection lower-saddle.

($\mathrm{v}_{2\mathrm{b}}$) From case ($\mathrm{v}_{2\mathrm{a}}$), for $\overline{x}_{j_1} = a_{j_1j_1 1}$, the equilibrium of $(x^*_{j_1}, x^*_{j_2}) = (a_{j_1j_1 1}, a_{j_1j_2 1})$ with $a_{j_2j_2 1} = a_{j_1j_2 1}$ has the following properties as in Eqs. (1.103)–(1.106):

- For $a_{j_1j_1 0} > 0$ and $a_{j_2j_2 0}(a_{j_1j_1 1} - a_{j_2j_1 1}) > 0$,

$$(a_{j_1j_1 1}, a_{j_1j_2 1}) = \underbrace{(\mathrm{DI} : \mathrm{II}, \mathrm{US})}_{\text{inflection upper-saddle}}.$$

The equilibrium of $(x^*_{j_1}, x^*_{j_2}) = (a_{j_1j_1 1}, a_{j_1j_2 1})$ is a (DI:II,US)-inflection upper-saddle.

- For $a_{j_1j_1 0} < 0$ and $a_{j_2j_2 0}(a_{j_1j_1 1} - a_{j_2j_1 1}) > 0$,

$$(a_{j_1j_1 1}, a_{j_1j_2 1}) = \underbrace{(\mathrm{II} : \mathrm{DI}, \mathrm{US})}_{\text{inflection upper-saddle}}.$$

The equilibrium of $(x^*_{j_1}, x^*_{j_2}) = (a_{j_1j_1 1}, a_{j_1j_2 1})$ is an (II:DI,US)-inflection upper-saddle.

- For $a_{j_1j_1 0} > 0$ and $a_{j_2j_2 0}(a_{j_1j_1 1} - a_{j_2j_1 1}) < 0$,

$$(a_{j_1j_1 1}, a_{j_1j_2 1}) = \underbrace{(\mathrm{II} : \mathrm{DI}, \mathrm{LS})}_{\text{inflection lower-saddle}}.$$

The equilibrium of $(x^*_{j_1}, x^*_{j_2}) = (a_{j_1j_1 1}, a_{j_1j_2 1})$ is an (II:DI,LS)-inflection lower-saddle.

- For $a_{j_1j_10}<0$ and $a_{j_2j_20}(a_{j_1j_11}-a_{j_2j_11})<0$,

$$(a_{j_1j_11},a_{j_1j_21})=\underbrace{(\text{DI}:\text{II},\text{LS})}_{\text{inflection lower-saddle}}.$$

The equilibrium of $(x^*_{j_1},x^*_{j_2})=(a_{j_1j_11},a_{j_1j_21})$ is a (DI:II,LS)-inflection lower-saddle.

(v_{2c}) For $x^*_{j_1}=a_{j_2j_11}$, in phase space, for $a_{j_2j_11}\neq a_{j_1j_11}$,

$$\frac{dx_{j_2}}{dx_{j_1}}\bigg|_{x^*_{j_1}=a_{j_2j_11}}=\frac{a_{j_2j_20}}{a_{j_1j_10}}\frac{(x_{j_1}-a_{j_2j_11})(x_{j_2}-a_{j_2j_21})^2}{(x_{j_1}-a_{j_1j_11})(x_{j_2}-a_{j_1j_21})^2}\bigg|_{x^*_{j_1}=a_{j_2j_11}}=0,$$

If

$$\frac{d^2x_{j_2}}{dx^2_{j_1}}\bigg|_{x^*_{j_1}=a_{j_2j_11}}=\frac{a_{j_2j_20}}{a_{j_1j_10}}\frac{(\overline{x}_{j_2}-a_{j_2j_21})^2}{(a_{j_2j_11}-a_{j_1j_11})(\overline{x}_{j_2}-a_{j_1j_21})^2}>0,$$

there is an up-parabola flow. If

$$\frac{d^2x_{j_2}}{dx^2_{j_1}}\bigg|_{x^*_{j_1}=a_{j_2j_11}}=\frac{a_{j_2j_20}}{a_{j_1j_10}}\frac{(\overline{x}_{j_2}-a_{j_2j_21})^2}{(a_{j_2j_11}-a_{j_1j_11})(\overline{x}_{j_2}-a_{j_1j_21})^2}<0,$$

there is a down-parabola flow. Let

$$\dot{x}_{j_1}=a_{j_1j_10}(a_{j_2j_11}-a_{j_1j_11})(\overline{x}_{j_2}-a_{j_1j_21})^2,$$

and for $\overline{x}_{j_2}\neq a_{j_1j_21}$,

$$(\overline{x}_{j_2}-a_{j_1j_21})^2>0.$$

The parabola flows at $x^*_{j_1}=a_{j_2j_11}$ in the x_{j_2}-direction are positive and negative for $a_{j_1j_10}(a_{j_2j_11}-a_{j_1j_11})>0$ and $a_{j_1j_10}(a_{j_2j_11}-a_{j_1j_11})<0$, respectively. Thus,

- For $a_{j_1j_10}(a_{j_2j_11}-a_{j_1j_11})(\overline{x}_{j_2}-a_{j_1j_21})^2>0$ and $a_{j_2j_20}(\overline{x}_{j_2}-a_{j_2j_21})^2>0$,

$$(\dot{x}_{j_1},a_{j_2j_11})=\underbrace{(\text{pF},\text{UP})}_{\text{up-parabola flow }(+)}.$$

- For $a_{j_1j_10}(a_{j_2j_11}-a_{j_1j_11})(\overline{x}_{j_2}-a_{j_1j_21})^2<0$ and $a_{j_2j_20}(\overline{x}_{j_2}-a_{j_2j_21})^2>0$,

$$(\dot{x}_{j_1},a_{j_2j_11})=\underbrace{(\text{nF},\text{DP})}_{\text{down-parabola flow }(-)}.$$

- For $a_{j_1j_1 0}(a_{j_2j_1 1}-a_{j_1j_1 1})(\bar{x}_{j_2}-a_{j_1j_2 1})^2>0$ and $a_{j_2j_2 0}(\bar{x}_{j_2}-a_{j_2j_2 1})^2<0$,

$$(\dot{x}_{j_1}, a_{j_2j_1 1}) = \underbrace{(\text{pF}, \text{DP})}_{\text{down-parabola flow } (+)}.$$

- For $a_{j_1j_1 0}(a_{j_2j_1 1}-a_{j_1j_1 1})(\bar{x}_{j_2}-a_{j_1j_2 1})^2<0$ and $a_{j_2j_2 0}(\bar{x}_{j_2}-a_{j_2j_2 1})^2<0$,

$$(\dot{x}_{j_1}, a_{j_2j_1 1}) = \underbrace{(\text{nF}, \text{UP})}_{\text{up-parabola flow } (-)}.$$

From the above analysis, for $\bar{x}_{j_2}=a_{j_1j_2 1}$ with case ($\text{v}_{2\text{a}}$), the equilibrium of $(x^*_{j_2}, x^*_{j_1})=(a_{j_1j_2 1}, a_{j_2j_1 1})$ with $a_{j_2j_2 1}=a_{j_1j_2 1}$ has the following properties as in Eqs. (1.107)–(1.110):

- For $a_{j_1j_1 0}(a_{j_2j_1 1}-a_{j_2j_1 1})>0$ and $a_{j_2j_2 0}(a_{j_1j_2 1}-a_{j_2j_2 1})^2>0$,

$$(a_{j_1j_2 1}, a_{j_2j_1 1}) = \underbrace{(\text{US}, \text{UP})}_{\text{up-parabola upper-saddle}}.$$

  The equilibrium of $(x^*_{j_2}, x^*_{j_1})=(a_{j_1j_2 1}, a_{j_2j_1 1})$ is a (US,UP)-up-parabola upper-saddle.
- For $a_{j_1j_1 0}(a_{j_2j_1 1}-a_{j_2j_1 1})<0$ and $a_{j_2j_2 0}(a_{j_1j_2 1}-a_{j_2j_2 1})^2>0$,

$$(a_{j_1j_2 1}, a_{j_2j_1 1}) = \underbrace{(\text{LS}, \text{DP})}_{\text{down-parabola lower-saddle}}.$$

  The equilibrium of $(x^*_{j_2}, x^*_{j_1})=(a_{j_1j_2 1}, a_{j_2j_1 1})$ is an (LS,DP)-down-parabola lower-saddle.
- For $a_{j_1j_1 0}(a_{j_2j_1 1}-a_{j_2j_1 1})>0$ and $a_{j_2j_2 0}(a_{j_1j_2 1}-a_{j_2j_2 1})^2<0$,

$$(a_{j_1j_2 1}, a_{j_2j_1 1}) = \underbrace{(\text{US}, \text{DP})}_{\text{down-parabola upper-saddle}}.$$

  The equilibrium of $(x^*_{j_2}, x^*_{j_1})=(a_{j_1j_2 1}, a_{j_2j_1 1})$ is a (US,DP)-down-parabola upper-saddle.
- For $a_{j_1j_1 0}(a_{j_2j_1 1}-a_{j_2j_1 1})<0$ and $a_{j_2j_2 0}(a_{j_1j_2 1}-a_{j_2j_2 1})^2<0$,

$$(a_{j_1j_2 1}, a_{j_2j_1 1}) = \underbrace{(\text{LS}, \text{UP})}_{\text{up-parabola lower-saddle}}.$$

  The equilibrium of $(x^*_{j_2}, x^*_{j_1})=(a_{j_1j_2 1}, a_{j_2j_1 1})$ is an (LS,UP)-up-parabola lower-saddle.

(v$_3$) For $a_{j_1j_1 1}=a_{j_2j_1 1}$ and $a_{j_2j_2 1}\neq a_{j_1j_2 1}$, in phase space,

$$\frac{dx_{j_1}}{dx_{j_2}}=\frac{a_{j_1j_1 0}}{a_{j_2j_2 0}}\frac{(x_{j_2}-a_{j_1j_2 1})^2}{(x_{j_2}-a_{j_2j_2 1})^2},$$

and the deformation of the foregoing equation is

$$dx_{j_1}=\frac{a_{j_1j_1 0}}{a_{j_2j_2 0}}\Big[1+\frac{2(a_{j_2j_2 1}-a_{j_1j_2 1})}{(x_{j_2}-a_{j_2j_2 1})}+\frac{(a_{j_2j_2 1}-a_{j_1j_2 1})^2}{(x_{j_2}-a_{j_2j_2 1})^2}\Big]dx_{j_2}.$$

With an initial condition of $(x_{j_1 0},x_{j_2 0})$ at $t=t_0$, the integration of the above equation gives the first integral manifold for $a_{j_1j_1 1}=a_{j_2j_1 1}$ and $a_{j_2j_2 1}\neq a_{j_1j_2 1}$ as

$$\begin{aligned}\big[(x_{j_1}-a_{j_1j_1 1})-(x_{j_1 0}-a_{j_1j_1 1})\big]=\frac{a_{j_1j_1 0}}{a_{j_2j_2 0}}\Big\{&\big[(x_{j_2}-a_{j_2j_2 1})-(x_{j_2 0}-a_{j_2j_2 1})\big]\\&+2(a_{j_2j_2 1}-a_{j_1j_2 1})\ln\frac{|x_{j_2}-a_{j_2j_2 1}|}{|x_{j_2 0}-a_{j_2j_2 1}|}\\&-(a_{j_2j_2 1}-a_{j_1j_2 1})^2\Big(\frac{1}{x_{j_2}-a_{j_2j_2 1}}-\frac{1}{x_{j_2 0}-a_{j_2j_2 1}}\Big)\Big\}.\end{aligned}$$

(v$_{3a}$) In phase space, for $x_{j_1}^*=a_{j_1j_1 1}=a_{j_2j_1 1}$ with $\bar{x}_{j_2}\neq a_{j_1j_2 1},a_{j_2j_2 1}$, if

$$\frac{dx_{j_2}}{dx_{j_1}}\Big|_{x_{j_1}^*=a_{j_2j_1 1}}=\frac{a_{j_2j_2 0}}{a_{j_1j_1 0}}\frac{(\bar{x}_{j_2}-a_{j_2j_2 1})^2}{(\bar{x}_{j_2}-a_{j_1j_2 1})^2}>0,$$

there is an increasing-inflection flow, and if

$$\frac{dx_{j_2}}{dx_{j_1}}\Big|_{x_{j_1}^*=a_{j_2j_1 1}}=\frac{a_{j_2j_2 0}}{a_{j_1j_1 0}}\frac{(\bar{x}_{j_2}-a_{j_2j_2 1})^2}{(\bar{x}_{j_2}-a_{j_1j_2 1})^2}<0,$$

there is a decreasing-inflection flow. The variational equation is

$$\Delta\dot{x}_{j_1}=a_{j_1j_1 0}(\bar{x}_{j_2}-a_{j_1j_2 1})^2\Delta x_{j_1}.$$

The inflection flows for $a_{j_1j_1 0}(\bar{x}_{j_2}-a_{j_1j_2 1})^2>0$ and $a_{j_1j_1 0}(\bar{x}_{j_2}-a_{j_1j_2 1})^2<0$ are source and sink, respectively.

Therefore, the infinite-equilibrium of $x_{j_1}^*=a_{j_1j_1 1}=a_{j_2j_1 1}$ with $\bar{x}_{j_2}\neq a_{j_1j_2 1},a_{j_2j_2 1}$ for $a_{j_2j_2 1}\neq a_{j_1j_2 1}$ has the following properties as in Eqs. (1.112)–(1.115):

- For $a_{j_1j_10}(\overline{x}_{j_2}-a_{j_1j_21})^2>0$ and $a_{j_2j_20}(\overline{x}_{j_2}-a_{j_2j_21})^2>0$,

$$(a_{j_1j_11},\overline{x}_{j_2})=\underbrace{(\text{SO},\text{II})}_{\text{increasing-inflection source}}.$$

The infinite-equilibrium of $x^*_{j_1}=a_{j_1j_11}=a_{j_2j_11}$ is an (SO,II)-increasing-inflection source.

- For $a_{j_1j_10}(\overline{x}_{j_2}-a_{j_1j_21})^2<0$ and $a_{j_2j_20}(\overline{x}_{j_2}-a_{j_2j_21})^2>0$,

$$(a_{j_1j_11},\overline{x}_{j_2})=\underbrace{(\text{SI},\text{DI})}_{\text{increasing-inflection sink}}.$$

The infinite-equilibrium of $x^*_{j_1}=a_{j_1j_11}=a_{j_2j_11}$ is an (SI,DI)-decreasing-inflection sink.

- For $a_{j_1j_10}(\overline{x}_{j_2}-a_{j_1j_21})^2>0$ and $a_{j_2j_20}(\overline{x}_{j_2}-a_{j_2j_21})^2<0$,

$$(a_{j_1j_11},\overline{x}_{j_2})=\underbrace{(\text{SO},\text{DI})}_{\text{increasing-inflection source}}.$$

The infinite-equilibrium of $x^*_{j_1}=a_{j_1j_11}=a_{j_2j_11}$ is an (SO,DI)-decreasing-inflection source.

- For $a_{j_1j_10}(\overline{x}_{j_2}-a_{j_1j_21})^2<0$ and $a_{j_2j_20}(\overline{x}_{j_2}-a_{j_2j_21})^2<0$,

$$(a_{j_1j_11},\overline{x}_{j_2})=\underbrace{(\text{SI},\text{II})}_{\text{increasing-inflection sink}}.$$

The infinite-equilibrium of $x^*_{j_1}=a_{j_1j_11}=a_{j_2j_11}$ is an (SI,II)-increasing-inflection sink.

(v_{3b}) For $\overline{x}_{j_2}=a_{j_1j_21}$, from case ($v_{3a}$), the equilibrium of $(x^*_{j_1},x^*_{j_2})=(a_{j_1j_11},a_{j_1j_21})$ with $a_{j_1j_11}=a_{j_2j_11}$ has the following properties as in Eqs. (1.116)–(1.119):

- For $a_{j_1j_10}>0$ and $a_{j_2j_20}(a_{j_1j_21}-a_{j_2j_21})^2>0$,

$$(a_{j_1j_11},a_{j_1j_21})=\underbrace{(\text{SO},\text{II}:\text{II})}_{\text{increasing-inflection source}}.$$

The equilibrium of $(x^*_{j_1},x^*_{j_2})=(a_{j_1j_11},a_{j_1j_21})$ is an (SO,II:II)-increasing-inflection source of the first kind.

- For $a_{j_1j_10}<0$ and $a_{j_2j_20}(a_{j_1j_21}-a_{j_2j_21})^2>0$,

$$(a_{j_1j_11},a_{j_1j_21})=\underbrace{(\text{SI},\text{DI}:\text{DI})}_{\text{decreasing-inflection sink}}.$$

The equilibrium of $(x_{j_1}^*, x_{j_2}^*) = (a_{j_1j_1}1, a_{j_1j_2}1)$ is an (SI,DI:DI)-decreasing-inflection sink of the first kind.

- For $a_{j_1j_10} > 0$ and $a_{j_2j_20}(a_{j_1j_2}1 - a_{j_2j_2}1)^2 < 0$,

$$(a_{j_1j_1}1, a_{j_1j_2}1) = \underbrace{(\mathrm{SO}, \mathrm{DI} : \mathrm{DI})}_{\text{decreasing-inflection source}}.$$

The equilibrium of $(x_{j_1}^*, x_{j_2}^*) = (a_{j_1j_1}1, a_{j_1j_2}1)$ is an (SO,DI:DI)-decreasing-inflection source of the first kind.

- For $a_{j_1j_10} < 0$ and $a_{j_2j_20}(a_{j_1j_2}1 - a_{j_2j_2}1)^2 < 0$,

$$(a_{j_1j_1}1, a_{j_1j_2}1) = \underbrace{(\mathrm{SI}, \mathrm{II} : \mathrm{II})}_{\text{increasing-inflection sink}}.$$

The equilibrium of $(x_{j_1}^*, x_{j_2}^*) = (a_{j_1j_1}1, a_{j_1j_2}1)$ is an (SI,II:II)-increasing-inflection sink of the first kind.

(v_{3c}) For $\bar{x}_{j_2} = a_{j_2j_2}1$, from case ($\mathrm{v}_{3a}$), the equilibrium of $(x_{j_2}^*, x_{j_1}^*) = (a_{j_1j_1}1, a_{j_2j_2}1)$ with $a_{j_1j_1}1 = a_{j_2j_2}1$ has the following properties as in Eqs. (1.120)–(1.123):

- For $a_{j_1j_10}(a_{j_2j_2}1 - a_{j_1j_2}1)^2 > 0$ and $a_{j_2j_20} > 0$,

$$(a_{j_1j_1}1, a_{j_2j_2}1) = \underbrace{(\mathrm{SO}, \mathrm{II} : \mathrm{II})}_{\text{increasing-inflection source}}.$$

The equilibrium of $(x_{j_2}^*, x_{j_1}^*) = (a_{j_1j_1}1, a_{j_2j_2}1)$ is an (SO,II:II)-increasing-inflection source of the second kind.

- For $a_{j_1j_10}(a_{j_2j_2}1 - a_{j_1j_2}1)^2 < 0$ and $a_{j_2j_20} > 0$,

$$(a_{j_1j_1}1, a_{j_2j_2}1) = \underbrace{(\mathrm{SI}, \mathrm{DI} : \mathrm{DI})}_{\text{decreasing-inflection sink}}.$$

The equilibrium of $(x_{j_2}^*, x_{j_1}^*) = (a_{j_1j_1}1, a_{j_2j_2}1)$ is an (SI,DI:DI)-decreasing-inflection sink of the second kind.

- For $a_{j_1j_10}(a_{j_2j_2}1 - a_{j_1j_2}1)^2 > 0$ and $a_{j_2j_20} < 0$,

$$(a_{j_1j_1}1, a_{j_2j_2}1) = \underbrace{(\mathrm{SO}, \mathrm{DI} : \mathrm{DI})}_{\text{decreasing-inflection source}}.$$

The equilibrium of $(x_{j_2}^*, x_{j_1}^*) = (a_{j_1j_1}1, a_{j_2j_2}1)$ is an (SO,DI:DI)-decreasing-inflection source of the second kind.

- For $a_{j_1j_10}(a_{j_2j_2}1 - a_{j_1j_2}1)^2 < 0$ and $a_{j_2j_20} < 0$,

$$(a_{j_1j_11}, a_{j_2j_21}) = \underbrace{(\text{SI}, \text{II} : \text{II})}_{\text{increasing-inflection sink}}.$$

The equilibrium of $(x^*_{j_2}, x^*_{j_1}) = (a_{j_1j_11}, a_{j_2j_21})$ is an (SI,II:II)-increasing-inflection sink of the second kind.

(v_4) For $a_{j_1j_11} = a_{j_2j_11}$ and $a_{j_2j_21} = a_{j_1j_21}$, in phase space,

$$\frac{dx_{j_1}}{dx_{j_2}} = \frac{a_{j_1j_10}}{a_{j_2j_20}},$$

and the deformation of the foregoing equation is

$$dx_{j_1} = \frac{a_{j_1j_10}}{a_{j_2j_20}} dx_{j_2}.$$

With an initial condition of (x_{j_10}, x_{j_20}) at $t = t_0$, the integration of the above equation gives the first integral manifold for $a_{j_1j_11} = a_{j_2j_11}$ and $a_{j_2j_21} = a_{j_1j_21}$ as

$$\left[(x_{j_1} - a_{j_1j_11}) - (x_{j_10} - a_{j_1j_11})\right] = \frac{a_{j_1j_10}}{a_{j_2j_20}} \left[(x_{j_2} - a_{j_2j_21}) - (x_{j_20} - a_{j_2j_21})\right].$$

($\text{v}_{4\text{a}}$) In phase space, for $x^*_{j_1} = a_{j_1j_11} = a_{j_2j_11}$ with $\bar{x}_{j_2} \neq a_{j_1j_21} = a_{j_2j_21}$, if

$$\left.\frac{dx_{j_2}}{dx_{j_1}}\right|_{x^*_{j_1} = a_{j_2j_11}} = \frac{a_{j_2j_20}}{a_{j_1j_10}} > 0,$$

there is an increasing-inflection flow, and if

$$\left.\frac{dx_{j_2}}{dx_{j_1}}\right|_{x^*_{j_1} = a_{j_2j_11}} = \frac{a_{j_2j_20}}{a_{j_1j_10}} < 0,$$

there is a decreasing-inflection flow. The variational equation is

$$\Delta\dot{x}_{j_1} = a_{j_1j_10}(\bar{x}_{j_2} - a_{j_1j_21})^2 \Delta x_{j_1}.$$

The inflection flows for $a_{j_1j_10}(\bar{x}_{j_2} - a_{j_1j_21})^2 > 0$ and $a_{j_1j_10}(\bar{x}_{j_2} - a_{j_1j_21})^2 < 0$ are source and sink, respectively.

Therefore, the infinite-equilibrium of $x^*_{j_1} = a_{j_1j_11} = a_{j_2j_11}$ with $\bar{x}_{j_2} \neq a_{j_1j_21}, a_{j_2j_21}$ for $a_{j_1j_11} = a_{j_2j_11}$ and $a_{j_2j_21} = a_{j_1j_21}$ has the following properties as in Eqs. (1.125)–(1.128):

- For $a_{j_1j_10}(\bar{x}_{j_2} - a_{j_1j_21})^2 > 0$ and $a_{j_2j_20}(\bar{x}_{j_2} - a_{j_2j_21})^2 > 0$,

$$(a_{j_1j_1 1}, \overline{x}_{j_2}) = \underbrace{(\text{SO, II})}_{\text{increasing-inflection source}}.$$

The infinite-equilibrium of $x^*_{j_1} = a_{j_1j_1 1} = a_{j_2j_1 1}$ is an (SO,II)-increasing-inflection source.

- For $a_{j_1j_1 0}(\overline{x}_{j_2} - a_{j_1j_2 1})^2 < 0$ and $a_{j_2j_2 0}(\overline{x}_{j_2} - a_{j_2j_2 1})^2 > 0$,

$$(a_{j_1j_1 1}, \overline{x}_{j_2}) = \underbrace{(\text{SI, DI})}_{\text{increasing-inflection sink}}.$$

The infinite-equilibrium of $x^*_{j_1} = a_{j_1j_1 1} = a_{j_2j_1 1}$ is an (SI,DI)-decreasing-inflection sink.

- For $a_{j_1j_1 0}(\overline{x}_{j_2} - a_{j_1j_2 1})^2 > 0$ and $a_{j_2j_2 0}(\overline{x}_{j_2} - a_{j_2j_2 1})^2 < 0$,

$$(a_{j_1j_1 1}, \overline{x}_{j_2}) = \underbrace{(\text{SO, DI})}_{\text{increasing-inflection source}}.$$

The infinite-equilibrium of $x^*_{j_1} = a_{j_1j_1 1} = a_{j_2j_1 1}$ is an (SO,DI)-decreasing-inflection source.

- For $a_{j_1j_1 0}(\overline{x}_{j_2} - a_{j_1j_2 1})^2 < 0$ and $a_{j_2j_2 0}(\overline{x}_{j_2} - a_{j_2j_2 1})^2 < 0$,

$$(a_{j_1j_1 1}, \overline{x}_{j_2}) = \underbrace{(\text{SI, II})}_{\text{increasing-inflection sink}}.$$

The infinite-equilibrium of $x^*_{j_1} = a_{j_1j_1 1} = a_{j_2j_1 1}$ is an (SI,II)-increasing-inflection sink.

(v_{4b}) In phase space, for $x^*_{j_2} = a_{j_2j_2 1} = a_{j_1j_2 1}$ with $\overline{x}_{j_1} \neq a_{j_1j_1 1} = a_{j_2j_1 1}$, if

$$\left.\frac{dx_{j_1}}{dx_{j_2}}\right|_{x^*_{j_2} = a_{j_2j_2 1}} = \frac{a_{j_1j_1 0}}{a_{j_2j_2 0}} > 0,$$

there is an increasing-inflection flow, and if

$$\left.\frac{dx_{j_1}}{dx_{j_2}}\right|_{x^*_{j_2} = a_{j_2j_2 1}} = \frac{a_{j_1j_1 0}}{a_{j_2j_2 0}} < 0,$$

there is a decreasing-inflection flow. The variational equation is

$$\Delta\dot{x}_{j_2} = a_{j_2j_2 0}(\overline{x}_{j_1} - a_{j_2j_1 1})(\Delta x_{j_2})^2.$$

The inflection flows for $a_{j_2j_2 0}(\overline{x}_{j_1} - a_{j_2j_1 1}) > 0$ and $a_{j_2j_2 0}(\overline{x}_{j_1} - a_{j_2j_1 1}) < 0$ are upper-saddle and lower-saddle, respectively.

Therefore, the infinite-equilibrium of $x_{j_2}^* = a_{j_2 j_2 1} = a_{j_1 j_2 1}$ with $\overline{x}_{j_1} \neq a_{j_1 j_1 1}, a_{j_2 j_1 1}$ for $a_{j_1 j_1 1} = a_{j_2 j_1 1}$ and $a_{j_2 j_2 1} = a_{j_1 j_2 1}$ has the following properties as in Eqs. (1.129)–(1.132):

- For $a_{j_1 j_1 0}(\overline{x}_{j_1} - a_{j_1 j_1 1}) > 0$ and $a_{j_2 j_2 0}(\overline{x}_{j_1} - a_{j_2 j_1 1}) > 0$,

$$(\overline{x}_{j_1}, a_{j_2 j_2 1}) = \underbrace{(\text{II}, \text{US})}_{\text{increasing-inflection upper-saddle}} .$$

The infinite-equilibrium of $x_{j_2}^* = a_{j_2 j_2 1} = a_{j_1 j_2 1}$ is an (II,US)-increasing-inflection upper-saddle.

- For $a_{j_1 j_1 0}(\overline{x}_{j_1} - a_{j_1 j_1 1}) < 0$ and $a_{j_2 j_2 0}(\overline{x}_{j_1} - a_{j_2 j_1 1}) > 0$,

$$(\overline{x}_{j_1}, a_{j_2 j_2 1}) = \underbrace{(\text{DI}, \text{US})}_{\text{decreasing-inflection upper-saddle}} .$$

The infinite-equilibrium of $x_{j_2}^* = a_{j_2 j_2 1} = a_{j_1 j_2 1}$ is a (DI,US)-decreasing-inflection upper-saddle.

- For $a_{j_1 j_1 0}(\overline{x}_{j_1} - a_{j_1 j_1 1}) > 0$ and $a_{j_2 j_2 0}(\overline{x}_{j_1} - a_{j_2 j_1 1}) < 0$,

$$(\overline{x}_{j_1}, a_{j_2 j_2 1}) = \underbrace{(\text{DI}, \text{LS})}_{\text{decreasing-inflection lower-saddle}} .$$

The infinite-equilibrium of $x_{j_2}^* = a_{j_2 j_2 1} = a_{j_1 j_2 1}$ is a (DI,US)-decreasing-inflection lower-saddle.

- For $a_{j_1 j_1 0}(\overline{x}_{j_1} - a_{j_1 j_1 1}) < 0$ and $a_{j_2 j_2 0}(\overline{x}_{j_1} - a_{j_2 j_1 1}) < 0$,

$$(\overline{x}_{j_1}, a_{j_2 j_2 1}) = \underbrace{(\text{II}, \text{LS})}_{\text{increasing-inflection lower-saddle}} .$$

The infinite-equilibrium of $x_{j_2}^* = a_{j_2 j_2 1} = a_{j_1 j_2 1}$ is an (II,LS)-increasing-inflection lower-saddle.

($\text{v}_{4\text{c}}$) From ($\text{v}_{4\text{a}}$), for $\overline{x}_{j_2} = a_{j_1 j_2 1} = a_{j_2 j_2 1}$, the equilibrium of $(x_{j_1}^*, x_{j_2}^*) = (a_{j_2 j_1 1}, a_{j_2 j_2 1}) = (a_{j_1 j_1 1}, a_{j_1 j_2 1})$ with $a_{j_1 j_1 1} = a_{j_2 j_1 1}$ and $a_{j_2 j_2 1} = a_{j_1 j_2 1}$ has the following properties as in Eqs. (1.133)–(1.136):

- For $a_{j_1 j_1 0} > 0$ and $a_{j_2 j_2 0} > 0$,

$$(a_{j_2 j_1 1}, a_{j_2 j_2 1}) = (a_{j_1 j_1 1}, a_{j_1 j_2 1}) = \underbrace{({}_{\text{II:II}}\text{SO}, {}_{\text{LS:US}}\text{II})}_{\text{increasing-inflection source}} .$$

The equilibrium of $(x_{j_1}^*, x_{j_2}^*) = (a_{j_2 j_1 1}, a_{j_2 j_2 1}) = (a_{j_1 j_1 1}, a_{j_1 j_2 1})$ is an $({}_{\text{II:II}}\text{SO}, {}_{\text{LS:US}}\text{II})$-increasing-inflection source.

- For $a_{j_1 j_1 0} < 0$ and $a_{j_2 j_2 0} > 0$,

$$(a_{j_2 j_1 1}, a_{j_2 j_2 1}) = (a_{j_1 j_1 1}, a_{j_1 j_2 1}) = \underbrace{({}_{\text{DI:DI}}\text{SI}, {}_{\text{LS:US}}\text{DI})}_{\text{decreasing-inflection sink}}.$$

The equilibrium of $(x_{j_1}^*, x_{j_2}^*) = (a_{j_2 j_1 1}, a_{j_2 j_2 1}) = (a_{j_1 j_1 1}, a_{j_1 j_2 1})$ is an $({}_{\text{DI:DI}}\text{SI}, {}_{\text{US:US}}\text{DI})$-decreasing-inflection sink.

- For $a_{j_1 j_1 0} > 0$ and $a_{j_2 j_2 0} < 0$,

$$(a_{j_2 j_1 1}, a_{j_2 j_2 1}) = (a_{j_1 j_1 1}, a_{j_1 j_2 1}) = \underbrace{({}_{\text{DI:DI}}\text{SO}, {}_{\text{US:LS}}\text{DI})}_{\text{decreasing-inflection source}}.$$

The equilibrium of $(x_{j_1}^*, x_{j_2}^*) = (a_{j_2 j_1 1}, a_{j_2 j_2 1}) = (a_{j_1 j_1 1}, a_{j_1 j_2 1})$ is an $({}_{\text{DI:DI}}\text{SO}, {}_{\text{US:LS}}\text{DI})$-decreasing-inflection source.

- For $a_{j_1 j_1 0} < 0$ and $a_{j_2 j_2 0} < 0$,

$$(a_{j_2 j_1 1}, a_{j_2 j_2 1}) = (a_{j_1 j_1 1}, a_{j_1 j_2 1}) = \underbrace{({}_{\text{II:II}}\text{SI}, {}_{\text{US:LS}}\text{II})}_{\text{increasing-inflection sink}}.$$

The equilibrium of $(x_{j_1}^*, x_{j_2}^*) = (a_{j_2 j_1 1}, a_{j_2 j_2 1}) = (a_{j_1 j_1 1}, a_{j_1 j_2 1})$ is an $({}_{\text{II:II}}\text{SI}, {}_{\text{US:LS}}\text{II})$-increasing-inflection sink.

The above results can also be obtained from case (v_{4b}) for $\bar{x}_{j_1} = a_{j_1 j_1 1} = a_{j_2 j_1 1}$.

(vi) For $\Delta_{j_1 j_2} = 0$ and $\Delta_{j_2 j_2} > 0$, the standard form is

$$\begin{aligned}
\dot{x}_{j_2} &= a_{j_2 j_2 0}(x_{j_1} - a_{j_2 j_1 1})(x_{j_2} - a_{j_2 j_2 1})(x_{j_2} - a_{j_2 j_2 2}),\\
\dot{x}_{j_1} &= a_{j_1 j_1 0}(x_{j_1} - a_{j_1 j_1 1})(x_{j_2} - a_{j_1 j_2 1})^2
\end{aligned}$$

where

$$\begin{aligned}
a_{j_1 j_2 1} &= -\frac{1}{2} B_{j_1 j_2},\\
b_{j_2 j_2 1}, b_{j_2 j_2 2} &= -\frac{1}{2}(B_{j_2 j_2} \pm \sqrt{\Delta_{j_2 j_2}}),\\
\{a_{j_2 j_2 1}, a_{j_2 j_2 2}\} &= \text{sort}\{b_{j_2 j_2 1}, b_{j_2 j_2 2}\}, a_{j_2 j_2 1} < a_{j_2 j_2 2}.
\end{aligned}$$

(vi_1) In phase space, $a_{j_2 j_2 s_1} \neq a_{j_1 j_2 1}$, we have

$$\frac{dx_{j_1}}{dx_{j_2}} = \frac{a_{j_1j_10}}{a_{j_2j_20}} \frac{(x_{j_1} - a_{j_1j_1}1)(x_{j_2} - a_{j_1j_2}1)^2}{(x_{j_1} - a_{j_2j_1}1)(x_{j_2} - a_{j_2j_2}1)(x_{j_2} - a_{j_2j_2}2)},$$

and the deformation of the foregoing equation is

$$\begin{aligned}\left[1 + \frac{(a_{j_1j_1}1 - a_{j_2j_1}1)}{(x_{j_1} - a_{j_1j_1}1)}\right]dx_{j_1} = \frac{a_{j_1j_10}}{a_{j_2j_20}}\Big[1 &+ \frac{(a_{j_2j_2}1 - a_{j_1j_2}1)^2}{(a_{j_2j_2}1 - a_{j_2j_2}2)(x_{j_2} - a_{j_2j_2}1)} \\ &+ \frac{(a_{j_2j_2}2 - a_{j_1j_2}1)^2}{(a_{j_2j_2}2 - a_{j_2j_2}1)(x_{j_2} - a_{j_2j_2}2)}\Big]dx_{j_2}.\end{aligned}$$

With an initial condition of (x_{j_10}, x_{j_20}) at $t = t_0$, the integration of the above equation gives the first integral manifold $(a_{j_2j_2s_1} \neq a_{j_1j_2}1)$ as

$$\begin{aligned}&\left[(x_{j_1} - a_{j_1j_1}1) - (x_{j_10} - a_{j_1j_1}1)\right] + (a_{j_1j_1}1 - a_{j_2j_1}1)\ln\frac{|x_{j_1} - a_{j_1j_1}1|}{|x_{j_10} - a_{j_1j_1}1|} \\ &= \frac{a_{j_1j_10}}{a_{j_2j_20}}\left[(x_{j_1} - x_{j_10}) + \sum\nolimits_{s_1=1, s_2=s_1}^{2} \frac{(a_{j_2j_2s_1} - a_{j_1j_2}1)^2}{(a_{j_2j_2s_1} - a_{j_2j_2s_2})}\ln\frac{|x_{j_2} - a_{j_2j_2s_1}|}{|x_{j_20} - a_{j_2j_2s_1}|}\right].\end{aligned}$$

(vi$_{1a}$) At $(x^*_{j_1}, x^*_{j_2}) = (a_{j_1j_1}1, a_{j_2j_2s_1})$, the variational equations are

$$\begin{aligned}\Delta\dot{x}_{j_2} &= a_{j_2j_20}(a_{j_1j_1}1 - a_{j_2j_1}1)(a_{j_2j_2s_1} - a_{j_2j_2s_2})\Delta x_{j_2}, \\ \Delta\dot{x}_{j_1} &= a_{j_1j_10}(a_{j_2j_2s_1} - a_{j_1j_2}1)^2\Delta x_{j_1}.\end{aligned}$$

The flows $(x^*_{j_1}, x^*_{j_2}) = (a_{j_1j_1}1, a_{j_2j_2s_1})$ are source and sink for $a_{j_1j_10}(a_{j_2j_2s_1} - a_{j_1j_2}1)^2 > 0$ and $a_{j_1j_10}(a_{j_2j_2s_1} - a_{j_1j_2}1)^2 < 0$ in the x_{j_1}-direction, respectively. The flows $(x^*_{j_1}, x^*_{j_2}) = (a_{j_1j_1}1, a_{j_2j_2s_1})$ are source and sink for $a_{j_2j_20}(a_{j_1j_1}1 - a_{j_2j_1}1)(a_{j_2j_2s_1} - a_{j_2j_2s_2}) > 0$ and $a_{j_2j_20}(a_{j_1j_1}1 - a_{j_2j_1}1)(a_{j_2j_2s_1} - a_{j_2j_2s_2}) < 0$ in the x_{j_2}-direction, respectively.

Thus, the equilibrium of $(x^*_{j_1}, x^*_{j_2}) = (a_{j_1j_1}1, a_{j_2j_2s_1})$ $(s_1, s_2 \in \{1, 2\}, s_1 \neq s_2)$ has the following properties as in Eqs. (1.140)–(1.143):

- For $a_{j_1j_10}(a_{j_2j_2s_1} - a_{j_1j_2}1)^2 > 0$ and $a_{j_2j_20}(a_{j_1j_1}1 - a_{j_2j_1}1)(a_{j_2j_2s_1} - a_{j_2j_2s_2}) > 0$,

$$(a_{j_1j_1}1, a_{j_1j_2s_1}) = \underbrace{(\text{SO}, \text{SO})}_{\text{source}}.$$

  The equilibrium of $(x^*_{j_1}, x^*_{j_2}) = (a_{j_1j_1}1, a_{j_2j_2s_1})$ is an (SO,SO)-source.

- For $a_{j_1j_10}(a_{j_2j_2s_1} - a_{j_1j_2}1)^2 < 0$ and $a_{j_2j_20}(a_{j_1j_1}1 - a_{j_2j_1}1)(a_{j_2j_2s_1} - a_{j_2j_2s_2}) > 0$,

$$(a_{j_1j_11}, a_{j_2j_2s_1}) = \underbrace{(\text{SI}, \text{SO})}_{\text{saddle}}.$$

The equilibrium of $(x^*_{j_1}, x^*_{j_2}) = (a_{j_1j_11}, a_{j_2j_2s_1})$ is an (SI,SO)-saddle.

- For $a_{j_1j_10}(a_{j_2j_2s_1} - a_{j_1j_21})^2 > 0$ and $a_{j_2j_20}(a_{j_1j_11} - a_{j_2j_11})(a_{j_2j_2s_1} - a_{j_2j_2s_2}) < 0$,

$$(a_{j_1j_11}, a_{j_2j_2s_1}) = \underbrace{(\text{SO}, \text{SI})}_{\text{saddle}}.$$

The equilibrium of $(x^*_{j_1}, x^*_{j_2}) = (a_{j_1j_11}, a_{j_2j_2s_1})$ is an (SO,SI)-saddle.

- For $a_{j_1j_10}(a_{j_2j_2s_1} - a_{j_1j_21})^2 < 0$ and $a_{j_2j_20}(a_{j_1j_11} - a_{j_2j_11})(a_{j_2j_2s_1} - a_{j_2j_2s_2}) < 0$,

$$(a_{j_1j_11}, a_{j_2j_2s_1}) = \underbrace{(\text{SI}, \text{SI})}_{\text{sink}}.$$

The equilibrium of $(x^*_{j_1}, x^*_{j_2}) = (a_{j_1j_11}, a_{j_2j_2s_1})$ is an (SI,SI)-sink.

(vi_{1b}) Consider two cases (I) $x^*_{j_2} = a_{j_1j_21}$ and (II) $x^*_{j_1} = a_{j_1j_11}$.
(I) In phase space, at $x^*_{j_2} = a_{j_1j_21}$ with $\overline{x}_{j_1} \neq a_{j_1j_11}, a_{j_2j_11}$

$$\begin{aligned}
&\frac{dx_{j_1}}{dx_{j_2}}\Big|_{x^*_{j_2}=a_{j_1j_21}} = \frac{a_{j_1j_10}}{a_{j_2j_20}} \frac{(\overline{x}_{j_1} - a_{j_1j_11})(x_{j_2} - a_{j_1j_21})^2}{(\overline{x}_{j_1} - a_{j_2j_11})(x_{j_2} - a_{j_2j_21})(x_{j_2} - a_{j_2j_22})}\Big|_{x^*_{j_2}=a_{j_1j_21}} = 0,\\
&\frac{d^2x_{j_1}}{dx^2_{j_2}}\Big|_{x^*_{j_2}=a_{j_1j_21}} = 0.
\end{aligned}$$

If

$$\frac{d^3x_{j_1}}{dx^3_{j_2}}\Big|_{x^*_{j_2}=a_{j_1j_21}} = \frac{a_{j_1j_10}}{a_{j_2j_20}} \frac{2(\overline{x}_{j_1} - a_{j_1j_11})}{(\overline{x}_{j_1} - a_{j_2j_11})(a_{j_1j_21} - a_{j_2j_21})(a_{j_1j_21} - a_{j_2j_22})} > 0,$$

there is a third-order increasing-inflection flow at $x^*_{j_2} = a_{j_1j_21}$ with $\overline{x}_{j_1} \neq a_{j_1j_11}, a_{j_2j_11}$. If

$$\frac{d^3x_{j_1}}{dx^3_{j_2}}\Big|_{x^*_{j_2}=a_{j_1j_21}} = \frac{a_{j_1j_10}}{a_{j_2j_20}} \frac{2(\overline{x}_{j_1} - a_{j_1j_11})}{(\overline{x}_{j_1} - a_{j_2j_11})(a_{j_1j_21} - a_{j_2j_21})(a_{j_1j_21} - a_{j_2j_22})} < 0,$$

there is a third-order decreasing-inflection flow at $x^*_{j_2} = a_{j_1j_21}$ with $\overline{x}_{j_1} \neq a_{j_1j_11}, a_{j_2j_11}$.

Let

$$\begin{aligned}
\dot{x}_{j_2} &= a_{j_2j_20}(\overline{x}_{j_1} - a_{j_2j_11})(a_{j_1j_21} - a_{j_2j_21})(a_{j_1j_21} - a_{j_2j_22})\\
&= a_{j_2j_20}(\overline{x}_{j_1} - a_{j_2j_11})\prod\nolimits_{s_1=1}^{2}(a_{j_1j_21} - a_{j_2j_2s_1}) \neq 0.
\end{aligned}$$

The third-order inflection flows at $x_{j_2}^* = a_{j_1j_2 1}$ with $\overline{x}_{j_1} \neq a_{j_1j_1 1}, a_{j_2j_1 1}$ are positive and negative for $\dot{x}_{j_2} > 0$ and $\dot{x}_{j_2} < 0$, respectively.

Thus, the equilibrium of $x_{j_2}^* = a_{j_1j_2 1}$ with $\overline{x}_{j_1} \neq a_{j_1j_1 1}$ has the following properties:

- For $a_{j_1j_1 0}(\overline{x}_{j_1} - a_{j_1j_1 1}) > 0$ and $a_{j_2j_2 0}(\overline{x}_{j_1} - a_{j_2j_1 1})\prod_{s_1=1}^{2}(a_{j_1j_2 1} - a_{j_2j_2 s_1}) > 0$,

$$(a_{j_1j_2 1}, \dot{x}_{j_2}) = \underbrace{(\text{II}, \text{pF})}_{\text{increasing-inflection flow }(+)}.$$

- For $a_{j_1j_1 0}(\overline{x}_{j_1} - a_{j_1j_1 1}) < 0$ and $a_{j_2j_2 0}(\overline{x}_{j_1} - a_{j_2j_1 1})\prod_{s_1=1}^{2}(a_{j_1j_2 1} - a_{j_2j_2 s_1}) > 0$,

$$(a_{j_1j_2 1}, \dot{x}_{j_2}) = \underbrace{(\text{DI}, \text{pF})}_{\text{decreasing-inflection flow }(+)}.$$

- For $a_{j_1j_1 0}(\overline{x}_{j_1} - a_{j_1j_1 1}) > 0$ and $a_{j_2j_2 0}(\overline{x}_{j_1} - a_{j_2j_1 1})\prod_{s_1=1}^{2}(a_{j_1j_2 1} - a_{j_2j_2 s_1}) < 0$,

$$(a_{j_1j_2 1}, \dot{x}_{j_2}) = \underbrace{(\text{DI}, \text{nF})}_{\text{decreasing-inflection flow }(-)}.$$

- For $a_{j_1j_1 0}(\overline{x}_{j_1} - a_{j_1j_1 1}) < 0$ and $a_{j_2j_2 0}(\overline{x}_{j_1} - a_{j_2j_1 1})\prod_{s_1=1}^{2}(a_{j_1j_2 1} - a_{j_2j_2 s_1}) < 0$,

$$(a_{j_1j_2 1}, \dot{x}_{j_2}) = \underbrace{(\text{II}, \text{nF})}_{\text{increasing-inflection flow }(-)}.$$

(II) The variational equation at $x_{j_1}^* = a_{j_1j_1 1}$ with $\overline{x}_{j_2} \neq a_{j_1j_2 1}$ is

$$\Delta\dot{x}_{j_1} = a_{j_1j_1 0}(\overline{x}_{j_2} - a_{j_1j_2 1})^2 \Delta x_{j_1}$$

and

$$\dot{x}_{j_2} = a_{j_2j_2 0}(a_{j_1j_1 1} - a_{j_2j_1 1})\prod_{s_1=1}^{2}(\overline{x}_{j_2} - a_{j_2j_2 s_1}).$$

Thus, the equilibrium of $x_{j_1}^* = a_{j_1j_1 1}$ with $\overline{x}_{j_2} \neq a_{j_1j_2 1}, a_{j_2j_2 s_1}$ $(s_1 = 1, 2)$ has the following properties:

- For $a_{j_1j_1 0}(\overline{x}_{j_2} - a_{j_1j_2 1})^2 > 0$ and $a_{j_2j_2 0}(a_{j_1j_1 1} - a_{j_2j_1 1})\prod_{s_1=1}^{2}(\overline{x}_{j_2} - a_{j_2j_2 s_1}) > 0$,

$$(a_{j_1j_1}1, \dot{x}_{j_2}) = \underbrace{(\text{SO}, \text{pF})}_{\text{source flow }(+)}.$$

- For $a_{j_1j_1 0}(\bar{x}_{j_2} - a_{j_1j_2}1)^2 < 0$ and $a_{j_2j_2 0}(a_{j_1j_1}1 - a_{j_2j_1}1)\prod_{s_1=1}^{2}(\bar{x}_{j_2} - a_{j_2j_2 s_1}) > 0$,

$$(a_{j_1j_1}1, \dot{x}_{j_2}) = \underbrace{(\text{SI}, \text{pF})}_{\text{sink flow }(+)}.$$

- For $a_{j_1j_1 0}(\bar{x}_{j_2} - a_{j_1j_2}1)^2 > 0$ and $a_{j_2j_2 0}(a_{j_1j_1}1 - a_{j_2j_1}1)\prod_{s_1=1}^{2}(\bar{x}_{j_2} - a_{j_2j_2 s_1}) < 0$,

$$(a_{j_1j_1}1, \dot{x}_{j_2}) = \underbrace{(\text{SO}, \text{nF})}_{\text{source flow }(-)}.$$

- For $a_{j_1j_1 0}(\bar{x}_{j_2} - a_{j_1j_2}1)^2 < 0$ and $a_{j_2j_2 0}(a_{j_1j_1}1 - a_{j_2j_1}1)\prod_{s_1=1}^{2}(\bar{x}_{j_2} - a_{j_2j_2 s_1}) < 0$,

$$(a_{j_1j_1}1, \dot{x}_{j_2}) = \underbrace{(\text{SI}, \text{nF})}_{\text{sink flow }(-)}.$$

From the above analysis, the equilibrium of $(x_{j_1}^*, x_{j_2}^*) = (a_{j_1j_1}1, a_{j_1j_2}1)$ with $a_{j_1j_2}1 \neq a_{j_2j_2}1, a_{j_2j_2}2$ has the following properties as in Eqs. (1.144)–(1.147):

- For $a_{j_1j_1 0} > 0$ and $a_{j_2j_2 0}(a_{j_1j_1}1 - a_{j_2j_1}1)\prod_{s_1=1}^{2}(a_{j_1j_2}1 - a_{j_2j_2 s_1}) > 0$,

$$(a_{j_1j_1}1, a_{j_1j_2}1) = \underbrace{(_{\text{DI:II}}\text{SO}, \text{pF})}_{\text{inflection-source flow }(+)}.$$

  The equilibrium of $(x_{j_1}^*, x_{j_2}^*) = (a_{j_1j_1}1, a_{j_1j_2}1)$ is a $(_{\text{DI:II}}\text{SO}, \text{pF})$-positive inflection-source flow.
- For $a_{j_1j_1 0} < 0$ and $a_{j_2j_2 0}(a_{j_1j_1}1 - a_{j_2j_1}1)\prod_{s_1=1}^{2}(a_{j_1j_2}1 - a_{j_2j_2 s_1}) > 0$,

$$(a_{j_1j_1}1, a_{j_1j_2}1) = \underbrace{(_{\text{II:DI}}\text{SI}, \text{pF})}_{\text{inflection-sink flow }(+)}.$$

  The equilibrium of $(x_{j_1}^*, x_{j_2}^*) = (a_{j_1j_1}1, a_{j_1j_2}1)$ is an $(_{\text{II:DI}}\text{SI}, \text{pF})$-positive inflection-sink flow.

- For $a_{j_1j_10}>0$ and $a_{j_2j_20}(a_{j_1j_11}-a_{j_2j_11})\prod_{s_1=1}^{2}(a_{j_1j_21}-a_{j_2j_2s_1})<0$,

$$(a_{j_1j_11},a_{j_1j_21})=\underbrace{({}_{\text{II:DI}}\text{SO, nF})}_{\text{inflection source flow }(-)}.$$

The equilibrium of $(x_{j_1}^*,x_{j_2}^*)=(a_{j_1j_11},a_{j_1j_21})$ is an (${}_{\text{II:DI}}$SO, nF)-negative inflection-source flow.

- For $a_{j_1j_10}<0$ and $a_{j_2j_20}(a_{j_1j_11}-a_{j_2j_11})\prod_{s_1=1}^{2}(a_{j_1j_21}-a_{j_2j_2s_1})<0$,

$$(a_{j_1j_11},a_{j_1j_21})=\underbrace{({}_{\text{DI:II}}\text{SI, nF})}_{\text{inflection sink flow }(-)}.$$

The equilibrium of $(x_{j_1}^*,x_{j_2}^*)=(a_{j_1j_11},a_{j_1j_21})$ is a (${}_{\text{DI:II}}$SI, nF)-negative inflection-sink flow.

(vi$_{1c}$) For $x_{j_1}^*=a_{j_2j_11}$, in phase space, for $a_{j_2j_11}\neq a_{j_1j_11}$,

$$\frac{dx_{j_2}}{dx_{j_1}}\bigg|_{x_{j_1}^*=a_{j_2j_11}}=\frac{a_{j_2j_20}}{a_{j_1j_10}}\frac{(x_{j_1}-a_{j_2j_11})\prod_{s_1=1}^{2}(\bar{x}_{j_2}-a_{j_2j_2s_1})}{(x_{j_1}-a_{j_1j_11})(\bar{x}_{j_2}-a_{j_1j_21})^2}\bigg|_{x_{j_1}^*=a_{j_2j_11}}=0.$$

If

$$\frac{d^2x_{j_2}}{dx_{j_1}^2}\bigg|_{x_{j_1}^*=a_{j_2j_11}}=\frac{a_{j_2j_20}}{a_{j_1j_10}}\frac{\prod_{s_1=1}^{2}(\bar{x}_{j_2}-a_{j_2j_2s_1})}{(a_{j_2j_11}-a_{j_1j_11})(\bar{x}_{j_2}-a_{j_1j_21})^2}>0,$$

there is an up-parabola flow. If

$$\frac{d^2x_{j_2}}{dx_{j_1}^2}\bigg|_{x_{j_1}^*=a_{j_2j_11}}=\frac{a_{j_2j_20}}{a_{j_1j_10}}\frac{\prod_{s_1=1}^{2}(\bar{x}_{j_2}-a_{j_2j_2s_1})}{(a_{j_2j_11}-a_{j_1j_11})(\bar{x}_{j_2}-a_{j_1j_21})^2}<0,$$

there is a down-parabola flow. Let

$$\dot{x}_{j_1}=a_{j_1j_10}(a_{j_2j_11}-a_{j_1j_11})(\bar{x}_{j_2}-a_{j_1j_21})^2,$$

and for $\bar{x}_{j_2}\neq a_{j_1j_21}$,

$$(\bar{x}_{j_2}-a_{j_1j_21})^2>0.$$

The parabola flows at $x_{j_1}^*=a_{j_2j_11}$ in the x_{j_2}-direction are positive and negative for $a_{j_1j_10}(a_{j_2j_11}-a_{j_1j_11})>0$ and $a_{j_1j_10}(a_{j_2j_11}-a_{j_1j_11})<0$, respectively. Thus,

- For $a_{j_1j_10}(a_{j_2j_11}-a_{j_1j_11})(\overline{x}_{j_2}-a_{j_1j_21})^2>0$ and $a_{j_2j_20}\prod_{s_1=1}^{2}(\overline{x}_{j_2}-a_{j_2j_2s_1})>0$,

$$(\dot{x}_{j_1},a_{j_2j_11})=\underbrace{(\text{pF, UP})}_{\text{up-parabola flow }(+)}.$$

- For $a_{j_1j_10}(a_{j_2j_11}-a_{j_1j_11})(\overline{x}_{j_2}-a_{j_1j_21})^2<0$ and $a_{j_2j_20}\prod_{s_1=1}^{2}(\overline{x}_{j_2}-a_{j_2j_2s_1})>0$,

$$(\dot{x}_{j_1},a_{j_2j_11})=\underbrace{(\text{nF, DP})}_{\text{down-parabola flow }(-)}.$$

- For $a_{j_1j_10}(a_{j_2j_11}-a_{j_1j_11})(\overline{x}_{j_2}-a_{j_1j_21})^2>0$ and $a_{j_2j_20}\prod_{s_1=1}^{2}(\overline{x}_{j_2}-a_{j_2j_2s_1})<0$,

$$(\dot{x}_{j_1},a_{j_2j_11})=\underbrace{(\text{pF, DP})}_{\text{down-parabola flow }(+)}.$$

- For $a_{j_1j_10}(a_{j_2j_11}-a_{j_1j_11})(\overline{x}_{j_2}-a_{j_1j_21})^2<0$ and $a_{j_2j_20}\prod_{s_1=1}^{2}(\overline{x}_{j_2}-a_{j_2j_2s_1})<0$,

$$(\dot{x}_{j_1},a_{j_2j_11})=\underbrace{(\text{nF, UP})}_{\text{up-parabola flow }(-)}.$$

From the above analysis, for $\overline{x}_{j_2}=a_{j_1j_21}$ with case (vi_{1b}), the equilibrium of $(x^*_{j_2},x^*_{j_1})=(a_{j_1j_21},a_{j_2j_11})$ has the following properties as in Eqs. (1.148)–(1.151):

- For $a_{j_1j_10}(a_{j_2j_10}-a_{j_2j_11})>0$ and $a_{j_2j_20}\prod_{s_1=1}^{2}(a_{j_1j_21}-a_{j_2j_2s_1})>0$,

$$(a_{j_1j_21},a_{j_2j_11})=\underbrace{(\text{US, UP})}_{\text{up-parabola upper-saddle}}.$$

  The equilibrium of $(x^*_{j_2},x^*_{j_1})=(a_{j_1j_21},a_{j_2j_11})$ is a (US,UP)-up-parabola upper-saddle.
- For $a_{j_1j_10}(a_{j_2j_11}-a_{j_2j_11})<0$ and $a_{j_2j_20}\prod_{s_1=1}^{2}(a_{j_1j_21}-a_{j_2j_2s_1})>0$,

$$(a_{j_1j_21},a_{j_2j_11})=\underbrace{(\text{LS, DP})}_{\text{down-parabola lower-saddle}}.$$

  The equilibrium of $(x^*_{j_2},x^*_{j_1})=(a_{j_1j_21},a_{j_2j_11})$ is an (LS,DP)-down-parabola lower-saddle.
- For $a_{j_1j_10}(a_{j_2j_11}-a_{j_2j_11})>0$ and $a_{j_2j_20}\prod_{s_1=1}^{2}(a_{j_1j_21}-a_{j_2j_2s_1})<0$,

$$(a_{j_1j_2 1}, a_{j_2j_1 1}) = \underbrace{(\text{US}, \text{DP})}_{\text{down-parabola upper-saddle}}.$$

The equilibrium of $(x_{j_2}^*, x_{j_1}^*) = (a_{j_1j_2 1}, a_{j_2j_1 1})$ is a (US,DP)-down-parabola upper-saddle.

- For $a_{j_1j_1 0}(a_{j_2j_1 1} - a_{j_2j_1 1}) < 0$ and $a_{j_2j_2 0}\prod_{s_1=1}^{2}(a_{j_1j_2 1} - a_{j_2j_2 s_1}) < 0$,

$$(a_{j_1j_2 1}, a_{j_2j_1 1}) = \underbrace{(\text{LS}, \text{UP})}_{\text{up-parabola lower-saddle}}.$$

The equilibrium of $(x_{j_2}^*, x_{j_1}^*) = (a_{j_1j_2 1}, a_{j_2j_1 1})$ is an (LS,UP)-up-parabola lower-saddle.

(vi$_{1d}$) For $\bar{x}_{j_2} = a_{j_2j_2 s_1}$, from the proof of (vi$_{1c}$), the equilibrium of $(x_{j_1}^*, x_{j_2}^*) = (a_{j_2j_1 1}, a_{j_2j_2 s_1})$ $(s_1, s_2 \in \{1, 2\}; s_1 \neq s_2)$ with $a_{j_1j_2 1} \neq a_{j_2j_2 1}, a_{j_2j_2 2}$ has the following properties as in Eqs. (1.152)–(1.155):

- For $a_{j_1j_1 0}(a_{j_2j_1 1} - a_{j_1j_1 1})(a_{j_2j_2 s_1} - a_{j_1j_2 1})^2 > 0$ and $a_{j_2j_2 0}(a_{j_2j_2 s_1} - a_{j_2j_2 s_2}) > 0$,

$$(a_{j_2j_1 1}, a_{j_2j_2 s_1}) = \underbrace{(\text{pF}, \text{DP} : \text{UP})}_{\text{hyperbolic flow } (+)}.$$

The equilibrium of $(x_{j_1}^*, x_{j_2}^*) = (a_{j_2j_1 1}, a_{j_2j_2 s_1})$ is a (pF, DP:UP)-positive hyperbolic flow.

- For $a_{j_1j_1 0}(a_{j_2j_1 1} - a_{j_1j_1 1})(a_{j_2j_2 s_1} - a_{j_1j_2 1})^2 < 0$ and $a_{j_2j_2 0}(a_{j_2j_2 s_1} - a_{j_2j_2 s_2}) > 0$,

$$(a_{j_2j_1 1}, a_{j_2j_2 s_1}) = \underbrace{(\text{nF}, \text{UP} : \text{DP})}_{\text{hyperbolic-secant flow } (-)}.$$

The equilibrium of $(x_{j_1}^*, x_{j_2}^*) = (a_{j_2j_1 1}, a_{j_2j_2 s_1})$ is an (nF, UP:DP)-negative hyperbolic-secant flow.

- For $a_{j_1j_1 0}(a_{j_2j_1 1} - a_{j_1j_1 1})(a_{j_2j_2 s_1} - a_{j_1j_2 1})^2 > 0$ and $a_{j_2j_2 0}(a_{j_2j_2 s_1} - a_{j_2j_2 s_2}) < 0$,

$$(a_{j_2j_1 1}, a_{j_2j_2 s_1}) = \underbrace{(\text{pF}, \text{UP} : \text{DP})}_{\text{hyperbolic-secant flow } (+)}.$$

The equilibrium of $(x_{j_1}^*, x_{j_2}^*) = (a_{j_2j_1 1}, a_{j_2j_2 s_1})$ is a (pF, UP:DP)-negative hyperbolic-secant flow.

- For $a_{j_1j_1 0}(a_{j_2j_1 1} - a_{j_1j_1 1})(a_{j_2j_2 s_1} - a_{j_1j_2 1})^2 < 0$ and $a_{j_2j_2 0}(a_{j_2j_2 s_1} - a_{j_2j_2 s_2}) < 0$,

$$(a_{j_2j_1 1}, a_{j_2j_2s_1}) = \underbrace{(\text{nF}, \text{DP}:\text{UP})}_{\text{hyperbolic flow } (-)}.$$

The equilibrium of $(x^*_{j_1}, x^*_{j_2}) = (a_{j_2j_1 1}, a_{j_2j_2s_1})$ is an (nF, DP:UP)-negative hyperbolic flow.

(vi$_2$) At $a_{j_2j_2s_1} = a_{j_1j_2 1}$ $(s_1, s_2 \in \{1,2\}, s_1 \neq s_2)$, we have

$$\frac{dx_{j_1}}{dx_{j_2}} = \frac{a_{j_1j_1 0}}{a_{j_2j_2 0}} \frac{(x_{j_1} - a_{j_1j_1 1})(x_{j_2} - a_{j_1j_2 1})}{(x_{j_1} - a_{j_2j_1 1})(x_{j_2} - a_{j_2j_2s_2})},$$

and the deformation of the foregoing equation is

$$[1 + \frac{(a_{j_1j_1 1} - a_{j_2j_1 1})}{(x_{j_1} - a_{j_1j_1 1})}]dx_{j_1} = \frac{a_{j_1j_1 0}}{a_{j_2j_2 0}}[1 + \frac{(a_{j_2j_2s_2} - a_{j_1j_2 1})}{(x_{j_2} - a_{j_2j_2s_2})}]dx_{j_2}.$$

With an initial condition of $(x_{j_1 0}, x_{j_2 0})$ at $t = t_0$, the integration of the above equation gives the first integral manifold $(a_{j_2j_2s_1} = a_{j_1j_2 1})$ as

$$\begin{aligned}&[(x_{j_1} - a_{j_1j_1 1}) - (x_{j_1 0} - a_{j_1j_1 1})] + (a_{j_1j_1 1} - a_{j_2j_1 1}) \ln \frac{|x_{j_1} - a_{j_1j_1 1}|}{|x_{j_1 0} - a_{j_1j_1 1}|}\\ &= \frac{a_{j_1j_1 0}}{a_{j_2j_2 0}}[(x_{j_2} - x_{j_2 0}) + (a_{j_2j_2s_2} - a_{j_1j_2 1}) \ln \frac{|x_{j_2} - a_{j_2j_2s_2}|}{|x_{j_2 0} - a_{j_2j_2s_2}|}].\end{aligned}$$

(vi$_{2a}$) At $x^*_{j_2} = a_{j_1j_2 1} = a_{j_2j_2s_1}$ with $\bar{x}_{j_1} \neq a_{j_1j_1 1}, a_{j_2j_1 1}$,

$$\frac{dx_{j_1}}{dx_{j_2}}\Big|_{x^*_{j_2} = a_{j_1j_2 1}} = \frac{a_{j_1j_1 0}}{a_{j_2j_2 0}} \frac{(\bar{x}_{j_1} - a_{j_1j_1 1})(x_{j_2} - a_{j_1j_2 1})}{(\bar{x}_{j_1} - a_{j_2j_1 1})(x_{j_2} - a_{j_2j_2s_2})}\Big|_{x^*_{j_2} = a_{j_1j_2 1}} = 0.$$

If

$$\frac{d^2x_{j_1}}{dx^2_{j_2}}\Big|_{x^*_{j_2} = a_{j_1j_2 1}} = \frac{a_{j_1j_1 0}}{a_{j_2j_2 0}} \frac{(\bar{x}_{j_1} - a_{j_1j_1 1})}{(\bar{x}_{j_1} - a_{j_2j_1 1})(a_{j_1j_2 1} - a_{j_2j_2s_2})} > 0,$$

there is an up-parabola flow at $x^*_{j_2} = a_{j_1j_2 1} = a_{j_2j_2s_1}$ with $\bar{x}_{j_1} \neq a_{j_1j_1 1}$. If

$$\frac{d^2x_{j_1}}{dx^2_{j_2}}\Big|_{x^*_{j_2} = a_{j_1j_2 1}} = \frac{a_{j_1j_1 0}}{a_{j_2j_2 0}} \frac{(\bar{x}_{j_1} - a_{j_1j_1 1})}{(\bar{x}_{j_1} - a_{j_2j_1 1})(a_{j_1j_2 1} - a_{j_2j_2s_2})} < 0,$$

there is a down-parabola flow at $x^*_{j_2} = a_{j_1j_2 1} = a_{j_2j_2s_1}$ with $\bar{x}_{j_1} \neq a_{j_1j_1 1}$. The corresponding variational equation is

$$\Delta\dot{x}_{j_2} = a_{j_2j_20}(\overline{x}_{j_1} - a_{j_2j_11})(a_{j_2j_2s_1} - a_{j_2j_2s_2})\Delta x_{j_2},$$

The parabola flows at $x_{j_2}^* = a_{j_1j_21} = a_{j_2j_2s_1}$ with $\overline{x}_{j_1} \neq a_{j_1j_11}, a_{j_2j_11}$ are source and sink for $a_{j_2j_20}(\overline{x}_{j_1} - a_{j_2j_11})(a_{j_2j_2s_1} - a_{j_2j_2s_2}) > 0$ and $a_{j_2j_20}(\overline{x}_{j_1} - a_{j_2j_11})(a_{j_2j_2s_1} - a_{j_2j_2s_2}) < 0$, respectively.

Therefore, the infinite-equilibrium of $x_{j_2}^* = a_{j_1j_21} = a_{j_2j_2s_1}$ with $\overline{x}_{j_1} \neq a_{j_1j_11}$ has the following properties as in Eqs. (1.157)–(1.160):

- For $a_{j_1j_10}(\overline{x}_{j_1} - a_{j_1j_11}) > 0$ and $a_{j_2j_20}(\overline{x}_{j_1} - a_{j_2j_11})(a_{j_2j_2s_1} - a_{j_2j_2s_2}) > 0$,

$$(\overline{x}_{j_1}, a_{j_2j_11}) = \underbrace{(\text{UP}, \text{SO})}_{\text{up-parabola-source}}.$$

  The infinite-equilibrium of $x_{j_2}^* = a_{j_1j_21} = a_{j_2j_2s_1}$ is a (UP,SO)-up-parabola-source.
- For $a_{j_1j_10}(\overline{x}_{j_1} - a_{j_1j_11}) < 0$ and $a_{j_2j_20}(\overline{x}_{j_1} - a_{j_2j_11})(a_{j_2j_2s_1} - a_{j_2j_2s_2}) > 0$,

$$(\overline{x}_{j_1}, a_{j_2j_2s_1}) = \underbrace{(\text{DP}, \text{SO})}_{\text{down-parabola-source}}.$$

  The infinite-equilibrium of $x_{j_2}^* = a_{j_1j_21} = a_{j_2j_2s_1}$ is a (DP,SO)-down-parabola-source.
- For $a_{j_1j_10}(\overline{x}_{j_1} - a_{j_1j_11}) > 0$ and $a_{j_2j_20}(\overline{x}_{j_1} - a_{j_2j_11})(a_{j_2j_2s_1} - a_{j_2j_2s_2}) < 0$,

$$(\overline{x}_{j_1}, a_{j_2j_2s_1}) = \underbrace{(\text{DP}, \text{SI})}_{\text{down-parabola-sink}}.$$

  The infinite-equilibrium of $x_{j_2}^* = a_{j_1j_21} = a_{j_2j_2s_1}$ is a (DP,SI)-down-parabola-sink.
- For $a_{j_1j_10}(\overline{x}_{j_1} - a_{j_1j_11}) < 0$ and $a_{j_2j_20}(\overline{x}_{j_1} - a_{j_2j_11})(a_{j_2j_2s_1} - a_{j_2j_2s_2}) < 0$,

$$(\overline{x}_{j_1}, a_{j_2j_2s_1}) = \underbrace{(\text{UP}, \text{SI})}_{\text{up-parabola-sink}}.$$

  The infinite-equilibrium of $x_{j_2}^* = a_{j_1j_21} = a_{j_2j_2s_1}$ is a (UP,SI)-up-parabola-sink.

($\text{vi}_{2\text{b}}$) From case ($\text{vi}_{2\text{a}}$), for $\overline{x}_{j_1} = a_{j_1j_11}$, thus, the equilibrium of $(x_{j_1}^*, x_{j_2}^*) = (a_{j_1j_11}, a_{j_1j_21})$ with $a_{j_2j_2s_1} = a_{j_1j_21}$ has the following properties as in Eqs. (1.161)–(1.164):

- For $a_{j_1j_10} > 0$ and $a_{j_2j_20}(a_{j_1j_11} - a_{j_2j_11})(a_{j_2j_2s_1} - a_{j_2j_2s_2}) > 0$,

$$(a_{j_1j_1}1, a_{j_1j_2}1) = \underbrace{(\text{DP}:\text{UP},\text{SO})}_{\text{hyperbolic source}}.$$

The equilibrium of $(x^*_{j_1}, x^*_{j_2}) = (a_{j_1j_1}1, a_{j_1j_2}1)$ is a (DP:UP,SO)-hyperbolic source.

- For $a_{j_1j_1}0 < 0$ and $a_{j_2j_2}0(a_{j_1j_1}1 - a_{j_2j_1}1)(a_{j_2j_2s_1} - a_{j_2j_2s_2}) > 0$,

$$(a_{j_1j_1}1, a_{j_1j_2}1) = \underbrace{(\text{UP}:\text{DP},\text{SO})}_{\text{hyperbolic-secant source}}.$$

The equilibrium of $(x^*_{j_1}, x^*_{j_2}) = (a_{j_1j_1}1, a_{j_1j_2}1)$ is a (UP:DP,SO)-hyperbolic-secant source.

- For $a_{j_1j_1}0 > 0$ and $a_{j_2j_2}0(a_{j_1j_1}1 - a_{j_2j_1}1)(a_{j_2j_2s_1} - a_{j_2j_2s_2}) < 0$,

$$(a_{j_1j_1}1, a_{j_1j_2}1) = \underbrace{(\text{UP}:\text{DP},\text{SI})}_{\text{hyperbolic-secant sink}}.$$

The equilibrium of $(x^*_{j_1}, x^*_{j_2}) = (a_{j_1j_1}1, a_{j_1j_2}1)$ is a (UP:DP,SI)-hyperbolic-secant sink.

- For $a_{j_1j_1}0 < 0$ and $a_{j_2j_2}0(a_{j_1j_1}1 - a_{j_2j_1}1)(a_{j_2j_2s_1} - a_{j_2j_2s_2}) < 0$,

$$(a_{j_1j_1}1, a_{j_1j_2}1) = \underbrace{(\text{DP}:\text{UP},\text{SI})}_{\text{hyperbolic sink}}.$$

The equilibrium of $(x^*_{j_1}, x^*_{j_2}) = (a_{j_1j_1}1, a_{j_1j_2}1)$ is a (DP:UP,SI)-hyperbolic sink.

(vi$_{2c}$) From case (vi$_{2a}$), for $\bar{x}_{j_1} = a_{j_2j_1}1$, thus, the equilibrium of $(x^*_{j_2}, x^*_{j_1}) = (a_{j_1j_2}1, a_{j_2j_1}1)$ with $a_{j_2j_2s_1} = a_{j_1j_2}1$ has the following properties as in Eqs. (1.165)–(1.168):

- For $a_{j_1j_1}0(a_{j_2j_1}1 - a_{j_1j_1}1) > 0$ and $a_{j_2j_2}0(a_{j_2j_2s_1} - a_{j_2j_2s_2}) > 0$,

$$(a_{j_1j_2}1, a_{j_2j_1}1) = \underbrace{(\text{DP}_+ : \text{UP}_+, {}_{\text{DP}}\text{SI}:{}_{\text{UP}}\text{SO})}_{\text{hyperbolic sink-to-source}}.$$

The equilibrium of $(x^*_{j_2}, x^*_{j_1}) = (a_{j_1j_2}1, a_{j_2j_1}1)$ is a (DP$_+$:UP$_+$, $_{\text{DP}}$SI:$_{\text{UP}}$SO)-hyperbolic sink-to-source.

- For $a_{j_1j_1}0(a_{j_2j_1}1 - a_{j_1j_1}1) < 0$ and $a_{j_2j_2}0(a_{j_2j_2s_1} - a_{j_2j_2s_2}) > 0$,

$$(a_{j_1j_21}, a_{j_2j_11}) = \underbrace{(\mathrm{UP}_- : \mathrm{DP}_-, {}_{\mathrm{UP}}\mathrm{SI}{:}_{\mathrm{DP}}\mathrm{SO})}_{\text{circular sink-to-source}}.$$

The equilibrium of $(x^*_{j_2}, x^*_{j_1}) = (a_{j_1j_21}, a_{j_2j_11})$ is a $(\mathrm{UP}_-{:}\mathrm{DP}_-, {}_{\mathrm{UP}}\mathrm{SI}{:}_{\mathrm{DP}}\mathrm{SO})$-circular sink-to-source.

- For $a_{j_1j_10}(a_{j_2j_11} - a_{j_1j_11}) > 0$ and $a_{j_2j_20}(a_{j_2j_2s_1} - a_{j_2j_2s_2}) < 0$,

$$(a_{j_1j_21}, a_{j_2j_11}) = \underbrace{(\mathrm{UP}_+ : \mathrm{DP}_+, {}_{\mathrm{UP}}\mathrm{SO}{:}_{\mathrm{DP}}\mathrm{SI})}_{\text{circular source-to-sink}}.$$

The equilibrium of $(x^*_{j_2}, x^*_{j_1}) = (a_{j_1j_21}, a_{j_2j_11})$ is a $(\mathrm{UP}_+{:}\mathrm{DP}_+, {}_{\mathrm{UP}}\mathrm{SO}{:}_{\mathrm{DP}}\mathrm{SI})$-circular source-to-sink.

- For $a_{j_1j_10}(a_{j_2j_11} - a_{j_1j_11}) < 0$ and $a_{j_2j_20}(a_{j_2j_2s_1} - a_{j_2j_2s_2}) < 0$,

$$(a_{j_1j_21}, a_{j_2j_11}) = \underbrace{(\mathrm{DP}_- : \mathrm{UP}_-, {}_{\mathrm{DP}}\mathrm{SO}{:}_{\mathrm{UP}}\mathrm{SI})}_{\text{hyperbolic source-to-sink}}.$$

The equilibrium of $(x^*_{j_2}, x^*_{j_1}) = (a_{j_1j_21}, a_{j_2j_11})$ is a $(\mathrm{DP}_-{:}\mathrm{UP}_-, {}_{\mathrm{DP}}\mathrm{SO}{:}_{\mathrm{UP}}\mathrm{SI})$-hyperbolic source-to-sink.

(vi$_3$) For $a_{j_1j_11} = a_{j_2j_11}$ and $a_{j_2j_2s_1} \neq a_{j_1j_21}$, in phase space,

$$\frac{dx_{j_1}}{dx_{j_2}} = \frac{a_{j_1j_10}}{a_{j_2j_20}} \frac{(x_{j_2} - a_{j_1j_21})^2}{(x_{j_2} - a_{j_2j_21})(x_{j_2} - a_{j_2j_22})},$$

and the deformation of the foregoing equation is

$$dx_{j_1} = \frac{a_{j_1j_10}}{a_{j_2j_20}} \Big[1 + \frac{(a_{j_2j_21} - a_{j_1j_21})^2}{(a_{j_2j_21} - a_{j_2j_22})(x_{j_2} - a_{j_2j_21})} + \frac{(a_{j_2j_22} - a_{j_1j_21})^2}{(a_{j_2j_22} - a_{j_2j_21})(x_{j_2} - a_{j_2j_22})}\Big] dx_{j_2}.$$

With an initial condition of (x_{j_10}, x_{j_20}) at $t = t_0$, the integration of the above equation gives the first integral manifold ($a_{j_1j_11} = a_{j_2j_11}$ and $a_{j_2j_2s_1} \neq a_{j_1j_21}$) as

$$\big[(x_{j_1} - a_{j_1j_11}) - (x_{j_10} - a_{j_1j_11})\big] = \frac{a_{j_1j_10}}{a_{j_2j_20}} \Big\{ \big[(x_{j_2} - a_{j_2j_21}) - (x_{j_20} - a_{j_2j_21})\big] + \sum\nolimits_{s_1=1}^{2} \frac{(a_{j_2j_2s_1} - a_{j_1j_21})^2}{(a_{j_2j_2s_1} - a_{j_2j_2s_2})} \ln \frac{|x_{j_2} - a_{j_2j_2s_1}|}{|x_{j_20} - a_{j_2j_2s_1}|} \Big\}.$$

(vi$_{3a}$) For $x_{j_1}^* = a_{j_1j_11} = a_{j_2j_11}$ with $\overline{x}_{j_2} \neq a_{j_2j_2s_1}, a_{j_1j_21}$, if

$$\left.\frac{dx_{j_2}}{dx_{j_1}}\right|_{x_{j_1}^*=a_{j_1j_11}} = \frac{a_{j_2j_20}}{a_{j_1j_10}}\frac{(\overline{x}_{j_2}-a_{j_2j_21})(\overline{x}_{j_2}-a_{j_2j_22})}{(\overline{x}_{j_2}-a_{j_1j_21})^2} > 0,$$

there is an increasing-inflection flow, if

$$\left.\frac{dx_{j_2}}{dx_{j_1}}\right|_{x_{j_1}^*=a_{j_1j_11}} = \frac{a_{j_2j_20}}{a_{j_1j_10}}\frac{(\overline{x}_{j_2}-a_{j_2j_21})(\overline{x}_{j_2}-a_{j_2j_22})}{(\overline{x}_{j_2}-a_{j_1j_21})^2} < 0,$$

there is a decreasing-inflection flow. The variational equation is

$$\Delta\dot{x}_{j_1} = a_{j_1j_10}(\overline{x}_{j_2}-a_{j_1j_21})^2\Delta x_{j_1}.$$

The inflection flows for $a_{j_1j_10}(\overline{x}_{j_2}-a_{j_1j_21})^2 > 0$ and $a_{j_1j_10}(\overline{x}_{j_2}-a_{j_1j_21})^2 < 0$ are source and sink, respectively.

Therefore, the infinite-equilibrium of $x_{j_1}^* = a_{j_1j_11} = a_{j_2j_11}$ with $\overline{x}_{j_2} \neq a_{j_2j_2s_1}, a_{j_1j_21}$ has the following properties as in Eqs. (1.170)–(1.173):

- For $a_{j_1j_10}(\overline{x}_{j_2}-a_{j_1j_21})^2 > 0$ and $a_{j_2j_20}\prod_{s_1=1}^2(\overline{x}_{j_2}-a_{j_2j_2s_1}) > 0$,

$$(a_{j_1j_11}, \overline{x}_{j_2}) = \underbrace{(\text{SO}, \text{II})}_{\text{increasing-inflection source}}.$$

  The infinite-equilibrium of $x_{j_1}^* = a_{j_1j_11} = a_{j_2j_11}$ is an (SO,II)-increasing-inflection source.
- For $a_{j_1j_10}(\overline{x}_{j_2}-a_{j_1j_21})^2 < 0$ and $a_{j_2j_20}\prod_{s_1=1}^2(\overline{x}_{j_2}-a_{j_2j_2s_1}) > 0$,

$$(a_{j_1j_11}, \overline{x}_{j_2}) = \underbrace{(\text{SI}, \text{II})}_{\text{decreasing-inflection sink}}.$$

  The infinite-equilibrium of $x_{j_1}^* = a_{j_1j_11} = a_{j_2j_11}$ is an (SI,DI)-decreasing-inflection sink.
- For $a_{j_1j_10}(\overline{x}_{j_2}-a_{j_1j_21})^2 > 0$ and $a_{j_2j_20}\prod_{s_1=1}^2(\overline{x}_{j_2}-a_{j_2j_2s_1}) < 0$,

$$(a_{j_1j_11}, \overline{x}_{j_2}) = \underbrace{(\text{SO}, \text{DI})}_{\text{decreasing-inflection source}}.$$

  The infinite-equilibrium of $x_{j_2}^* = a_{j_1j_21} = a_{j_2j_2s_1}$ is an (SO,DI)-decreasing-inflection source.
- For $a_{j_1j_10}(\overline{x}_{j_2}-a_{j_1j_21})^2 < 0$ and $a_{j_2j_20}\prod_{s_1=1}^2(\overline{x}_{j_2}-a_{j_2j_2s_1}) < 0$,

$$(a_{j_1j_11}, \bar{x}_{j_2}) = \underbrace{(\text{SI}, \text{II})}_{\text{increasing-inflection sink}}.$$

The infinite-equilibrium of $x^*_{j_1} = a_{j_1j_11} = a_{j_2j_11}$ is an (SI,II)-increasing-inflection sink.

($\text{vi}_{3\text{b}}$) From ($\text{vi}_{3\text{a}}$), for $\bar{x}_{j_2} = a_{j_1j_21}$, the equilibrium of $(x^*_{j_1}, x^*_{j_2}) = (a_{j_1j_11}, a_{j_1j_21})$ with $a_{j_1j_11} = a_{j_2j_11}$ has the following properties as in Eqs. (1.174)–(1.177):

- For $a_{j_1j_10} > 0$ and $a_{j_2j_20}\prod_{s_1=1}^{2}(a_{j_1j_21} - a_{j_2j_2s_1}) > 0$,

$$(a_{j_1j_11}, a_{j_1j_21}) = \underbrace{(\text{SO}, \ \text{II} : \text{II})}_{\text{increasing-inflection source}}.$$

  The equilibrium of $(x^*_{j_1}, x^*_{j_2}) = (a_{j_1j_11}, a_{j_1j_21})$ is an (SO,II:II)-increasing-inflection source of the first kind.
- For $a_{j_1j_10} < 0$ and $a_{j_2j_20}\prod_{s_1=1}^{2}(a_{j_1j_21} - a_{j_2j_2s_1}) > 0$,

$$(a_{j_1j_11}, a_{j_1j_21}) = \underbrace{(\text{SI}, \ \text{DI} : \text{DI})}_{\text{decreasing-inflection sink}}.$$

  The equilibrium of $(x^*_{j_1}, x^*_{j_2}) = (a_{j_1j_11}, a_{j_1j_21})$ is an (SI,DI:DI)-decreasing-inflection sink of the first kind.
- For $a_{j_1j_10} > 0$ and $a_{j_2j_20}\prod_{s_1=1}^{2}(a_{j_1j_21} - a_{j_2j_2s_1}) < 0$,

$$(a_{j_1j_11}, a_{j_1j_21}) = \underbrace{(\text{SO}, \ \text{DI} : \text{DI})}_{\text{decreasing-inflection source}}.$$

  The equilibrium of $(x^*_{j_1}, x^*_{j_2}) = (a_{j_1j_11}, a_{j_1j_21})$ is an (SO,DI:DI)-decreasing-inflection source of the first kind.
- For $a_{j_1j_10} < 0$ and $a_{j_2j_20}\prod_{s_1=1}^{2}(a_{j_1j_21} - a_{j_2j_2s_1}) < 0$,

$$(a_{j_1j_11}, a_{j_1j_21}) = \underbrace{(\text{SI}, \ \text{II} : \text{II})}_{\text{increasing-inflection sink}}.$$

  The equilibrium of $(x^*_{j_1}, x^*_{j_2}) = (a_{j_1j_11}, a_{j_1j_21})$ is an (SI,II:II)-increasing-inflection sink of the first kind.

($\text{vi}_{3\text{c}}$) From case ($\text{vi}_{3\text{a}}$), for $\bar{x}_{j_2} = a_{j_2j_2s_1}$, the equilibrium of $(x^*_{j_1}, x^*_{j_2}) = (a_{j_1j_11}, a_{j_2j_2s_1})$ with $a_{j_2j_11} = a_{j_1j_11}$ has the following properties as in Eqs. (1.178)–(1.181):

- For $a_{j_1j_10}(a_{j_2j_2s_1}-a_{j_1j_21})^2>0$ and $a_{j_2j_20}(a_{j_2j_2s_1}-a_{j_2j_2s_2})>0$,

$$(a_{j_1j_11},a_{j_2j_2s_1})=\underbrace{(\mathrm{SO},\mathrm{DI}:\mathrm{II})}_{\text{inflection source}}.$$

The equilibrium of $(x_{j_1}^*,x_{j_2}^*)=(a_{j_1j_11},a_{j_2j_2s_1})$ is an (SO,DI:II)-inflection source.

- For $a_{j_1j_10}(a_{j_2j_2s_1}-a_{j_1j_21})^2<0$ and $a_{j_2j_20}(a_{j_2j_2s_1}-a_{j_2j_2s_2})>0$,

$$(a_{j_1j_11},a_{j_2j_2s_1})=\underbrace{(\mathrm{SI},\mathrm{II}:\mathrm{DI})}_{\text{inflection sink}}.$$

The equilibrium of $(x_{j_1}^*,x_{j_2}^*)=(a_{j_1j_11},a_{j_2j_2s_1})$ is an (SI,II:DI)-inflection sink.

- For $a_{j_1j_10}(a_{j_2j_2s_1}-a_{j_1j_21})^2>0$ and $a_{j_2j_20}(a_{j_2j_2s_1}-a_{j_2j_2s_2})<0$,

$$(a_{j_1j_11},a_{j_2j_2s_1})=\underbrace{(\mathrm{SO},\mathrm{II}:\mathrm{DI})}_{\text{inflection source}}.$$

The equilibrium of $(x_{j_1}^*,x_{j_2}^*)=(a_{j_1j_11},a_{j_2j_2s_1})$ is an (SO,II:DI)-inflection source.

- For $a_{j_1j_10}(a_{j_2j_2s_1}-a_{j_1j_21})^2<0$ and $a_{j_2j_20}(a_{j_2j_2s_1}-a_{j_2j_2s_2})<0$,

$$(a_{j_1j_11},a_{j_2j_2s_1})=\underbrace{(\mathrm{SI},\mathrm{DI}:\mathrm{II})}_{\text{inflection sink}}.$$

The equilibrium of $(x_{j_1}^*,x_{j_2}^*)=(a_{j_1j_11},a_{j_2j_2s_1})$ is an (SI,DI:II)-inflection sink.

(vi$_4$) For $a_{j_1j_11}=a_{j_2j_11}$ and $a_{j_2j_2s_1}=a_{j_1j_21}$, in phase space,

$$\frac{dx_{j_1}}{dx_{j_2}}=\frac{a_{j_1j_10}}{a_{j_2j_20}}\frac{(x_{j_2}-a_{j_1j_21})}{(x_{j_2}-a_{j_2j_2s_2})},$$

and the deformation of the foregoing equation is

$$dx_{j_1}=\frac{a_{j_1j_10}}{a_{j_2j_20}}\left[1+\frac{(a_{j_2j_2s_2}-a_{j_1j_21})}{(x_{j_2}-a_{j_2j_2s_2})}\right]dx_{j_2}.$$

With an initial condition of (x_{j_10},x_{j_20}) at $t=t_0$, the integration of the above equation gives the first integral manifold ($a_{j_1j_11}=a_{j_2j_11}$ and $a_{j_2j_2s_1}=a_{j_1j_21}$) as

$$\begin{aligned}\left[(x_{j_1}-a_{j_1j_11})-(x_{j_10}-a_{j_1j_11})\right]=\frac{a_{j_1j_10}}{a_{j_2j_20}}\Big\{&\left[(x_{j_2}-a_{j_2j_21})-(x_{j_20}-a_{j_2j_21})\right]\\&+(a_{j_2j_2s_2}-a_{j_1j_21})\ln\frac{|x_{j_2}-a_{j_2j_2s_2}|}{|x_{j_20}-a_{j_2j_2s_2}|}\Big\}.\end{aligned}$$

From (vi$_{2a}$), for $\overline{x}_{j_1} = a_{j_1j_1 1} = a_{j_2j_1 1}$, the equilibrium of $(x^*_{j_1}, x^*_{j_2}) = (a_{j_1j_1 1}, a_{j_1j_2 1})$ with $a_{j_1j_1 1} = a_{j_2j_1 1}$ and $a_{j_1j_2 1} = a_{j_2j_2 s_1}$ has the following properties as in Eqs. (1.183)–(1.186)

- For $a_{j_1j_1 0} > 0$ and $a_{j_2j_2 0}(a_{j_2j_2 s_1} - a_{j_2j_2 s_2}) > 0$,

$$(a_{j_1j_1 1}, a_{j_1j_2 1}) = \underbrace{({}_{\mathrm{DI:II}}\mathrm{SO},\ {}_{\mathrm{SI:SO}}\mathrm{UP})}_{\text{up-parabola-source}}.$$

  The equilibrium of $(x^*_{j_1}, x^*_{j_2}) = (a_{j_1j_1 1}, a_{j_1j_2 1})$ is a $({}_{\mathrm{DI:II}}\mathrm{SO},\ {}_{\mathrm{SI:SO}}\mathrm{UP})$-up-parabola-source.
- For $a_{j_1j_1 0} < 0$ and $a_{j_2j_2 0}(a_{j_2j_2 s_1} - a_{j_2j_2 s_2}) > 0$,

$$(a_{j_1j_1 1}, a_{j_1j_2 1}) = \underbrace{({}_{\mathrm{II:DI}}\mathrm{SI},\ {}_{\mathrm{SO:SI}}\mathrm{DP})}_{\text{down-parabola-sink}}.$$

  The equilibrium of $(x^*_{j_1}, x^*_{j_2}) = (a_{j_1j_1 1}, a_{j_1j_2 1})$ is an $({}_{\mathrm{II:DI}}\mathrm{SI},\ {}_{\mathrm{SO:SI}}\mathrm{DP})$-down-parabola-sink.
- For $a_{j_1j_1 0} > 0$ and $a_{j_2j_2 0}(a_{j_2j_2 s_1} - a_{j_2j_2 s_2}) < 0$,

$$(a_{j_1j_1 1}, a_{j_1j_2 1}) = \underbrace{({}_{\mathrm{II:DI}}\mathrm{SO},\ {}_{\mathrm{SI:SO}}\mathrm{DP})}_{\text{down-parabola-source}}.$$

  The equilibrium of $(x^*_{j_1}, x^*_{j_2}) = (a_{j_1j_1 1}, a_{j_1j_2 1})$ is an $({}_{\mathrm{II:DI}}\mathrm{SO},\ {}_{\mathrm{SI:SO}}\mathrm{DP})$-down-parabola-source.
- For $a_{j_1j_1 0} < 0$ and $a_{j_2j_2 0}(a_{j_2j_2 s_1} - a_{j_2j_2 s_2}) < 0$,

$$(a_{j_1j_1 1}, a_{j_1j_2 1}) = \underbrace{({}_{\mathrm{DI:II}}\mathrm{SI},\ {}_{\mathrm{SI:SO}}\mathrm{UP})}_{\text{up-parabola-sink}}.$$

  The equilibrium of $(x^*_{j_1}, x^*_{j_2}) = (a_{j_1j_1 1}, a_{j_1j_2 1})$ is a $({}_{\mathrm{DI:II}}\mathrm{SI},\ {}_{\mathrm{SO:SI}}\mathrm{UP})$-up-parabola-sink.

  The above conclusion can be obtained from case (vi$_{3a}$) for $\overline{x}_{j_2} = a_{j_2j_2 s_1} = a_{j_1j_2 1}$.

(vii) For $\Delta_{j_1j_2} > 0$ and $\Delta_{j_2j_2} < 0$, the standard form is

$$\begin{aligned}
\dot{x}_{j_2} &= a_{j_2j_2 0}(x_{j_1} - a_{j_2j_1 1})\left[(x_{j_2} - a_{j_2j_2})^2 + b_{j_2j_2}\right],\\
\dot{x}_{j_1} &= a_{j_1j_1 0}(x_{j_1} - a_{j_1j_1 1})(x_{j_2} - a_{j_1j_2 1})(x_{j_2} - a_{j_1j_2 2})
\end{aligned}$$

where

$$b_{j_1j_21}, b_{j_1j_22} = -\frac{1}{2}(B_{j_1j_2} \pm \sqrt{\Delta_{j_1j_2}}),$$
$$\{a_{j_1j_21}, a_{j_1j_22}\} = \mathrm{sort}\{b_{j_1j_21}, b_{j_1j_22}\}, a_{j_1j_21} < a_{j_1j_22};$$
$$a_{j_2j_2} = -\frac{1}{2}B_{j_2j_2}, b_{j_2j_2} = \frac{1}{4}(-\Delta_{j_2j_2}).$$

(vii$_1$) In phase space,

$$\frac{dx_{j_1}}{dx_{j_2}} = \frac{a_{j_1j_10}}{a_{j_2j_20}} \frac{(x_{j_1} - a_{j_1j_11})(x_{j_2} - a_{j_1j_21})(x_{j_2} - a_{j_1j_22})}{(x_{j_1} - a_{j_2j_11})\left[(x_{j_2} - a_{j_2j_2})^2 + b_{j_2j_2}\right]},$$

and the deformation of the foregoing equation is

$$\begin{aligned}\left[1 + \frac{(a_{j_1j_11} - a_{j_2j_11})}{(x_{j_1} - a_{j_1j_11})}\right] dx_{j_1} = \frac{a_{j_1j_10}}{a_{j_2j_20}} \Big[& 1 + \frac{(a_{j_2j_2} - a_{j_1j_21})(x_{j_2} - a_{j_2j_2})}{(x_{j_2} - a_{j_2j_2})^2 + b_{j_2j_2}} \\ & + \frac{(a_{j_2j_2} - a_{j_1j_22})(x_{j_2} - a_{j_2j_2})}{(x_{j_2} - a_{j_2j_2})^2 + b_{j_2j_2}} \\ & + \frac{(a_{j_2j_2} - a_{j_1j_21})(a_{j_2j_2} - a_{j_1j_22}) - b_{j_2j_2}}{(x_{j_2} - a_{j_2j_2})^2 + b_{j_2j_2}} \Big] dx_{j_2}.\end{aligned}$$

With an initial condition of (x_{j_10}, x_{j_20}) at $t = t_0$, the integration of the above equation gives the first integral manifold as

$$\begin{aligned}&\left[(x_{j_1} - a_{j_1j_11}) - (x_{j_10} - a_{j_1j_11})\right] + (a_{j_1j_11} - a_{j_2j_11}) \ln \frac{|x_{j_1} - a_{j_1j_11}|}{|x_{j_10} - a_{j_1j_11}|} \\ &= \frac{a_{j_1j_10}}{a_{j_2j_20}} \Big\{ \left[(x_{j_2} - a_{j_2j_2}) - (x_{j_20} - a_{j_2j_2})\right] \\ &\quad + \sum\nolimits_{s_1=1}^{2} (a_{j_2j_2} - a_{j_1j_2s_1}) \ln \frac{|(x_{j_2} - a_{j_2j_2})^2 + b_{j_2j_2}|}{|(x_{j_20} - a_{j_2j_2})^2 + b_{j_2j_2}|} \\ &\quad + \left[\prod\nolimits_{s_1=1}^{2} (a_{j_2j_2} - a_{j_1j_2s_1}) - b_{j_2j_2}\right] \\ &\quad \times \frac{1}{\sqrt{b_{j_2j_2}}} \left(\arctan \frac{x_{j_2} - a_{j_2j_2}}{\sqrt{b_{j_2j_2}}} - \arctan \frac{x_{j_20} - a_{j_2j_2}}{\sqrt{b_{j_2j_2}}} \right) \Big\}.\end{aligned}$$

(vii$_{1a}$) For $x_{j_2}^* = a_{j_1j_2s_1}$ and $\overline{x}_{j_1} \neq a_{j_1j_1 1}$ $(s_1, s_2 \in \{1,2\}, s_1 \neq s_2)$, in phase space,

$$\frac{dx_{j_1}}{dx_{j_2}}\bigg|_{x_{j_2}^*=a_{j_1j_2s_1}} = \frac{a_{j_1j_10}}{a_{j_2j_20}} \frac{(\overline{x}_{j_1} - a_{j_1j_11})(x_{j_2} - a_{j_1j_2s_1})(x_{j_2} - a_{j_1j_2s_2})}{(\overline{x}_{j_1} - a_{j_2j_11})(x_{j_2} - a_{j_2j_2})^2 + b_{j_2j_2})}\bigg|_{x_{j_2}^*=a_{j_1j_2s_1}} = 0.$$

If

$$\frac{d^2x_{j_1}}{dx_{j_2}^2}\bigg|_{x_{j_2}^*=a_{j_1j_2s_1}} = \frac{a_{j_1j_10}}{a_{j_2j_20}} \frac{(\overline{x}_{j_1} - a_{j_1j_11})(a_{j_1j_2s_1} - a_{j_1j_2s_2})}{(\overline{x}_{j_1} - a_{j_2j_11})\left[(x_{j_2} - a_{j_2j_2})^2 + b_{j_2j_2}\right]} > 0,$$

there is an up-parabola flow at $x_{j_2}^* = a_{j_1j_2s_1}$ with $\overline{x}_{j_1} \neq a_{j_1j_11}$. If

$$\frac{d^2x_{j_1}}{dx_{j_2}^2}\bigg|_{x_{j_2}^*=a_{j_1j_2s_1}} = \frac{a_{j_1j_10}}{a_{j_2j_20}} \frac{(\overline{x}_{j_1} - a_{j_1j_11})(a_{j_1j_2s_1} - a_{j_1j_2s_2})}{(\overline{x}_{j_1} - a_{j_2j_11})\left[(x_{j_2} - a_{j_2j_2})^2 + b_{j_2j_2}\right]} < 0,$$

there is a down-parabola flow at $x_{j_2}^* = a_{j_1j_2s_1}$ with $\overline{x}_{j_1} \neq a_{j_1j_11}$. Let

$$\dot{x}_{j_2} = a_{j_2j_20}(\overline{x}_{j_1} - a_{j_2j_11})\left[(a_{j_1j_2s_1} - a_{j_2j_2})^2 + b_{j_2j_2}\right],$$

with

$$(a_{j_1j_2s_1} - a_{j_2j_2})^2 + b_{j_2j_2} > 0.$$

The inflection flows at $x_{j_2}^* = a_{j_1j_2s_1}$ with $\overline{x}_{j_1} \neq a_{j_1j_11}$ are positive and negative for $a_{j_2j_20} > 0$ and $a_{j_2j_20} < 0$, respectively.

- For $a_{j_1j_10}(\overline{x}_{j_1} - a_{j_1j_11})(a_{j_1j_2s_1} - a_{j_1j_2s_2}) > 0$ and $a_{j_2j_20}(\overline{x}_{j_1} - a_{j_2j_11}) > 0$,

$$(a_{j_1j_2s_1}, \dot{x}_{j_2}) = \underbrace{(\mathrm{UP}, \mathrm{pF})}_{\text{up-parabola flow }(+)}.$$

- For $a_{j_1j_10}(\overline{x}_{j_1} - a_{j_1j_11})(a_{j_1j_2s_1} - a_{j_1j_2s_2}) < 0$ and $a_{j_2j_20}(\overline{x}_{j_1} - a_{j_2j_11}) > 0$,

$$(a_{j_1j_2s_1}, \dot{x}_{j_2}) = \underbrace{(\mathrm{DP}, \mathrm{pF})}_{\text{down-parabola flow }(+)}.$$

- For $a_{j_1j_10}(\overline{x}_{j_1} - a_{j_1j_11})(a_{j_1j_2s_1} - a_{j_1j_2s_2}) > 0$ and $a_{j_2j_20}(\overline{x}_{j_1} - a_{j_2j_11}) < 0$,

$$(a_{j_1j_2s_1}, \dot{x}_{j_2}) = \underbrace{(\mathrm{DP}, \mathrm{nF})}_{\text{down-parabola flow }(-)}.$$

- For $a_{j_1j_1 0}(\overline{x}_{j_1} - a_{j_1j_1 1})(a_{j_1j_2s_1} - a_{j_1j_2s_2}) < 0$ and $a_{j_2j_2 0}(\overline{x}_{j_1} - a_{j_2j_1 1}) < 0$,

$$(a_{j_1j_2s_1}, \dot{x}_{j_2}) = \underbrace{(\text{UP, nF})}_{\text{up-parabola flow }(-)}.$$

From the above analysis, thus, the equilibrium of $(x^*_{j_1}, x^*_{j_2}) = (a_{j_1j_1 1}, a_{j_1j_2s_1})$ $(s_1, s_2 \in \{1, 2\}, s_1 \neq s_2)$ has the following properties as in Eqs. (1.190)–(1.193):

- For $a_{j_1j_1 0}(a_{j_1j_2s_1} - a_{j_1j_2s_2}) > 0$ and $a_{j_2j_2 0}(a_{j_1j_1 1} - a_{j_2j_1 1}) > 0$,

$$(a_{j_1j_1 1}, a_{j_1j_2s_1}) = \underbrace{(\text{DP : UP, pF})}_{\text{hyperbolic flow }(+)}.$$

  The equilibrium of $(x^*_{j_1}, x^*_{j_2}) = (a_{j_1j_1 1}, a_{j_1j_2s_1})$ is a (DP:UP,pF)-positive hyperbolic flow.
- For $a_{j_1j_1 0}(a_{j_1j_2s_1} - a_{j_1j_2s_2}) < 0$ and $a_{j_2j_2 0}(a_{j_1j_1 1} - a_{j_2j_1 1}) > 0$,

$$(a_{j_1j_1 1}, a_{j_1j_2s_1}) = \underbrace{(\text{UP : DP, pF})}_{\text{hyperbolic-secant flow }(+)}.$$

  The equilibrium of $(x^*_{j_1}, x^*_{j_2}) = (a_{j_1j_1 1}, a_{j_1j_2s_1})$ is a (UP:DP,pF)-positive hyperbolic-secant flow.
- For $a_{j_1j_1 0}(a_{j_1j_2s_1} - a_{j_1j_2s_2}) > 0$ and $a_{j_2j_2 0}(a_{j_1j_1 1} - a_{j_2j_1 1}) < 0$,

$$(a_{j_1j_1 1}, a_{j_1j_2s_1}) = \underbrace{(\text{UP : DP, nF})}_{\text{hyperbolic-secant flow }(-)}.$$

  The equilibrium of $(x^*_{j_1}, x^*_{j_2}) = (a_{j_1j_1 1}, a_{j_1j_2s_1})$ is a (UP:DP,nF)-negative hyperbolic-secant flow.
- For $a_{j_1j_1 0}(a_{j_1j_2s_1} - a_{j_1j_2s_2}) < 0$ and $a_{j_2j_2 0}(a_{j_1j_1 1} - a_{j_2j_1 1}) < 0$,

$$(a_{j_1j_1 1}, a_{j_1j_2s_1}) = \underbrace{(\text{DP : UP, nF})}_{\text{hyperbolic flow }(-)}.$$

  The equilibrium of $(x^*_{j_1}, x^*_{j_2}) = (a_{j_1j_1 1}, a_{j_1j_2s_1})$ is a (DP:UP,nF)-negative hyperbolic flow.

(vii$_{1b}$) For $x^*_{j_1} = a_{j_2j_1 1}$ and $\overline{x}_{j_2} \neq a_{j_1j_2s_1}$ $(s_1, s_2 \in \{1, 2\}, s_1 \neq s_2)$, in phase space,

$$\frac{dx_{j_2}}{dx_{j_1}}\bigg|_{x^*_{j_1} = a_{j_2j_1 1}} = \frac{a_{j_2j_2 0}}{a_{j_1j_1 0}} \frac{(x_{j_1} - a_{j_2j_1 1})\left[(\overline{x}_{j_2} - a_{j_2j_2})^2 + b_{j_2j_2}\right]}{(x_{j_1} - a_{j_1j_1 1})\prod_{s_1=1}^{2}(\overline{x}_{j_2} - a_{j_1j_2s_1})}\bigg|_{x^*_{j_1} = a_{j_2j_1 1}} = 0.$$

If

$$\left.\frac{d^2x_{j_2}}{dx_{j_1}^2}\right|_{x_{j_2}^*=a_{j_1j_2s_1}}=\frac{a_{j_2j_20}}{a_{j_1j_10}}\frac{(\overline{x}_{j_2}-a_{j_2j_2})^2+b_{j_2j_2}}{(a_{j_2j_11}-a_{j_1j_11})\prod_{s_1=1}^{2}(\overline{x}_{j_2}-a_{j_2j_2s_1})}>0,$$

there is an up-parabola flow at $x_{j_2}^*=a_{j_1j_2s_1}$ with $\overline{x}_{j_1}\neq a_{j_1j_11}$. If

$$\left.\frac{d^2x_{j_2}}{dx_{j_1}^2}\right|_{x_{j_2}^*=a_{j_1j_2s_1}}=\frac{a_{j_2j_20}}{a_{j_1j_10}}\frac{(\overline{x}_{j_2}-a_{j_2j_2})^2+b_{j_2j_2}}{(a_{j_2j_11}-a_{j_1j_11})\prod_{s_1=1}^{2}(\overline{x}_{j_2}-a_{j_2j_2s_1})}<0,$$

there is a down-parabola flow at $x_{j_2}^*=a_{j_1j_2s_1}$ with $\overline{x}_{j_1}\neq a_{j_1j_11}$. Let

$$\dot{x}_{j_1}=a_{j_1j_10}(a_{j_2j_11}-a_{j_1j_11})\prod_{s_1=1}^{2}(\overline{x}_{j_2}-a_{j_1j_2s_1}).$$

The inflection flows at $x_{j_1}^*=a_{j_2j_11}$ with $\overline{x}_{j_1}\neq a_{j_1j_11}$ are positive and negative for $\dot{x}_{j_1}>0$ and $\dot{x}_{j_1}<0$, respectively.

- For $a_{j_1j_10}(a_{j_2j_11}-a_{j_1j_11})\prod_{s_1=1}^{2}(\overline{x}_{j_2}-a_{j_1j_2s_1})>0$ and $a_{j_2j_20}>0$,

$$(\dot{x}_{j_1},a_{j_2j_11})=\underbrace{(\text{pF, UP})}_{\text{up-parabola flow }(+)}.$$

- For $a_{j_1j_10}(a_{j_2j_11}-a_{j_1j_11})\prod_{s_1=1}^{2}(\overline{x}_{j_2}-a_{j_1j_2s_1})<0$ and $a_{j_2j_20}>0$,

$$(\dot{x}_{j_1},a_{j_2j_11})=\underbrace{(\text{nF, DP})}_{\text{down-parabola flow }(-)}.$$

- For $a_{j_1j_10}(a_{j_2j_11}-a_{j_1j_11})\prod_{s_1=1}^{2}(\overline{x}_{j_2}-a_{j_1j_2s_1})>0$ and $a_{j_2j_20}<0$,

$$(\dot{x}_{j_1},a_{j_2j_11})=\underbrace{(\text{pF, DP})}_{\text{down-parabola flow }(+)}.$$

- For $a_{j_1j_10}(a_{j_2j_11}-a_{j_1j_11})\prod_{s_1=1}^{2}(\overline{x}_{j_2}-a_{j_1j_2s_1})<0$ and $a_{j_2j_20}<0$,

$$(\dot{x}_{j_1},a_{j_2j_11})=\underbrace{(\text{pF, UP})}_{\text{up-parabola flow }(+)}.$$

From the above analysis with the proof of (vii$_{1a}$), the equilibrium of $(x^*_{j_2}, x^*_{j_1}) = (a_{j_1j_2s_1}, a_{j_2j_1}1)$ $(s_1,\ s_2 \in \{1,\ 2\},\ s_1 \neq s_2)$ has the following properties as in Eqs. (1.194)–(1.197):

- For $a_{j_1j_10}(a_{j_2j_1}1 - a_{j_1j_1}1)(a_{j_1j_2s_1} - a_{j_1j_2s_2}) > 0$ and $a_{j_2j_20} > 0$,

$$(a_{j_1j_2s_1}, a_{j_2j_1}1) = \underbrace{(\mathrm{UP}_+, \mathrm{UP}_+)}_{\text{positive saddle}}.$$

  The equilibrium of $(x^*_{j_2}, x^*_{j_1}) = (a_{j_1j_2s_1}, a_{j_2j_1}1)$ is a $(\mathrm{UP}_+,\mathrm{UP}_+)$-positive saddle.
- For $a_{j_1j_10}(a_{j_2j_1}1 - a_{j_1j_1}1)(a_{j_1j_2s_1} - a_{j_1j_2s_2}) < 0$ and $a_{j_2j_20} > 0$,

$$(a_{j_1j_2s_1}, a_{j_2j_1}1) = \underbrace{(\mathrm{DP}_+, \mathrm{DP}_-)}_{\text{CCW center}}.$$

  The equilibrium of $(x^*_{j_2}, x^*_{j_1}) = (a_{j_1j_2s_1}, a_{j_2j_1}1)$ is a $(\mathrm{DP}_+,\mathrm{DP}_-)$-counterclockwise center.
- For $a_{j_1j_10}(a_{j_2j_1}1 - a_{j_1j_1}1)(a_{j_1j_2s_1} - a_{j_1j_2s_2}) > 0$ and $a_{j_2j_20} < 0$,

$$(a_{j_1j_2s_1}, a_{j_2j_1}1) = \underbrace{(\mathrm{DP}_-, \mathrm{DP}_+)}_{\text{CW center}}.$$

  The equilibrium of $(x^*_{j_2}, x^*_{j_1}) = (a_{j_1j_2s_1}, a_{j_2j_1}1)$ is a $(\mathrm{DP}_-,\mathrm{DP}_+)$-clockwise center.
- For $a_{j_1j_10}(a_{j_2j_1}1 - a_{j_1j_1}1)(a_{j_1j_2s_1} - a_{j_1j_2s_2}) < 0$ and $a_{j_2j_20} < 0$,

$$(a_{j_1j_2s_1}, a_{j_2j_1}1) = \underbrace{(\mathrm{UP}_-, \mathrm{UP}_-)}_{\text{negative saddle}}.$$

  The equilibrium of $(x^*_{j_2}, x^*_{j_1}) = (a_{j_1j_2s_1}, a_{j_2j_1}1)$ is a $(\mathrm{UP}_-,\mathrm{UP}_-)$-negative saddle.

(vii$_2$) For $a_{j_1j_1}1 = a_{j_2j_1}1$, in phase space,

$$\frac{dx_{j_1}}{dx_{j_2}} = \frac{a_{j_1j_10}}{a_{j_2j_20}} \frac{(x_{j_2} - a_{j_1j_2}1)(x_{j_2} - a_{j_1j_2}2)}{(x_{j_2} - a_{j_2j_2})^2 + b_{j_2j_2}},$$

and the deformation of the foregoing equation is

$$\begin{aligned} dx_{j_1} = & \frac{a_{j_1j_10}}{a_{j_2j_20}} \left[1 + \frac{(a_{j_2j_2} - a_{j_1j_2}1)(x_{j_2} - a_{j_2j_2})}{(x_{j_2} - a_{j_2j_2})^2 + b_{j_2j_2}} + \frac{(a_{j_2j_2} - a_{j_1j_2}2)(x_{j_2} - a_{j_2j_2})}{(x_{j_2} - a_{j_2j_2})^2 + b_{j_2j_2}} \right. \\ & \left. + \frac{(a_{j_2j_2} - a_{j_1j_2}1)(a_{j_2j_2} - a_{j_1j_2}2) - b_{j_2j_2}}{(x_{j_2} - a_{j_2j_2})^2 + b_{j_2j_2}} \right] dx_{j_2}. \end{aligned}$$

With an initial condition of $(x_{j_1 0}, x_{j_2 0})$ at $t = t_0$, the integration of the above equation gives the first integral manifold for $a_{j_1 j_1 1} = a_{j_2 j_1 1}$ as

$$
\begin{aligned}
&\left[(x_{j_1} - a_{j_1 j_1 1}) - (x_{j_1 0} - a_{j_1 j_1 1})\right] \\
&= \frac{a_{j_1 j_1 0}}{a_{j_2 j_2 0}} \left\{ \left[(x_{j_2} - a_{j_2 j_2}) - (x_{j_2 0} - a_{j_2 j_2})\right] \right. \\
&\quad + \sum\nolimits_{s_1=1}^{2} (a_{j_2 j_2} - a_{j_1 j_2 s_1}) \ln \frac{|(x_{j_2} - a_{j_2 j_2})^2 + b_{j_2 j_2}|}{|(x_{j_2 0} - a_{j_2 j_2})^2 + b_{j_2 j_2}|} \\
&\quad + \left[\prod\nolimits_{s_1=1}^{2} (a_{j_2 j_2} - a_{j_1 j_2 s_1}) - b_{j_2 j_2}\right] \\
&\quad \left. \times \frac{1}{\sqrt{b_{j_2 j_2}}} \left(\arctan \frac{x_{j_2} - a_{j_2 j_2}}{\sqrt{b_{j_2 j_2}}} - \arctan \frac{x_{j_2 0} - a_{j_2 j_2}}{\sqrt{b_{j_2 j_2}}}\right) \right\}.
\end{aligned}
$$

(vii$_{2a}$) For $x_{j_1}^* = a_{j_1 j_1 1} = a_{j_2 j_1 1}$ with $\overline{x}_{j_2} \neq a_{j_1 j_2 s_1}$ $(s_1, s_2 \in \{1, 2\}, s_1 \neq s_2)$, in phase space, if

$$
\left.\frac{dx_{j_2}}{dx_{j_1}}\right|_{x_{j_1}^* = a_{j_2 j_1 1}} = \frac{a_{j_2 j_2 0}}{a_{j_1 j_1 0}} \frac{(\overline{x}_{j_2} - a_{j_2 j_2})^2 + b_{j_2 j_2}}{(\overline{x}_{j_2} - a_{j_1 j_2 1})(\overline{x}_{j_2} - a_{j_1 j_2 2})} > 0,
$$

there is an increasing-inflection flow, and if

$$
\left.\frac{dx_{j_2}}{dx_{j_1}}\right|_{x_{j_1}^* = a_{j_2 j_1 1}} = \frac{a_{j_2 j_2 0}}{a_{j_1 j_1 0}} \frac{(\overline{x}_{j_2} - a_{j_2 j_2})^2 + b_{j_2 j_2}}{(\overline{x}_{j_2} - a_{j_1 j_2 1})(\overline{x}_{j_2} - a_{j_1 j_2 2})} < 0,
$$

there is a decreasing-inflection flow. The variational equation is

$$
\Delta \dot{x}_{j_1} = a_{j_1 j_1 0}(\overline{x}_{j_2} - a_{j_1 j_2 s_1})(\overline{x}_{j_2} - a_{j_1 j_2 s_2}) \Delta x_{j_1}.
$$

The inflection flows at $x_{j_1}^* = a_{j_1 j_1 1} = a_{j_2 j_1 1}$ for $a_{j_1 j_1 0}(\overline{x}_{j_2} - a_{j_1 j_2 s_1})(\overline{x}_{j_2} - a_{j_1 j_2 s_2}) > 0$ and $a_{j_1 j_1 0}(\overline{x}_{j_2} - a_{j_1 j_2 s_1})(\overline{x}_{j_2} - a_{j_1 j_2 s_2}) < 0$ are source and sink, respectively.

Therefore, the infinite-equilibrium of $x_{j_1}^* = a_{j_1 j_1 1} = a_{j_2 j_1 1}$ with $\overline{x}_{j_2} \neq a_{j_1 j_2 s_1}$ $(s_1, s_2 \in \{1, 2\}, s_1 \neq s_2)$ for $a_{j_1 j_1 1} = a_{j_2 j_1 1}$ has the following properties as in Eqs. (1.199)–(1.202):

- For $a_{j_1 j_1 0}(\overline{x}_{j_2} - a_{j_1 j_2 s_1})(\overline{x}_{j_2} - a_{j_1 j_2 s_2}) > 0$ and $a_{j_2 j_2 0} > 0$,

$$
(a_{j_1 j_1 1}, \overline{x}_{j_2}) = \underbrace{(\mathrm{SO}, \mathrm{II})}_{\text{increasing-inflection source}}.
$$

  The infinite-equilibrium of $x_{j_1}^* = a_{j_1 j_1 1} = a_{j_2 j_1 1}$ is an (SO,II)-increasing-inflection source.

- For $a_{j_1j_10}(\overline{x}_{j_2}-a_{j_1j_2s_1})(\overline{x}_{j_2}-a_{j_1j_2s_2})<0$ and $a_{j_2j_20}>0$,

$$(a_{j_1j_11},\overline{x}_{j_2})=\underbrace{(\mathrm{SI},\mathrm{DI})}_{\text{decreasing-inflection sink}}.$$

  The infinite-equilibrium of $x^*_{j_1}=a_{j_1j_11}=a_{j_2j_11}$ is an (SI,DI)-decreasing-inflection sink.
- For $a_{j_1j_10}(\overline{x}_{j_2}-a_{j_1j_2s_1})(\overline{x}_{j_2}-a_{j_1j_2s_2})>0$ and $a_{j_2j_20}<0$,

$$(a_{j_1j_11},\overline{x}_{j_2})=\underbrace{(\mathrm{SO},\mathrm{DI})}_{\text{decreasing-inflection source}}.$$

  The infinite-equilibrium of $x^*_{j_1}=a_{j_1j_11}=a_{j_2j_11}$ is an (SO,DI)-decreasing-inflection source.
- For $a_{j_1j_10}(\overline{x}_{j_2}-a_{j_1j_2s_1})(\overline{x}_{j_2}-a_{j_1j_2s_2})<0$ and $a_{j_2j_20}<0$,

$$(a_{j_1j_11},\overline{x}_{j_2})=\underbrace{(\mathrm{SI},\mathrm{II})}_{\text{increasing-inflection sink}}.$$

  The infinite-equilibrium of $x^*_{j_1}=a_{j_1j_11}=a_{j_2j_11}$ is an (SI,II)-increasing-inflection sink.

(vii$_{2b}$) For $x^*_{j_2}=a_{j_1j_2s_1}$ with $\overline{x}_{j_1}\neq a_{j_1j_11}=a_{j_2j_11}$, in phase space,

$$\left.\frac{dx_{j_1}}{dx_{j_2}}\right|_{x^*_{j_2}=a_{j_1j_2s_1}}=\left.\frac{a_{j_1j_10}}{a_{j_2j_20}}\frac{(x_{j_2}-a_{j_1j_2s_1})(x_{j_2}-a_{j_1j_2s_2})}{(x_{j_2}-a_{j_2j_2})^2+b_{j_2j_2}}\right|_{x^*_{j_2}=a_{j_1j_2s_1}}=0.$$

If

$$\left.\frac{d^2x_{j_1}}{dx^2_{j_2}}\right|_{x^*_{j_2}=a_{j_1j_2s_1}}=\frac{a_{j_1j_10}}{a_{j_2j_20}}\frac{(a_{j_1j_2s_1}-a_{j_1j_2s_2})}{(a_{j_1j_2s_1}-a_{j_2j_2})^2+b_{j_2j_2}}>0,$$

there is an up-parabola flow. If

$$\left.\frac{d^2x_{j_1}}{dx^2_{j_2}}\right|_{x^*_{j_2}=a_{j_1j_2s_1}}=\frac{a_{j_1j_10}}{a_{j_2j_20}}\frac{(a_{j_1j_2s_1}-a_{j_1j_2s_2})}{(a_{j_1j_2s_1}-a_{j_2j_2})^2+b_{j_2j_2}}<0,$$

there is a down-parabola flow. The variational equation is

$$\dot{x}_{j_2} = a_{j_2 j_2 0}(\overline{x}_{j_1} - a_{j_2 j_1 1})\left[(a_{j_1 j_2 s_1} - a_{j_2 j_2})^2 + b_{j_2 j_2}\right].$$

The parabola flows at $x_{j_2}^* = a_{j_1 j_2 s_1}$ are positive and negative for $a_{j_2 j_2 0}(\overline{x}_{j_1} - a_{j_2 j_1 1}) > 0$ and $a_{j_2 j_2 0}(\overline{x}_{j_1} - a_{j_2 j_1 1}) < 0$, respectively.

From the above analysis with the proof of (vii$_{2a}$), the equilibrium of $(x_{j_2}^*, x_{j_1}^*) = (a_{j_1 j_2 s_1}, a_{j_2 j_1 1})$ $(s_1, s_2 \in \{1, 2\}, s_1 \neq s_2)$ *with* $a_{j_1 j_1 1} = a_{j_2 j_1 1}$ has the following properties as in Eqs. (1.203)–(1.206):

- For $a_{j_1 j_1 0}(a_{j_1 j_2 s_1} - a_{j_1 j_2 s_2}) > 0$ and $a_{j_2 j_2 0} > 0$,

$$(a_{j_1 j_2 s_1}, a_{j_2 j_1 1}) = \underbrace{({}_{\mathrm{DI:II}}\mathrm{UP},\ {}_{\mathrm{SI:SO}}\mathrm{US})}_{\text{up-parabola upper-saddle}}.$$

  The equilibrium of $(x_{j_2}^*, x_{j_1}^*) = (a_{j_1 j_2 s_1}, a_{j_2 j_1 1})$ is a $({}_{\mathrm{DI:II}}\mathrm{UP}, {}_{\mathrm{SI:SO}}\mathrm{US})$-up-parabola upper-saddle.

- For $a_{j_1 j_1 0}(a_{j_1 j_2 s_1} - a_{j_1 j_2 s_2}) < 0$ and $a_{j_2 j_2 0} > 0$,

$$(a_{j_1 j_2 s_1}, a_{j_2 j_1 1}) = \underbrace{({}_{\mathrm{II:DI}}\mathrm{DP},\ {}_{\mathrm{SO:SI}}\mathrm{LS})}_{\text{down-parabola lower-saddle}}.$$

  The equilibrium of $(x_{j_2}^*, x_{j_1}^*) = (a_{j_1 j_2 s_1}, a_{j_2 j_1 1})$ is a $({}_{\mathrm{II:DI}}\mathrm{DP}, {}_{\mathrm{SO:SI}}\mathrm{LS})$-down-parabola lower-saddle.

- For $a_{j_1 j_1 0}(a_{j_1 j_2 s_1} - a_{j_1 j_2 s_2}) > 0$ and $a_{j_2 j_2 0} < 0$,

$$(a_{j_1 j_2 s_1}, a_{j_2 j_1 1}) = \underbrace{({}_{\mathrm{II:DI}}\mathrm{DP},\ {}_{\mathrm{SI:SO}}\mathrm{US})}_{\text{down-parabola lower-saddle}}.$$

  The equilibrium of $(x_{j_1}^*, x_{j_2}^*) = (a_{j_1 j_1 1}, a_{j_1 j_2 s_1})$ is a $({}_{\mathrm{II:DI}}\mathrm{DP}, {}_{\mathrm{SI:SO}}\mathrm{US})$-down-parabola upper-saddle.

- For $a_{j_1 j_1 0}(a_{j_1 j_2 s_1} - a_{j_1 j_2 s_2}) < 0$ and $a_{j_2 j_2 0} < 0$,

$$(a_{j_1 j_2 s_1}, a_{j_2 j_1 1}) = \underbrace{({}_{\mathrm{DI:II}}\mathrm{UP},\ {}_{\mathrm{SO:SI}}\mathrm{LS})}_{\text{up-parabola lower-saddle}}.$$

  The equilibrium of $(x_{j_1}^*, x_{j_2}^*) = (a_{j_1 j_1 1}, a_{j_1 j_2 s_1})$ is a $({}_{\mathrm{DI:II}}\mathrm{UP}, {}_{\mathrm{SO:SI}}\mathrm{LS})$-up-parabola lower-saddle.

(viii) For $\Delta_{j_1 j_2} > 0$ and $\Delta_{j_2 j_2} = 0$, the standard form is

$$\dot{x}_{j_2} = a_{j_2 j_2 0}(x_{j_1} - a_{j_2 j_1 1})(x_{j_2} - a_{j_2 j_2 1})^2,$$
$$\dot{x}_{j_1} = a_{j_1 j_1 0}(x_{j_1} - a_{j_1 j_1 1})(x_{j_2} - a_{j_1 j_2 1})(x_{j_2} - a_{j_1 j_2 2})$$

where

$$
\begin{aligned}
& b_{j_1 j_2 1}, b_{j_1 j_2 2} = -\frac{1}{2}(B_{j_1 j_2} + \sqrt{\Delta_{j_2 j_2}}), \\
& \{a_{j_1 j_2 1}, a_{j_1 j_2 2}\} = \mathrm{sort}\{b_{j_1 j_2 1}, b_{j_1 j_2 2}\}, a_{j_1 j_2 1} < a_{j_1 j_2 2}; \\
& a_{j_2 j_2 1} = -\frac{1}{2} B_{j_2 j_2}.
\end{aligned}
$$

(viii$_1$) For $a_{j_2 j_2 1} \neq a_{j_1 j_2 1}, a_{j_1 j_2 2}$, in phase space,

$$
\frac{dx_{j_1}}{dx_{j_2}} = \frac{a_{j_1 j_1 0}}{a_{j_2 j_2 0}} \frac{(x_{j_1} - a_{j_1 j_1 1})(x_{j_2} - a_{j_1 j_2 1})(x_{j_2} - a_{j_1 j_2 2})}{(x_{j_1} - a_{j_2 j_1 1})(x_{j_2} - a_{j_2 j_2 1})^2},
$$

and the deformation of the foregoing equation is

$$
\begin{aligned}
\left[1 + \frac{(a_{j_1 j_1 1} - a_{j_2 j_1 1})}{(x_{j_1} - a_{j_1 j_1 1})}\right] dx_{j_1} = & \frac{a_{j_1 j_1 0}}{a_{j_2 j_2 0}} \left[1 + \sum\nolimits_{s_1=1}^{2} (a_{j_2 j_2 1} - a_{j_1 j_2 s_1}) \frac{1}{(x_{j_2} - a_{j_2 j_2 1})} \right. \\
& \left. + \prod\nolimits_{s_1=1}^{2} (a_{j_2 j_2 1} - a_{j_1 j_2 s_1}) \frac{1}{(x_{j_2} - a_{j_2 j_2 1})^2} \right] dx_{j_2}.
\end{aligned}
$$

With an initial condition of $(x_{j_1 0}, x_{j_2 0})$ at $t = t_0$, the integration of the above equation gives the first integral manifold for $a_{j_2 j_2 1} \neq a_{j_1 j_2 1}, a_{j_1 j_2 2}$ as

$$
\begin{aligned}
& \left[(x_{j_1} - a_{j_1 j_1 1}) - (x_{j_1 0} - a_{j_1 j_1 1})\right] + (a_{j_1 j_1 1} - a_{j_2 j_1 1}) \ln \frac{|x_{j_1} - a_{j_1 j_1 1}|}{|x_{j_1 0} - a_{j_1 j_1 1}|} \\
& = \frac{a_{j_1 j_1 0}}{a_{j_2 j_2 0}} \left\{ \left[(x_{j_2} - a_{j_2 j_2 1}) - (x_{j_2 0} - a_{j_2 j_2 1})\right] \right. \\
& \quad + \sum\nolimits_{s_1=1}^{2} (a_{j_2 j_2 1} - a_{j_1 j_2 s_1}) \ln \frac{|x_{j_2} - a_{j_2 j_2 1}|}{|x_{j_2 0} - a_{j_2 j_2 1}|} \\
& \quad \left. - \prod\nolimits_{s_1=1}^{2} (a_{j_2 j_2 1} - a_{j_1 j_2 s_1}) \left(\frac{1}{x_{j_2} - a_{j_2 j_2 1}} - \frac{1}{x_{j_2 0} - a_{j_2 j_2 1}}\right) \right\}.
\end{aligned}
$$

(viii$_{1a}$) At $(x_{j_1}^*, x_{j_2}^*) = (a_{j_1 j_1 1}, a_{j_2 j_2 1})$, the variational equations are

$$
\begin{aligned}
\Delta \dot{x}_{j_2} &= a_{j_2 j_2 0} (a_{j_1 j_1 1} - a_{j_2 j_1 1})(\Delta x_{j_2})^2, \\
\Delta \dot{x}_{j_1} &= a_{j_1 j_1 0} \prod\nolimits_{s_1=1}^{2} (a_{j_2 j_2 1} - a_{j_1 j_2 s_1}) \Delta x_{j_1}.
\end{aligned}
$$

The flows at $(x_{j_1}^*, x_{j_2}^*) = (a_{j_1 j_1 1}, a_{j_2 j_2 1})$ in the x_{j_1} -direction are source and sink for $a_{j_1 j_1 0} \prod_{s_1=1}^{2} (a_{j_2 j_2 1} - a_{j_1 j_2 s_1}) > 0$ and $a_{j_1 j_1 0} \prod_{s_1=1}^{2} (a_{j_2 j_2 1} - a_{j_1 j_2 s_1}) < 0$, respectively. The flows at $(x_{j_1}^*, x_{j_2}^*) = (a_{j_1 j_1 1}, a_{j_2 j_2 1})$ in the x_{j_2}-direction are upper-saddle and lower-saddle for $a_{j_2 j_2 0}(a_{j_1 j_1 1} - a_{j_2 j_1 1}) > 0$ and $a_{j_2 j_2 0}(a_{j_1 j_1 1} - a_{j_2 j_1 1}) < 0$, respectively.

Therefore, the equilibrium of $(x_{j_1}^*, x_{j_2}^*) = (a_{j_1 j_1 1}, a_{j_2 j_2 1})$ has the following properties as in Eqs. (1.210)–(1.213):

- For $a_{j_1 j_1 0}\prod_{s_1=1}^{2}(a_{j_2 j_2 1} - a_{j_1 j_2 s_1}) > 0$ and $a_{j_2 j_2 0}(a_{j_1 j_1 1} - a_{j_2 j_1 1}) > 0$,
$$(a_{j_1 j_1 1}, a_{j_2 j_2 1}) = \underbrace{(\mathrm{SO}, \mathrm{US})}_{\text{upper-saddle-source}}.$$
The equilibrium of $(x_{j_1}^*, x_{j_2}^*) = (a_{j_1 j_1 1}, a_{j_2 j_2 1})$ is an (SO,US)-upper-saddle-source.
- For $a_{j_1 j_1 0}\prod_{s_1=1}^{2}(a_{j_2 j_2 1} - a_{j_1 j_2 s_1}) < 0$ and $a_{j_2 j_2 0}(a_{j_1 j_1 1} - a_{j_2 j_1 1}) > 0$,
$$(a_{j_1 j_1 1}, a_{j_2 j_2 1}) = \underbrace{(\mathrm{SI}, \mathrm{US})}_{\text{upper-saddle-sink}}.$$
The equilibrium of $(x_{j_1}^*, x_{j_2}^*) = (a_{j_1 j_1 1}, a_{j_2 j_2 1})$ is an (SI,US)-upper-saddle-sink.
- For $a_{j_1 j_1 0}\prod_{s_1=1}^{2}(a_{j_2 j_2 1} - a_{j_1 j_2 s_1}) > 0$ and $a_{j_2 j_2 0}(a_{j_1 j_1 1} - a_{j_2 j_1 1}) < 0$,
$$(a_{j_1 j_1 1}, a_{j_2 j_2 1}) = \underbrace{(\mathrm{SO}, \mathrm{LS})}_{\text{lower-saddle-source}}.$$
The equilibrium of $(x_{j_1}^*, x_{j_2}^*) = (a_{j_1 j_1 1}, a_{j_2 j_2 1})$ is an (SO,LS)-lower-saddle-source.
- For $a_{j_1 j_1 0}\prod_{s_1=1}^{2}(a_{j_2 j_2 1} - a_{j_1 j_2 s_1}) < 0$ and $a_{j_2 j_2 0}(a_{j_1 j_1 1} - a_{j_2 j_1 1}) < 0$,
$$(a_{j_1 j_1 1}, a_{j_2 j_2 1}) = \underbrace{(\mathrm{SI}, \mathrm{LS})}_{\text{lower-saddle-sink}}.$$
The equilibrium of $(x_{j_1}^*, x_{j_2}^*) = (a_{j_1 j_1 1}, a_{j_2 j_2 1})$ is an (SI,LS)-lower-saddle-sink.

(viii$_{1b}$) For $x_{j_2}^* = a_{j_1 j_2 s_1} \neq a_{j_2 j_2 1}$ and $\overline{x}_{j_1} \neq a_{j_1 j_1 1}$ ($s_1, s_2 \in \{1, \ 2\}, s_1 \neq s_2$), in phase space,

$$\left.\frac{dx_{j_1}}{dx_{j_2}}\right|_{x_{j_2}^* = a_{j_1 j_2 s_1}} = \frac{a_{j_1 j_1 0}}{a_{j_2 j_2 0}} \left.\frac{(\overline{x}_{j_1} - a_{j_1 j_1 1})(x_{j_2} - a_{j_1 j_2 s_1})(x_{j_2} - a_{j_1 j_2 s_2})}{(\overline{x}_{j_1} - a_{j_2 j_1 1})(x_{j_2} - a_{j_2 j_2 1})^2}\right|_{x_{j_2}^* = a_{j_1 j_2 s_1}} = 0.$$

If

$$\left.\frac{d^2 x_{j_1}}{dx_{j_2}^2}\right|_{x_{j_2}^* = a_{j_1 j_2 s_1}} = \frac{a_{j_1 j_1 0}}{a_{j_2 j_2 0}} \frac{(\overline{x}_{j_1} - a_{j_1 j_1 1})(a_{j_1 j_2 s_1} - a_{j_1 j_2 s_2})}{(\overline{x}_{j_1} - a_{j_2 j_1 1})(a_{j_1 j_2 s_1} - a_{j_1 j_2 1})^2} > 0,$$

there is an up-parabola flow at $x_{j_2}^* = a_{j_1 j_2 s_1} \neq a_{j_2 j_2 1}$ with $\overline{x}_{j_1} \neq a_{j_1 j_1 1}$. If

$$\left.\frac{d^2x_{j_1}}{dx_{j_2}^2}\right|_{x_{j_2}^*=a_{j_1j_2s_1}} = \frac{a_{j_1j_10}}{a_{j_2j_20}}\frac{(\bar{x}_{j_1}-a_{j_1j_11})(a_{j_1j_2s_1}-a_{j_1j_2s_2})}{(\bar{x}_{j_1}-a_{j_2j_11})(a_{j_1j_2s_1}-a_{j_1j_21})^2}<0,$$

there is a down-parabola flow at $x_{j_2}^*=a_{j_1j_2s_1}$ with $\bar{x}_{j_1}\neq a_{j_1j_11}$. Let

$$\dot{x}_{j_2}=a_{j_2j_20}(\bar{x}_{j_1}-a_{j_2j_11})(a_{j_1j_2s_1}-a_{j_2j_21})^2\neq 0.$$

The parabola flows at $x_{j_2}^*=a_{j_1j_2s_1}\neq a_{j_2j_21}$ with $\bar{x}_{j_1}\neq a_{j_1j_11}$ are positive and negative for $a_{j_2j_20}(\bar{x}_{j_1}-a_{j_2j_11})>0$ and $a_{j_2j_20}(\bar{x}_{j_1}-a_{j_2j_11})<0$, respectively.

- For $a_{j_1j_10}(\bar{x}_{j_1}-a_{j_1j_11})(a_{j_1j_2s_1}-a_{j_1j_2s_2})>0$ and $a_{j_2j_20}(\bar{x}_{j_1}-a_{j_2j_11})(a_{j_1j_2s_1}-a_{j_2j_21})^2>0$,

$$(a_{j_1j_2s_1},\dot{x}_{j_2})=\underbrace{(\text{UP},\text{pF})}_{\text{up-parabola flow }(+)}.$$

- For $a_{j_1j_10}(\bar{x}_{j_1}-a_{j_1j_11})(a_{j_1j_2s_1}-a_{j_1j_2s_2})<0$ and $a_{j_2j_20}(\bar{x}_{j_1}-a_{j_2j_11})(a_{j_1j_2s_1}-a_{j_2j_21})^2>0$,

$$(a_{j_1j_2s_1},\dot{x}_{j_2})=\underbrace{(\text{DP},\text{pF})}_{\text{down-parabola flow }(+)}.$$

- For $a_{j_1j_10}(\bar{x}_{j_1}-a_{j_1j_11})(a_{j_1j_2s_1}-a_{j_1j_2s_2})>0$ and $a_{j_2j_20}(\bar{x}_{j_1}-a_{j_2j_11})(a_{j_1j_2s_1}-a_{j_2j_21})^2<0$,

$$(a_{j_1j_2s_1},\dot{x}_{j_2})=\underbrace{(\text{DP},\text{nF})}_{\text{down-parabola flow }(-)}.$$

- For $a_{j_1j_10}(\bar{x}_{j_1}-a_{j_1j_11})(a_{j_1j_2s_1}-a_{j_1j_2s_2})<0$ and $a_{j_2j_20}(\bar{x}_{j_1}-a_{j_2j_11})(a_{j_1j_2s_1}-a_{j_2j_21})^2<0$,

$$(a_{j_1j_2s_1},\dot{x}_{j_2})=\underbrace{(\text{UP},\text{nF})}_{\text{up-parabola flow }(-)}.$$

Therefore, from the above analysis, for $\bar{x}_{j_1}=a_{j_1j_11}$, the equilibrium of $(x_{j_1}^*,x_{j_2}^*)=(a_{j_1j_11},a_{j_1j_2s_1})$ with $a_{j_1j_2s_1}\neq a_{j_2j_21}$ $(s_1,s_2\in\{1,\ 2\},\ s_1\neq s_2)$ has the following properties as in Eqs. (1.214)–(1.217):

- For $a_{j_1j_10}(a_{j_1j_2s_1}-a_{j_1j_2s_2})>0$ and $a_{j_2j_20}(a_{j_1j_11}-a_{j_2j_11})(a_{j_1j_2s_1}-a_{j_2j_21})^2>0$,

$$(a_{j_1j_11},a_{j_1j_2s_1})=\underbrace{(\text{DP}:\text{UP},\text{pF})}_{\text{hyperbolic flow }(+)}.$$

The equilibrium of $(x_{j_1}^*,x_{j_2}^*)=(a_{j_1j_11},a_{j_1j_2s_1})$ is a (DP:UP, pF)-positive hyperbolic flow.

- For $a_{j_1j_10}(a_{j_1j_2s_1}-a_{j_1j_2s_2})<0$ and $a_{j_2j_20}(a_{j_1j_11}-a_{j_2j_11})(a_{j_1j_2s_1}-a_{j_2j_21})^2>0$,

$$(a_{j_1j_11},a_{j_1j_2s_1})=\underbrace{(\mathrm{UP}:\mathrm{DP},\mathrm{pF})}_{\text{hyperbolic-secant flow }(+)}.$$

The equilibrium of $(x^*_{j_1},x^*_{j_2})=(a_{j_1j_11},a_{j_1j_2s_1})$ is a (UP:DP, pF)-positive hyperbolic-secant flow.

- For $a_{j_1j_10}(a_{j_1j_2s_1}-a_{j_1j_2s_2})>0$ and $a_{j_2j_20}(a_{j_1j_11}-a_{j_2j_11})(a_{j_1j_2s_1}-a_{j_2j_21})^2<0$,

$$(a_{j_1j_11},a_{j_1j_2s_1})=\underbrace{(\mathrm{UP}:\mathrm{DP},\mathrm{nF})}_{\text{hyperbolic-secant flow }(-)}.$$

The equilibrium of $(x^*_{j_1},x^*_{j_2})=(a_{j_1j_11},a_{j_1j_2s_1})$ is a (UP:DP,nF)-negative hyperbolic-secant flow.

- For $a_{j_1j_10}(a_{j_1j_2s_1}-a_{j_1j_2s_2})<0$ and $a_{j_2j_20}(a_{j_1j_11}-a_{j_2j_11})(a_{j_1j_2s_1}-a_{j_2j_21})^2<0$,

$$(a_{j_1j_11},a_{j_1j_2s_1})=\underbrace{(\mathrm{DP}:\mathrm{UP},\mathrm{nF})}_{\text{hyperbolic flow }(-)}.$$

The equilibrium of $(x^*_{j_1},x^*_{j_2})=(a_{j_1j_11},a_{j_1j_21}$)is a (DP:UP,nF)-negative hyperbolic flow.

(viii$_{1c}$) For $x^*_{j_1}=a_{j_2j_11}$ and $\overline{x}_{j_2}\neq a_{j_1j_2s_1},a_{j_2j_21}$ ($s_1=1,2$), in phase space,

$$\left.\frac{dx_{j_2}}{dx_{j_1}}\right|_{x^*_{j_1}=a_{j_2j_11}}=\frac{a_{j_2j_20}}{a_{j_1j_10}}\left.\frac{(x_{j_1}-a_{j_2j_11})(\overline{x}_{j_2}-a_{j_2j_21})^2}{(x_{j_1}-a_{j_1j_11})\prod_{s_1=1}^{2}(\overline{x}_{j_2}-a_{j_1j_2s_1})}\right|_{x^*_{j_1}=a_{j_2j_11}}=0.$$

If

$$\left.\frac{d^2x_{j_2}}{dx^2_{j_1}}\right|_{x^*_{j_1}=a_{j_2j_11}}=\frac{a_{j_2j_20}}{a_{j_1j_10}}\frac{(\overline{x}_{j_2}-a_{j_2j_21})^2}{(a_{j_2j_11}-a_{j_1j_11})\prod_{s_1=1}^{2}(\overline{x}_{j_2}-a_{j_1j_2s_1})}>0,$$

there is an up-parabola flow at $x^*_{j_1}=a_{j_2j_11}$ with $\overline{x}_{j_2}\neq a_{j_1j_21},a_{j_1j_22},a_{j_2j_21}$. If

$$\left.\frac{d^2x_{j_2}}{dx^2_{j_1}}\right|_{x^*_{j_1}=a_{j_2j_11}}=\frac{a_{j_2j_20}}{a_{j_1j_10}}\frac{(\overline{x}_{j_2}-a_{j_2j_21})^2}{(a_{j_2j_11}-a_{j_1j_11})\prod_{s_1=1}^{2}(\overline{x}_{j_2}-a_{j_1j_2s_1})}<0,$$

there is a down-parabola flow at $x^*_{j_1}=a_{j_2j_11}$ with $\overline{x}_{j_2}\neq a_{j_1j_21},a_{j_1j_22},a_{j_2j_21}$. Let

$$\dot{x}_{j_1} = a_{j_1j_10}(a_{j_2j_11} - a_{j_1j_11})\prod_{s_1=1}^{2}(\overline{x}_{j_2} - a_{j_1j_2s_1}).$$

The inflection flows at $x^*_{j_1} = a_{j_2j_11}$ with $\overline{x}_{j_2} \neq a_{j_1j_21}, a_{j_1j_22}, a_{j_2j_21}$ are positive and negative for $\dot{x}_{j_1} > 0$ and $\dot{x}_{j_1} < 0$, respectively.

- For $a_{j_1j_10}(a_{j_2j_11} - a_{j_1j_11})\prod_{s_1=1}^{2}(\overline{x}_{j_2} - a_{j_1j_2s_1}) > 0$ and $a_{j_2j_20}(\overline{x}_{j_2} - a_{j_2j_21})^2 > 0$,

$$(\dot{x}_{j_1}, a_{j_2j_11}) = \underbrace{(\text{pF}, \text{UP})}_{\text{up-parabola flow }(+)}.$$

- For $a_{j_1j_10}(a_{j_2j_11} - a_{j_1j_11})\prod_{s_1=1}^{2}(\overline{x}_{j_2} - a_{j_1j_2s_1}) < 0$ and $a_{j_2j_20}(\overline{x}_{j_2} - a_{j_2j_21})^2 > 0$,

$$(\dot{x}_{j_1}, a_{j_2j_11}) = \underbrace{(\text{nF}, \text{DP})}_{\text{down-parabola flow }(-)}.$$

- For $a_{j_1j_10}(a_{j_2j_11} - a_{j_1j_11})\prod_{s_1=1}^{2}(\overline{x}_{j_2} - a_{j_1j_2s_1}) > 0$ and $a_{j_2j_20}(\overline{x}_{j_2} - a_{j_2j_21})^2 < 0$,

$$(\dot{x}_{j_1}, a_{j_2j_11}) = \underbrace{(\text{pF}, \text{DP})}_{\text{down-parabola flow }(+)}.$$

- For $a_{j_1j_10}(a_{j_2j_11} - a_{j_1j_11})\prod_{s_1=1}^{2}(\overline{x}_{j_2} - a_{j_1j_2s_1}) < 0$ and $a_{j_2j_20}(\overline{x}_{j_2} - a_{j_2j_21})^2 < 0$,

$$(\dot{x}_{j_1}, a_{j_2j_11}) = \underbrace{(\text{nF}, \text{UP})}_{\text{up-parabola flow }(-)}.$$

From the above analysis of $\overline{x}_{j_2} = a_{j_1j_2s_1}$ and the proof of ($\text{viii}_{1\text{b}}$) with $x_{j_2} = a_{j_2j_11}$, the equilibrium of $(x^*_{j_2}, x^*_{j_1}) = (a_{j_1j_2s_1}, a_{j_2j_11})$ $(s_1, s_2 \in \{1,\ 2\}, s_1 \neq s_2)$ has the following properties as in Eqs. (1.218)–(1.221):

- For $a_{j_1j_10}(a_{j_2j_11} - a_{j_1j_11})(a_{j_1j_2s_1} - a_{j_1j_2s_2}) > 0$ and $a_{j_2j_20}(a_{j_1j_2s_1} - a_{j_2j_21})^2 > 0$,

$$(a_{j_1j_2s_1}, a_{j_2j_11}) = \underbrace{(\text{UP}_+, \text{UP}_+)}_{\text{positive saddle}}.$$

The equilibrium of $(x^*_{j_2}, x^*_{j_1}) = (a_{j_1j_2s_1}, a_{j_2j_11})$ is a $(\text{UP}_+,\text{UP}_+)$-positive saddle.

- For $a_{j_1j_10}(a_{j_2j_11}-a_{j_1j_11})(a_{j_1j_2s_1}-a_{j_1j_2s_2})<0$ and $a_{j_2j_20}(a_{j_1j_2s_1}-a_{j_2j_21})^2>0$,

$$(a_{j_1j_2s_1},a_{j_2j_11})=\underbrace{(\mathrm{DP}_+,\mathrm{DP}_-)}_{\text{CCW center}}.$$

The equilibrium of $(x^*_{j_2},x^*_{j_1})=(a_{j_1j_2s_1},a_{j_2j_11})$ is a (DP$_+$,DP$_-$)-counterclockwise center.

- For $a_{j_1j_10}(a_{j_2j_11}-a_{j_1j_11})(a_{j_1j_2s_1}-a_{j_1j_2s_2})>0$ and $a_{j_2j_20}(a_{j_1j_2s_1}-a_{j_2j_21})^2<0$,

$$(a_{j_1j_2s_1},a_{j_2j_11})=\underbrace{(\mathrm{DP}_-,\mathrm{DP}_+)}_{\text{CW center}}.$$

The equilibrium of $(x^*_{j_2},x^*_{j_1})=(a_{j_1j_2s_1},a_{j_2j_11})$ is a (DP$_-$,DP$_+$)-clockwise center.

- For $a_{j_1j_10}(a_{j_2j_11}-a_{j_1j_11})(a_{j_1j_2s_1}-a_{j_1j_2s_2})<0$ and $a_{j_2j_20}(a_{j_1j_2s_1}-a_{j_2j_21})^2<0$,

$$(a_{j_1j_2s_1},a_{j_2j_11})=\underbrace{(\mathrm{UP}_-,\mathrm{UP}_-)}_{\text{negative saddle}}.$$

The equilibrium of $(x^*_{j_2},x^*_{j_1})=(a_{j_1j_2s_1},a_{j_2j_11})$ is a (UP$_-$,UP$_-$)-negative saddle.

(viii$_{1d}$) For the above analysis of $\overline{x}_{j_2}=a_{j_2j_21}$, the equilibrium of $(x^*_{j_1},x^*_{j_2})=(a_{j_2j_11},a_{j_2j_21})$ with $a_{j_2j_21}\neq a_{j_1j_21}$ has the following properties as in Eqs. (1.222)–(1.225):

- For $a_{j_1j_10}(a_{j_2j_11}-a_{j_1j_11})\prod_{s_1=1}^{2}(a_{j_2j_21}-a_{j_1j_2s_1})>0$ and $a_{j_2j_20}>0$,

$$(a_{j_2j_11},a_{j_2j_21})=\underbrace{(\mathrm{pF},\mathrm{UP}:\mathrm{UP})}_{\text{hyperbolic-secant-to-hyperbolic flow }(+)}.$$

The equilibrium of $(x^*_{j_1},x^*_{j_2})=(a_{j_2j_11},a_{j_2j_21})$ is a (pF, UP:UP)-positive hyperbolic-secant-to-hyperbolic flow.

- For $a_{j_1j_10}(a_{j_2j_11}-a_{j_1j_11})\prod_{s_1=1}^{2}(a_{j_2j_21}-a_{j_1j_2s_1})<0$ and $a_{j_2j_20}>0$,

$$(a_{j_2j_11},a_{j_2j_21})=\underbrace{(\mathrm{nF},\mathrm{DP}:\mathrm{DP})}_{\text{hyperbolic-to-hyperbolic-secant flow }(-)}.$$

The equilibrium of $(x^*_{j_1},x^*_{j_2})=(a_{j_2j_11},a_{j_2j_21})$ is an (nF,DP:DP)-negative hyperbolic-to-hyperbolic-secant flow.

- For $a_{j_1j_10}(a_{j_2j_11}-a_{j_1j_11})\prod_{s_1=1}^{2}(a_{j_2j_21}-a_{j_1j_2s_1})>0$ and $a_{j_2j_20}<0$,

$$(a_{j_2j_11},a_{j_2j_21})=\underbrace{(\text{pF},\text{DP}:\text{DP})}_{\text{hyperbolic-to-hyperbolic-secant flow }(+)}.$$

  The equilibrium of $(x_{j_1}^*,x_{j_2}^*)=(a_{j_1j_11},a_{j_1j_21})$ is a (pF,UP:DP)-positive hyperbolic-to-hyperbolic-secant flow.
- For $a_{j_1j_10}(a_{j_2j_11}-a_{j_1j_11})\prod_{s_1=1}^{2}(a_{j_2j_21}-a_{j_1j_2s_1})<0$ and $a_{j_2j_20}<0$,

$$(a_{j_2j_11},a_{j_2j_21})=\underbrace{(\text{nF},\text{UP}:\text{UP})}_{\text{hyperbolic-secant-to-hyperbolic flow }(-)}.$$

  The equilibrium of $(x_{j_1}^*,x_{j_2}^*)=(a_{j_1j_11},a_{j_1j_21})$ is an (nF, DP:UP)-negative hyperbolic-secant-to-hyperbolic flow.

(viii$_2$) For $a_{j_2j_21}=a_{j_1j_2s_1}$ $(s_1,s_2\in\{1,\ 2\},\ s_1\neq s_2)$,

$$\frac{dx_{j_1}}{dx_{j_2}}=\frac{a_{j_1j_10}}{a_{j_2j_20}}\frac{(x_{j_1}-a_{j_1j_11})(x_{j_2}-a_{j_1j_2s_1})(x_{j_2}-a_{j_1j_2s_2})}{(x_{j_1}-a_{j_2j_11})(x_{j_2}-a_{j_2j_21})^2},$$

and the deformation of the foregoing equation is

$$\Big[1+\frac{(a_{j_1j_11}-a_{j_2j_11})}{(x_{j_1}-a_{j_1j_11})}\Big]dx_{j_1}=\frac{a_{j_1j_10}}{a_{j_2j_20}}\Big[1+(a_{j_2j_21}-a_{j_1j_2s_2})\frac{1}{(x_{j_2}-a_{j_2j_21})}\Big]dx_{j_2}.$$

With an initial condition of (x_{j_10},x_{j_20}) at $t=t_0$, the integration of the above equation gives the first integral manifold for $a_{j_2j_21}=a_{j_1j_2s_1}$ $(s_1,s_2\in\{1,\ 2\},s_1\neq s_2)$ is

$$\begin{aligned}&\big[(x_{j_1}-a_{j_1j_11})-(x_{j_10}-a_{j_1j_11})\big]+(a_{j_1j_11}-a_{j_2j_11})\ln\frac{|x_{j_1}-a_{j_1j_11}|}{|x_{j_10}-a_{j_1j_11}|}\\&=\frac{a_{j_1j_10}}{a_{j_2j_20}}\Big\{\big[(x_{j_2}-a_{j_2j_21})-(x_{j_20}-a_{j_2j_21})\big]+(a_{j_2j_21}-a_{j_1j_2s_2})\ln\frac{|x_{j_2}-a_{j_2j_21}|}{|x_{j_20}-a_{j_2j_21}|}\Big\}.\end{aligned}$$

(viii$_{2a}$) At $x_{j_2}^*=a_{j_1j_2s_1}=a_{j_2j_21}$ with $\bar{x}_{j_1}\neq a_{j_1j_11},a_{j_2j_11}$, from the first integral manifold, the flow is an asymptotic flow. In addition,

$$\frac{dx_{j_1}}{dx_{j_2}}\bigg|_{x_{j_2}^*=a_{j_2j_21}\pm\varepsilon}=\pm\frac{a_{j_1j_10}}{a_{j_2j_20}}\frac{(\bar{x}_{j_1}-a_{j_1j_11})(a_{j_1j_2s_1}-a_{j_1j_2s_2})}{(\bar{x}_{j_1}-a_{j_2j_11})\varepsilon}.$$

If

$$\frac{a_{j_1j_10}}{a_{j_2j_20}}\frac{(\overline{x}_{j_1}-a_{j_1j_11})(a_{j_1j_2s_1}-a_{j_1j_2s_2})}{(\overline{x}_{j_1}-a_{j_2j_11})}>0,$$

the flow at $x_{j_2}^*=a_{j_1j_2s_1}=a_{j_2j_21}$ is a down-up flow. If

$$\frac{a_{j_1j_10}}{a_{j_2j_20}}\frac{(\overline{x}_{j_1}-a_{j_1j_11})(a_{j_1j_2s_1}-a_{j_1j_2s_2})}{(\overline{x}_{j_1}-a_{j_2j_11})}<0,$$

the flow at $x_{j_2}^*=a_{j_1j_2s_1}=a_{j_2j_21}$ is an up-down flow. The variational equation at $x_{j_2}^*=a_{j_1j_2s_1}=a_{j_2j_21}$ is

$$\Delta\dot{x}_{j_2}=a_{j_2j_20}(\overline{x}_{j_1}-a_{j_2j_11})(\Delta x_{j_2})^2.$$

The asymptotic flows in the x_{j_2}-direction are upper-saddle and lower-saddle for $a_{j_2j_20}>0$ and $a_{j_2j_20}<0$, respectively.

Therefore, the infinite-equilibrium of $x_{j_2}^*=a_{j_1j_2s_1}=a_{j_2j_21}$ with $\overline{x}_{j_1}\neq a_{j_1j_11}$ has the following properties as in Eqs. (1.227)–(1.230):

- For $a_{j_1j_10}(\overline{x}_{j_1}-a_{j_1j_11})(a_{j_1j_2s_1}-a_{j_1j_2s_2})>0$ and $a_{j_2j_20}(\overline{x}_{j_1}-a_{j_2j_11})>0$,

$$(\overline{x}_{j_1},a_{j_2j_21})=\underbrace{(\text{DU},\text{US})}_{\text{down-up upper-saddle}}.$$

  The infinite-equilibrium of $x_{j_2}^*=a_{j_1j_2s_1}=a_{j_2j_21}$ is an (DU,US)-down-up upper-saddle.
- For $a_{j_1j_10}(\overline{x}_{j_1}-a_{j_1j_11})(a_{j_1j_2s_1}-a_{j_1j_2s_2})<0$ and $a_{j_2j_20}(\overline{x}_{j_1}-a_{j_2j_11})>0$,

$$(\overline{x}_{j_1},a_{j_2j_21})=\underbrace{(\text{UD},\text{US})}_{\text{up-down upper-saddle}}.$$

  The infinite-equilibrium of $x_{j_2}^*=a_{j_1j_2s_1}=a_{j_2j_21}$ is a (UD,US)-up-down upper-saddle.
- For $a_{j_1j_10}(\overline{x}_{j_1}-a_{j_1j_11})(a_{j_1j_2s_1}-a_{j_1j_2s_2})>0$ and $a_{j_2j_20}(\overline{x}_{j_1}-a_{j_2j_11})<0$,

$$(\overline{x}_{j_1},a_{j_2j_21})=\underbrace{(\text{UD},\text{LS})}_{\text{up-down lower-saddle}}.$$

  The infinite-equilibrium of $x_{j_2}^*=a_{j_1j_2s_1}=a_{j_2j_21}$ is a (UD,LS)-up-down lower-saddle.

- For $a_{j_1j_10}(\bar{x}_{j_1} - a_{j_1j_11})(a_{j_1j_2s_1} - a_{j_1j_2s_2}) < 0$ and $a_{j_2j_20}(\bar{x}_{j_1} - a_{j_2j_11}) < 0$,

$$(\bar{x}_{j_1}, a_{j_2j_21}) = \underbrace{(\mathrm{DU}, \mathrm{LS})}_{\text{down-up lower-saddle}}.$$

The infinite-equilibrium of $x^*_{j_2} = a_{j_1j_2s_1} = a_{j_2j_21}$ is a (DU,LS)-down-up lower-saddle.

(viii$_{2b}$) From case (viii$_{2a}$), for $\bar{x}_{j_1} = a_{j_1j_11}$, the equilibrium of $(x^*_{j_1}, x^*_{j_2}) = (a_{j_1j_11}, a_{j_2j_21})$ with $a_{j_2j_21} = a_{j_1j_2s_1}$ has the following properties as in Eqs. (1.231)–(1.234):

- For $a_{j_1j_10}(a_{j_1j_2s_1} - a_{j_1j_2s_2}) > 0$ and $a_{j_2j_20}(a_{j_1j_11} - a_{j_2j_11}) > 0$,

$$(a_{j_1j_11}, a_{j_2j_21}) = \underbrace{(\mathrm{DU} : \mathrm{UD}, \mathrm{US})}_{\text{down-up-to-up-down upper-saddle}}.$$

The equilibrium of $(x^*_{j_1}, x^*_{j_2}) = (a_{j_1j_11}, a_{j_2j_21})$ is a (DU:UD,US)-down-up-to-up-down upper-saddle.

- For $a_{j_1j_10}(a_{j_1j_2s_1} - a_{j_1j_2s_2}) < 0$ and $a_{j_2j_20}(a_{j_1j_11} - a_{j_2j_11}) > 0$,

$$(a_{j_1j_11}, a_{j_2j_21}) = \underbrace{(\mathrm{UD} : \mathrm{DU}, \mathrm{US})}_{\text{up-down-to-down-up upper-saddle}}.$$

The equilibrium of $(x^*_{j_1}, x^*_{j_2}) = (a_{j_1j_11}, a_{j_2j_21})$ is a (UD:DU,US)-up-down-to-down-up upper-saddle.

- For $a_{j_1j_10}(a_{j_1j_2s_1} - a_{j_1j_2s_2}) > 0$ and $a_{j_2j_20}(a_{j_1j_11} - a_{j_2j_11}) < 0$,

$$(a_{j_1j_11}, a_{j_2j_21}) = \underbrace{(\mathrm{UD} : \mathrm{DU}, \mathrm{LS})}_{\text{up-down-to-down-up lower-saddle}}.$$

The equilibrium of $(x^*_{j_1}, x^*_{j_2}) = (a_{j_1j_11}, a_{j_2j_21})$ is a (UD:DU,LS)-up-down-to-down-up lower-saddle.

- For $a_{j_1j_10}(a_{j_1j_2s_1} - a_{j_1j_2s_2}) < 0$ and $a_{j_2j_20}(a_{j_1j_11} - a_{j_2j_11}) < 0$,

$$(a_{j_1j_11}, a_{j_2j_21}) = \underbrace{(\mathrm{DU} : \mathrm{UD}, \mathrm{LS})}_{\text{down-up-to-up-down lower-saddle}}.$$

The equilibrium of $(x^*_{j_1}, x^*_{j_2}) = (a_{j_1j_11}, a_{j_2j_21})$ is a (DU:UD,LS)-down-up-to-up-down lower-saddle.

(viii$_{2c}$) At $x_{j_1}^* = a_{j_2j_1 1}$ with $\bar{x}_{j_2} \neq a_{j_2j_2 1} = a_{j_1j_2s_1}, a_{j_1j_2s_2}$, from the first integral manifold, the flow is an asymptotic flow. In addition,

$$\left.\frac{dx_{j_2}}{dx_{j_1}}\right|_{x_{j_1}^* = a_{j_2j_1 1}} = \left.\frac{a_{j_2j_2 0}}{a_{j_1j_1 0}} \frac{(x_{j_1} - a_{j_2j_1 1})(\bar{x}_{j_2} - a_{j_2j_2 1})}{(x_{j_1} - a_{j_1j_1 1})(\bar{x}_{j_2} - a_{j_1j_2s_2})}\right|_{x_{j_1}^* = a_{j_2j_1 1}} = 0.$$

If

$$\left.\frac{d^2x_{j_2}}{dx_{j_1}^2}\right|_{x_{j_1}^* = a_{j_2j_1 1}} = \frac{a_{j_2j_2 0}}{a_{j_1j_1 0}} \frac{(\bar{x}_{j_2} - a_{j_2j_2 1})^2}{(a_{j_2j_1 1} - a_{j_1j_1 1})(\bar{x}_{j_2} - a_{j_1j_2s_1})(\bar{x}_{j_2} - a_{j_1j_2s_2})} > 0,$$

the flow at $x_{j_1}^* = a_{j_2j_1 1}$ is an up-parabola flow. If

$$\left.\frac{d^2x_{j_2}}{dx_{j_1}^2}\right|_{x_{j_1}^* = a_{j_2j_1 1}} = \frac{a_{j_2j_2 0}}{a_{j_1j_1 0}} \frac{(\bar{x}_{j_2} - a_{j_2j_2 1})^2}{(a_{j_2j_1 1} - a_{j_1j_1 1})(\bar{x}_{j_2} - a_{j_1j_2s_1})(\bar{x}_{j_2} - a_{j_1j_2s_2})} < 0,$$

the flow at $x_{j_1}^* = a_{j_2j_1 1}$ is a down-parabola flow. At $x_{j_1}^* = a_{j_2j_1 1}$, let

$$\dot{x}_{j_1} = a_{j_1j_1 0}(a_{j_2j_1 1} - a_{j_1j_1 1})(\bar{x}_{j_2} - a_{j_1j_2s_1})(\bar{x}_{j_2} - a_{j_1j_2s_2}).$$

The parabola flows are positive and negative for $\dot{x}_{j_1} > 0$ and $\dot{x}_{j_1} < 0$, respectively.

- For $a_{j_1j_1 0}(a_{j_2j_1 1} - a_{j_1j_1 1})(\bar{x}_{j_2} - a_{j_1j_2s_1})(\bar{x}_{j_2} - a_{j_1j_2s_2}) < 0$ and $a_{j_2j_2 0}(\bar{x}_{j_2} - a_{j_2j_2 1})^2 > 0$,

$$(\dot{x}_{j_1}, a_{j_2j_1 1}) = \underbrace{(\text{pF}, \text{UP})}_{\text{up-parabola flow }(+)}.$$

- For $a_{j_1j_1 0}(a_{j_2j_1 1} - a_{j_1j_1 1})(\bar{x}_{j_2} - a_{j_1j_2s_1})(\bar{x}_{j_2} - a_{j_1j_2s_2}) < 0$ and $a_{j_2j_2 0}(\bar{x}_{j_2} - a_{j_2j_2 1})^2 > 0$,

$$(\dot{x}_{j_1}, a_{j_2j_1 1}) = \underbrace{(\text{nF}, \text{DP})}_{\text{down-parabola flow }(-)}.$$

- For $a_{j_1j_1 0}(a_{j_2j_1 1} - a_{j_1j_1 1})(\bar{x}_{j_2} - a_{j_1j_2s_1})(\bar{x}_{j_2} - a_{j_1j_2s_2}) > 0$ and $a_{j_2j_2 0}(\bar{x}_{j_2} - a_{j_2j_2 1})^2 < 0$,

$$(\dot{x}_{j_1}, a_{j_2j_1 1}) = \underbrace{(\text{pF}, \text{DP})}_{\text{down-parabola flow }(+)}.$$

- For $a_{j_1j_10}(a_{j_2j_11}-a_{j_1j_11})(\bar{x}_{j_2}-a_{j_1j_2s_1})(\bar{x}_{j_2}-a_{j_1j_2s_2})<0$ and $a_{j_2j_20}(\bar{x}_{j_2}-a_{j_2j_21})^2>0$,

$$(\dot{x}_{j_1}, a_{j_2j_11}) = \underbrace{(\text{nF, UP})}_{\text{up-parabola flow }(-)}.$$

From the above analysis with $\bar{x}_{j_2}=a_{j_1j_2s_1}$ and the proof of (viii$_{2a}$) for $\bar{x}_{j_1}=a_{j_2j_11}$, the equilibrium of $(x^*_{j_2}, x^*_{j_1})=(a_{j_1j_2s_1}, a_{j_2j_11})$ with $a_{j_2j_21}=a_{j_1j_2s_1}$ has the following properties as in Eqs. (1.235)–(1.238):

- For $a_{j_1j_10}(a_{j_2j_11}-a_{j_1j_11})(a_{j_1j_2s_1}-a_{j_1j_2s_2})>0$ and $a_{j_2j_20}>0$,

$$(a_{j_1j_2s_1}, a_{j_2j_11}) = \underbrace{({}_{\text{UD}}\text{LS:}{}_{\text{DU}}\text{US, DP}_- : \text{UP}_+)}_{\text{hyperbolic lower-to-upper saddle}}.$$

The equilibrium of $(x^*_{j_2}, x^*_{j_1})=(a_{j_1j_2s_1}, a_{j_2j_11})$ is a (${}_{\text{UD}}$LS:${}_{\text{DU}}$US, DP$_-$:UP$_+$)-hyperbolic lower-to-upper saddle.

- For $a_{j_1j_10}(a_{j_2j_11}-a_{j_1j_11})(a_{j_1j_2s_1}-a_{j_1j_2s_2})<0$ and $a_{j_2j_20}>0$,

$$(a_{j_1j_2s_1}, a_{j_2j_11}) = \underbrace{({}_{\text{DU}}\text{US:}{}_{\text{UD}}\text{LS, UP}_- : \text{DP}_+)}_{\text{hyperbolic-secant upper-to-lower saddle}}.$$

The equilibrium of $(x^*_{j_2}, x^*_{j_1})=(a_{j_1j_2s_1}, a_{j_2j_11})$ is a (${}_{\text{DU}}$US:${}_{\text{UD}}$LS, UP$_-$:DP$_+$)-hyperbolic-secant upper-to-lower saddle.

- For $a_{j_1j_10}(a_{j_2j_11}-a_{j_1j_11})(a_{j_1j_2s_1}-a_{j_1j_2s_2})>0$ and $a_{j_2j_20}<0$,

$$(a_{j_1j_2s_1}, a_{j_2j_11}) = \underbrace{({}_{\text{DU}}\text{LS:}{}_{\text{UD}}\text{US, UP}_+ : \text{DP}_-)}_{\text{hyperbolic-secant lower-to-upper saddle}}.$$

The equilibrium of $(x^*_{j_2}, x^*_{j_1})=(a_{j_1j_2s_1}, a_{j_2j_11})$ is a (${}_{\text{DU}}$LS:${}_{\text{UD}}$US, UP$_+$:DP$_-$)-hyperbolic-secant lower-to-upper saddle.

- For $a_{j_1j_10}(a_{j_2j_11}-a_{j_1j_11})(a_{j_1j_2s_1}-a_{j_1j_2s_2})<0$ and $a_{j_2j_20}<0$,

$$(a_{j_1j_2s_1}, a_{j_2j_11}) = \underbrace{({}_{\text{UD}}\text{US : }{}_{\text{DU}}\text{LS, DP}_+ : \text{UP}_-)}_{\text{hypebolic upper-to-lower saddle}}.$$

The equilibrium of $(x^*_{j_2}, x^*_{j_1})=(a_{j_1j_2s_1}, a_{j_2j_11})$ is a (${}_{\text{UD}}$US:${}_{\text{DU}}$LS, DP$_+$:UP$_-$)-hyperbolic-upper-to-lower saddle.

(viii$_3$) For $a_{j_1j_11}=a_{j_2j_11}$ and $a_{j_2j_21}\neq a_{j_2j_2s_1}$ ($s_1, s_2\in\{1, 2\}$, $s_1\neq s_2$), in phase space,

$$\frac{dx_{j_1}}{dx_{j_2}}=\frac{a_{j_1j_10}}{a_{j_2j_20}}\frac{(x_{j_2}-a_{j_1j_21})(x_{j_2}-a_{j_1j_22})}{(x_{j_2}-a_{j_2j_21})^2},$$

and the deformation of the foregoing equation is

$$dx_{j_1} = \frac{a_{j_1j_10}}{a_{j_2j_20}}\Big[1 + \sum\nolimits_{s_1=1}^{2}(a_{j_2j_21} - a_{j_1j_2s_1})\frac{1}{(x_{j_2} - a_{j_2j_21})} + \prod\nolimits_{s_1=1}^{2}(a_{j_2j_21} - a_{j_1j_2s_1})\frac{1}{(x_{j_2} - a_{j_2j_21})}\Big]dx_{j_2}.$$

With an initial condition of (x_{j_10}, x_{j_20}) at $t = t_0$, the integration of the above equation gives the first integral manifold for $a_{j_1j_11} = a_{j_2j_11}$ and $a_{j_2j_21} \neq a_{j_1j_2s_1}$ ($s_1, s_2 \in \{1,\ 2\}$, $s_1 \neq s_2$) as

$$\begin{aligned}
&\big[(x_{j_1} - a_{j_1j_11}) - (x_{j_10} - a_{j_1j_11})\big]\\
&= \frac{a_{j_1j_10}}{a_{j_2j_20}}\Big\{\big[(x_{j_2} - a_{j_2j_21}) - (x_{j_20} - a_{j_2j_21})\big]\\
&+\sum\nolimits_{s_1=1}^{2}(a_{j_2j_21} - a_{j_1j_2s_1})\ln\frac{|x_{j_2} - a_{j_2j_21}|}{|x_{j_20} - a_{j_2j_21}|}\\
&-\prod\nolimits_{s_1=1}^{2}(a_{j_2j_21} - a_{j_1j_2s_1})\Big(\frac{1}{x_{j_2} - a_{j_2j_21}} - \frac{1}{x_{j_20} - a_{j_2j_21}}\Big)\Big\}.
\end{aligned}$$

(viii$_{3a}$) For $x_{j_1}^* = a_{j_1j_11} = a_{j_2j_11}$ with $\overline{x}_{j_2} \neq a_{j_2j_21}, a_{j_1j_2s_1}$ ($s_1 = 1, 2$; $a_{j_1j_2s_1} \neq a_{j_2j_21}$), if

$$\frac{dx_{j_2}}{dx_{j_1}}\bigg|_{x_{j_1}^* = a_{j_2j_11}} = \frac{a_{j_2j_20}}{a_{j_1j_10}}\frac{(\overline{x}_{j_2} - a_{j_2j_21})^2}{(\overline{x}_{j_2} - a_{j_1j_21})(\overline{x}_{j_2} - a_{j_1j_22})} > 0,$$

there is an increasing-inflection flow, and if

$$\frac{dx_{j_2}}{dx_{j_1}}\bigg|_{x_{j_1}^* = a_{j_2j_11}} = \frac{a_{j_2j_20}}{a_{j_1j_10}}\frac{(\overline{x}_{j_2} - a_{j_2j_21})^2}{(\overline{x}_{j_2} - a_{j_1j_21})(\overline{x}_{j_2} - a_{j_1j_22})} < 0,$$

there is a decreasing-inflection flow. The variational equation is

$$\Delta\dot{x}_{j_1} = a_{j_1j_10}(\overline{x}_{j_2} - a_{j_1j_2s_1})(\overline{x}_{j_2} - a_{j_1j_2s_2})\Delta x_{j_1}.$$

The inflection-flow is source and sink for $a_{j_1j_10}(\overline{x}_{j_2} - a_{j_1j_2s_1})(\overline{x}_{j_2} - a_{j_1j_2s_2}) > 0$ and $a_{j_1j_10}(\overline{x}_{j_2} - a_{j_1j_2s_1})(\overline{x}_{j_2} - a_{j_1j_2s_2}) < 0$, respectively.

Therefore, the infinite-equilibrium of $x_{j_1}^* = a_{j_1j_11} = a_{j_2j_11}$ with $\overline{x}_{j_2} \neq a_{j_2j_21}, a_{j_1j_2s_1}$ ($a_{j_1j_2s_1} \neq a_{j_1j_21}$) has the following properties as in Eqs. (1.240)–(1.243):

- For $a_{j_1j_10}(\overline{x}_{j_2}-a_{j_1j_2s_1})(\overline{x}_{j_2}-a_{j_1j_2s_2})>0$ and $a_{j_2j_20}(\overline{x}_{j_2}-a_{j_2j_21})^2>0$,

$$(a_{j_1j_11},\overline{x}_{j_2})=\underbrace{(\mathrm{SO},\mathrm{II})}_{\text{increasing-inflection source}}.$$

  The infinite-equilibrium of $x^*_{j_1}=a_{j_1j_11}=a_{j_2j_11}$ is an (SO,II)-increasing-inflection source.
- For $a_{j_1j_10}(\overline{x}_{j_2}-a_{j_1j_2s_1})(\overline{x}_{j_2}-a_{j_1j_2s_2})<0$ and $a_{j_2j_20}(\overline{x}_{j_2}-a_{j_2j_21})^2>0$,

$$(a_{j_1j_11},\overline{x}_{j_2})=\underbrace{(\mathrm{SI},\mathrm{DI})}_{\text{decreasing-inflection sink}}.$$

  The infinite-equilibrium of $x^*_{j_1}=a_{j_1j_11}=a_{j_2j_11}$ is an (SI,DI)-decreasing-inflection sink.
- For $a_{j_1j_10}(\overline{x}_{j_2}-a_{j_1j_2s_1})(\overline{x}_{j_2}-a_{j_1j_2s_2})>0$ and $a_{j_2j_20}(\overline{x}_{j_2}-a_{j_2j_21})^2<0$,

$$(a_{j_1j_11},\overline{x}_{j_2})=\underbrace{(\mathrm{SO},\mathrm{DI})}_{\text{decreasing-inflection source}}.$$

  The infinite-equilibrium of $x^*_{j_1}=a_{j_1j_11}=a_{j_2j_11}$ is an (SO,DI)-decreasing-inflection source.
- For $a_{j_1j_10}(\overline{x}_{j_2}-a_{j_1j_2s_1})(\overline{x}_{j_2}-a_{j_1j_2s_2})<0$ and $a_{j_2j_20}(\overline{x}_{j_2}-a_{j_2j_21})^2<0$,

$$(a_{j_1j_11},\overline{x}_{j_2})=\underbrace{(\mathrm{SI},\mathrm{II})}_{\text{increasing-inflection sink}}.$$

  The infinite-equilibrium of $x^*_{j_1}=a_{j_1j_11}=a_{j_2j_11}$ is an (SI,II)-increasing-inflection sink.

(viii$_{3b}$) For $x^*_{j_2}=a_{j_1j_2s_1}$ $(a_{j_1j_2s_1}\neq a_{j_2j_21})$ with $\overline{x}_{j_1}\neq a_{j_1j_11},a_{j_2j_11}$,

$$\left.\frac{dx_{j_1}}{dx_{j_2}}\right|_{x^*_{j_2}=a_{j_1j_2s_1}}=\left.\frac{a_{j_1j_10}}{a_{j_2j_20}}\frac{(x_{j_2}-a_{j_1j_2s_1})(x_{j_2}-a_{j_1j_2s_2})}{(x_{j_2}-a_{j_2j_21})^2}\right|_{x^*_{j_2}=a_{j_1j_2s_1}}=0.$$

If

$$\left.\frac{d^2x_{j_1}}{dx^2_{j_2}}\right|_{x^*_{j_2}=a_{j_1j_2s_1}}=\frac{a_{j_1j_10}}{a_{j_2j_20}}\frac{(a_{j_1j_2s_1}-a_{j_1j_2s_2})}{(a_{j_1j_2s_1}-a_{j_2j_21})^2}>0,$$

there is an up-parabola flow. If

$$\left.\frac{d^2x_{j_1}}{dx_{j_2}^2}\right|_{x_{j_2}^*=a_{j_1j_2s_1}}=\frac{a_{j_1j_10}}{a_{j_2j_20}}\frac{(a_{j_1j_2s_1}-a_{j_1j_2s_2})}{(a_{j_1j_2s_1}-a_{j_2j_21})^2}<0,$$

there is a down-parabola flow. Let

$$\dot{x}_{j_2}=a_{j_2j_20}(\overline{x}_{j_2}-a_{j_2j_11})(a_{j_1j_2s_1}-a_{j_2j_21})^2.$$

The parabola flows are positive and negative for $\dot{x}_{j_2}>0$ and $\dot{x}_{j_2}<0$, respectively.

- For $a_{j_1j_10}(\overline{x}_{j_1}-a_{j_1j_11})(a_{j_1j_2s_1}-a_{j_1j_2s_2})>0$ and $a_{j_2j_20}(\overline{x}_{j_1}-a_{j_2j_11})(a_{j_1j_2s_1}-a_{j_2j_21})^2>0$,

$$(a_{j_1j_2s_1},\dot{x}_{j_2})=\underbrace{(\text{UP},\text{pF})}_{\text{up-parabola flow }(+)}.$$

- For $a_{j_1j_10}(\overline{x}_{j_1}-a_{j_1j_11})(a_{j_1j_2s_1}-a_{j_1j_2s_2})<0$ and $a_{j_2j_20}(\overline{x}_{j_1}-a_{j_2j_11})(a_{j_1j_2s_1}-a_{j_2j_21})^2>0$,

$$(a_{j_1j_2s_1},\dot{x}_{j_2})=\underbrace{(\text{DP},\text{pF})}_{\text{down-parabola flow }(+)}.$$

- For $a_{j_1j_10}(\overline{x}_{j_1}-a_{j_1j_11})(a_{j_1j_2s_1}-a_{j_1j_2s_2})>0$ and $a_{j_2j_20}(\overline{x}_{j_1}-a_{j_2j_11})(a_{j_1j_2s_1}-a_{j_2j_21})^2<0$,

$$(a_{j_1j_2s_1},\dot{x}_{j_2})=\underbrace{(\text{DP},\text{nF})}_{\text{down-parabola flow }(-)}.$$

- For $a_{j_1j_10}(\overline{x}_{j_1}-a_{j_1j_11})(a_{j_1j_2s_1}-a_{j_1j_2s_2})<0$ and $a_{j_2j_20}(\overline{x}_{j_1}-a_{j_2j_11})(a_{j_1j_2s_1}-a_{j_2j_21})^2<0$,

$$(a_{j_1j_2s_1},\dot{x}_{j_2})=\underbrace{(\text{UP},\text{nF})}_{\text{up-parabola flow }(-)}.$$

From the above analysis of $\overline{x}_{j_1}=a_{j_2j_11}=a_{j_1j_11}$ and the case of (viii$_{3b}$) with $\overline{x}_{j_2}=a_{j_1j_2s_1}$, the equilibrium of $(x_{j_2}^*,x_{j_1}^*)=(a_{j_1j_2s_1},a_{j_2j_11})$ with $a_{j_1j_11}=a_{j_2j_11}$ has the following properties as in Eqs. (1.244)–(1.247):

- For $a_{j_1j_10}(a_{j_1j_2s_1}-a_{j_1j_2s_2})>0$ and $a_{j_2j_20}(a_{j_1j_2s_1}-a_{j_2j_21})^2>0$,

$$(a_{j_1j_2s_1},a_{j_2j_11})=\underbrace{({}_{\text{DI:II}}\text{UP},\ {}_{\text{SI:SO}}\text{US})}_{\text{up-parabola upper-saddle}}.$$

  The equilibrium of $(x_{j_2}^*,x_{j_1}^*)=(a_{j_1j_2s_1},a_{j_2j_11})$ is a $({}_{\text{DI:II}}\text{UP},{}_{\text{SI:SO}}\text{US})$-up-parabola upper-saddle.
- For $a_{j_1j_10}(a_{j_1j_2s_1}-a_{j_1j_2s_2})<0$ and $a_{j_2j_20}(a_{j_1j_2s_1}-a_{j_2j_21})^2>0$,

$$(a_{j_1j_2s_1}, a_{j_2j_1 1}) = \underbrace{({}_{\text{II:DI}}\text{DP},\ {}_{\text{SO:SI}}\text{LS})}_{\text{down-parabola lower-saddle}}.$$

The equilibrium of $(x^*_{j_2}, x^*_{j_1}) = (a_{j_1j_2s_1}, a_{j_2j_1 1})$ is a $({}_{\text{II:DI}}\text{DP},\ {}_{\text{SO:SI}}\text{LS})$-down-parabola lower-saddle.

- For $a_{j_1j_1 0}(a_{j_1j_2s_1} - a_{j_1j_2s_2}) > 0$ and $a_{j_2j_2 0}(a_{j_1j_2s_1} - a_{j_2j_2 1})^2 < 0$,

$$(a_{j_1j_2s_1}, a_{j_2j_1 1}) = \underbrace{({}_{\text{II:DI}}\text{DP},\ {}_{\text{SI:SO}}\text{US})}_{\text{down-parabola upper-saddle}}.$$

The equilibrium of $(x^*_{j_2}, x^*_{j_1}) = (a_{j_1j_2s_1}, a_{j_2j_1 1})$ is a $({}_{\text{II:DI}}\text{DP}, {}_{\text{SI:SO}}\text{US})$-down-parabola upper-saddle.

- For $a_{j_1j_1 0}(a_{j_1j_2s_1} - a_{j_1j_2s_2}) < 0$ and $a_{j_2j_2 0}(a_{j_1j_2s_1} - a_{j_2j_2 1})^2 < 0$,

$$(a_{j_1j_2s_1}, a_{j_2j_1 1}) = \underbrace{({}_{\text{DI:II}}\text{UP},\ {}_{\text{SO:SI}}\text{LS})}_{\text{up-parabola lower-saddle}}.$$

The equilibrium of $(x^*_{j_2}, x^*_{j_1}) = (a_{j_1j_2s_1}, a_{j_2j_1 1})$ is a $({}_{\text{DI:II}}\text{UP}, {}_{\text{SO:SI}}\text{LS})$-up-parabola lower-saddle.

(viii_{3c}) From (viii_{3a}), for $\bar{x}_{j_2} = a_{j_2j_2 1}$, the equilibrium of $(x^*_{j_1}, x^*_{j_2}) = (a_{j_1j_1 1}, a_{j_2j_2 1})$ with $a_{j_1j_1 1} = a_{j_2j_1 1}$ has the following properties as in Eqs. (1.248)–(1.251):

- For $a_{j_1j_1 0}(a_{j_2j_2 1} - a_{j_1j_2s_1})(a_{j_2j_2 1} - a_{j_1j_2s_2}) > 0$ and $a_{j_2j_2 0} > 0$,

$$(a_{j_1j_1 1}, a_{j_2j_2 1}) = \underbrace{(\text{SO}, \text{II} : \text{II})}_{\text{increasing-inflection source}}.$$

The equilibrium of $(x^*_{j_1}, x^*_{j_2}) = (a_{j_1j_1 1}, a_{j_2j_2 1})$ is an (SO,II:II)-increasing-inflection source of the second kind.

- For $a_{j_1j_1 0}(a_{j_2j_2 1} - a_{j_1j_2s_1})(a_{j_2j_2 1} - a_{j_1j_2s_2}) < 0$ and $a_{j_2j_2 0} > 0$,

$$(a_{j_1j_1 1}, a_{j_2j_2 1}) = \underbrace{(\text{SI}, \text{DI} : \text{DI})}_{\text{dereasing-inflection sink}}.$$

The equilibrium of $(x^*_{j_1}, x^*_{j_2}) = (a_{j_1j_1 1}, a_{j_2j_2 1})$ is an (SI,DI:DI)-deceasing-inflection sink of the second kind.

- For $a_{j_1j_1 0}(a_{j_2j_2 1} - a_{j_1j_2s_1})(a_{j_2j_2 1} - a_{j_1j_2s_2}) > 0$ and $a_{j_2j_2 0} < 0$,

$$(a_{j_1j_11}, a_{j_2j_21}) = \underbrace{(\mathrm{SO}, \mathrm{DI}:\mathrm{DI})}_{\text{dereasing-inflection source}}.$$

The equilibrium of $(x_{j_1}^*, x_{j_2}^*) = (a_{j_1j_11}, a_{j_2j_21})$ is an (SO,DI:DI)-decreasing-inflecting source of the second kind.

- For $a_{j_1j_10}(a_{j_2j_21} - a_{j_1j_2s_1})(a_{j_2j_21} - a_{j_1j_2s_2}) < 0$ and $a_{j_2j_20} < 0$,

$$(a_{j_1j_11}, a_{j_2j_21}) = \underbrace{(\mathrm{SI}, \mathrm{II}:\mathrm{II})}_{\text{increasing-inflection sink}}.$$

The equilibrium of $(x_{j_1}^*, x_{j_2}^*) = (a_{j_1j_11}, a_{j_2j_21})$ is an (SI,II:II)-increasing-inflection sink of the second kind.

(viii$_4$) For $a_{j_2j_21} = a_{j_1j_2s_1}$ $(s_1, s_2 \in \{1,\ 2\}, s_1 \neq s_2)$ and $a_{j_1j_11} = a_{j_2j_11}$,

$$\frac{dx_{j_1}}{dx_{j_2}} = \frac{a_{j_1j_10}}{a_{j_2j_20}} \frac{(x_{j_2} - a_{j_1j_2s_2})}{(x_{j_2} - a_{j_2j_21})},$$

and the deformation of the foregoing equation is

$$dx_{j_1} = \frac{a_{j_1j_10}}{a_{j_2j_20}} \Big[1 + \frac{(a_{j_2j_21} - a_{j_1j_2s_2})}{(x_{j_2} - a_{j_2j_21})}\Big] dx_{j_2}.$$

With an initial condition of (x_{j_10}, x_{j_20}) at $t = t_0$, the integration of the above equation gives the first integral manifold for $a_{j_1j_11} = a_{j_2j_11}$ and $a_{j_2j_21} = a_{j_1j_2s_1}$ $(s_1, s_2 \in \{1,\ 2\}, s_1 \neq s_2)$ as

$$\begin{aligned}\big[(x_{j_1} - a_{j_1j_11}) - (x_{j_10} - a_{j_1j_11})\big] = \frac{a_{j_1j_10}}{a_{j_2j_20}} \Big\{ &\big[(x_{j_2} - a_{j_2j_21}) - (x_{j_20} - a_{j_2j_21})\big] \\ &+ (a_{j_2j_21} - a_{j_1j_2s_1}) \ln \frac{|x_{j_2} - a_{j_2j_21}|}{|x_{j_20} - a_{j_2j_21}|} \Big\}.\end{aligned}$$

(viii$_{4a}$) From (viii$_{3a}$), for $\bar{x}_{j_2} = a_{j_1j_2s_1} = a_{j_2j_21}$, the equilibrium of $(x_{j_1}^*, x_{j_2}^*) = (a_{j_1j_11}, a_{j_1j_2s_1})$ with $a_{j_1j_11} = a_{j_2j_11}$ and $a_{j_1j_2s_1} = a_{j_2j_21}$ has the following properties as in Eqs. (1.253)–(1.256):

- For $a_{j_1j_10}(a_{j_1j_2s_1} - a_{j_1j_2s_2}) > 0$ and $a_{j_2j_20}(a_{j_1j_2s_1} - a_{j_2j_21})^2 > 0$,

$$(a_{j_1j_11}, a_{j_1j_2s_1}) = \underbrace{({}_{\mathrm{LS:US}}\mathrm{DU}, {}_{\mathrm{DI}}\mathrm{SI:}_{\mathrm{II}}\mathrm{SO})}_{\text{down-up inflection sink-to-source}}.$$

The equilibrium of $(x_{j_1}^*, x_{j_2}^*) = (a_{j_1j_11}, a_{j_1j_2s_1})$ is an (${}_{\mathrm{LS:US}}\mathrm{DU}$,${}_{\mathrm{DI}}\mathrm{SI:}_{\mathrm{II}}\mathrm{SO}$)-down-up inflection sink-to-source.

- For $a_{j_1j_10}(a_{j_1j_2s_1} - a_{j_1j_2s_2}) < 0$ and $a_{j_2j_20}(a_{j_1j_2s_1} - a_{j_2j_21})^2 > 0$,

$$(a_{j_1j_11}, a_{j_1j_2s_1}) = \underbrace{({}_{\mathrm{LS:US}}\mathrm{UD}, {}_{\mathrm{II}}\mathrm{SO:}{}_{\mathrm{DI}}\mathrm{SI})}_{\text{up-down inflection source-to-sink}}.$$

The equilibrium of $(x^*_{j_1}, x^*_{j_2}) = (a_{j_1j_11}, a_{j_1j_2s_1})$ is an $({}_{\mathrm{LS:US}}\mathrm{UD}, {}_{\mathrm{II}}\mathrm{SO:}{}_{\mathrm{DI}}\mathrm{SI})$ inflection source-to-sink.

- For $a_{j_1j_10}(a_{j_1j_2s_1} - a_{j_1j_2s_2}) > 0$ and $a_{j_2j_20}(a_{j_1j_2s_1} - a_{j_2j_21})^2 < 0$,

$$(a_{j_1j_11}, a_{j_1j_2s_1}) = \underbrace{({}_{\mathrm{US:LS}}\mathrm{UD}, {}_{\mathrm{II}}\mathrm{SI:}{}_{\mathrm{DI}}\mathrm{SO})}_{\text{up-down inflection sink-to-source}}.$$

The equilibrium of $(x^*_{j_1}, x^*_{j_2}) = (a_{j_1j_11}, a_{j_1j_2s_1})$ is a $({}_{\mathrm{US:LS}}\mathrm{UD}, {}_{\mathrm{II}}\mathrm{SI:}{}_{\mathrm{DI}}\mathrm{SO})$-up-down inflection sink-to-source.

- For $a_{j_1j_10}(a_{j_1j_2s_1} - a_{j_1j_2s_2}) < 0$ and $a_{j_2j_20}(a_{j_1j_2s_1} - a_{j_2j_21})^2 < 0$,

$$(a_{j_1j_11}, a_{j_1j_2s_1}) = \underbrace{({}_{\mathrm{US:LS}}\mathrm{DU}, {}_{\mathrm{DI}}\mathrm{SO:}{}_{\mathrm{II}}\mathrm{SI})}_{\text{down-up inflection source-to-sink}}.$$

The equilibrium of $(x^*_{j_1}, x^*_{j_2}) = (a_{j_1j_11}, a_{j_1j_2s_1})$ is a $({}_{\mathrm{US:LS}}\mathrm{DU}, {}_{\mathrm{DI}}\mathrm{SI:}{}_{\mathrm{II}}\mathrm{SO})$-down-up inflection sink-to-source.

(viii$_{4b}$) From (viii$_{3b}$), let $a_{j_1j_2s_1} = a_{j_1j_2s_2}$. Thus, the equilibrium of $(x^*_{j_2}, x^*_{j_1}) = (a_{j_1j_2s_2}, a_{j_2j_11})$ with $a_{j_1j_11} = a_{j_2j_11}$ and $a_{j_1j_2s_1} \neq a_{j_2j_21}$ has the following properties as in Eqs. (1.257)–(1.260):

- For $a_{j_1j_10}(a_{j_1j_2s_2} - a_{j_1j_2s_1}) > 0$ and $a_{j_2j_20}(a_{j_1j_2s_2} - a_{j_2j_21})^2 > 0$,

$$(a_{j_1j_2s_2}, a_{j_2j_11}) = \underbrace{({}_{\mathrm{DI:II}}\mathrm{UP}, {}_{\mathrm{SI:ISO}}\mathrm{US})}_{\text{up-parabola upper-saddle}}.$$

The equilibrium of $(x^*_{j_2}, x^*_{j_1}) = (a_{j_1j_2s_2}, a_{j_2j_11})$ is a $({}_{\mathrm{DI:II}}\mathrm{UP}, {}_{\mathrm{SI:SO}}\mathrm{US})$-up-parabola upper-saddle.

- For $a_{j_1j_10}(a_{j_1j_2s_2} - a_{j_1j_2s_1}) < 0$ and $a_{j_2j_20}(a_{j_1j_2s_2} - a_{j_2j_21})^2 > 0$,

$$(a_{j_1j_2s_2}, a_{j_2j_11}) = \underbrace{({}_{\mathrm{II:DI}}\mathrm{DP}, {}_{\mathrm{SO:SI}}\mathrm{LS})}_{\text{down-parabola lower-saddle}}.$$

The equilibrium of $(x^*_{j_2}, x^*_{j_1}) = (a_{j_1j_2s_2}, a_{j_2j_11})$ is an $({}_{\mathrm{II:DI}}\mathrm{DP}, {}_{\mathrm{SO:SI}}\mathrm{LS})$-down-parabola lower-saddle.

- For $a_{j_1j_10}(a_{j_1j_2s_2} - a_{j_1j_2s_1}) > 0$ and $a_{j_2j_20}(a_{j_1j_2s_2} - a_{j_2j_21})^2 < 0$,

$$(a_{j_1j_2s_2}, a_{j_2j_1 1}) = \underbrace{({}_{\text{II:DI}}\text{DP}, {}_{\text{SI:SO}}\text{US})}_{\text{down-parabola upper-saddle}}.$$

The equilibrium of $(x^*_{j_2}, x^*_{j_1}) = (a_{j_1j_2s_2}, a_{j_2j_1 1})$ is an (${}_{\text{II:DI}}$DP, ${}_{\text{SI:SO}}$US)-down-parabola upper-saddle.

- For $a_{j_1j_1 0}(a_{j_1j_2s_2} - a_{j_1j_2s_1}) < 0$ and $a_{j_2j_2 0}(a_{j_1j_2s_2} - a_{j_2j_2 1})^2 < 0$,

$$(a_{j_1j_2s_2}, a_{j_2j_1 1}) = \underbrace{({}_{\text{DI:II}}\text{UP}, {}_{\text{SO:SI}}\text{LS})}_{\text{up-parabola lower-saddle}}.$$

The equilibrium of $(x^*_{j_2}, x^*_{j_1}) = (a_{j_1j_2s_2}, a_{j_2j_1 1})$ is a (${}_{\text{DI:II}}$UP, ${}_{\text{SO:SI}}$LS)-up-parabola lower-saddle.

(ix) For $\Delta_{j_1j_2} > 0$ and $\Delta_{j_2j_2} = 0$, the standard form is

$$\begin{aligned}
\dot{x}_{j_2} &= a_{j_2j_2 0}(x_{j_1} - a_{j_2j_1 1})(x_{j_2} - a_{j_2j_2 1})(x_{j_2} - a_{j_2j_2 2}),\\
\dot{x}_{j_1} &= a_{j_1j_1 0}(x_{j_1} - a_{j_1j_1 1})(x_{j_2} - a_{j_1j_2 1})(x_{j_2} - a_{j_1j_2 2})
\end{aligned}$$

where

$$\begin{aligned}
&b_{j_1j_2 1}, b_{j_1j_2 2} = -\frac{1}{2}(B_{j_1j_2} \pm \sqrt{\Delta_{j_1j_2}}),\\
&\{a_{j_1j_2 1}, a_{j_1j_2 2}\} = \text{sort}\{b_{j_1j_2 1}, b_{j_1j_2 2}\}, a_{j_1j_2 1} < a_{j_1j_2 2};\\
&b_{j_2j_2 1}, b_{j_2j_2 2} = -\frac{1}{2}(B_{j_2j_2} \pm \sqrt{\Delta_{j_2j_2}}),\\
&\{a_{j_2j_2 1}, a_{j_2j_2 2}\} = \text{sort}\{b_{j_2j_2 1}, b_{j_2j_2 2}\}, a_{j_2j_2 1} < a_{j_2j_2 2}.
\end{aligned}$$

(ix_1) For $a_{j_1j_2s_1} \neq a_{j_2j_2 1}, a_{j_2j_2 2}$ $(s_1 = 1, 2)$ and $a_{j_1j_1 1} \neq a_{j_2j_1 1}$, in phase space,

$$\frac{dx_{j_1}}{dx_{j_2}} = \frac{a_{j_1j_1 0}}{a_{j_2j_2 0}} \frac{(x_{j_1} - a_{j_1j_1 1})(x_{j_2} - a_{j_1j_2 1})(x_{j_2} - a_{j_1j_2 2})}{(x_{j_1} - a_{j_2j_1 1})(x_{j_2} - a_{j_2j_2 1})(x_{j_2} - a_{j_2j_2 2})},$$

and the deformation of the foregoing equation is

$$\begin{aligned}
&[1 + \frac{(a_{j_1j_1 1} - a_{j_2j_1 1})}{(x_{j_1} - a_{j_1j_1 1})}]dx_{j_1}\\
&= \frac{a_{j_1j_1 0}}{a_{j_2j_2 0}}[1 + \sum\nolimits_{l_1=1}^{2} \frac{\prod_{s_1=1}^{2}(a_{j_2j_2l_1} - a_{j_1j_2s_1})}{(a_{j_2j_2l_1} - a_{j_2j_2l_2})} \frac{1}{(x_{j_2} - a_{j_2j_2l_1})}]dx_{j_2}.
\end{aligned}$$

With an initial condition of $(x_{j_1 0}, x_{j_2 0})$ at $t = t_0$, the integration of the above equation gives the first integral manifold as

$$\left[(x_{j_1}-a_{j_1j_1 1})-(x_{j_1 0}-a_{j_1j_1 1})\right]+(a_{j_1j_1 1}-a_{j_2j_1 1})\ln\frac{|x_{j_1}-a_{j_1j_1 1}|}{|x_{j_1 0}-a_{j_1j_1 1}|}$$
$$=\frac{a_{j_1j_1 0}}{a_{j_2j_2 0}}\left[(x_{j_2}-x_{j_2 0})+\sum_{l_1=1}^{2}\frac{\prod_{s_1=1}^{2}(a_{j_2j_2l_1}-a_{j_1j_2s_1})}{(a_{j_2j_2l_1}-a_{j_2j_2l_2})}\ln\frac{|x_{j_2}-a_{j_2j_2l_1}|}{|x_{j_2 0}-a_{j_2j_2l_1}|}\right].$$

(ix$_{1a}$) For $a_{j_1j_1 1}=a_{j_2j_1 1}$, $(s_1, s_2, l_1, l_2 \in \{1,\ 2\};\ s_1\neq s_2,\ l_1\neq l_2)$ and $a_{j_1j_1 1}\neq a_{j_2j_1 1}$, at $(x^*_{j_1},x^*_{j_2})=(a_{j_1j_1 1},a_{j_2j_2s_1})$, the variational equations are

$$\Delta\dot{x}_{j_2}=a_{j_2j_2 0}(a_{j_1j_1 1}-a_{j_2j_1 1})(a_{j_2j_2s_1}-a_{j_2j_2s_2})\Delta x_{j_2},$$
$$\Delta\dot{x}_{j_1}=a_{j_1j_1 0}\prod_{l_1=1}^{2}(a_{j_2j_2s_1}-a_{j_1j_2l_1})\Delta x_{j_1}.$$

The flows in the x_{j_1}-direction are source and sink for $a_{j_1j_1 0}\prod_{l_1=1}^{2}(a_{j_2j_2s_1}-a_{j_1j_2l_1})>0$ and $a_{j_1j_1 0}\prod_{l_1=1}^{2}(a_{j_2j_2s_1}-a_{j_1j_2l_1})<0$, respectively. The flows in the x_{j_2}-direction for $a_{j_2j_2 0}(a_{j_1j_1 1}-a_{j_2j_1 1})(a_{j_2j_2s_1}-a_{j_2j_2s_2})>0$ and $a_{j_2j_2 0}(a_{j_1j_1 1}-a_{j_2j_1 1})(a_{j_2j_2s_1}-a_{j_2j_2s_2})<0$ are source and sink, respectively.

Therefore, the equilibrium of $(x^*_{j_1},x^*_{j_2})=(a_{j_1j_1 1},a_{j_2j_2s_1})\,(s_1,s_2\in\{1,\ 2\},\ s_1\neq s_2)$ has the following properties as in Eqs. (1.264)–(1.267):

- For $a_{j_1j_1 0}\prod_{l_1=1}^{2}(a_{j_2j_2s_1}-a_{j_1j_2l_1})>0$ and $a_{j_2j_2 0}(a_{j_1j_1 1}-a_{j_2j_1 1})(a_{j_2j_2s_1}-a_{j_2j_2s_2})>0$,

$$(a_{j_1j_1 1},a_{j_2j_2s_1})=\underbrace{(\text{SO, SO})}_{\text{source}}.$$

The equilibrium of $(x^*_{j_1},x^*_{j_2})=(a_{j_1j_1 1},a_{j_2j_2s_1})$ is an (SO,SO)-source.

- For $a_{j_1j_1 0}\prod_{l_1=1}^{2}(a_{j_2j_2s_1}-a_{j_1j_2l_1})<0$ and $a_{j_2j_2 0}(a_{j_1j_1 1}-a_{j_2j_1 1})(a_{j_2j_2s_1}-a_{j_2j_2s_2})>0$,

$$(a_{j_1j_1 1},a_{j_2j_2s_1})=\underbrace{(\text{SI, SO})}_{\text{saddle}}.$$

The equilibrium of $(x^*_{j_1},x^*_{j_2})=(a_{j_1j_1 1},a_{j_2j_2s_1})$ is an (SI,SO)-saddle.

- For $a_{j_1j_1 0}\prod_{l_1=1}^{2}(a_{j_2j_2s_1}-a_{j_1j_2l_1})>0$ and $a_{j_2j_2 0}(a_{j_1j_1 1}-a_{j_2j_1 1})(a_{j_2j_2s_1}-a_{j_2j_2s_2})<0$,

$$(a_{j_1j_1 1},a_{j_2j_2s_1})=\underbrace{(\text{SO, SI})}_{\text{saddle}}.$$

The equilibrium of $(x^*_{j_1},x^*_{j_2})=(a_{j_1j_1 1},a_{j_2j_2s_1})$ is an (SO,SI)-saddle.

- For $a_{j_1j_1 0}\prod_{l_1=1}^{2}(a_{j_2j_2s_1}-a_{j_1j_2l_1})<0$ and $a_{j_2j_2 0}(a_{j_1j_1 1}-a_{j_2j_1 1})(a_{j_2j_2s_1}-a_{j_2j_2s_2})<0$,

$$(a_{j_1j_11}, a_{j_2j_2s_1}) = \underbrace{(\mathrm{SI}, \mathrm{SI})}_{\text{sink}}.$$

The equilibrium of $(x_{j_1}^*, x_{j_2}^*) = (a_{j_1j_11}, a_{j_2j_2s_1})$ is an (SI,SI)-sink.

(ix$_{1b}$) For $x_{j_2}^* = a_{j_1j_2l_1} \neq a_{j_2j_21}, \ a_{j_2j_22}$ and $\overline{x}_{j_1} \neq a_{j_1j_11}, \ a_{j_2j_11}$ $(s_1, s_2 \in \{1, \ 2\}, s_1 \neq s_2)$, in phase space,

$$\left.\frac{dx_{j_1}}{dx_{j_2}}\right|_{x_{j_2}^* = a_{j_1j_2l_1}} = \left.\frac{a_{j_1j_10}}{a_{j_2j_20}} \frac{(\overline{x}_{j_1} - a_{j_1j_11})(x_{j_2} - a_{j_2j_2l_1})(x_{j_2} - a_{j_1j_2l_2})}{(\overline{x}_{j_1} - a_{j_2j_11})\prod_{s_1=1}^{2}(x_{j_2} - a_{j_2j_2s_1})}\right|_{x_{j_2}^* = a_{j_1j_2l_1}} = 0.$$

If

$$\left.\frac{d^2x_{j_1}}{dx_{j_2}^2}\right|_{x_{j_2}^* = a_{j_1j_2l_1}} = \frac{a_{j_1j_10}}{a_{j_2j_20}} \frac{(\overline{x}_{j_1} - a_{j_1j_11})(a_{j_1j_2l_1} - a_{j_1j_2l_2})}{(\overline{x}_{j_1} - a_{j_2j_11})\prod_{s_1=1}^{2}(a_{j_1j_2l_1} - a_{j_2j_2s_1})} > 0,$$

there is an up-parabola flow at $x_{j_2}^* = a_{j_1j_2l_1} \neq a_{j_2j_21}, a_{j_2j_22}$ with $\overline{x}_{j_1} \neq a_{j_1j_11}$. If

$$\left.\frac{d^2x_{j_1}}{dx_{j_2}^2}\right|_{x_{j_2}^* = a_{j_1j_2l_1}} = \frac{a_{j_1j_10}}{a_{j_2j_20}} \frac{(\overline{x}_{j_1} - a_{j_1j_11})(a_{j_1j_2l_1} - a_{j_1j_2l_2})}{(\overline{x}_{j_1} - a_{j_2j_11})\prod_{s_1=1}^{2}(a_{j_1j_2l_1} - a_{j_2j_2s_1})} < 0,$$

there is a down-parabola flow at $x_{j_2}^* = a_{j_1j_2l_1}$ with $\overline{x}_{j_1} \neq a_{j_1j_11}$. Let

$$\dot{x}_{j_2} = a_{j_2j_20}(\overline{x}_{j_1} - a_{j_2j_11})\prod_{s_1=1}^{2}(a_{j_1j_2l_1} - a_{j_2j_2s_1}) \neq 0.$$

The parabola flows at $x_{j_2}^* = a_{j_1j_2s_1} \neq a_{j_2j_21}$ with $\overline{x}_{j_1} \neq a_{j_1j_11}$ are positive and negative for $a_{j_2j_20} > 0$ and $a_{j_2j_20} < 0$, respectively.

- For

$$a_{j_1j_10}(\overline{x}_{j_1} - a_{j_1j_11})(a_{j_1j_2l_1} - a_{j_1j_2l_2}) > 0 \text{ and}$$
$$a_{j_2j_20}(\overline{x}_{j_1} - a_{j_2j_11})\prod_{s_1=1}^{2}(a_{j_1j_2l_1} - a_{j_2j_2s_1}) > 0,$$

we have

$$(a_{j_1j_2s_1}, \dot{x}_{j_2}) = \underbrace{(\mathrm{UP}, \mathrm{pF})}_{\text{up-parabola flow }(+)}.$$

- For

$$a_{j_1j_10}(\overline{x}_{j_1} - a_{j_1j_11})(a_{j_1j_2l_1} - a_{j_1j_2l_2}) < 0 \text{ and}$$
$$a_{j_2j_20}(\overline{x}_{j_1} - a_{j_2j_11})\prod\nolimits_{s_1=1}^{2}(a_{j_1j_2l_1} - a_{j_2j_2s_1}) > 0,$$

we have

$$(a_{j_1j_2s_1}, \dot{x}_{j_2}) = \underbrace{(\text{DP}, \text{pF})}_{\text{down-parabola flow }(+)}.$$

- For

$$a_{j_1j_10}(\overline{x}_{j_1} - a_{j_1j_11})(a_{j_1j_2l_1} - a_{j_1j_2l_2}) > 0 \text{ and}$$
$$a_{j_2j_20}(\overline{x}_{j_1} - a_{j_2j_11})\prod\nolimits_{s_1=1}^{2}(a_{j_1j_2l_1} - a_{j_2j_2s_1}) < 0,$$

we have

$$(a_{j_1j_2s_1}, \dot{x}_{j_2}) = \underbrace{(\text{DP}, \text{nF})}_{\text{down-parabola flow }(-)}.$$

- For

$$a_{j_1j_10}(\overline{x}_{j_1} - a_{j_1j_11})(a_{j_1j_2l_1} - a_{j_2j_2l_2}) < 0 \text{ and}$$
$$a_{j_2j_20}(\overline{x}_{j_1} - a_{j_2j_11})\prod\nolimits_{s_1=1}^{2}(a_{j_1j_2l_1} - a_{j_2j_2s_1}) < 0,$$

we have

$$(a_{j_1j_2s_1}, \dot{x}_{j_2}) = \underbrace{(\text{UP}, \text{nF})}_{\text{up-parabola flow }(-)}.$$

From the above analysis, for $\overline{x}_{j_1} = a_{j_1j_11}$, therefore, the equilibrium of $(x^*_{j_1}, x^*_{j_2}) = (a_{j_1j_11}, a_{j_1j_2l_1})$ $(l_1,\ l_2 \in \{1,\ 2\},\ l_1 \neq l_2)$ with $a_{j_1j_2l_1} \neq a_{j_2j_21}, a_{j_2j_22}$ has the following properties as in Eqs. (1.268)–(1.271):

- For $a_{j_1j_10}(a_{j_1j_2l_1} - a_{j_1j_2l_2}) > 0$ and $a_{j_2j_20}(a_{j_1j_11} - a_{j_2j_11})\prod_{s_1=1}^{2}(a_{j_1j_2l_1} - a_{j_2j_2s_1}) > 0$,

$$(a_{j_1j_11}, a_{j_1j_2l_1}) = \underbrace{(\text{DP} : \text{UP}, \text{pF})}_{\text{hyperbolic flow }(+)}.$$

  The equilibrium of $(x^*_{j_1}, x^*_{j_2}) = (a_{j_1j_11}, a_{j_1j_2l_1})$ is a (DP:UP,pF)-positive hyperbolic flow.
- For $a_{j_1j_10}(a_{j_1j_2l_1} - a_{j_1j_2l_2}) < 0$ and $a_{j_2j_20}(a_{j_1j_11} - a_{j_2j_11})\prod_{s_1=1}^{2}(a_{j_1j_2l_1} - a_{j_2j_2s_1}) > 0$,

$$(a_{j_1j_11}, a_{j_1j_2l_1}) = \underbrace{(\text{UP}:\text{DP},\text{pF})}_{\text{hyperbolic-secant flow }(+)}.$$

The equilibrium of $(x^*_{j_1}, x^*_{j_2}) = (a_{j_1j_11}, a_{j_1j_2l_1})$ is a (UP:DP,pF)-positive hyperbolic-secant flow.

- For $a_{j_1j_10}(a_{j_1j_2l_1} - a_{j_1j_2l_2}) > 0$ and $a_{j_2j_20}(a_{j_1j_11} - a_{j_2j_11})\prod_{s_1=1}^{2}(a_{j_1j_2l_1} - a_{j_2j_2s_1}) < 0$,

$$(a_{j_1j_11}, a_{j_1j_2l_1}) = \underbrace{(\text{UP}:\text{DP},\text{nF})}_{\text{hyperbolic-secant flow }(-)}.$$

The equilibrium of $(x^*_{j_1}, x^*_{j_2}) = (a_{j_1j_11}, a_{j_1j_2l_1})$ is a (UP:DP,nF)-negative hyperbolic-secant flow

- For $a_{j_1j_10}(a_{j_1j_2l_1} - a_{j_1j_2l_2}) < 0$ and $a_{j_2j_20}(a_{j_1j_11} - a_{j_2j_11})\prod_{s_1=1}^{2}(a_{j_1j_2l_1} - a_{j_2j_2s_1}) < 0$,

$$(a_{j_1j_11}, a_{j_1j_2l_1}) = \underbrace{(\text{DP}:\text{UP},\text{nF})}_{\text{hyperbolic flow }(-)}.$$

The equilibrium of $(x^*_{j_1}, x^*_{j_2}) = (a_{j_1j_11}, a_{j_1j_2l_1})$ is a (DP:UP,nF)-negative hyperbolic flow.

(ix$_{1c}$) Consider two cases (I) $x^*_{j_1} = a_{j_2j_11}$ and (II) $x^*_{j_2} = a_{j_1j_2l_1}$.

(I) For $x^*_{j_1} = a_{j_2j_11}$ with $a_{j_2j_11} \neq a_{j_1j_11}$ and $\bar{x}_{j_2} \neq a_{j_2j_2s_1}, a_{j_1j_2l_1}$ (s_1, l_1 = 1, 2), in phase space,

$$\left.\frac{dx_{j_2}}{dx_{j_1}}\right|_{x^*_{j_1} = a_{j_2j_11}} = \left.\frac{a_{j_2j_20}}{a_{j_1j_10}}\frac{(x_{j_1} - a_{j_2j_11})\prod_{s_1=1}^{2}(\bar{x}_{j_2} - a_{j_2j_2s_1})}{(x_{j_1} - a_{j_1j_11})\prod_{l_1=1}^{2}(\bar{x}_{j_2} - a_{j_1j_2l_1})}\right|_{x^*_{j_1} = a_{j_2j_11}} = 0.$$

If

$$\left.\frac{d^2x_{j_2}}{dx^2_{j_1}}\right|_{x^*_{j_1} = a_{j_2j_11}} = \frac{a_{j_2j_20}}{a_{j_1j_10}}\frac{\prod_{s_1=1}^{2}(\bar{x}_{j_2} - a_{j_2j_2s_1})}{(a_{j_2j_11} - a_{j_1j_11})\prod_{l_1=1}^{2}(\bar{x}_{j_2} - a_{j_1j_2l_1})} > 0,$$

there is an up-parabola flow. If

$$\left.\frac{d^2x_{j_2}}{dx^2_{j_1}}\right|_{x^*_{j_1} = a_{j_2j_11}} = \frac{a_{j_2j_20}}{a_{j_1j_10}}\frac{\prod_{s_1=1}^{2}(\bar{x}_{j_2} - a_{j_2j_2s_1})}{(a_{j_2j_11} - a_{j_1j_11})\prod_{l_1=1}^{2}(\bar{x}_{j_2} - a_{j_1j_2l_1})} < 0,$$

there is a down-parabola flow. Let

$$\dot{x}_{j_1} = a_{j_1 j_1 0}(a_{j_2 j_1 1} - a_{j_1 j_1 1})\prod\nolimits_{l_1=1}^{2}(\overline{x}_{j_2} - a_{j_1 j_2 l_1}).$$

The parabola flows at $x_{j_1}^* = a_{j_2 j_1 1}$ in the x_{j_2}-direction are positive and negative for $\dot{x}_{j_1} > 0$ and $\dot{x}_{j_1} < 0$, respectively. Thus,

- For

$$a_{j_1 j_1 0}(a_{j_2 j_1 1} - a_{j_1 j_1 1})\prod\nolimits_{l_1=1}^{2}(\overline{x}_{j_2} - a_{j_1 j_2 l_1}) > 0 \text{ and}$$
$$a_{j_2 j_2 0}\prod\nolimits_{s_1=1}^{2}(\overline{x}_{j_2} - a_{j_2 j_2 s_1}) > 0,$$

  we have

$$(\dot{x}_{j_1}, a_{j_2 j_1 1}) = \underbrace{(\text{pF}, \text{UP})}_{\text{up-parabola flow } (+)}.$$

- For

$$a_{j_1 j_1 0}(a_{j_2 j_1 1} - a_{j_1 j_1 1})\prod\nolimits_{l_1=1}^{2}(\overline{x}_{j_2} - a_{j_1 j_2 l_1}) < 0 \text{ and}$$
$$a_{j_2 j_2 0}\prod\nolimits_{s_1=1}^{2}(\overline{x}_{j_2} - a_{j_2 j_2 s_1}) > 0,$$

  we have

$$(\dot{x}_{j_1}, a_{j_2 j_1 1}) = \underbrace{(\text{nF}, \text{DP})}_{\text{down-parabola flow } (-)}.$$

- For

$$a_{j_1 j_1 0}(a_{j_2 j_1 1} - a_{j_1 j_1 1})\prod\nolimits_{l_1=1}^{2}(\overline{x}_{j_2} - a_{j_1 j_2 l_1}) > 0 \text{ and}$$
$$a_{j_2 j_2 0}\prod\nolimits_{s_1=1}^{2}(\overline{x}_{j_2} - a_{j_2 j_2 s_1}) < 0,$$

  we have

$$(\dot{x}_{j_1}, a_{j_2 j_1 1}) = \underbrace{(\text{pF}, \text{DP})}_{\text{down-parabola flow } (+)}.$$

- For

$$a_{j_1j_10}(a_{j_2j_11}-a_{j_1j_11})\prod\nolimits_{l_1=1}^{2}(\bar{x}_{j_2}-a_{j_1j_2l_1})<0 \text{ and}$$
$$a_{j_2j_20}\prod\nolimits_{s_1=1}^{2}(\bar{x}_{j_2}-a_{j_2j_2s_1})<0,$$

we have

$$(\dot{x}_{j_1}, a_{j_2j_11}) = \underbrace{(\text{nF}, \text{UP})}_{\text{up-parabola flow } (-)}.$$

(II) For $x^*_{j_2}=a_{j_1j_2l_1}$ and $\bar{x}_{j_1}\neq a_{j_1j_11}, a_{j_2j_11}$ (l_1, $l_2\in\{1,\ 2\}$, $l_1\neq l_2$), in phase space,

$$\left.\frac{dx_{j_1}}{dx_{j_2}}\right|_{x^*_{j_2}=a_{j_1j_2l_1}} = \left.\frac{a_{j_1j_10}}{a_{j_2j_20}}\frac{(\bar{x}_{j_1}-a_{j_1j_11})(x_{j_2}-a_{j_1j_2l_1})(x_{j_2}-a_{j_1j_2l_2})}{(\bar{x}_{j_1}-a_{j_2j_11})\prod_{s_1=1}^{2}(x_{j_2}-a_{j_2j_2s_1})}\right|_{x^*_{j_2}=a_{j_1j_2l_1}}=0.$$

If

$$\left.\frac{d^2x_{j_1}}{dx^2_{j_2}}\right|_{x^*_{j_2}=a_{j_1j_2l_1}} = \frac{a_{j_1j_10}}{a_{j_2j_20}}\frac{(\bar{x}_{j_1}-a_{j_1j_11})(a_{j_1j_2l_1}-a_{j_1j_2l_2})}{(\bar{x}_{j_1}-a_{j_2j_11})\prod_{s_1=1}^{2}(a_{j_1j_2l_1}-a_{j_2j_2s_1})}>0,$$

there is an up-parabola flow at $x^*_{j_2}=a_{j_1j_2s_1}$ with $\bar{x}_{j_1}\neq a_{j_1j_11}$. If

$$\left.\frac{d^2x_{j_1}}{dx^2_{j_2}}\right|_{x^*_{j_2}=a_{j_1j_2l_1}} = \frac{a_{j_1j_10}}{a_{j_2j_20}}\frac{(\bar{x}_{j_1}-a_{j_1j_11})(a_{j_1j_2l_1}-a_{j_1j_2l_2})}{(\bar{x}_{j_1}-a_{j_2j_11})\prod_{s_1=1}^{2}(a_{j_1j_2l_1}-a_{j_2j_2s_1})}<0,$$

there is a down-parabola flow at $x^*_{j_2}=a_{j_1j_2l_1}$ with $\bar{x}_{j_1}\neq a_{j_1j_11}$. Let

$$\dot{x}_{j_2}=(\bar{x}_{j_1}-a_{j_2j_11})\prod\nolimits_{s_1=1}^{2}(a_{j_1j_2l_1}-a_{j_2j_2s_1}).$$

The parabola flows at $x^*_{j_2}=a_{j_1j_2l_1}$ with $\bar{x}_{j_1}\neq a_{j_1j_11}$ are positive and negative for $\dot{x}_{j_2}>0$ and $\dot{x}_{j_2}<0$, respectively.

- For

$$a_{j_1j_10}(\bar{x}_{j_1}-a_{j_1j_11})(a_{j_1j_2l_1}-a_{j_1j_2l_2})>0 \text{ and}$$
$$(\bar{x}_{j_1}-a_{j_2j_11})\prod\nolimits_{s_1=1}^{2}(a_{j_1j_2l_1}-a_{j_2j_2s_1})>0,$$

we have

$$(a_{j_1j_2l_1}, \dot{x}_{j_2}) = \underbrace{(\text{UP, pF})}_{\text{up-parabola flow }(+)}.$$

- For

$$a_{j_1j_1 0}(\overline{x}_{j_1} - a_{j_1j_1 1})(a_{j_1j_2l_1} - a_{j_1j_2l_2}) < 0 \text{ and}$$
$$(\overline{x}_{j_1} - a_{j_2j_1 1})\prod\nolimits_{s_1=1}^{2}(a_{j_1j_2l_1} - a_{j_2j_2s_1}) > 0,$$

we have

$$(a_{j_1j_2l_1}, \dot{x}_{j_2}) = \underbrace{(\text{DP, pF})}_{\text{down-parabola flow }(+)}.$$

- For

$$a_{j_1j_1 0}(\overline{x}_{j_1} - a_{j_1j_1 1})(a_{j_1j_2l_1} - a_{j_1j_2l_2}) > 0 \text{ and}$$
$$(\overline{x}_{j_1} - a_{j_2j_1 1})\prod\nolimits_{s_1=1}^{2}(a_{j_1j_2l_1} - a_{j_2j_2s_1}) < 0,$$

we have

$$(a_{j_1j_2l_1}, \dot{x}_{j_2}) = \underbrace{(\text{DP, nF})}_{\text{down-parabola flow }(-)}.$$

- For

$$a_{j_1j_1 0}(\overline{x}_{j_1} - a_{j_1j_1 1})(a_{j_1j_2l_1} - a_{j_1j_2l_2}) < 0 \text{ and}$$
$$(\overline{x}_{j_1} - a_{j_2j_1 1})\prod\nolimits_{s_1=1}^{2}(a_{j_1j_2l_1} - a_{j_2j_2s_1}) < 0,$$

we have

$$(a_{j_1j_2l_1}, \dot{x}_{j_2}) = \underbrace{(\text{UP, nF})}_{\text{up-parabola flow }(-)}.$$

Therefore, the equilibrium of $(x^*_{j_2}, x^*_{j_1}) = (a_{j_1j_2l_1}, a_{j_2j_1 1})$ $(l_1, l_2 \in \{1, 2\}, l_1 \neq l_2)$ has the following properties as in Eqs. (1.272)–(1.275):

- For $a_{j_1j_1 0}(a_{j_2j_1 1} - a_{j_1j_1 1})(a_{j_1j_2l_1} - a_{j_1j_2l_2}) > 0$ and $a_{j_2j_2 0}\prod_{s_1=1}^{2}(a_{j_1j_2l_1} - a_{j_2j_2s_1}) > 0$,

$$(a_{j_1j_2l_1}, a_{j_2j_11}) = \underbrace{(\mathrm{UP}_+, \mathrm{UP}_+)}_{\text{positive saddle}}.$$

The equilibrium of $(x^*_{j_2}, x^*_{j_1}) = (a_{j_1j_2l_1}, a_{j_2j_11})$ is a (UP$_+$,UP$_+$)-positive saddle.

- For $a_{j_1j_10}(a_{j_2j_11} - a_{j_1j_11})(a_{j_1j_2l_1} - a_{j_1j_2l_2}) < 0$ and $a_{j_2j_20}\prod_{s_1=1}^{2}(a_{j_1j_2l_1} - a_{j_2j_2s_1}) > 0$,

$$(a_{j_1j_2l_1}, a_{j_2j_11}) = \underbrace{(\mathrm{DP}_+, \mathrm{DP}_-)}_{\text{CCW center}}.$$

The equilibrium of $(x^*_{j_2}, x^*_{j_1}) = (a_{j_1j_2l_1}, a_{j_2j_11})$ is a (DP$_+$,DP$_-$)-counterclockwise center.

- For $a_{j_1j_10}(a_{j_2j_11} - a_{j_1j_11})(a_{j_1j_2l_1} - a_{j_1j_2l_2}) > 0$ and $a_{j_2j_20}\prod_{s_1=1}^{2}(a_{j_1j_2l_1} - a_{j_2j_2s_1}) < 0$,

$$(a_{j_1j_2l_1}, a_{j_2j_11}) = \underbrace{(\mathrm{DP}_-, \mathrm{DP}_+)}_{\text{CW center}}.$$

The equilibrium of $(x^*_{j_2}, x^*_{j_1}) = (a_{j_1j_2l_1}, a_{j_2j_11})$ is a (DP$_-$,DP$_+$)-clockwise center.

- For $a_{j_1j_10}(a_{j_2j_11} - a_{j_1j_11})(a_{j_1j_2l_1} - a_{j_1j_2l_2}) < 0$ and $a_{j_2j_20}\prod_{s_1=1}^{2}(a_{j_1j_2l_1} - a_{j_2j_2s_1}) < 0$,

$$(a_{j_1j_2l_1}, a_{j_2j_11}) = \underbrace{(\mathrm{UP}_-, \mathrm{UP}_-)}_{\text{negatives addle}}.$$

The equilibrium of $(x^*_{j_2}, x^*_{j_1}) = (a_{j_1j_1l_1}, a_{j_2j_11})$ is a (UP$_-$,UP$_-$)-negative saddle.

(ix$_{1d}$) From the case (I) of (ix$_{1c}$) with $\overline{x}_{j_2} = a_{j_2j_2s_1}$, the equilibrium of $(x^*_{j_1}, x^*_{j_2}) = (a_{j_2j_11}, a_{j_2j_2s_1})$ ($l_1, l_2 \in \{1, 2\}, l_1 \neq l_2$) with $a_{j_1j_2l_1} \neq a_{j_2j_21}, a_{j_2j_22}$ has the following properties as in Eqs. (1.276)–(1.279):

- For $a_{j_1j_10}(a_{j_2j_11} - a_{j_1j_11})\prod_{l_1=1}^{2}(a_{j_2j_2s_1} - a_{j_1j_2l_1}) > 0$ and $a_{j_2j_20}(a_{j_2j_2s_1} - a_{j_2j_2s_2}) > 0$,

$$(a_{j_2j_11}, a_{j_2j_2s_1}) = \underbrace{(\mathrm{pF}, \mathrm{DP} : \mathrm{UP})}_{\text{hyperbolic flow } (+)}.$$

The equilibrium of $(x^*_{j_1}, x^*_{j_2}) = (a_{j_2j_11}, a_{j_2j_2s_1})$ is a (pF, DP:UP)-positive hyperbolic flow.

- For $a_{j_1j_10}(a_{j_2j_11} - a_{j_1j_11})\prod_{l_1=1}^{2}(a_{j_2j_2s_1} - a_{j_1j_2l_1}) < 0$ and $a_{j_2j_20}(a_{j_2j_2s_1} - a_{j_2j_2s_2}) > 0$,

$$(a_{j_2j_11}, a_{j_2j_2s_1}) = \underbrace{(\mathrm{nF}, \mathrm{DP} : \mathrm{UP})}_{\text{hyperbolic-secant flow } (-)}.$$

The equilibrium of $(x^*_{j_1}, x^*_{j_2}) = (a_{j_2j_11}, a_{j_2j_2s_1})$ is an (nF,UP:DP)-negative hyperbolic-secant flow.

- For $a_{j_1j_10}(a_{j_2j_11} - a_{j_1j_11})\prod_{l_1=1}^{2}(a_{j_2j_2s_1} - a_{j_1j_2l_1}) > 0$ and $a_{j_2j_20}(a_{j_2j_2s_1} - a_{j_2j_2s_2}) < 0$,

$$(a_{j_2j_1 1}, a_{j_2j_2s_1}) = \underbrace{(\text{nF}, \text{UP}:\text{DP})}_{\text{hyperbolic-secant flow }(+)}.$$

The equilibrium of $(x_{j_1}^*, x_{j_2}^*) = (a_{j_2j_1 1}, a_{j_2j_2s_1})$ is an (pF,UP:DP)-negative hyperbolic-secant flow.

- For $a_{j_1j_1 0}(a_{j_2j_1 1} - a_{j_1j_1 1})\prod_{l_1=1}^{2}(a_{j_2j_2s_1} - a_{j_1j_2l_1}) < 0$ and $a_{j_2j_2 0}(a_{j_2j_2s_1} - a_{j_2j_2s_2}) < 0$,

$$(a_{j_2j_1 1}, a_{j_2j_2s_1}) = \underbrace{(\text{nF}, \text{DP}:\text{UP})}_{\text{hyperbolic flow }(-)}.$$

The equilibrium of $(x_{j_1}^*, x_{j_2}^*) = (a_{j_2j_1 1}, a_{j_2j_2s_1})$ is an (nF,DP:UP)-negative hyperbolic flow.

(ix$_2$) For $a_{j_2j_2s_1} = a_{j_1j_2l_1}$ and $a_{j_2j_2s_2} \neq a_{j_1j_2l_2}$ (s_1, s_2, l_1, $l_2 \in \{1,\ 2\}$; $s_1 \neq s_2$, $l_1 \neq l_2$), in phase space

$$\begin{aligned}\frac{dx_{j_1}}{dx_{j_2}} &= \frac{a_{j_1j_1 0}}{a_{j_2j_2 0}} \frac{(x_{j_1} - a_{j_1j_1 1})(x_{j_2} - a_{j_1j_2l_1})(x_{j_2} - a_{j_1j_2l_2})}{(x_{j_1} - a_{j_2j_1 1})(x_{j_2} - a_{j_2j_2s_1})(x_{j_2} - a_{j_2j_2s_2})}\\ &= \frac{a_{j_1j_1 0}}{a_{j_2j_2 0}} \frac{(x_{j_1} - a_{j_1j_1 1})(x_{j_2} - a_{j_1j_2l_2})}{(x_{j_1} - a_{j_2j_1 1})(x_{j_2} - a_{j_2j_2s_2})}\end{aligned}$$

and the deformation of the foregoing equation is

$$\Big[1 + \frac{(a_{j_1j_1 1} - a_{j_2j_1 1})}{(x_{j_1} - a_{j_1j_1 1})}\Big]dx_{j_1} = \frac{a_{j_1j_1 0}}{a_{j_2j_2 0}}\Big[1 + \frac{(a_{j_2j_2s_2} - a_{j_1j_2l_2})}{(x_{j_2} - a_{j_2j_2s_2})}\Big]dx_{j_2}.$$

With an initial condition of $(x_{j_1 0}, x_{j_2 0})$ at $t = t_0$, the integration of the above equation gives the first integral manifold as

$$\begin{aligned}&\big[(x_{j_1} - a_{j_1j_1 1}) - (x_{j_1 0} - a_{j_1j_1 1})\big] + (a_{j_1j_1 1} - a_{j_2j_1 1}) \ln \frac{|x_{j_1} - a_{j_1j_1 1}|}{|x_{j_1 0} - a_{j_1j_1 1}|}\\ &= \frac{a_{j_1j_1 0}}{a_{j_2j_2 0}} \Big\{ \big[(x_{j_2} - a_{j_2j_2s_2}) - (x_{j_2 0} - a_{j_2j_2s_2})\big] + (a_{j_2j_2s_2} - a_{j_1j_2l_2}) \ln \frac{|x_{j_2} - a_{j_2j_2s_2}|}{|x_{j_2 0} - a_{j_2j_2s_2}|}\Big\},\end{aligned}$$

and the first integral manifold ($a_{j_2j_2s_1} = a_{j_1j_2l_1}$ and $a_{j_2j_2s_2} = a_{j_1j_2l_2}$) is

$$(x_{j_1} - x_{j_1 0}) + (a_{j_1j_1 1} - a_{j_2j_1 1}) \ln \frac{|x_{j_1} - a_{j_1j_1 1}|}{|x_{j_1 0} - a_{j_1j_1 1}|} = \frac{a_{j_1j_1 0}}{a_{j_2j_2 0}}(x_{j_2} - x_{j_2 0}).$$

(ix$_{2a}$) For $x_{j_2}^* = a_{j_1j_2l_1} = a_{j_2j_2s_1}$ with $\bar{x}_{j_1} \neq a_{j_1j_1 1}$, if

$$\left.\frac{dx_{j_1}}{dx_{j_2}}\right|_{x_{j_2}^*=a_{j_1j_2l_1}} = \frac{a_{j_1j_10}}{a_{j_2j_20}}\frac{(\bar{x}_{j_1}-a_{j_1j_11})(a_{j_1j_2l_1}-a_{j_1j_2l_2})}{(\bar{x}_{j_1}-a_{j_2j_11})(a_{j_2j_2s_1}-a_{j_2j_2s_2})} > 0,$$

there is an increasing-inflection flow, and if

$$\left.\frac{dx_{j_1}}{dx_{j_2}}\right|_{x_{j_2}^*=a_{j_1j_2l_1}} = \frac{a_{j_1j_10}}{a_{j_2j_20}}\frac{(\bar{x}_{j_1}-a_{j_1j_11})(a_{j_1j_2l_1}-a_{j_1j_2l_2})}{(\bar{x}_{j_1}-a_{j_2j_11})(a_{j_2j_2s_1}-a_{j_2j_2s_2})} < 0,$$

there is a decreasing-inflection flow. The variational equation is

$$\Delta\dot{x}_{j_2} = a_{j_2j_20}(\bar{x}_{j_1}-a_{j_2j_11})(a_{j_2j_2s_1}-a_{j_2j_2s_2})\Delta x_{j_2}.$$

For $a_{j_2j_20}(\bar{x}_{j_1}-a_{j_2j_11})(a_{j_2j_2s_1}-a_{j_2j_2s_2})>0$ and $a_{j_2j_20}(\bar{x}_{j_1}-a_{j_2j_11})(a_{j_2j_2s_1}-a_{j_2j_2s_2})<0$, the inflection flows are source and sink, respectively.

Therefore, the infinite-equilibrium of $x_{j_2}^*=a_{j_1j_2l_1}=a_{j_2j_2s_1}$ with $\bar{x}_{j_1}\neq a_{j_1j_11}$ has the following properties as in Eqs. (1.282)–(1.285):

- For $a_{j_1j_10}(\bar{x}_{j_1}-a_{j_1j_11})(a_{j_1j_2l_1}-a_{j_1j_2l_2})>0$ and $a_{j_2j_20}(\bar{x}_{j_1}-a_{j_2j_11})(a_{j_2j_2s_1}-a_{j_2j_2s_2})>0$,

$$(\bar{x}_{j_1},a_{j_2j_2s_1}) = \underbrace{(\text{II},\text{SO})}_{\text{increasing-inflection source}}.$$

  The infinite-equilibrium of $x_{j_2}^*=a_{j_1j_2l_1}=a_{j_2j_2s_1}$ is an (II,SO)-increasing-inflection source.
- For $a_{j_1j_10}(\bar{x}_{j_1}-a_{j_1j_11})(a_{j_1j_2l_1}-a_{j_1j_2l_2})<0$ and $a_{j_2j_20}(\bar{x}_{j_1}-a_{j_2j_11})(a_{j_2j_2s_1}-a_{j_2j_2s_2})>0$,

$$(\bar{x}_{j_1},a_{j_2j_2s_1}) = \underbrace{(\text{DI},\text{SO})}_{\text{decreasing-inflection source}}.$$

  The infinite-equilibrium of $x_{j_2}^*=a_{j_1j_2l_1}=a_{j_2j_2s_1}$ is a (DI,SO)-decreasing-inflection source.
- For $a_{j_1j_10}(\bar{x}_{j_1}-a_{j_1j_11})(a_{j_1j_2l_1}-a_{j_1j_2l_2})>0$ and $a_{j_2j_20}(\bar{x}_{j_1}-a_{j_2j_11})(a_{j_2j_2s_1}-a_{j_2j_2s_2})<0$,

$$(\bar{x}_{j_1},a_{j_2j_2s_1}) = \underbrace{(\text{DI},\text{SI})}_{\text{decreasing-inflection sink}}.$$

  The infinite-equilibrium of $x_{j_2}^*=a_{j_1j_2l_1}=a_{j_2j_2s_1}$ is a (DI,SI)-decreasing-inflection sink.
- For $a_{j_1j_10}(\bar{x}_{j_1}-a_{j_1j_11})(a_{j_1j_2l_1}-a_{j_1j_2l_2})<0$ and $a_{j_2j_20}(\bar{x}_{j_1}-a_{j_2j_11})(a_{j_2j_2s_1}-a_{j_2j_2s_2})<0$,

$$(\bar{x}_{j_1},a_{j_2j_2s_1}) = \underbrace{(\text{II},\text{SI})}_{\text{increasing-inflection sink}}.$$

The infinite-equilibrium of $x_{j_2}^* = a_{j_1j_2l_1} = a_{j_2j_2s_1}$ is an (II,SI)-increasing-inflection sink.

(ix$_{2b}$) From case (ix$_{2a}$), for $\bar{x}_{j_1} = a_{j_1j_1 1}$, therefore, the equilibrium of $(x_{j_1}^*, x_{j_2}^*) = (a_{j_1j_1 1}, a_{j_2j_2s_1})$ with $a_{j_2j_2s_1} = a_{j_1j_2l_1}$ has the following properties as in Eqs. (1.286)–(1.289):

- For $a_{j_1j_10}(a_{j_1j_2l_1} - a_{j_1j_2l_2}) > 0$ and $a_{j_2j_20}(a_{j_1j_11} - a_{j_2j_11})(a_{j_2j_2s_1} - a_{j_2j_2s_2}) > 0$,

$$(a_{j_1j_11}, a_{j_2j_2s_1}) = \underbrace{(\text{DI} : \text{II}, \text{SO})}_{\text{inflection source}}.$$

The equilibrium of $(x_{j1}^*, x_{j2}^*) = (a_{j_1j_11}, a_{j_2j_2s_1})$ is a (DI:II,SO)-inflection source.

- For $a_{j_1j_10}(a_{j_1j_2l_1} - a_{j_1j_2l_2}) < 0$ and $a_{j_2j_20}(a_{j_1j_11} - a_{j_2j_11})(a_{j_2j_2s_1} - a_{j_2j_2s_2}) > 0$,

$$(a_{j_1j_11}, a_{j_2j_2s_1}) = \underbrace{(\text{II} : \text{DI}, \text{SO})}_{\text{inflection source}}.$$

The equilibrium of $(x_{j_1}^*, x_{j_2}^*) = (a_{j_1j_11}, a_{j_2j_2s_1})$ is an (II:DI,SO)-inflection source.

- For $a_{j_1j_10}(a_{j_1j_2l_1} - a_{j_1j_2l_2}) > 0$ and $a_{j_2j_20}(a_{j_1j_11} - a_{j_2j_11})(a_{j_2j_2s_1} - a_{j_2j_2s_2}) < 0$,

$$(a_{j_1j_11}, a_{j_2j_2s_1}) = \underbrace{(\text{II} : \text{DI}, \text{SI})}_{\text{inflection sink}}.$$

The equilibrium of $(x_{j_1}^*, x_{j_2}^*) = (a_{j_1j_11}, a_{j_2j_2s_1})$ is an (II:DI,SI)-inflection sink.

- For $a_{j_1j_10}(a_{j_1j_2l_1} - a_{j_1j_2l_2}) < 0$ and $a_{j_2j_20}(a_{j_1j_11} - a_{j_2j_11})(a_{j_2j_2s_1} - a_{j_2j_2s_2}) < 0$,

$$(a_{j_1j_11}, a_{j_2j_2s_1}) = \underbrace{(\text{DI} : \text{II}, \text{SI})}_{\text{inflection sink}}.$$

The equilibrium of $(x_{j_1}^*, x_{j_2}^*) = (a_{j_1j_11}, a_{j_2s_1})$ is a (DI:II,SI)-inflection sink.

(ix$_{2c}$) From the case (II) for the proof of (ix$_{1c}$) with $\bar{x}_{j_1} = a_{j_2j_11}$ and (ix$_{2a}$) with $\bar{x}_{j_1} = a_{j_2j_11}$, the equilibrium of $(x_{j2}^*, x_{j1}^*) = (a_{j_1j_2l_1}, a_{j_2j_11})$ with $a_{j_2j_2s_1} = a_{j_1j_2l_1}$ has the following properties as in Eqs. (1.290)–(1.293):

- For $a_{j_1j_10}(a_{j_2j_11} - a_{j_1j_11})(a_{j_1j_2l_1} - a_{j_1j_2l_2}) > 0$ and $a_{j_2j_20}(a_{j_2j_2s_1} - a_{j_2j_2s_2}) > 0$,

$$(a_{j_1j_2l_1}, a_{j_2j_11}) = \underbrace{({}_{\text{SI:SO}}\text{US}, {}_{\text{DI:II}}\text{UP})}_{\text{up-parabola upper-saddle}}.$$

The equilibrium of $(x_{j_2}^*, x_{j_1}^*) = (a_{j_1j_2l_1}, a_{j_2j_11})$ is an (${}_{\text{SI:SO}}$US, ${}_{\text{DI:II}}$UP)-up-parabola upper-saddle.

- For $a_{j_1j_10}(a_{j_2j_11} - a_{j_1j_11})(a_{j_1j_2l_1} - a_{j_1j_2l_2}) < 0$ and $a_{j_2j_20}(a_{j_2j_2s_1} - a_{j_2j_2s_2}) > 0$,

$$(a_{j_1j_2l_1}, a_{j_2j_1 1}) = \underbrace{({}_{\text{SI:SO}}\text{US}, {}_{\text{II:DI}}\text{DP})}_{\text{down-parabola upper-saddle}}.$$

The equilibrium of $(x^*_{j_2}, x^*_{j_1}) = (a_{j_1j_2l_1}, a_{j_2j_1 1})$ is an $({}_{\text{SI:SO}}\text{US}, {}_{\text{II:DI}}\text{DP})$-down-parabola upper-saddle.

- For $a_{j_1j_1 0}(a_{j_2j_1 1} - a_{j_1j_1 1})(a_{j_1j_2l_1} - a_{j_1j_2l_2}) > 0$ and $a_{j_2j_2 0}(a_{j_2j_2s_1} - a_{j_2j_2s_2}) < 0$,

$$(a_{j_1j_2l_1}, a_{j_2j_1 1}) = \underbrace{({}_{\text{SO:SI}}\text{LS}, {}_{\text{II:DI}}\text{DP})}_{\text{down-parabola lower-saddle}}.$$

The equilibrium of $(x^*_{j_2}, x^*_{j_1}) = (a_{j_1j_2l_1}, a_{j_2j_1 1})$ is an $({}_{\text{SO:SI}}\text{LS}, {}_{\text{II:DI}}\text{DP})$-down-parabola lower-saddle.

- For $a_{j_1j_1 0}(a_{j_2j_1 1} - a_{j_1j_1 1})(a_{j_1j_2l_1} - a_{j_1j_2l_2}) < 0$ and $a_{j_2j_2 0}(a_{j_2j_2s_1} - a_{j_2j_2s_2}) < 0$,

$$(a_{j_1j_2l_1}, a_{j_2j_1 1}) = \underbrace{({}_{\text{SO:SI}}\text{LS}, {}_{\text{DI:II}}\text{UP})}_{\text{up-parabola lower-saddle}}.$$

The equilibrium of $(x^*_{j_2}, x^*_{j_1}) = (a_{j_1j_2l_1}, a_{j_2j_1 1})$ is an $({}_{\text{SO:SI}}\text{LS}, {}_{\text{DI:II}}\text{UP})$-up-parabola lower-saddle.

(ix$_3$) For $a_{j_2j_2s_1} \neq a_{j_1j_2l_1}$ $(s_1, s_2, l_1, l_2 \in \{1, 2\}, s_1 \neq s_2, l_1 \neq l_2)$ and $a_{j_1j_1 1} = a_{j_2j_1 1}$, in phase space,

$$\frac{dx_{j_1}}{dx_{j_2}} = \frac{a_{j_1j_1 0}}{a_{j_2j_2 0}} \frac{(x_{j_2} - a_{j_1j_2l_1})(x_{j_2} - a_{j_1j_2l_2})}{(x_{j_2} - a_{j_2j_2s_1})(x_{j_2} - a_{j_2j_2s_2})}$$

and the deformation of the foregoing equation is

$$dx_{j_1} = \frac{a_{j_1j_1 0}}{a_{j_2j_2 0}} \Big[1 + \sum\nolimits_{s_1=1, s_2 \neq s_1}^{2} \frac{\prod_{l_1=0}^{2}(a_{j_2j_2s_1} - a_{j_1j_2l_1})}{(a_{j_2j_2s_1} - a_{j_2j_2s_2})} \frac{1}{(x_{j_2} - a_{j_2j_2s_1})}\Big] dx_{j_2}.$$

With an initial condition of $(x_{j_1 0}, x_{j_2 0})$ at $t = t_0$, the integration of the above equation gives the first integral manifold for $a_{j_2j_2s_1} \neq a_{j_1j_2l_1}$ $(s_1, s_2, l_1, l_2 \in \{1, 2\}, s_1 \neq s_2, l_1 \neq l_2)$ and $a_{j_1j_1 1} = a_{j_2j_1 1}$ as

$$\begin{aligned}
&\big[(x_{j_1} - a_{j_1j_1 1}) - (x_{j_1 0} - a_{j_1j_1 1})\big] \\
&= \frac{a_{j_1j_1 0}}{a_{j_2j_2 0}} \Big[(x_{j_2} - x_{j_2 0}) + \sum\nolimits_{s_1=1, s_2 \neq s_1}^{2} \frac{\prod_{l_1=1}^{2}(a_{j_2j_2s_1} - a_{j_1j_2l_1})}{(a_{j_2j_2s_1} - a_{j_2j_2s_2})} \ln \frac{|x_{j_2} - a_{j_2j_2s_1}|}{|x_{j_2 0} - a_{j_2j_2s_1}|}\Big].
\end{aligned}$$

(ix$_{3a}$) For $x^*_{j_1} = a_{j_1j_1 1} = a_{j_2j_1 1}$ with $\bar{x}_{j_2} \neq a_{j_2j_2s_1}, a_{j_1j_2l_1}$ $(a_{j_1j_2s_1}, a_{j_1j_2s_2} \neq a_{j_2j_2l_1}, a_{j_2j_2l_2})$ if

$$\left.\frac{dx_{j_2}}{dx_{j_1}}\right|_{x^*_{j_1}=a_{j_2j_11}} = \frac{a_{j_2j_20}}{a_{j_1j_10}}\frac{\prod_{s_1=1}^{2}(\bar{x}_{j_2}-a_{j_2j_2s_1})}{\prod_{l_1=1}^{2}(\bar{x}_{j_2}-a_{j_1j_2l_1})} > 0,$$

there is an increasing-inflection flow, and if

$$\left.\frac{dx_{j_2}}{dx_{j_1}}\right|_{x^*_{j_1}=a_{j_2j_11}} = \frac{a_{j_2j_20}}{a_{j_1j_10}}\frac{\prod_{s_1=1}^{2}(\bar{x}_{j_2}-a_{j_2j_2s_1})}{\prod_{l_1=1}^{2}(\bar{x}_{j_2}-a_{j_1j_2l_1})} < 0,$$

there is a decreasing-inflection flow. The variational equation is

$$\Delta\dot{x}_{j_1} = a_{j_1j_10}\prod\nolimits_{l_1=1}^{2}(\bar{x}_{j_2}-a_{j_1j_2l_1})\Delta x_{j_1}.$$

The inflection flows for $a_{j_1j_10}\prod_{l_1=1}^{2}(\bar{x}_{j_2}-a_{j_1j_2l_1}) > 0$ and $a_{j_1j_10}\prod_{l_1=1}^{2}(\bar{x}_{j_2}-a_{j_1j_2l_1}) < 0$ are source and sink, respectively.

Therefore, the infinite-equilibrium of $x^*_{j_1} = a_{j_1j_11} = a_{j_2j_11}$ with $\bar{x}_{j_2} \neq a_{j_2j_2s_1}, a_{j_1j_2l_1}$ $(a_{j_1j_2s_1}, a_{j_1j_2s_2} \neq a_{j_2j_2l_1}, a_{j_2j_2l_2})$ has the following properties as in Eqs. (1.295)–(1.298):

- For $a_{j_1j_10}\prod_{l_1=1}^{2}(\bar{x}_{j_2}-a_{j_1j_2l_1}) > 0$ and $a_{j_2j_20}\prod_{s_1=1}^{2}(\bar{x}_{j_2}-a_{j_2j_2s_1}) > 0$,

$$(a_{j_1j_11}, \bar{x}_{j_2}) = \underbrace{(\text{SO}, \text{II})}_{\text{increasing-inflection source}}.$$

  The infinite-equilibrium of $x^*_{j_1} = a_{j_1j_11} = a_{j_2j_11}$ is an (SO,II)-increasing-inflection source.
- For $a_{j_1j_10}\prod_{l_1=0}^{2}(\bar{x}_{j_2}-a_{j_1j_2l_1}) < 0$ and $a_{j_2j_20}\prod_{s_1=0}^{2}(\bar{x}_{j_2}-a_{j_2j_2s_1}) > 0$,

$$(a_{j_1j_11}, \bar{x}_{j_2}) = \underbrace{(\text{SI}, \text{DI})}_{\text{decreasing-inflection sink}}.$$

  The infinite-equilibrium of $x^*_{j_1} = a_{j_1j_11} = a_{j_2j_11}$ is an (SI,DI)-decreasing-inflection sink.
- For $a_{j_1j_10}\prod_{l_1=0}^{2}(\bar{x}_{j_2}-a_{j_1j_2l_1}) > 0$ and $a_{j_2j_20}\prod_{s_1=1}^{2}(\bar{x}_{j_2}-a_{j_2j_2s_1}) < 0$,

$$(a_{j_1j_11}, \bar{x}_{j_2}) = \underbrace{(\text{SO}, \text{DI})}_{\text{decreasing-inflection source}}.$$

  The infinite-equilibrium of $x^*_{j_1} = a_{j_1j_11} = a_{j_2j_11}$ is an (SO,DI)-decreasing-inflection source.
- For $a_{j_1j_10}\prod_{l_1=1}^{2}(\bar{x}_{j_2}-a_{j_1j_2l_1}) < 0$ and $a_{j_2j_20}\prod_{s_1=1}^{2}(x_{j_2}-a_{j_2j_2s_1}) < 0$,

$$(a_{j_1j_11}, \bar{x}_{j_2}) = \underbrace{(\text{SI}, \text{II})}_{\text{increasing-inflection sink}}.$$

The infinite-equilibrium of $x^*_{j_1} = a_{j_1j_11} = a_{j_2j_11}$ is an (SI,II)-increasing-inflection sink.

($\text{ix}_{3\text{b}}$) From the case (I) of ($\text{ix}_{1\text{a}}$) with $\bar{x}_{j_2} = a_{j_1j_2l_1}$ and ($\text{ix}_{3\text{a}}$) with $\bar{x}_{j_2} = a_{j_1j_2l_1}$, the equilibrium of $(x^*_{j_2}, x^*_{j_1}) = (a_{j_1j_2l_1}, a_{j_2j_11})$ with $a_{j_1j_11} = a_{j_2j_11}$ has the following properties as in Eqs. (1.299)–(1.302).

- For $a_{j_1j_10}(a_{j_1j_2l_1} - a_{j_1j_2l_2}) > 0$ and $a_{j_2j_20}\prod_{s_1=1}^{2}(a_{j_1j_2l_1} - a_{j_2j_2s_1}) > 0$,

$$(a_{j_1j_2l_1}, a_{j_2j_11}) = \underbrace{({}_{\text{DI:II}}\text{UP}, {}_{\text{SI:SO}}\text{US})}_{\text{up-parabola upper-saddle}}.$$

  The equilibrium of $(x^*_{j_2}, x^*_{j_1}) = (a_{j_1j_2l_1}, a_{j_2j_11})$ is a (${}_{\text{DI:II}}$UP, ${}_{\text{SI:SO}}$US)-up-parabola upper-saddle.

- For $a_{j_1j_10}(a_{j_1j_2l_1} - a_{j_1j_2l_2}) < 0$ and $a_{j_2j_20}\prod_{s_1=1}^{2}(a_{j_1j_2l_1} - a_{j_2j_2s_1}) > 0$,

$$(a_{j_1j_2l_1}, a_{j_2j_11}) = \underbrace{({}_{\text{II:DI}}\text{DP}, {}_{\text{SO:SI}}\text{LS})}_{\text{down-parabola lower-saddle}}.$$

  The equilibrium of $(x^*_{j_2}, x^*_{j_1}) = (a_{j_1j_2l_1}, a_{j_2j_11})$ is an (${}_{\text{II:DI}}$DP, ${}_{\text{SO:SI}}$LS)-down-parabola lower-saddle.

- For $a_{j_1j_10}(a_{j_1j_2l_1} - a_{j_1j_2l_2}) > 0$ and $a_{j_2j_20}\prod_{s_1=1}^{2}(a_{j_1j_2l_1} - a_{j_2j_2s_1}) < 0$,

$$(a_{j_1j_2l_1}, a_{j_2j_11}) = \underbrace{({}_{\text{II:DI}}\text{DP}, {}_{\text{SI:SO}}\text{US})}_{\text{down-parabola upper-saddle}}.$$

  The equilibrium of $(x^*_{j_2}, x^*_{j_1}) = (a_{j_1j_2l_1}, a_{j_2j_11})$ is an (${}_{\text{II:DI}}$DP, ${}_{\text{SI:SO}}$US)-down-parabola upper-saddle.

- For $a_{j_1j_10}(a_{j_1j_2l_1} - a_{j_1j_2l_2}) < 0$ and $a_{j_2j_20}\prod_{s_1=1}^{2}(a_{j_1j_2l_1} - a_{j_2j_2s_1}) < 0$,

$$(a_{j_1j_2l_1}, a_{j_2j_11}) = \underbrace{({}_{\text{DI:II}}\text{UP}, {}_{\text{SI:SO}}\text{LS})}_{\text{up-parabola lower-saddle}}.$$

  The equilibrium of $(x^*_{j_2}, x^*_{j_1}) = (a_{j_1j_2l_1}, a_{j_2j_11})$ is a (${}_{\text{DI:II}}$UP, ${}_{\text{SO:SI}}$LS)-up-parabola lower-saddle.

(ix$_{3c}$) From (xi$_{3a}$) with $\overline{x}_{j_2} = a_{j_2 j_2 s_1}$, the equilibrium of $(x_{j_1}^*, x_{j_2}^*) = (a_{j_1 j_1 1}, a_{j_2 j_2 s_1})$ with $a_{j_1 j_1 1} = a_{j_2 j_1 1}$ has the following properties as in Eqs. (1.303)–(1.306):

- For $a_{j_1 j_1 0} \prod_{l_1=1}^{2} (a_{j_2 j_2 s_1} - a_{j_1 j_2 l_1}) > 0$ and $a_{j_2 j_2 0}(a_{j_2 j_2 s_1} - a_{j_2 j_2 s_2}) > 0$,

$$(a_{j_1 j_1 1}, a_{j_2 j_2 s_1}) = \underbrace{(\text{SO}, \text{DI} : \text{II})}_{\text{inflection source}}.$$

The equilibrium of $(x_{j_1}^*, x_{j_2}^*) = (a_{j_1 j_1 1}, a_{j_2 j_2 s_1})$ is an (SO,DI:II)-inflection source.

- For $a_{j_1 j_1 0} \prod_{l_1=1}^{2} (a_{j_2 j_2 s_1} - a_{j_1 j_2 l_1}) > 0$ and $a_{j_2 j_2 0}(a_{j_2 j_2 s_1} - a_{j_2 j_2 s_2}) > 0$,

$$(a_{j_1 j_1 1}, a_{j_2 j_2 s_1}) = \underbrace{(\text{SI}, \text{II} : \text{DI})}_{\text{inflection sink}}.$$

The equilibrium of $(x_{j_1}^*, x_{j_2}^*) = (a_{j_1 j_1 1}, a_{j_2 j_2 s_1})$ is an (SI,II:DI)-inflection sink.

- For $a_{j_1 j_1 0} \prod_{l_1=1}^{2} (a_{j_2 j_2 s_1} - a_{j_1 j_2 l_1}) > 0$ and $a_{j_2 j_2 0}(a_{j_2 j_2 s_1} - a_{j_2 j_2 s_2}) < 0$,

$$(a_{j_1 j_1 1}, a_{j_2 j_2 s_1}) = \underbrace{(\text{SO}, \text{II} : \text{DI})}_{\text{inflection source}}.$$

The equilibrium of $(x_{j_1}^*, x_{j_2}^*) = (a_{j_1 j_1 1}, a_{j_2 j_2 s_1})$ is an (SO,II:DI)-inflection source.

- For $a_{j_1 j_1 0} \prod_{l_1=1}^{2} (a_{j_2 j_2 s_1} - a_{j_1 j_2 l_1}) < 0$ and $a_{j_2 j_2 0}(a_{j_2 j_2 s_1} - a_{j_2 j_2 s_2}) < 0$,

$$(a_{j_1 j_1 1}, a_{j_2 j_2 s_1}) = \underbrace{(\text{SI}, \text{DI} : \text{II})}_{\text{inflection sink}}.$$

The equilibrium of $(x_{j_1}^*, x_{j_2}^*) = (a_{j_1 j_1 1}, a_{j_2 j_2 s_1})$ is an (SI,DI:II)-inflection sink.

(ix$_4$) For $a_{j_2 j_2 s_1} = a_{j_1 j_2 l_1}$ ($s_1, s_2, l_1, l_2 \in \{1, \ 2\}, s_1 \neq s_2, l_1 \neq l_2$) and $a_{j_1 j_1 1} = a_{j_2 j_1 1}$, in phase space,

$$\frac{dx_{j_1}}{dx_{j_2}} = \frac{a_{j_1 j_1 0}}{a_{j_2 j_2 0}} \frac{(x_{j_2} - a_{j_1 j_2 l_2})}{(x_{j_2} - x_{j_2 j_2 s_2})}$$

and the deformation of the foregoing equation is

$$dx_{j_1} = \frac{a_{j_1 j_1 0}}{a_{j_2 j_2 0}} \left[1 + \frac{(a_{j_2 j_2 s_2} - a_{j_1 j_2 l_2})}{(x_{j_2} - a_{j_2 j_2 s_2})}\right] dx_{j_2}.$$

With an initial condition of $(x_{j_1 0}, x_{j_2 0})$ at $t = t_0$, the integration of the above equation gives the first integral manifold for $a_{j_2 j_2 s_1} = a_{j_1 j_2 l_1}$ ($s_1, s_2, l_1, l_2 \in \{1, \ 2\}, s_1 \neq s_2, l_1 \neq l_2$) and $a_{j_1 j_1 1} = a_{j_2 j_1 1}$ as

$$\begin{aligned}&\left[(x_{j_1}-a_{j_1j_11})-(x_{j_10}-a_{j_1j_11})\right]\\&=\frac{a_{j_1j_10}}{a_{j_2j_20}}\left[(x_{j_2}-x_{j_20})+(a_{j_2j_2s_2}-a_{j_1j_2l_2})\ln\frac{|x_{j_2}-a_{j_2j_2l_2}|}{|x_{j_20}-a_{j_2j_2l_2}|}\right].\end{aligned}$$

The first integral manifold for $a_{j_2j_21}=a_{j_1j_21}$ and $a_{j_2j_22}=a_{j_1j_22}$ with $a_{j_1j_11}=a_{j_2j_11}$ is

$$(x_{j_1}-x_{j_10})=\frac{a_{j_1j_10}}{a_{j_2j_20}}(x_{j_2}-x_{j_20}).$$

For $x^*_{j_1}=a_{j_1j_11}=a_{j_2j_11}$ with $a_{j_1j_11}=a_{j_2j_11}$ and $a_{j_2j_2s_1}=a_{j_1j_2s_1}$, if

$$\begin{aligned}\left.\frac{dx_{j_2}}{dx_{j_1}}\right|_{x^*_{j_1}=a_{j_2j_11}}&=\frac{a_{j_2j_20}\,(\overline{x}_{j_2}-a_{j_2j_2s_2})}{a_{j_1j_10}\,(\overline{x}_{j_2}-a_{j_1j_2l_2})}\\&=\frac{a_{j_2j_20}\,(\overline{x}_{j_2}-a_{j_1j_2l_2})(\overline{x}_{j_2}-a_{j_2j_2s_2})}{a_{j_1j_10}\,(\overline{x}_{j_2}-a_{j_1j_2l_2})(\overline{x}_{j_2}-a_{j_1j_2l_2})}>0,\end{aligned}$$

there is an increasing-inflection flow, and if

$$\begin{aligned}\left.\frac{dx_{j_2}}{dx_{j_1}}\right|_{x^*_{j_1}=a_{j_2j_11}}&=\frac{a_{j_2j_20}\,(\overline{x}_{j_2}-a_{j_2j_2s_2})}{a_{j_1j_10}\,(\overline{x}_{j_2}-a_{j_1j_2l_2})}\\&=\frac{a_{j_2j_20}\,(\overline{x}_{j_2}-a_{j_1j_2l_2})(\overline{x}_{j_2}-a_{j_2j_2s_2})}{a_{j_1j_10}\,(\overline{x}_{j_2}-a_{j_1j_2l_2})(\overline{x}_{j_2}-a_{j_1j_2l_2})}<0,\end{aligned}$$

there is a decreasing-inflection flow. The variational equation is

$$\Delta\dot{x}_{j_1}=a_{j_1j_10}(\overline{x}_{j_2}-a_{j_1j_2l_1})(\overline{x}_{j_2}-a_{j_1j_2l_2})\Delta x_{j_1}.$$

The inflection flows are source and sink for $a_{j_1j_10}(\overline{x}_{j_2}-a_{j_1j_2l_1})(\overline{x}_{j_2}-a_{j_1j_2l_2})>0$ and $a_{j_1j_10}(\overline{x}_{j_2}-a_{j_1j_2l_1})(\overline{x}_{j_2}-a_{j_1j_2l_2})<0$, respectively.

Thus, the infinite-equilibrium of $x^*_{j_2}=a_{j_1j_11}=a_{j_2j_11}$ has the following properties:

- For $a_{j_1j_10}(\overline{x}_{j_2}-a_{j_1j_2l_1})(\overline{x}_{j_2}-a_{j_1j_2l_2})>0$ and $a_{j_2j_20}(\overline{x}_{j_2}-a_{j_1j_2l_1})(\overline{x}_{j_2}-a_{j_2j_2s_2})>0$,

$$(a_{j_1j_11},\overline{x}_{j_2})=\underbrace{(\mathrm{SO},\mathrm{II})}_{\text{inflection source}}.$$

- For $a_{j_1j_10}(\overline{x}_{j_2}-a_{j_1j_2l_1})(\overline{x}_{j_2}-a_{j_1j_2l_2})<0$ and $a_{j_2j_20}(\overline{x}_{j_2}-a_{j_1j_2l_1})(\overline{x}_{j_2}-a_{j_2j_2s_2})>0$,

$$(a_{j_1j_11}, \overline{x}_{j_2}) = \underbrace{(\mathrm{SI, DI})}_{\text{inflection source}}.$$

- For $a_{j_1j_10}(\overline{x}_{j_2} - a_{j_1j_2l_1})(\overline{x}_{j_2} - a_{j_1j_2l_2}) > 0$ and $a_{j_2j_20}(\overline{x}_{j_2} - a_{j_1j_2l_1})(\overline{x}_{j_2} - a_{j_2j_2s_2}) < 0$,

$$(a_{j_1j_11}, \overline{x}_{j_2}) = \underbrace{(\mathrm{SO, DI})}_{\text{inflection source}}.$$

- For $a_{j_1j_10}(\overline{x}_{j_2} - a_{j_1j_2l_1})(\overline{x}_{j_2} - a_{j_1j_2l_2}) < 0$ and $a_{j_2j_20}(\overline{x}_{j_2} - a_{j_1j_2l_1})(\overline{x}_{j_2} - a_{j_2j_2s_2}) < 0$,

$$(a_{j_1j_11}, \overline{x}_{j_2}) = \underbrace{(\mathrm{SI, II})}_{\text{inflection sink}}.$$

From the above analysis, for $\overline{x}_{j_2} = a_{j_1j_2l_1} = a_{j_2j_2s_1}$, the equilibrium of $(x^*_{j_1}, x^*_{j_2}) = (a_{j_1j_11}, a_{j_2j_2s_1})$ with $a_{j_1j_11} = a_{j_2j_11}$ and $a_{j_2j_2s_1} = a_{j_1j_2l_1}$ $(s_1, s_2, l_1, l_2 \in \{1, 2\}; s_1, \neq s_2, l_1 \neq l_2)$ has the following properties as in Eqs. (1.309)–(1.312):

- For $a_{j_1j_10}(a_{j_1j_2l_1} - a_{j_1j_2l_2}) > 0$ and $a_{j_2j_20}(a_{j_2j_2s_1} - a_{j_2j_2s_2}) > 0$,

$$(a_{j_1j_11}, a_{j_2j_2s_1}) = \underbrace{({}_{\mathrm{SI:SO}}\mathrm{II}, {}_{\mathrm{SI:SO}}\mathrm{II})}_{\text{increasing-inflection saddle}}.$$

  The equilibrium of $(x^*_{j_1}, x^*_{j_2}) = (a_{j_1j_11}, a_{j_2j_2s_1})$ is an $({}_{\mathrm{SI:SO}}\mathrm{II}, {}_{\mathrm{SI:SO}}\mathrm{II})$-increasing-inflection saddle.
- For $a_{j_1j_10}(a_{j_1j_2s_1} - a_{j_1j_2s_2}) < 0$ and $a_{j_2j_20}(a_{j_2j_2s_1} - a_{j_2j_2s_2}) > 0$,

$$(a_{j_1j_11}, a_{j_2j_2s_1}) = \underbrace{({}_{\mathrm{SI:SO}}\mathrm{DI}, {}_{\mathrm{SO:SI}}\mathrm{DI})}_{\text{decreasing-inflection saddle}}.$$

  The equilibrium of $(x^*_{j_1}, x^*_{j_2}) = (a_{j_1j_11}, a_{j_2j_2s_1})$ is an $({}_{\mathrm{SI:SO}}\mathrm{DI}, {}_{\mathrm{SO:SI}}\mathrm{DI})$-decreasing-inflection saddle.
- For $a_{j_1j_10}(a_{j_1j_2s_1} - a_{j_1j_2s_2}) > 0$ and $a_{j_2j_20}(a_{j_2j_2s_1} - a_{j_2j_2s_2}) < 0$,

$$(a_{j_1j_11}, a_{j_2j_2s_1}) = \underbrace{({}_{\mathrm{SO:SI}}\mathrm{DI}, {}_{\mathrm{SI:SO}}\mathrm{DI})}_{\text{decreasing-inflection saddle}}.$$

  The equilibrium of $(x^*_{j_1}, x^*_{j_2}) = (a_{j_1j_11}, a_{j_2j_2s_1})$ is an $({}_{\mathrm{SO:SI}}\mathrm{DI}, {}_{\mathrm{SI:SO}}\mathrm{DI})$-decreasing-inflection saddle.
- For $a_{j_1j_10}(a_{j_1j_2s_1} - a_{j_1j_2s_2}) < 0$ and $a_{j_2j_20}(a_{j_2j_2s_1} - a_{j_2j_2s_2}) < 0$,

$$(a_{j_1 j_1 1}, a_{j_2 j_2 s_1}) = \underbrace{(\text{SO:SI}\text{II}, \text{SO:SI}\text{II})}_{\text{increasing-inflection saddle}}.$$

The equilibrium of $(x_{j_1}^*, x_{j_2}^*) = (a_{j_1 j_1 1}, a_{j_2 j_2 s_1})$ is an $(\text{SO:SI}\text{II}, \text{SO:SI}\text{II})$-increasing-inflection saddle.

In the end, this theorem is proved. ■

Chapter 2
Parabola-Saddle and Saddle-Source (Sink) Singularity

In this section, parabola-saddles, saddle-sources (sinks) singularities with the hyperbolic-to-hyperbolic-secant flows and inflection-source (sink) singularities in the two different product cubic systems are presented, and the corresponding bifurcation dynamics are discussed. The switching bifurcations of the parabola-saddle with hyperbolic-to-hyperbolic-secant flows and saddle-source (sink) with inflection-source (sink) flows are presented, and the switching bifurcations of saddle-source (sink) with hyperbolic-to-hyperbolic-secant flow and the parabola-saddle with inflection-source (sink) flow are also discussed.

2.1 Parabola-Saddle and Saddle-Node with Flow Singularity

Consider a dynamical system as

$$
\begin{aligned}
\dot{x}_1 &= a_{110}(x_1 - a_{111})(x_2 - a_{121})^2,\\
\dot{x}_2 &= a_{220}(x_1 - a_{211})(x_2 - a_{221})^2,
\end{aligned}
\tag{2.1}
$$

and the corresponding first integral manifold is

$$
\begin{aligned}
&[(x_1 - a_{111}) - (x_{10} - a_{111})] + (a_{111} - a_{211}) \ln \frac{|x_1 - a_{111}|}{|x_{10} - a_{111}|}\\
&= \frac{a_{110}}{a_{220}} \Big\{ [(x_2 - a_{221}) - (x_{20} - a_{221})] + 2(a_{221} - a_{121}) \ln \frac{|x_2 - a_{221}|}{|x_{20} - a_{221}|}\\
&\quad - (a_{221} - a_{121})^2 \Big(\frac{1}{x_2 - a_{221}} - \frac{1}{x_{20} - a_{221}}\Big) \Big\}.
\end{aligned}
\tag{2.2}
$$

A. C. J. Luo, *Two-dimensional Two-product Cubic Systems, Vol. I*,
https://doi.org/10.1007/978-3-031-48487-2_2

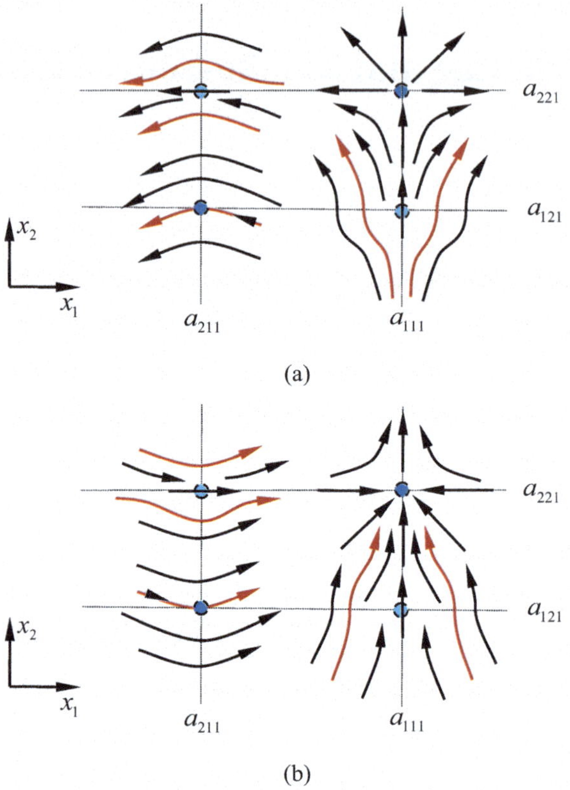

Fig. 2.1 Phase portraits ($a_{121} < a_{221}$, and $a_{221} < a_{111}$) for two-dimensional systems on the x_1-direction with $x_1^* = a_{111}, a_{211}$ and on the x_2-direction with $x_2^* = a_{121}, a_{221}$. The equilibriums and flows: (**a**) ($a_{110} > 0$, $a_{220} > 0$), (**b**) ($a_{110} < 0$, $a_{220} > 0$), (**c**) ($a_{110} > 0$, $a_{220} < 0$), and (**d**) ($a_{110} < 0$, $a_{220} < 0$)

(i) For $a_{121} < a_{221}$ and $a_{211} < a_{111}$, the equilibriums of $x_1^* = a_{111}, a_{211}$ with $x_2^* = a_{121}, a_{221}$ have the following properties, as shown in Fig. 2.1.

$$\left\{\begin{array}{cc} (a_{211}, a_{221}) & (a_{111}, a_{221}) \\ (a_{211}, a_{121}) & (a_{111}, a_{121}) \end{array}\right\} = \left\{\begin{array}{cc} \underbrace{(\text{nF}, \text{DP} : \text{DP})}_{\text{hyperbolic-to-hyperbolic-secant flow }(-)} & \underbrace{(\text{SO}, \text{US})}_{\text{upper-saddle-source}} \\ \underbrace{(\text{LS}, \text{DP})}_{\text{down-parabola lower-saddle}} & \underbrace{({}_{\text{DI:II}}\text{SO}, \text{pF})}_{\text{inflection-source flow }(+)} \end{array}\right\}$$

for $a_{110} > 0$ and $a_{220} > 0$; (2.3)

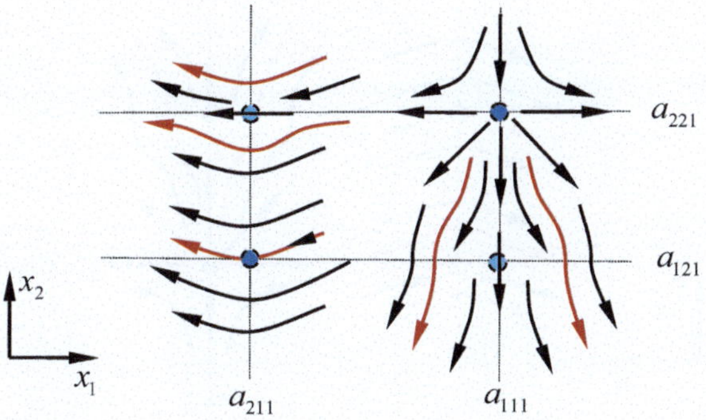

(c)

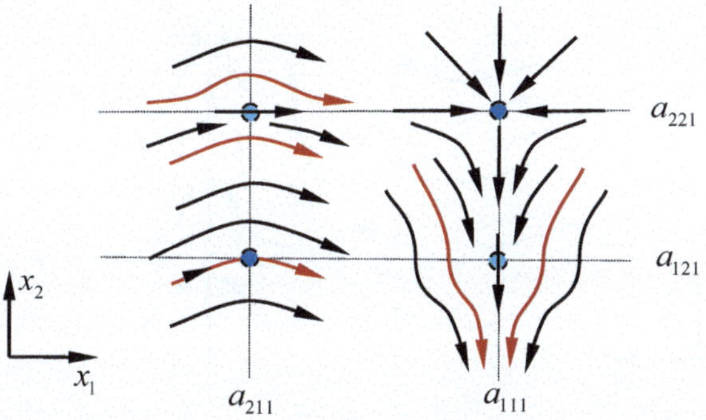

(d)

Fig. 2.1 (continued)

$$\left\{\begin{matrix} (a_{211}, a_{221}) & (a_{111}, a_{221}) \\ (a_{211}, a_{121}) & (a_{111}, a_{121}) \end{matrix}\right\} = \left\{\begin{matrix} \underbrace{(\text{pF}, \text{UP} : \text{UP})}_{\text{hyperbolic-secant-to-hyperbolic flow }(+)} & \underbrace{(\text{SI}, \text{US})}_{\text{upper-saddle-sink}} \\ \underbrace{(\text{US}, \text{UP})}_{\text{up-parabola upper-saddle}} & \underbrace{({}_{\text{II:DI}}\text{SI}, \text{pF})}_{\text{inflection-sink flow }(+)} \end{matrix}\right\}$$

for $a_{110} < 0$ and $a_{220} > 0$;

(2.4)

$$\left\{\begin{matrix} (a_{211}, a_{221}) & (a_{111}, a_{221}) \\ (a_{211}, a_{121}) & (a_{111}, a_{121}) \end{matrix}\right\} = \left\{\begin{matrix} \underbrace{(\text{pF}, \text{UP} : \text{UP})}_{\text{hyperbolic-secant-to-hyperbolic flow }(-)} & \underbrace{(\text{SO}, \text{LS})}_{\text{lower-saddle-source}} \\ \underbrace{(\text{LS}, \text{UP})}_{\text{up-parabola lower-saddle}} & \underbrace{({}_{\text{II:DI}}\text{SO}, \text{nF})}_{\text{inflection-source flow }(-)} \end{matrix}\right\}$$

for $a_{110} > 0$ and $a_{220} < 0$;

(2.5)

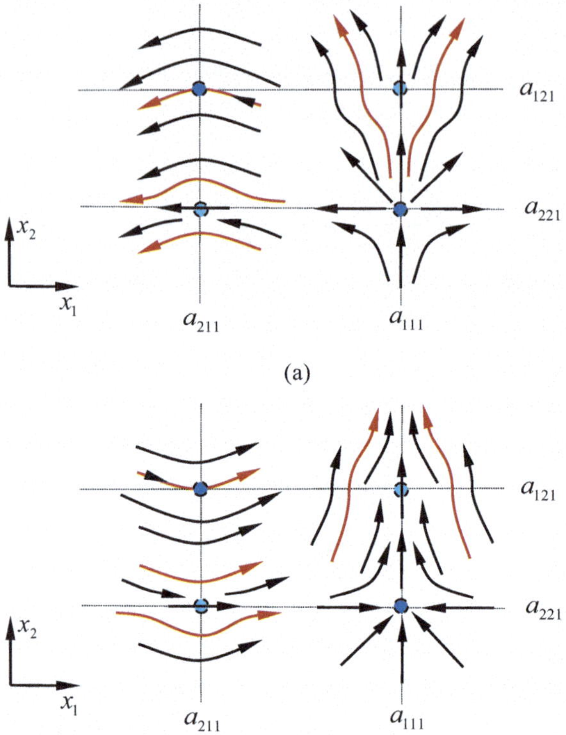

(a)

(b)

Fig. 2.2 Phase portraits ($a_{221} < a_{121}$) for two-dimensional systems on the x_1-direction with $x_1^* = a_{111}, a_{211}$ and on the x_2-direction with $x_2^* = a_{121}, a_{221}$. The equilibriums and flows: (**a**) ($a_{110} > 0, a_{220} > 0$), (**b**) ($a_{110} < 0, a_{220} > 0$), (**c**) ($a_{110} > 0, a_{220} < 0$), and (**d**) ($a_{110} < 0, a_{220} < 0$)

$$\left\{\begin{matrix} (a_{211}, a_{221}) & (a_{111}, a_{221}) \\ (a_{211}, a_{121}) & (a_{111}, a_{121}) \end{matrix}\right\} = \left\{\begin{matrix} \underbrace{(\text{PF}, \text{DP} : \text{DP})}_{\text{hyperbolic-to-hyperbolic-secant flow } (+)} & \underbrace{(\text{SI}, \text{LS})}_{\text{lower-saddle-sink}} \\ \underbrace{(\text{US}, \text{DP})}_{\text{down-parabola upper-saddle}} & \underbrace{({}_{\text{DI:II}}\text{SI}, \text{nF})}_{\text{inflection-sink flow } (-)} \end{matrix}\right\}$$

for $a_{110} < 0$ and $a_{220} < 0$.

(2.6)

(ii) For $a_{221} < a_{121}$ and $a_{211} < a_{111}$, the equilibriums of $x_1^* = a_{111}, a_{211}$ with $x_2^* = a_{121}, a_{221}$ have the properties as in Eqs. (2.3)–(2.6), and the corresponding phase portraits are also sketched in Fig. 2.2. Compared to the case of $a_{121} < a_{221}$, just the locations of equilibriums are exchanged, and the corresponding dynamical behaviors of the equilibriums are identical.

(iii) For $a_{121} < a_{221}$ and $a_{111} < a_{211}$, the equilibriums of $x_1^* = a_{111}, a_{211}$ with $x_2^* = a_{121}, a_{221}$ have the following properties, as shown in Fig. 2.3.

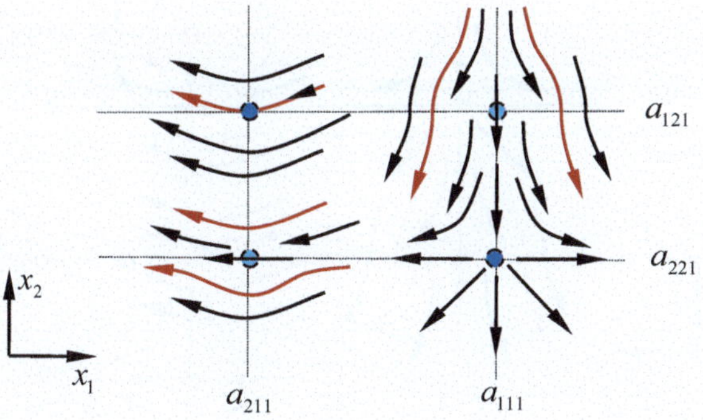

(c)

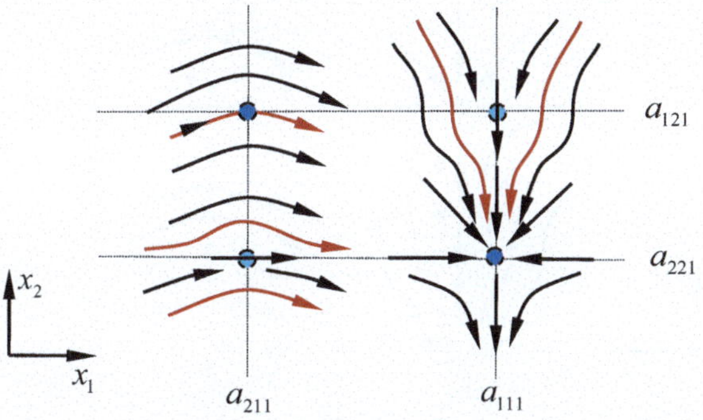

(d)

Fig. 2.2 (continued)

$$\begin{Bmatrix} (a_{111}, a_{221}) & (a_{211}, a_{221}) \\ (a_{111}, a_{121}) & (a_{211}, a_{121}) \end{Bmatrix} = \begin{Bmatrix} \underbrace{(\text{SO}, \text{LS})}_{\text{lower-saddle-source}} & \underbrace{(\text{pF}, \text{UP}:\text{UP})}_{\text{hyperbolic-secant-to-hyperbolic flow }(+)} \\ \underbrace{({}_{\text{II:DI}}\text{SO}, \text{nF})}_{\text{inflection-source flow }(-)} & \underbrace{(\text{US}, \text{UP})}_{\text{up-parabola upper-saddle}} \end{Bmatrix}$$

for $a_{110} > 0$ and $a_{220} > 0$;

(2.7)

$$\begin{Bmatrix} (a_{111}, a_{221}) & (a_{211}, a_{221}) \\ (a_{111}, a_{121}) & (a_{211}, a_{121}) \end{Bmatrix} = \begin{Bmatrix} \underbrace{(\text{SI}, \text{LS})}_{\text{lower-saddle-sink}} & \underbrace{(\text{nF}, \text{DP}:\text{DP})}_{\text{hyperbolic-to-hyperbolic-secant flow }(-)} \\ \underbrace{({}_{\text{DI:II}}\text{SI}, \text{nF})}_{\text{inflection-sink flow }(-)} & \underbrace{(\text{LS}, \text{DP})}_{\text{down-parabola lower-saddle}} \end{Bmatrix}$$

for $a_{110} < 0$ and $a_{220} > 0$;

(2.8)

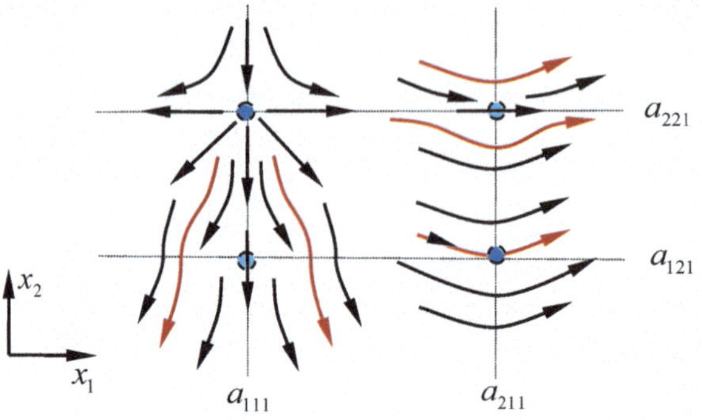

(a)

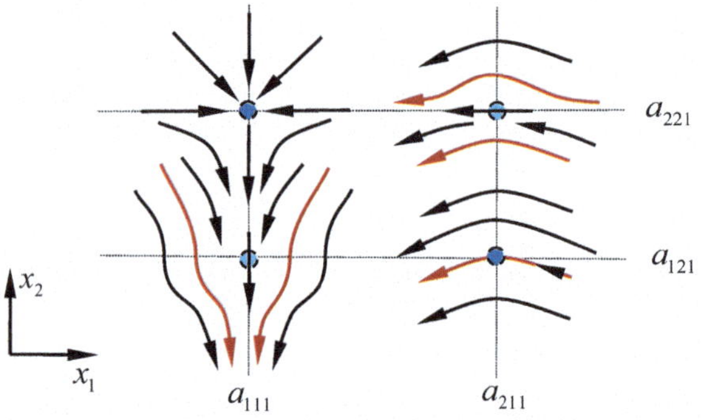

(b)

Fig. 2.3 Phase portraits ($a_{121} < a_{221}, a_{111} < a_{211}$) for two-dimensional systems on the x_1-direction with $x_1^* = a_{111}, a_{211}$ and on the x_2-direction with $x_2^* = a_{121}, a_{221}$. The equilibriums and flows: (**a**) ($a_{110} > 0, a_{220} > 0$), (**b**) ($a_{110} < 0, a_{220} > 0$), (**c**) ($a_{110} > 0, a_{220} < 0$), and (**d**) ($a_{110} < 0, a_{220} < 0$)

$$\left\{\begin{matrix} (a_{111}, a_{221}) & (a_{211}, a_{221}) \\ (a_{111}, a_{121}) & (a_{211}, a_{121}) \end{matrix}\right\} = \left\{\begin{matrix} \underbrace{(\text{SO}, \text{US})}_{\text{lower-saddle-source}} & \underbrace{(\text{pF}, \text{DP} : \text{DP})}_{\text{hyperbolic-to-hyperbolic-secant flow }(+)} \\ \underbrace{({}_{\text{DI:II}}\text{SO}, \text{pF})}_{\text{inflection-source flow }(+)} & \underbrace{(\text{US}, \text{DP})}_{\text{down-parabola upper-saddle}} \end{matrix}\right\}$$

for $a_{110} > 0$ and $a_{220} < 0$;

(2.9)

$$\left\{\begin{matrix} (a_{111}, a_{221}) & (a_{211}, a_{221}) \\ (a_{111}, a_{121}) & (a_{211}, a_{121}) \end{matrix}\right\} = \left\{\begin{matrix} \underbrace{(\text{SI}, \text{US})}_{\text{upper-saddle-sink}} & \underbrace{(\text{nF}, \text{UP} : \text{UP})}_{\text{hyperbolic-secant-to-hyperbolic flow }(-)} \\ \underbrace{({}_{\text{II:DI}}\text{SI}, \text{pF})}_{\text{inflection-sink flow }(+)} & \underbrace{(\text{LS}, \text{UP})}_{\text{up-parabola lower-saddle}} \end{matrix}\right\}$$

for $a_{110} < 0$ and $a_{220} < 0$.

(2.10)

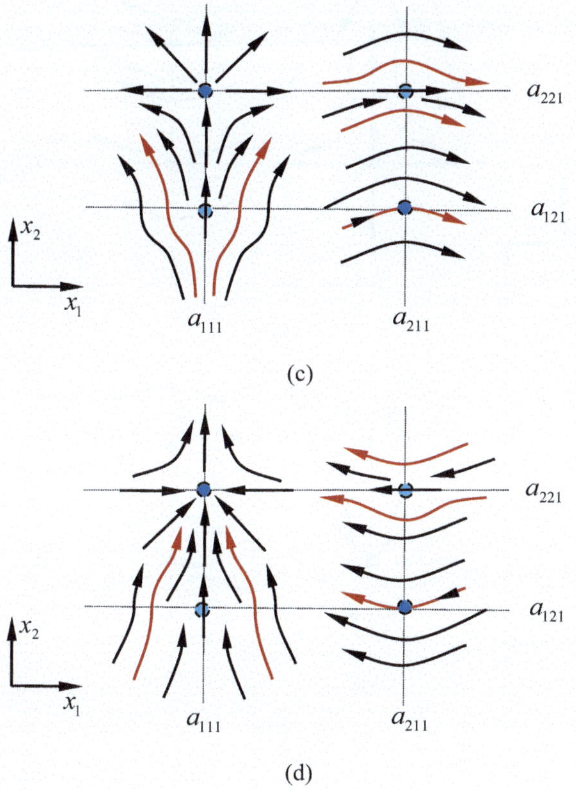

Fig. 2.3 (continued)

(iv) For $a_{221} < a_{121}$ and $a_{211} < a_{111}$, the equilibriums of $x_1^* = a_{111}, a_{211}$ with $x_2^* = a_{121}, a_{221}$ have the properties as in Eqs. (2.7)–(2.10), and the corresponding phase portraits are also sketched in Fig. 2.4. Compared to the case of $a_{121} < a_{221}$, just the locations of equilibriums are also exchanged, and the corresponding dynamical behaviors of the equilibriums are identical.

The upper-saddle-source and sink are the appearing bifurcations of saddle with source and sink, and the lower-saddle-source and sink are also the appearing bifurcations of saddle with source and sink. The up-parabola upper-saddle and down-parabola lower-saddle are the appearing bifurcations of positive and negative saddles with clockwise and counterclockwise centers, respectively. The inflection-source flows are the appearing bifurcations for the connected hyperbolic and hyperbolic-secant flows. The hyperbolic and hyperbolic-secant flows have the same directrix flows. The hyperbolic-to-hyperbolic-secant and hyperbolic-secant-to-hyperbolic flows are the appearing bifurcations for the separated hyperbolic and hyperbolic-secant flows. The hyperbolic and hyperbolic-secant flows have two directrix flows.

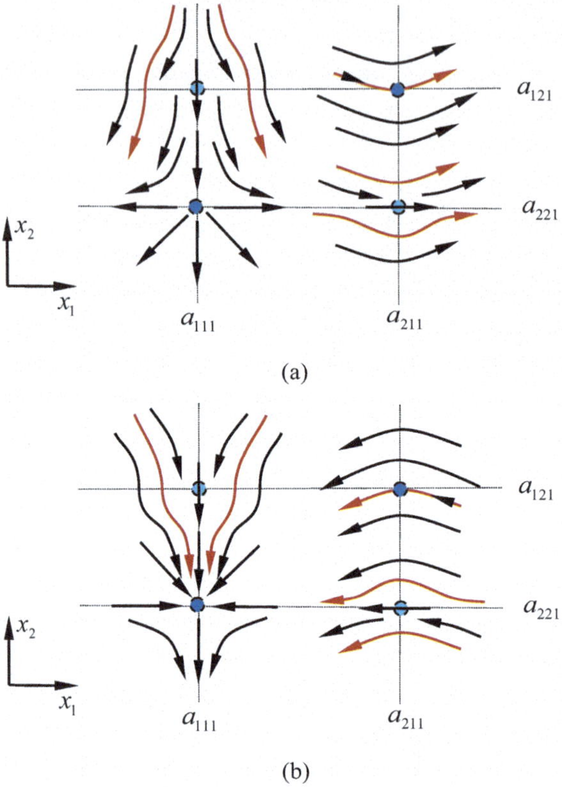

Fig. 2.4 Phase portraits ($a_{221} < a_{121}, a_{111} < a_{211}$) for two-dimensional systems on the x_1-direction with $x_1^* = a_{111}, a_{211}$ and on the x_2-direction with $x_2^* = a_{121}, a_{221}$. The equilibriums and flows: (**a**) ($a_{110} > 0, a_{220} > 0$), (**b**) ($a_{110} < 0, a_{220} > 0$), (**c**) ($a_{110} > 0, a_{220} < 0$), and (**d**) ($a_{110} < 0, a_{220} < 0$)

2.2 Inflection-Saddle and Parabola-Saddles

In this section, the inflection-saddle infinite-equilibriums will be presented for the switching bifurcations.

2.2.1 Horizontal Inflection-Saddle Infinite-Equilibriums

Consider a dynamical system for $a_{221} = a_{121}$ as

$$\begin{aligned}\dot{x}_1 &= a_{110}(x_1 - a_{111})(x_2 - a_{121})^2,\\ \dot{x}_2 &= a_{220}(x_1 - a_{211})(x_2 - a_{221})^2,\end{aligned} \tag{2.11}$$

and the corresponding first integral manifold for $a_{221} = a_{121}$ is

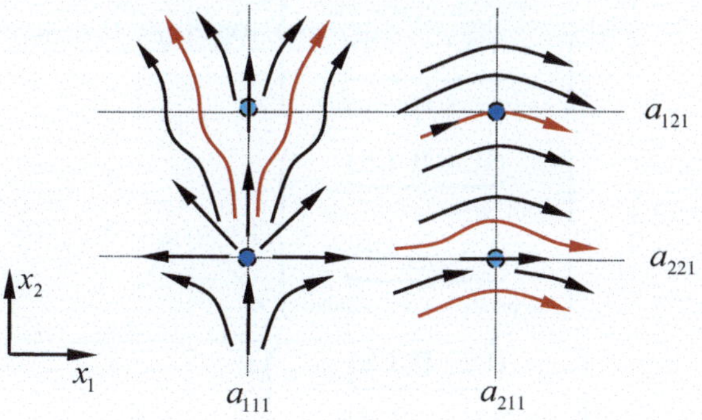

(c)

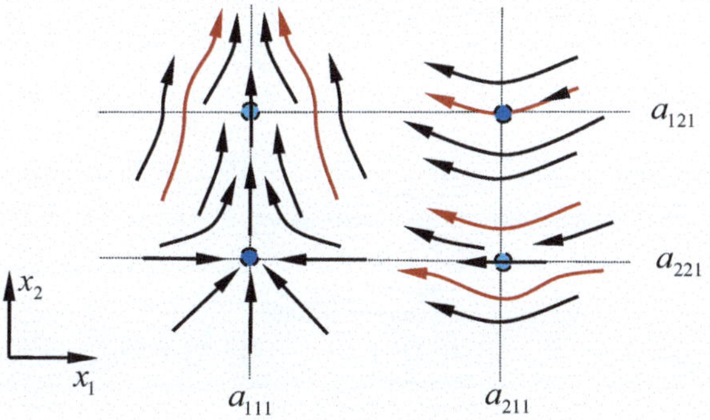

(d)

Fig. 2.4 (continued)

$$[(x_1 - a_{111}) - (x_{10} - a_{111})] + (a_{111} - a_{211}) \ln \frac{|x_1 - a_{111}|}{|x_{10} - a_{111}|} = \frac{a_{110}}{a_{220}}(x_2 - x_{20}). \quad (2.12)$$

(i) For $a_{221} = a_{121}, a_{111} > a_{211}$, the equilibriums of $(x_1^*, x_2^*) = (a_{111}, a_{221}), (a_{211}, a_{221})$ are

$$\left\{ \begin{array}{c} (a_{111}, a_{221}) \\ (a_{121}, a_{211}) \end{array} \right\} = \left\{ \begin{array}{c} \underbrace{(\text{DI}:\text{II}, \text{US})}_{\text{inflection upper-saddle}} \\ \underbrace{(\text{LS}, {}_{\text{II:DI}}\text{DP})}_{\text{down-parabola lower-saddle}} \end{array} \right\} \text{ for } a_{110} > 0, a_{220} > 0, \quad (2.13)$$

$$\left\{\begin{array}{l}(a_{111}, a_{221}) \\ (a_{121}, a_{211})\end{array}\right\} = \left\{\begin{array}{c}\underbrace{(\mathrm{II}:\mathrm{DI}, \mathrm{US})}_{\text{inflection upper-saddle}} \\ \underbrace{(\mathrm{US}, {}_{\mathrm{DI:II}}\mathrm{UP})}_{\text{up-parabola upper-saddle}}\end{array}\right\} \text{ for } a_{110} < 0, a_{220} > 0, \tag{2.14}$$

$$\left\{\begin{array}{l}(a_{111}, a_{221}) \\ (a_{121}, a_{211})\end{array}\right\} = \left\{\begin{array}{c}\underbrace{(\mathrm{II}:\mathrm{DI}, \mathrm{LS})}_{\text{inflection lower-saddle}} \\ \underbrace{(\mathrm{LS}, {}_{\mathrm{DI:II}}\mathrm{UP})}_{\text{up-parabola lower-saddle}}\end{array}\right\} \text{ for } a_{110} > 0, a_{220} < 0, \tag{2.15}$$

$$\left\{\begin{array}{l}(a_{111}, a_{221}) \\ (a_{121}, a_{211})\end{array}\right\} = \left\{\begin{array}{c}\underbrace{(\mathrm{DI}:\mathrm{II}, \mathrm{LS})}_{\text{inflection lower-saddle}} \\ \underbrace{(\mathrm{US}, {}_{\mathrm{II:DI}}\mathrm{DP})}_{\text{down-parabola upper-saddle}}\end{array}\right\} \text{ for } a_{110} < 0, a_{220} < 0. \tag{2.16}$$

and the infinite-equilibrium of $x_2^* = a_{121} = a_{221}$ is summarized as follows:

$$\left.\begin{array}{lll}(\bar{x}_1, a_{221}) = & \underbrace{(\mathrm{II}, \mathrm{LS})}_{\text{increasing-inflection lower-saddle}} & \text{if } \bar{x}_1 \in (-\infty, a_{211}) \\ (\bar{x}_1, a_{221}) = & \underbrace{(\mathrm{DI}, \mathrm{US})}_{\text{decreasing-inflection upper-saddle}} & \text{if } \bar{x}_1 \in (a_{211}, a_{111}), \\ (\bar{x}_1, a_{221}) = & \underbrace{(\mathrm{II}, \mathrm{US})}_{\text{increasing-inflection upper-saddle}} & \text{if } \bar{x}_1 \in (a_{111}, \infty), \\ \text{for } a_{110} > 0, a_{220} > 0; & & \end{array}\right. \tag{2.17}$$

$$\left.\begin{array}{lll}(\bar{x}_1, a_{221}) = & \underbrace{(\mathrm{DI}, \mathrm{LS})}_{\text{decreasing-inflection lower-saddle}} & \text{if } \bar{x}_1 \in (-\infty, a_{211}) \\ (\bar{x}_1, a_{221}) = & \underbrace{(\mathrm{II}, \mathrm{US})}_{\text{increasing-inflection upper-saddle}} & \text{if } \bar{x}_1 \in (a_{211}, a_{111}), \\ (\bar{x}_1, a_{221}) = & \underbrace{(\mathrm{DI}, \mathrm{US})}_{\text{decreasing-inflection upper-saddle}} & \text{if } \bar{x}_1 \in (a_{111}, \infty), \\ \text{for } a_{110} < 0, a_{220} > 0; & & \end{array}\right. \tag{2.18}$$

$$\left.\begin{array}{lll}(\bar{x}_1, a_{221}) = & \underbrace{(\mathrm{DI}, \mathrm{US})}_{\text{decreasing-inflection upper-saddle}} & \text{if } \bar{x}_1 \in (-\infty, a_{211}), \\ (\bar{x}_1, a_{221}) = & \underbrace{(\mathrm{II}, \mathrm{LS})}_{\text{increasing-inflection lower-saddle}} & \text{if } \bar{x}_1 \in (a_{211}, a_{111}), \\ (\bar{x}_1, a_{221}) = & \underbrace{(\mathrm{DI}, \mathrm{LS})}_{\text{decreasing-inflection lower-saddle}} & \text{if } \bar{x}_1 \in (a_{111}, \infty), \\ \text{for } a_{110} > 0, a_{220} < 0; & & \end{array}\right. \tag{2.19}$$

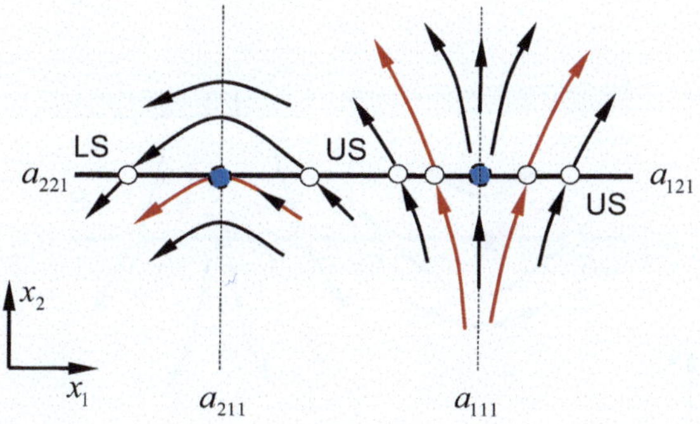

(a)

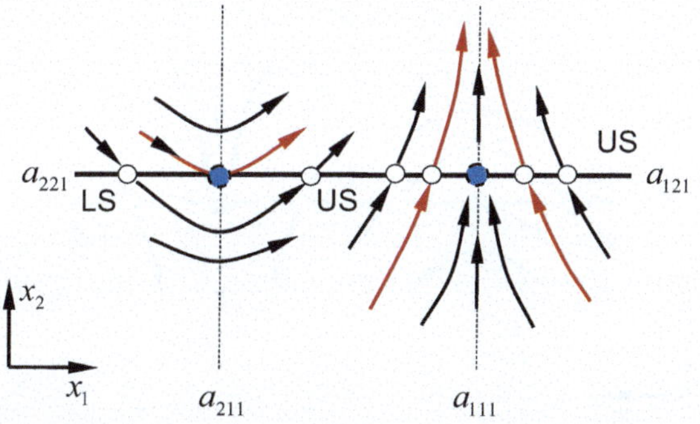

(b)

Fig. 2.5 Phase portraits of infinite-equilibrium ($a_{221} = a_{121}$, $a_{211} < a_{111}$) for two-dimensional systems on the x_1-direction with $x_1^* = a_{111}, a_{211}$ and on the x_2-direction with $x_2^* = a_{121} = a_{221}$. The infinite-equilibriums: (**a**) ($a_{110} > 0$, $a_{220} > 0$), (**b**) ($a_{110} < 0$, $a_{220} > 0$), (**c**) ($a_{110} > 0$, $a_{220} < 0$), and (**d**) ($a_{110} < 0$, $a_{220} < 0$)

$$
\begin{aligned}
(\bar{x}_1, a_{221}) &= \underbrace{(\mathrm{II}, \mathrm{US})}_{\text{increasing-inflection upper-saddle}} && \text{if } \bar{x}_1 \in (-\infty, a_{211}) \\
(\bar{x}_1, a_{221}) &= \underbrace{(\mathrm{DI}, \mathrm{LS})}_{\text{decreasing-inflection lower-saddle}} && \text{if } \bar{x}_1 \in (a_{211}, a_{111}), \\
(\bar{x}_1, a_{221}) &= \underbrace{(\mathrm{II}, \mathrm{LS})}_{\text{increasing-inflection lower-saddle}} && \text{if } \bar{x}_1 \in (a_{111}, \infty),
\end{aligned} \tag{2.20}
$$

for $a_{110} < 0, a_{220} < 0$.

The infinite-equilibrium of $x_2^* = a_{121} = a_{221}$ is increasing-inflection and decreasing-inflection upper-saddle and lower-saddle in Fig. 2.5. The infinite-equilibrium is the switching bifurcation for $a_{121} < a_{221}$ and $a_{121} > a_{221}$ with $a_{211} < a_{111}$.

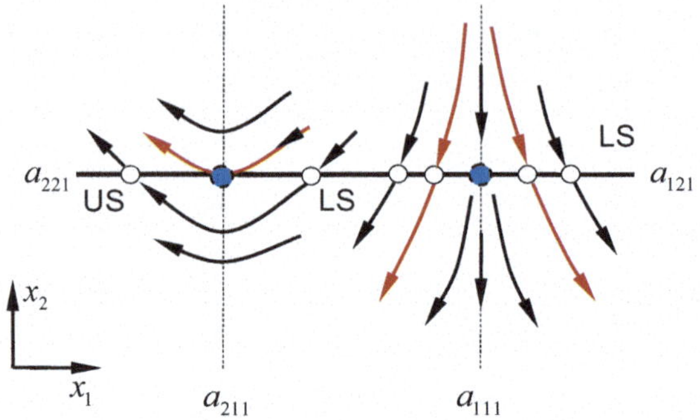

(c)

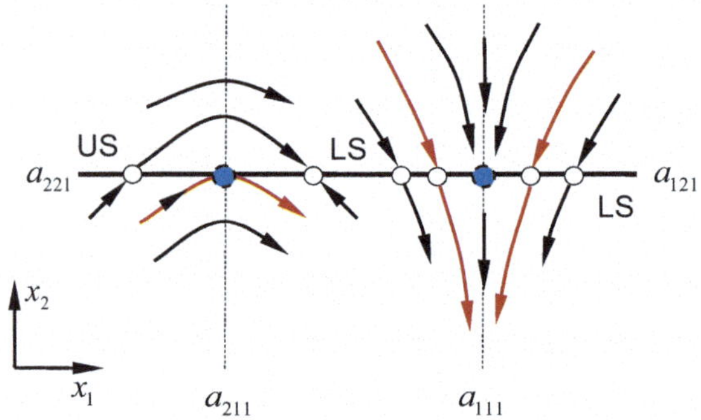

(d)

Fig. 2.5 (continued)

(ii) For $a_{221} = a_{121}$, $a_{111} < a_{211}$, the equilibriums of $(x_1^*, x_2^*) = (a_{111}, a_{221}), (a_{211}, a_{221})$ are

$$\begin{Bmatrix} (a_{121}, a_{211}) \\ (a_{111}, a_{221}) \end{Bmatrix} = \begin{Bmatrix} \underbrace{(\mathrm{US},{}_{\mathrm{DI:II}}\mathrm{UP})}_{\text{up-parabola upper-saddle}} \\ \underbrace{(\mathrm{II}:\mathrm{DI},\mathrm{LS})}_{\text{inflection lower-saddle}} \end{Bmatrix} \text{ for } a_{110} > 0, a_{220} > 0, \tag{2.21}$$

$$\begin{Bmatrix} (a_{121}, a_{211}) \\ (a_{111}, a_{221}) \end{Bmatrix} = \begin{Bmatrix} \underbrace{(\mathrm{LS},{}_{\mathrm{II:DI}}\mathrm{DP})}_{\text{down-parabola lower-saddle}} \\ \underbrace{(\mathrm{DI}:\mathrm{II},\mathrm{LS})}_{\text{inflection lower-saddle}} \end{Bmatrix} \text{ for } a_{110} < 0, a_{220} > 0, \tag{2.22}$$

$$\left\{\begin{array}{l}(a_{121},a_{211})\\(a_{111},a_{221})\end{array}\right\}=\left\{\begin{array}{c}\underbrace{(\text{US},{}_{\text{II:DI}}\text{DP})}_{\text{down-parabola upper-saddle}}\\\underbrace{(\text{DI : II, US})}_{\text{inflection upper-saddle}}\end{array}\right\}\text{ for } a_{110}>0, a_{220}<0, \tag{2.23}$$

$$\left\{\begin{array}{l}(a_{121},a_{211})\\(a_{111},a_{221})\end{array}\right\}=\left\{\begin{array}{c}\underbrace{(\text{LS},{}_{\text{DI:II}}\text{UP})}_{\text{up-parabola lower-saddle}}\\\underbrace{(\text{II : DI, US})}_{\text{inflection upper-saddle}}\end{array}\right\}\text{ for } a_{110}<0, a_{220}<0. \tag{2.24}$$

The infinite-equilibrium of $x_2^*=a_{121}=a_{221}$ for $a_{111}<a_{211}$ is summarized as follows:

$$\left.\begin{array}{ll}(\bar{x}_1,a_{221})=\underbrace{(\text{II, LS})}_{\text{increasing-inflection lower-saddle}} & \text{if } \bar{x}_1\in(-\infty,a_{111})\\(\bar{x}_1,a_{221})=\underbrace{(\text{DI, LS})}_{\text{decreasing-inflection lower-saddle}} & \text{if } \bar{x}_1\in(a_{111},a_{211}),\\(\bar{x}_1,a_{221})=\underbrace{(\text{II, US})}_{\text{increasing-inflection upper-saddle}} & \text{if } \bar{x}_1\in(a_{211},\infty),\\\text{for } a_{110}>0, a_{220}>0; & \end{array}\right. \tag{2.25}$$

$$\left.\begin{array}{ll}(\bar{x}_1,a_{221})=\underbrace{(\text{DI, LS})}_{\text{decreasing-inflection lower-saddle}} & \text{if } \bar{x}_1\in(-\infty,a_{111})\\(\bar{x}_1,a_{221})=\underbrace{(\text{II, LS})}_{\text{increasing-inflection lower-saddle}} & \text{if } \bar{x}_1\in(a_{111},a_{211}),\\(\bar{x}_1,a_{221})=\underbrace{(\text{DI, US})}_{\text{decreasing-inflection upper-saddle}} & \text{if } \bar{x}_1\in(a_{211},\infty),\\\text{for } a_{110}<0, a_{220}>0; & \end{array}\right. \tag{2.26}$$

$$\left.\begin{array}{ll}(\bar{x}_1,a_{221})=\underbrace{(\text{DI, US})}_{\text{decreasing-inflection upper-saddle}} & \text{if } \bar{x}_1\in(-\infty,a_{111})\\(\bar{x}_1,a_{221})=\underbrace{(\text{II, US})}_{\text{increasing-inflection upper-saddle}} & \text{if } \bar{x}_1\in(a_{111},a_{211}),\\(\bar{x}_1,a_{221})=\underbrace{(\text{DI, LS})}_{\text{decreasing-inflection lower-saddle}} & \text{if } \bar{x}_1\in(a_{211},\infty),\\\text{for } a_{110}>0, a_{220}<0; & \end{array}\right. \tag{2.27}$$

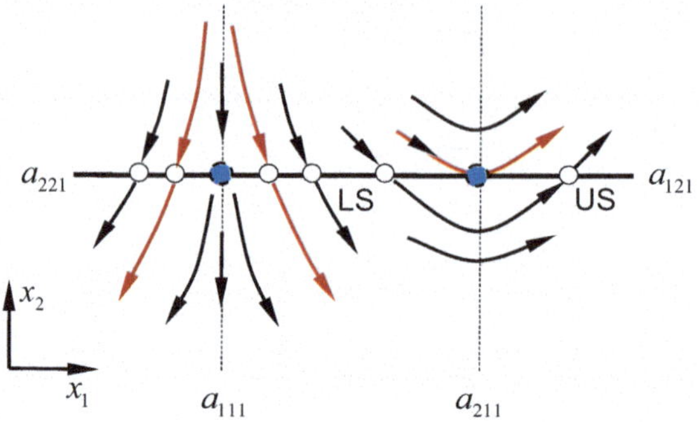

(a)

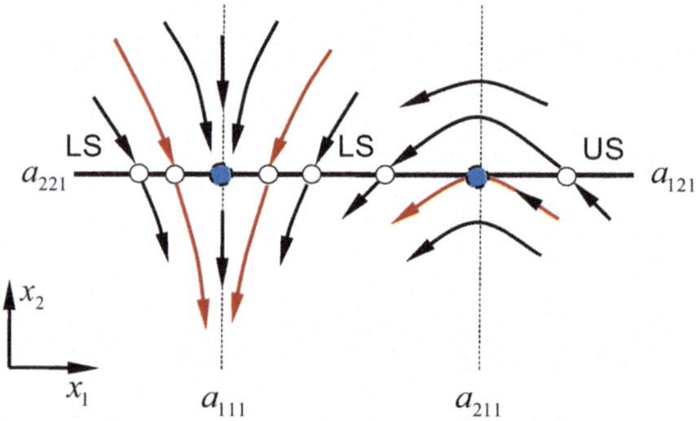

(b)

Fig. 2.6 Phase portraits of infinite-equilibrium ($a_{221} = a_{121}$, $a_{111} < a_{211}$) for two-dimensional systems on the x_1-direction with $x_1^* = a_{111}, a_{211}$ and on the x_2-direction with $x_2^* = a_{121} = a_{221}$. The infinite-equilibriums: (**a**) ($a_{110} > 0$, $a_{220} > 0$), (**b**) ($a_{110} < 0$, $a_{220} > 0$), (**c**) ($a_{110} > 0$, $a_{220} < 0$), and (**d**) ($a_{110} < 0$, $a_{220} < 0$)

$$
\begin{aligned}
(\bar{x}_1, a_{221}) &= \underbrace{(\mathrm{II}, \mathrm{US})}_{\text{increasing-inflection upper-saddle}} && \text{if } \bar{x}_1 \in (-\infty, a_{111}) \\
(\bar{x}_1, a_{221}) &= \underbrace{(\mathrm{DI}, \mathrm{US})}_{\text{decreasing-inflection upper-saddle}} && \text{if } \bar{x}_1 \in (a_{111}, a_{211}), \\
(\bar{x}_1, a_{221}) &= \underbrace{(\mathrm{II}, \mathrm{LS})}_{\text{increasing-inflection lower-saddle}} && \text{if } \bar{x}_1 \in (a_{211}, \infty),
\end{aligned} \tag{2.28}
$$

for $a_{110} < 0, a_{220} < 0$.

The infinite-equilibrium of $x_2^* = a_{121} = a_{221}$ is increasing-inflection and decreasing-inflection upper-saddle and lower-saddle in Fig. 2.6. The infinite-equilibrium is the switching bifurcation for $a_{121} < a_{221}$ and $a_{121} > a_{221}$ with $a_{111} < a_{211}$.

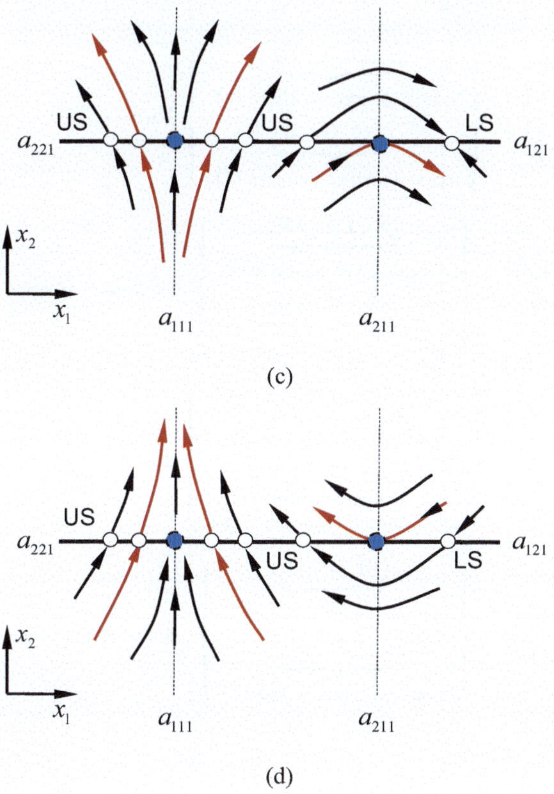

Fig. 2.6 (continued)

2.2.2 *Vertical Inflection-Source (Sink) Infinite-Equilibriums*

Consider a dynamical system for $a_{111} = a_{211}$ as

$$\begin{aligned}\dot{x}_1 &= a_{110}(x_1 - a_{111})(x_2 - a_{121})^2,\\ \dot{x}_2 &= a_{220}(x_1 - a_{211})(x_2 - a_{221})^2,\end{aligned} \tag{2.29}$$

and the corresponding first integral manifold for $a_{111} = a_{211}$ is

$$\begin{aligned}[(x_1 - a_{111}) - (x_{10} - a_{111})] = \frac{a_{110}}{a_{220}}\Big[&(x_2 - x_{20}) + 2(a_{221} - a_{121})\ln\frac{|x_2 - a_{221}|}{|x_{20} - a_{221}|}\\ &- (a_{221} - a_{121})^2\Big(\frac{1}{x_2 - a_{221}} - \frac{1}{x_{20} - a_{221}}\Big)\Big].\end{aligned} \tag{2.30}$$

(i) For $a_{111} = a_{211}, a_{121} < a_{221}$, the equilibrium of $(x_1^*, x_2^*) = (a_{111}, a_{121}),\ (a_{111}, a_{221})$ are

$$\left\{\begin{array}{l}(a_{111}, a_{221})\\ (a_{111}, a_{121})\end{array}\right\} = \left\{\begin{array}{c}\underbrace{(\mathrm{SO}, \mathrm{II} : \mathrm{II})}_{\text{increasing-inflection source (2)}}\\ \underbrace{(\mathrm{SO}, \mathrm{II} : \mathrm{II})}_{\text{increasing-inflection source (1)}}\end{array}\right\} \text{ for } a_{110} > 0, a_{220} > 0, \tag{2.31}$$

$$\left\{\begin{array}{l}(a_{111}, a_{221})\\ (a_{111}, a_{121})\end{array}\right\} = \left\{\begin{array}{c}\underbrace{(\mathrm{SI}, \mathrm{DI} : \mathrm{DI})}_{\text{decreasing-inflection sink (2)}}\\ \underbrace{(\mathrm{SI}, \mathrm{DI} : \mathrm{DI})}_{\text{decreasing-inflection sink (1)}}\end{array}\right\} \text{ for } a_{110} < 0, a_{220} > 0, \tag{2.32}$$

$$\left\{\begin{array}{l}(a_{111}, a_{221})\\ (a_{111}, a_{121})\end{array}\right\} = \left\{\begin{array}{c}\underbrace{(\mathrm{SO}, \mathrm{DI} : \mathrm{DI})}_{\text{decreasing-inflection source (2)}}\\ \underbrace{(\mathrm{SO}, \mathrm{DI} : \mathrm{DI})}_{\text{decreasing-inflection source (1)}}\end{array}\right\} \text{ for } a_{110} > 0, a_{220} < 0, \tag{2.33}$$

$$\left\{\begin{array}{l}(a_{111}, a_{221})\\ (a_{111}, a_{121})\end{array}\right\} = \left\{\begin{array}{c}\underbrace{(\mathrm{SI}, \mathrm{II} : \mathrm{II})}_{\text{increasing-inflection sink (2)}}\\ \underbrace{(\mathrm{SI}, \mathrm{II} : \mathrm{II})}_{\text{increasing inflection sink (1)}}\end{array}\right\} \text{ for } a_{110} < 0, a_{220} < 0. \tag{2.34}$$

The infinite-equilibrium of $x_2^* = a_{121} = a_{221}$ for $a_{111} < a_{211}$ is summarized as follows:

$$\begin{array}{lll}(a_{111}, \bar{x}_2) = \underbrace{(\mathrm{SO}, \mathrm{II})}_{\text{increasing-inflection source}} & \text{if } \bar{x}_1 \in (-\infty, a_{121}) \\ (a_{111}, \bar{x}_2) = \underbrace{(\mathrm{SO}, \mathrm{II})}_{\text{increasing-inflection source}} & \text{if } \bar{x}_1 \in (a_{121}, a_{221}), \\ (a_{111}, \bar{x}_2) = \underbrace{(\mathrm{SO}, \mathrm{II})}_{\text{increasing-inflection source}} & \text{if } \bar{x}_2 \in (a_{221}, \infty), \\ \text{for } a_{110} > 0, a_{220} > 0; \end{array} \tag{2.35}$$

$$\begin{array}{lll}(a_{111}, \bar{x}_2) = \underbrace{(\mathrm{SI}, \mathrm{DI})}_{\text{decreasing-inflection sink}} & \text{if } \bar{x}_1 \in (-\infty, a_{121}) \\ (a_{111}, \bar{x}_2) = \underbrace{(\mathrm{SI}, \mathrm{DI})}_{\text{decreasing-inflection sink}} & \text{if } \bar{x}_1 \in (a_{121}, a_{221}), \\ (a_{111}, \bar{x}_2) = \underbrace{(\mathrm{SI}, \mathrm{DI})}_{\text{decreasing-inflection sink}} & \text{if } \bar{x}_2 \in (a_{221}, \infty), \\ \text{for } a_{110} < 0, a_{220} > 0; \end{array} \tag{2.36}$$

$$\left.\begin{array}{ll}(a_{111},\bar{x}_2)=\underbrace{(\text{SO, DI})}_{\text{decreasing-inflection source}} & \text{if } \bar{x}_1\in(-\infty,a_{121})\\ (a_{111},\bar{x}_2)=\underbrace{(\text{SO, DI})}_{\text{decreasing-inflection source}} & \text{if } \bar{x}_1\in(a_{121},a_{221}),\\ (a_{111},\bar{x}_2)=\underbrace{(\text{SO, DI})}_{\text{decreasing-inflection source}} & \text{if } \bar{x}_2\in(a_{221},\infty),\\ \text{for } a_{110}>0, a_{220}<0; & \end{array}\right\} \tag{2.37}$$

$$\left.\begin{array}{ll}(a_{111},\bar{x}_2)=\underbrace{(\text{SI, II})}_{\text{increasing-inflection sink}} & \text{if } \bar{x}_1\in(-\infty,a_{121})\\ (a_{111},\bar{x}_2)=\underbrace{(\text{SI, II})}_{\text{increasing-inflection sink}} & \text{if } \bar{x}_1\in(a_{121},a_{221}),\\ (a_{111},\bar{x}_2)=\underbrace{(\text{SI, II})}_{\text{increasing-infllection sink}} & \text{if } \bar{x}_2\in(a_{221},\infty),\\ \text{for } a_{110}<0, a_{220}<0. & \end{array}\right\} \tag{2.38}$$

The infinite-equilibriums of $x_1^*=a_{111}=a_{211}$ are presented in Fig. 2.7, which are the inflection-sink and source. For $x_2^*=a_{221}$, the inflection-sink and source infinite-equilibrium are asymptotic. The infinite-equilibriums of $x_1^*=a_{111}=a_{211}$ are the switching bifurcations of the inflection-sink and source for cases of $a_{111}<a_{211}$ and $a_{111}>a_{211}$ with $a_{221}>a_{121}$.

(ii) For $a_{111}=a_{211}$ and $a_{221}<a_{121}$, the equilibriums of $(x_1^*,x_2^*)=(a_{111},a_{121})$, (a_{111},a_{221}) have the same properties as in Eqs. (2.31)–(2.34). The infinite-equilibriums of $x_1^*=a_{111}=a_{211}$ have inflection-sink and source as in Eqs. (2.35)–(2.38). The infinite-equilibrium of $x_1^*=a_{111}=a_{211}$ are the switching bifurcations of the inflection-sink and source for cases of $a_{111}<a_{211}$ and $a_{111}>a_{211}$ with $a_{221}<a_{121}$. Such infinite-equilibriums of $x_1^*=a_{111}=a_{211}$ are presented in Fig. 2.8.

2.2.3 Vertical and Horizontal Infinite-Equilibriums

Consider a dynamical system for $a_{111}=a_{211}$ and $a_{221}=a_{121}$ as

$$\begin{aligned}\dot{x}_1&=a_{110}(x_1-a_{111})(x_2-a_{121})^2,\\ \dot{x}_2&=a_{220}(x_1-a_{211})(x_2-a_{221})^2,\end{aligned} \tag{2.39}$$

and the corresponding first integral manifold for $a_{111}=a_{211}$ is

$$[(x_1-a_{111})-(x_{10}-a_{111})]=\frac{a_{110}}{a_{220}}(x_2-x_{20}). \tag{2.40}$$

The equilibrium of $(x_1^*,x_2^*)=(a_{111},a_{121})=(a_{211},a_{221})$ is

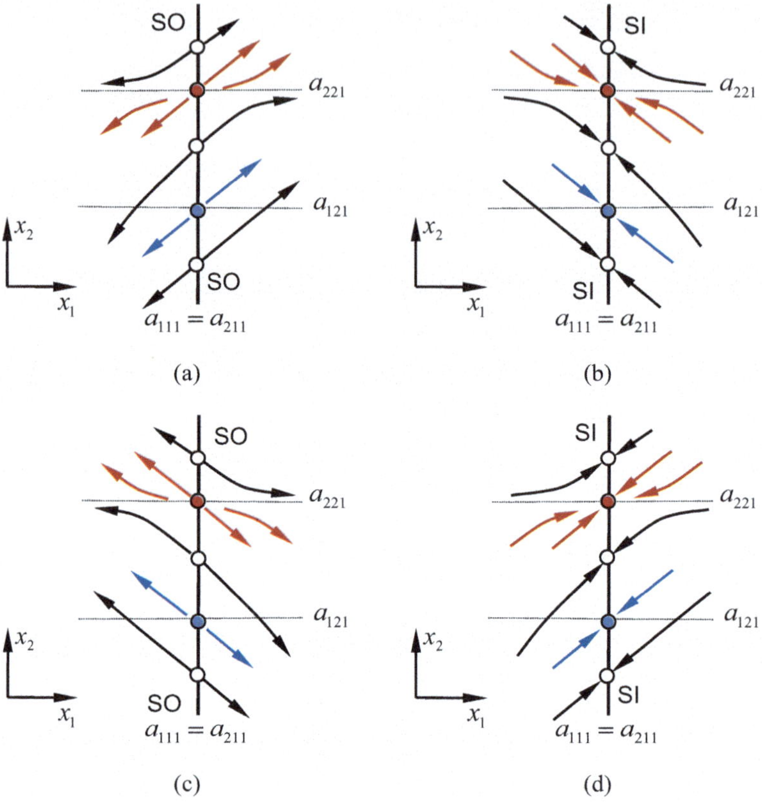

Fig. 2.7 Phase portraits ($a_{111} = a_{211}$, $a_{121} < a_{221}$) for two-dimensional systems on the x_1-direction with $x_1^* = a_{111}, a_{211}$ and on the x_2-direction with $x_2^* = a_{121}, a_{221}$. The equilibrium flows: (**a**) ($a_{110} > 0$, $a_{220} > 0$), (**b**) ($a_{110} < 0$, $a_{220} > 0$), (**c**) ($a_{110} > 0$, $a_{220} < 0$), and (**d**) ($a_{110} < 0$, $a_{220} < 0$)

$$
\begin{aligned}
(a_{111}, a_{121}) = (a_{211}, a_{221}) &= \underbrace{({}_{\text{II:II}}\text{SO}, {}_{\text{LS:US}}\text{II})}_{\text{increasing-inflection source}} \quad \text{for } a_{110} > 0, a_{220} > 0;\\
(a_{111}, a_{121}) = (a_{211}, a_{221}) &= \underbrace{({}_{\text{DI:DI}}\text{SI}, {}_{\text{LS:US}}\text{DI})}_{\text{decreasing-inflection sink}} \quad \text{for } a_{110} < 0, a_{220} > 0;\\
(a_{111}, a_{121}) = (a_{211}, a_{221}) &= \underbrace{({}_{\text{DI:DI}}\text{SO}, {}_{\text{US:LS}}\text{DI})}_{\text{decreasing-inflection source}} \quad \text{for } a_{110} > 0, a_{220} < 0;\\
(a_{111}, a_{121}) = (a_{211}, a_{221}) &= \underbrace{({}_{\text{II:II}}\text{SI}, {}_{\text{LS:US}}\text{II})}_{\text{increasing-inflection sink}} \quad \text{for } a_{110} < 0, a_{220} < 0.
\end{aligned}
\tag{2.41}
$$

The infinite-equilibriums of $x_2^* = a_{121} = a_{221}$ and $x_1^* = a_{111} = a_{211}$ are

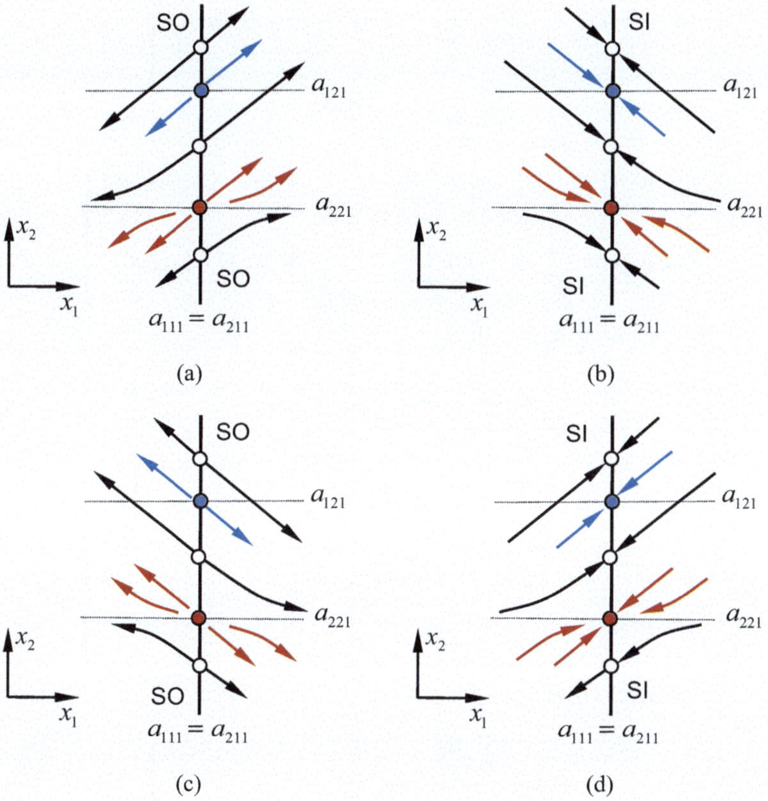

Fig. 2.8 Phase portraits ($a_{111} = a_{211}$, $a_{121} < a_{221}$) for two-dimensional systems on the x_1-direction with $x_1^* = a_{111}, a_{211}$ and on the x_2-direction with $x_2^* = a_{121}, a_{221}$. The equilibrium flows: **(a)** ($a_{110} > 0$, $a_{220} > 0$), **(b)** ($a_{110} < 0$, $a_{220} > 0$), **(c)** ($a_{110} > 0$, $a_{220} < 0$), and **(d)** ($a_{110} < 0$, $a_{220} < 0$)

$$
\left.
\begin{aligned}
&\left.
\begin{aligned}
(a_{111}, \bar{x}_2) &= \underbrace{(\mathrm{SO}, \mathrm{II})}_{\text{increasing-inflection source}} && \text{if } \bar{x}_2 \in (-\infty, a_{121}) \\
(a_{111}, \bar{x}_2) &= \underbrace{(\mathrm{SO}, \mathrm{II})}_{\text{increasing-inflection source}} && \text{if } \bar{x}_2 \in (a_{121}, \infty)
\end{aligned}
\right\}, \\
&\left.
\begin{aligned}
(\bar{x}_1, a_{221}) &= \underbrace{(\mathrm{II}, \mathrm{LS})}_{\text{increasing-inflection lower-saddle}} && \text{if } \bar{x}_1 \in (-\infty, a_{111}) \\
(\bar{x}_1, a_{221}) &= \underbrace{(\mathrm{II}, \mathrm{US})}_{\text{increasing-inflection upper-saddle}} && \text{if } \bar{x}_1 \in (a_{111}, \infty)
\end{aligned}
\right\}
\end{aligned}
\right. \tag{2.42}
$$

for $a_{110} > 0, a_{220} > 0$;

$$\left.\begin{array}{ll}(a_{111},\bar{x}_2)=\underbrace{(\mathrm{SI},\mathrm{DI})}_{\text{decreasing-inflection sink}} & \text{if } \bar{x}_2\in(-\infty,a_{121})\\ (a_{111},\bar{x}_2)=\underbrace{(\mathrm{SI},\mathrm{DI})}_{\text{decreasing-inflection sink}} & \text{if } \bar{x}_2\in(a_{121},\infty)\end{array}\right\},$$
$$\left.\begin{array}{ll}(\bar{x}_1,a_{221})=\underbrace{(\mathrm{DI},\mathrm{LS})}_{\text{decreasing-inflection lower-saddle}} & \text{if } \bar{x}_1\in(-\infty,a_{111})\\ (\bar{x}_1,a_{221})=\underbrace{(\mathrm{DI},\mathrm{US})}_{\text{decreasing-inflection upper-saddle}} & \text{if } \bar{x}_1\in(a_{111},\infty)\end{array}\right\} \tag{2.43}$$
for $a_{110}<0, a_{220}>0$;

$$\left.\begin{array}{ll}(a_{111},\bar{x}_2)=\underbrace{(\mathrm{SO},\mathrm{DI})}_{\text{decreasing-inflection source}} & \text{if } \bar{x}_2\in(-\infty,a_{121})\\ (a_{111},\bar{x}_2)=\underbrace{(\mathrm{SO},\mathrm{DI})}_{\text{decreasing-inflection source}} & \text{if } \bar{x}_2\in(a_{121},\infty)\end{array}\right\},$$
$$\left.\begin{array}{ll}(\bar{x}_1,a_{221})=\underbrace{(\mathrm{DI},\mathrm{US})}_{\text{decreasing-inflection upper-saddle}} & \text{if } \bar{x}_1\in(-\infty,a_{111})\\ (\bar{x}_1,a_{221})=\underbrace{(\mathrm{DI},\mathrm{LS})}_{\text{decreasing-inflection lower-saddle}} & \text{if } \bar{x}_1\in(a_{111},\infty)\end{array}\right\} \tag{2.44}$$
for $a_{110}>0, a_{220}<0$;

$$\left.\begin{array}{ll}(a_{111},\bar{x}_2)=\underbrace{(\mathrm{SI},\mathrm{II})}_{\text{increasing-inflection sink}} & \text{if } \bar{x}_2\in(-\infty,a_{121})\\ (a_{111},\bar{x}_2)=\underbrace{(\mathrm{SI},\mathrm{II})}_{\text{increasing-inflection sink}} & \text{if } \bar{x}_2\in(a_{121},\infty)\end{array}\right\},$$
$$\left.\begin{array}{ll}(\bar{x}_1,a_{221})=\underbrace{(\mathrm{II},\mathrm{US})}_{\text{increasing-inflection upper-saddle}} & \text{if } \bar{x}_1\in(-\infty,a_{111})\\ (\bar{x}_1,a_{221})=\underbrace{(\mathrm{II},\mathrm{LS})}_{\text{increasing-inflection lower-saddle}} & \text{if } \bar{x}_1\in(a_{111},\infty)\end{array}\right\} \tag{2.45}$$
for $a_{110}<0, a_{220}<0$.

The infinite-equilibriums of $x_1^*=a_{111}=a_{211}$ are the inflection-sink and source. The infinite-equilibriums of $x_2^*=a_{121}=a_{221}$ are the inflection upper-saddle and lower-saddle. The two infinite-equilibriums are presented in Fig. 2.9.

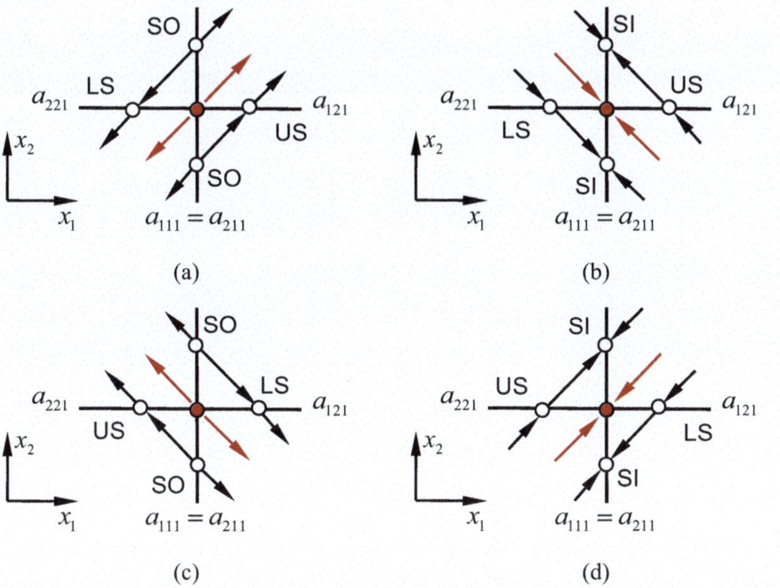

Fig. 2.9 Phase portraits ($a_{111} = a_{211}$, $a_{121} = a_{221}$) for two-dimensional systems on the x_1-direction with $x_1^* = a_{111}, a_{211}$ and on the x_2-direction with $x_2^* = a_{121}, a_{221}$. The equilibrium flows: (**a**) ($a_{110} > 0$, $a_{220} > 0$), (**b**) ($a_{110} < 0$, $a_{220} > 0$), (**c**) ($a_{110} > 0$, $a_{220} < 0$), and (**d**) ($a_{110} < 0$, $a_{220} < 0$)

Chapter 3
Inflection-Source (Sink) Flows and Parabola-Saddles

In this chapter, inflection-source (sink) flows and parabola-saddles with hyperbolic flows and simple equilibriums are discussed, and the corresponding networks are presented for inflection flows and parabola-saddles with simple flows and equilibriums. The corresponding switching bifurcations are discussed through the infinite-equilibriums, which are parabola-source (sink) infinite-equilibriums, inflection-source (sink) infinite-equilibriums, and two infinite-equilibriums of parabola-source (sink) and inflection-source (sink).

3.1 Networks of Inflection Flows with Saddles, Source, and Sink

Consider a dynamical system as

$$\begin{aligned}\dot{x}_1 &= a_{110}(x_1 - a_{111})(x_2 - a_{121})^2,\\ \dot{x}_2 &= a_{220}(x_1 - a_{211})(x_2 - a_{221})(x_2 - a_{222}),\end{aligned} \tag{3.1}$$

and the corresponding first integral manifold is

$$\begin{aligned}&[(x_1 - a_{111}) - (x_{10} - a_{111})] + (a_{111} - a_{211}) \ln \frac{|x_1 - a_{111}|}{|x_{10} - a_{111}|}\\ &= \frac{a_{110}}{a_{220}} \left[(x_2 - x_{20}) + \frac{(a_{221} - a_{121})^2}{a_{221} - a_{222}} \ln \frac{|x_2 - a_{221}|}{|x_{20} - a_{221}|} \right.\\ &\left. + \frac{(a_{222} - a_{121})^2}{a_{222} - a_{221}} \ln \frac{|x_2 - a_{222}|}{|x_{20} - a_{222}|} \right].\end{aligned} \tag{3.2}$$

A. C. J. Luo, *Two-dimensional Two-product Cubic Systems, Vol. I*,
https://doi.org/10.1007/978-3-031-48487-2_3

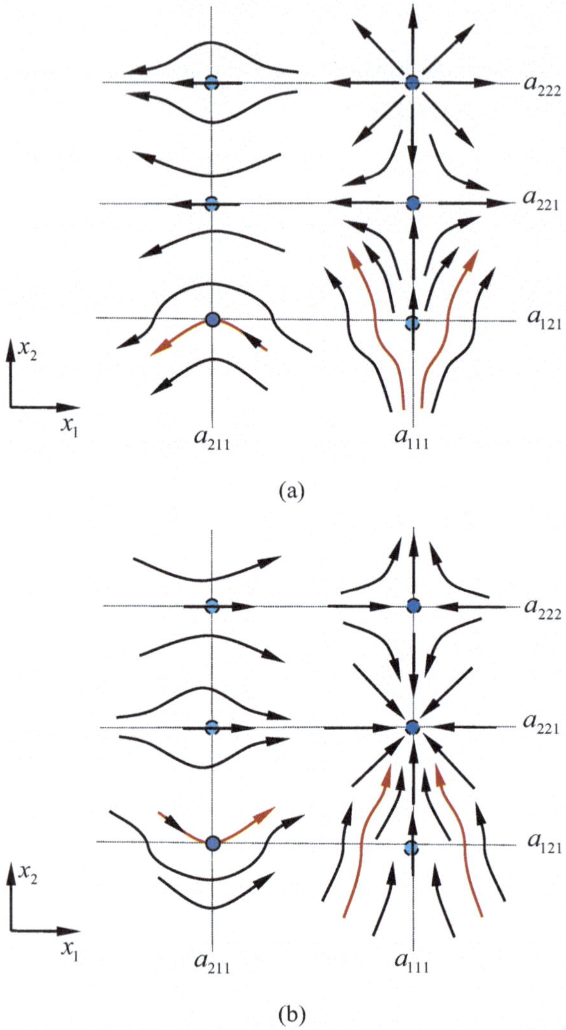

Fig. 3.1 Phase portraits ($a_{121} < a_{221}$ and $a_{211} < a_{111}$) for two-dimensional systems on the x_1-direction with $x_1^* = a_{111}, a_{211}$ and on the x_2-direction with $x_2^* = a_{121}, a_{221}, a_{222}$. The equilibrium flows: (**a**) ($a_{110} > 0, a_{220} > 0$), (**b**) ($a_{110} < 0, a_{220} > 0$), (**c**) ($a_{110} > 0, a_{220} < 0$), and (**d**) ($a_{110} < 0, a_{220} < 0$)

(i) For $a_{121} < a_{221}$ and $a_{222} < a_{121}$ with $a_{211} < a_{111}$, the equilibriums of $x_1^* = a_{111}, a_{211}$ with $x_2^* = a_{121}$, a_{221}, a_{222} have the following properties, as shown in Figs. 3.1 and 3.2.

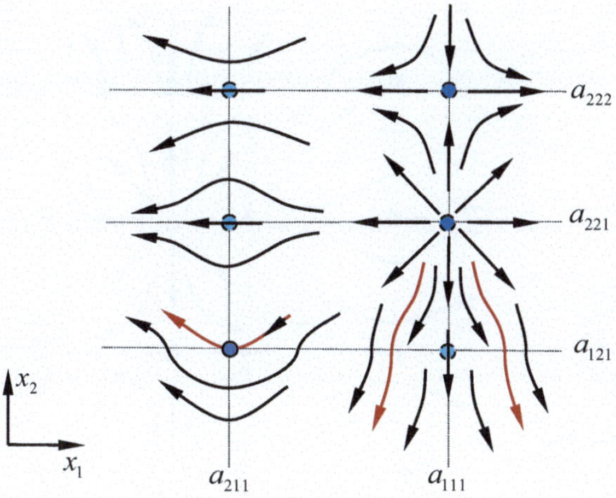

(c)

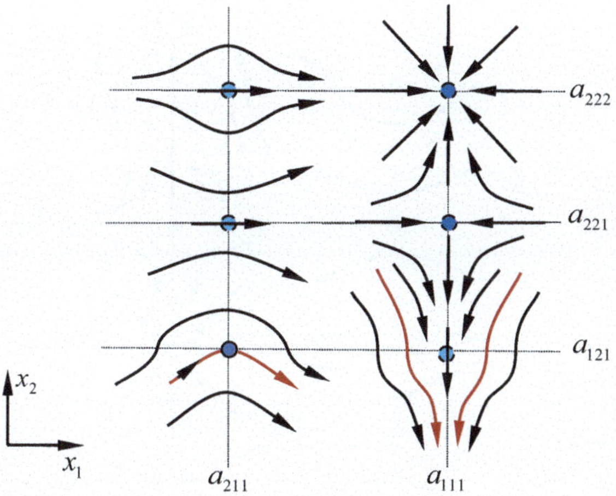

(d)

Fig. 3.1 (continued)

$$\left\{\begin{array}{cc} (a_{211}, a_{222}) & (a_{111}, a_{222}) \\ (a_{211}, a_{221}) & (a_{111}, a_{221}) \\ (a_{211}, a_{121}) & (a_{111}, a_{121}) \end{array}\right\} = \left\{\begin{array}{cc} \underbrace{(\text{nF}, \text{UP} : \text{DP})}_{\text{hyperbolic-secant flow } (-)} & \underbrace{(\text{SO}, \text{SO})}_{\text{source}} \\ \underbrace{(\text{nF}, \text{DP} : \text{UP})}_{\text{hyperbolic flow } (-)} & \underbrace{(\text{SO}, \text{SI})}_{\text{saddle}} \\ \underbrace{(\text{LS}, \text{DP})}_{\text{down-parabola lower-saddle}} & \underbrace{({}_{\text{DI:II}}\text{SO}, \text{pF})}_{\text{inflection-source flow } (+)} \end{array}\right\}$$

for $a_{110} > 0$ and $a_{220} > 0$;

(3.3)

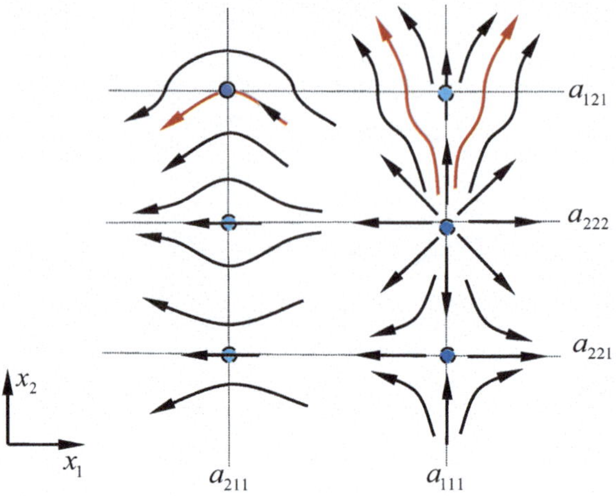

(a)

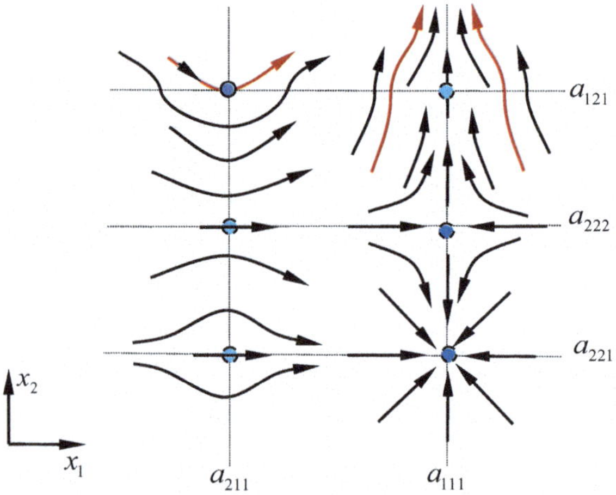

(b)

Fig. 3.2 Phase portraits ($a_{121} < a_{221}$ and $a_{211} < a_{111}$) for two-dimensional systems on the x_1-direction with $x_1^* = a_{111}, a_{211}$ and on the x_2-direction with $x_2^* = a_{121}, a_{221}, a_{222}$. The equilibrium flows: (**a**) ($a_{110} > 0$, $a_{220} > 0$), (**b**) ($a_{110} < 0$, $a_{220} > 0$), (**c**) ($a_{110} > 0$, $a_{220} < 0$), and (**d**) ($a_{110} < 0$, $a_{220} < 0$)

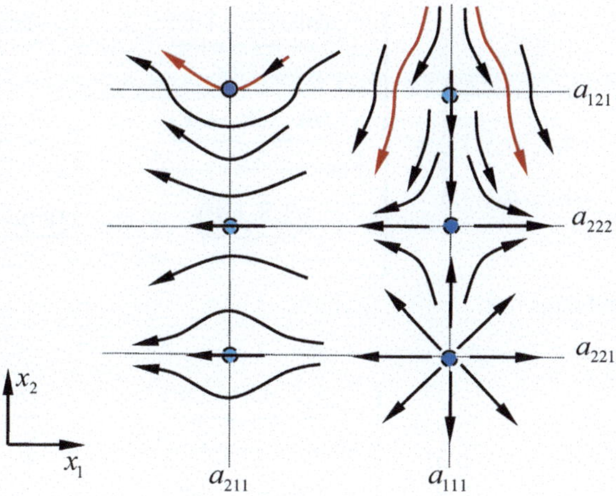

(c)

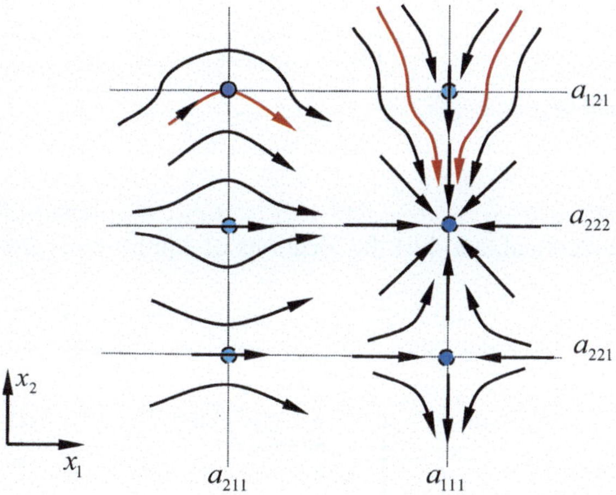

(d)

Fig. 3.2 (continued)

$$\left\{\begin{matrix} (a_{211}, a_{222}) & (a_{111}, a_{222}) \\ (a_{211}, a_{221}) & (a_{111}, a_{221}) \\ (a_{211}, a_{121}) & (a_{111}, a_{121}) \end{matrix}\right\} = \left\{\begin{matrix} \underbrace{(\text{pF}, \text{DP} : \text{UP})}_{\text{hyperbolic flow } (+)} & \underbrace{(\text{SI}, \text{SO})}_{\text{saddle}} \\ \underbrace{(\text{pF}, \text{UP} : \text{DP})}_{\text{hyperbolic-secant flow } (+)} & \underbrace{(\text{SI}, \text{SI})}_{\text{sink}} \\ \underbrace{(\text{US}, \text{UP})}_{\text{down-parabola upper-saddle}} & \underbrace{({}_{\text{II:DI}}\text{SI}, \text{pF})}_{\text{inflection-sink flow } (+)} \end{matrix}\right\}$$

for $a_{110} < 0$ and $a_{220} > 0$; (3.4)

$$\left\{\begin{array}{cc}(a_{211},a_{222}) & (a_{111},a_{222})\\(a_{211},a_{221}) & (a_{111},a_{221})\\(a_{211},a_{121}) & (a_{111},a_{121})\end{array}\right\}=\left\{\begin{array}{cc}\underbrace{(\text{nF},\text{DP}:\text{UP})}_{\text{hyperbolic flow }(-)} & \underbrace{(\text{SO},\text{SI})}_{\text{saddle}}\\\underbrace{(\text{nF},\text{UP}:\text{DP})}_{\text{hyperbolic-secant flow }(-)} & \underbrace{(\text{SO},\text{SO})}_{\text{source}}\\\underbrace{(\text{LS},\text{UP})}_{\text{down-parabola lower-saddle}} & \underbrace{({}_{\text{II:DI}}\text{SO},\text{nF})}_{\text{inflection-source flow }(-)}\end{array}\right\}$$

for $a_{110}>0$ and $a_{220}<0$;

(3.5)

$$\left\{\begin{array}{cc}(a_{211},a_{222}) & (a_{111},a_{222})\\(a_{211},a_{221}) & (a_{111},a_{221})\\(a_{211},a_{121}) & (a_{111},a_{121})\end{array}\right\}=\left\{\begin{array}{cc}\underbrace{(\text{pF},\text{UP}:\text{DP})}_{\text{hyperbolic-secant flow }(+)} & \underbrace{(\text{SI},\text{SI})}_{\text{sink}}\\\underbrace{(\text{pF},\text{DP}:\text{UP})}_{\text{hyperbolic flow }(+)} & \underbrace{(\text{SI},\text{SO})}_{\text{source}}\\\underbrace{(\text{US},\text{DP})}_{\text{down-parabola upper-saddle}} & \underbrace{({}_{\text{DI:II}}\text{SI},\text{nF})}_{\text{inflection-sink flow }(-)}\end{array}\right\}$$

for $a_{110}<0$ and $a_{220}<0$.

(3.6)

For both cases of $a_{121}<a_{221}$ and $a_{222}<a_{121}$, the dynamical behaviors of equilibriums are identical, and just the locations of equilibriums are switched.

(ii) For $a_{121}\in(a_{221},a_{222})$ with $a_{211}<a_{111}$, the equilibriums of $x_1^*=a_{111},a_{211}$ with $x_2^*=a_{121},a_{221},a_{222}$ have the following properties, as shown in Fig. 3.3.

$$\left\{\begin{array}{cc}(a_{211},a_{222}) & (a_{111},a_{222})\\(a_{211},a_{121}) & (a_{111},a_{221})\\(a_{211},a_{221}) & (a_{111},a_{121})\end{array}\right\}=\left\{\begin{array}{cc}\underbrace{(\text{nF},\text{UP}:\text{DP})}_{\text{hyperbolic-secant flow }(-)} & \underbrace{(\text{SO},\text{SO})}_{\text{source}}\\\underbrace{(\text{LS},\text{UP})}_{\text{up-parabola lower-saddle}} & \underbrace{({}_{\text{DI:II}}\text{SO},\text{pF})}_{\text{inflection-source flow }(+)}\\\underbrace{(\text{nF},\text{DP}:\text{UP})}_{\text{hyperbolic flow }(-)} & \underbrace{(\text{SO},\text{SI})}_{\text{saddle}}\end{array}\right\}$$

for $a_{110}>0$ and $a_{220}>0$;

(3.7)

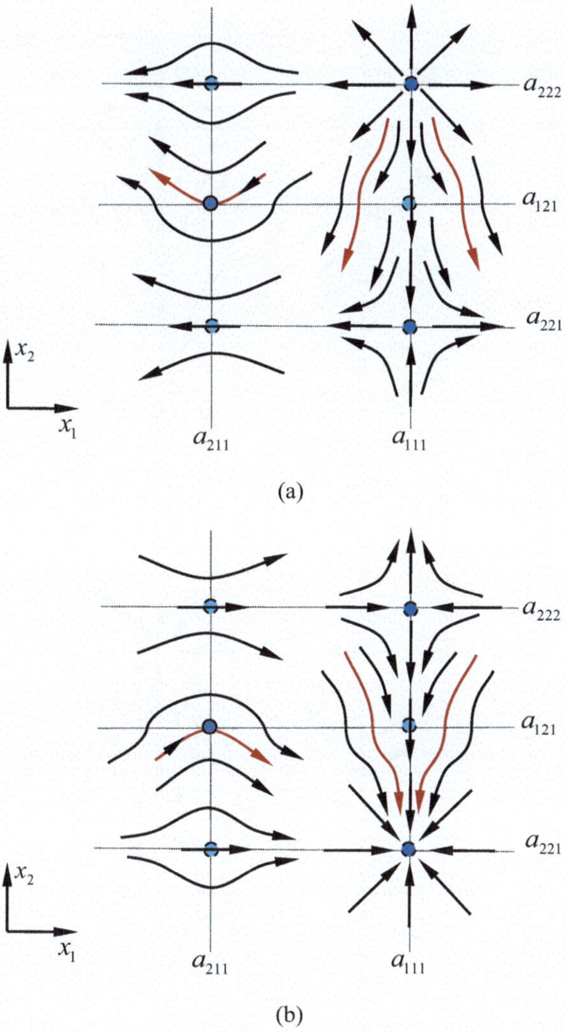

Fig. 3.3 Phase portraits ($a_{121} \in (a_{221}, a_{222})$ and $a_{211} < a_{111}$) for two-dimensional systems on the x_1-direction with $x_1^* = a_{111}, a_{211}$ and on the x_2-direction with $x_2^* = a_{121}, a_{221}, a_{222}$. The equilibrium flows: (**a**) ($a_{110} > 0$, $a_{220} > 0$), (**b**) ($a_{110} < 0$, $a_{220} > 0$), (**c**) ($a_{110} > 0$, $a_{220} < 0$), and (**d**) ($a_{110} < 0$, $a_{220} < 0$)

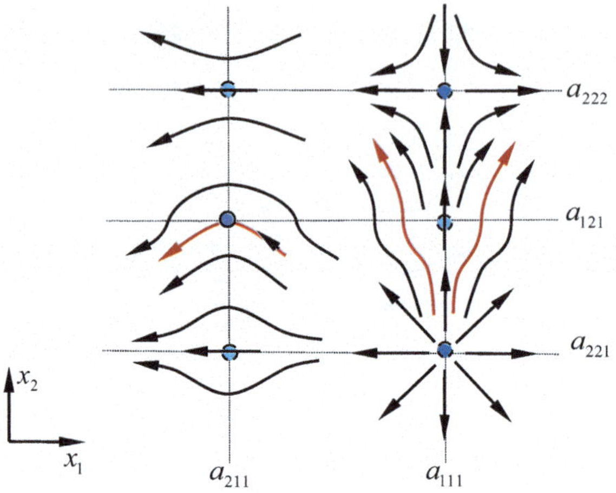

(c)

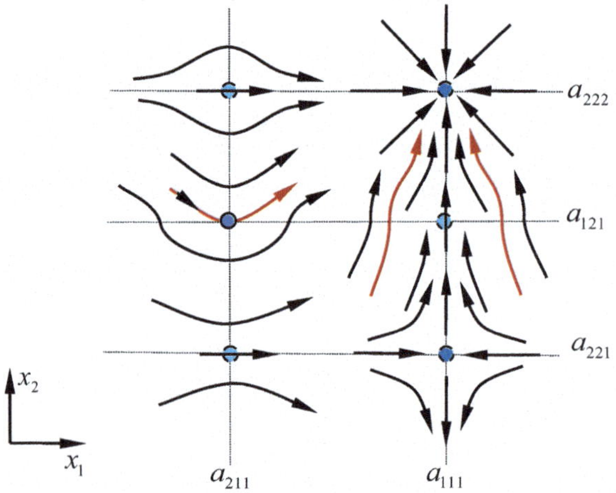

(d)

Fig. 3.3 (continued)

$$\left\{\begin{array}{cc}(a_{211},a_{222}) & (a_{111},a_{222})\\(a_{211},a_{121}) & (a_{111},a_{221})\\(a_{211},a_{221}) & (a_{111},a_{121})\end{array}\right\}=\left\{\begin{array}{cc}\underbrace{(\text{pF},\text{DP}:\text{UP})}_{\text{hyperbolic flow }(+)} & \underbrace{(\text{SI},\text{SO})}_{\text{saddle}}\\\underbrace{(\text{US},\text{DP})}_{\text{down-parabola upper-saddle}} & \underbrace{({}_{\text{DI:II}}\text{SI},\text{nF})}_{\text{inflection-sink flow }(-)}\\\underbrace{(\text{pF},\text{UP}:\text{DP})}_{\text{hyperbolic-secant flow }(+)} & \underbrace{(\text{SI},\text{SI})}_{\text{sink}}\end{array}\right\}$$

for $a_{110} < 0$ and $a_{220} > 0$;

(3.8)

$$\left\{\begin{array}{cc}(a_{211},a_{222}) & (a_{111},a_{222})\\(a_{211},a_{121}) & (a_{111},a_{221})\\(a_{211},a_{221}) & (a_{111},a_{121})\end{array}\right\}=\left\{\begin{array}{cc}\underbrace{(\text{nF},\text{DP}:\text{UP})}_{\text{hyperbolic flow }(-)} & \underbrace{(\text{SO},\text{SI})}_{\text{saddle}}\\\underbrace{(\text{LS},\text{DP})}_{\text{down-parabola lower-saddle}} & \underbrace{({}_{\text{DI:II}}\text{SO},\text{pF})}_{\text{inflection-source flow }(+)}\\\underbrace{(\text{nF},\text{UP}:\text{DP})}_{\text{hyperbolic-secant flow }(-)} & \underbrace{(\text{SO},\text{SO})}_{\text{source}}\end{array}\right\}$$

for $a_{110} > 0$ and $a_{220} < 0$;

(3.9)

$$\left\{\begin{array}{cc}(a_{211},a_{222}) & (a_{111},a_{222})\\(a_{211},a_{121}) & (a_{111},a_{221})\\(a_{211},a_{221}) & (a_{111},a_{121})\end{array}\right\}=\left\{\begin{array}{cc}\underbrace{(\text{pF},\text{UP}:\text{DP})}_{\text{hyperbolic-secant flow }(+)} & \underbrace{(\text{SI},\text{SI})}_{\text{sink}}\\\underbrace{(\text{US},\text{UP})}_{\text{up-parabola upper-saddle}} & \underbrace{({}_{\text{II:DI}}\text{SI},\text{pF})}_{\text{inflection-sink flow }(+)}\\\underbrace{(\text{pF},\text{DP}:\text{UP})}_{\text{hyperbolic flow }(+)} & \underbrace{(\text{SI},\text{SO})}_{\text{source}}\end{array}\right\}$$

for $a_{110} < 0$ and $a_{220} < 0$.

(3.10)

The dynamical behaviors for the equilibriums of (a_{111}, a_{221}) and (a_{111}, a_{222}) are unchanged for comparison of the cases of $a_{121} < a_{221}$ and $a_{222} < a_{121}$. However, the inflection flows are changed from positive to negative or from negative to positive.

3.2 Infinite-Equilibriums and Switching Bifurcations

Consider parabola-sink and source infinite-equilibriums and inflection-sink and source infinite-equilibriums, and the corresponding switching bifurcations will be presented herein.

3.2.1 Parabola-Source and Sink Infinite-Equilibriums

Consider a dynamical system for $a_{221} = a_{121}$ as

$$\begin{aligned}\dot{x}_1 &= a_{110}(x_1 - a_{111})(x_2 - a_{121})^2,\\ \dot{x}_2 &= a_{220}(x_1 - a_{211})(x_2 - a_{221})(x_2 - a_{222}),\end{aligned} \tag{3.11}$$

and the corresponding first integral manifold for $a_{221} = a_{121}$ is

$$\begin{aligned}&[(x_1 - a_{111}) - (x_{10} - a_{111})] + (a_{111} - a_{211}) \ln \frac{|x_1 - a_{111}|}{|x_{10} - a_{111}|}\\ &= \frac{a_{110}}{a_{220}}\left[(x_2 - x_{20}) + (a_{222} - a_{121}) \ln \frac{|x_2 - a_{222}|}{|x_{20} - a_{222}|}\right].\end{aligned} \tag{3.12}$$

(i) The equilibriums of $x_2^* = a_{221}, a_{222}, a_{121}$ with $x_1^* = a_{111}, a_{211}$ for $a_{221} = a_{121}$ with $a_{211} < a_{111}$ are

$$\left\{\begin{matrix}(a_{211}, a_{222}) & (a_{111}, a_{222})\\ (a_{121}, a_{211}) & (a_{111}, a_{221})\end{matrix}\right\} = \left\{\begin{matrix}\underbrace{(\text{nF}, \text{UP}:\text{DP})}_{\text{hyperbolic-secant flow }(-)} & \underbrace{(\text{SO}, \text{SO})}_{\text{source}}\\ \underbrace{(\text{DP}_-:\text{UP}_-, {}_{\text{DP}}\text{SO}:{}_{\text{UP}}\text{SI})}_{\text{hyperbolic source-to-sink}} & \underbrace{(\text{UP}:\text{DP}, \text{SI})}_{\text{hyperbolic-secant sink}}\end{matrix}\right\}$$

for $a_{110} > 0, a_{220} > 0,$

(3.13)

$$\left\{\begin{matrix}(a_{211}, a_{222}) & (a_{111}, a_{222})\\ (a_{121}, a_{211}) & (a_{111}, a_{221})\end{matrix}\right\} = \left\{\begin{matrix}\underbrace{(\text{pF}, \text{DP}:\text{UP})}_{\text{hyperbolic flow }(+)} & \underbrace{(\text{SI}, \text{SO})}_{\text{saddle}}\\ \underbrace{(\text{UP}_+:\text{DP}_+, {}_{\text{UP}}\text{SO}:{}_{\text{DP}}\text{SI})}_{\text{circular source-to-sink}} & \underbrace{(\text{DP}:\text{UP}, \text{SI})}_{\text{hyperbolic sink}}\end{matrix}\right\}$$

for $a_{110} < 0, a_{220} > 0,$

(3.14)

$$\left\{\begin{matrix}(a_{211}, a_{222}) & (a_{111}, a_{222})\\ (a_{121}, a_{211}) & (a_{111}, a_{221})\end{matrix}\right\} = \left\{\begin{matrix}\underbrace{(\text{nF}, \text{DP}:\text{UP})}_{\text{hyperbolic flow }(-)} & \underbrace{(\text{SO}, \text{SI})}_{\text{saddle}}\\ \underbrace{(\text{UP}_-:\text{DP}_-, {}_{\text{UP}}\text{SI}:{}_{\text{DP}}\text{SO})}_{\text{circular sink-to-source}} & \underbrace{(\text{DP}:\text{UP}, \text{SO})}_{\text{hyperbolic source}}\end{matrix}\right\}$$

for $a_{110} > 0, a_{220} < 0,$

(3.15)

$$\left\{\begin{array}{cc}(a_{211},a_{222}) & (a_{111},a_{222})\\ (a_{121},a_{211}) & (a_{111},a_{221})\end{array}\right\}\left\{\begin{array}{cc}\underbrace{(\mathrm{pF},\mathrm{UP}:\mathrm{DP})}_{\text{hyperbolic-secant flow }(+)} & \underbrace{(\mathrm{SI},\mathrm{SI})}_{\text{sink}}\\ \underbrace{(\mathrm{DP}_{+}:\mathrm{UP}_{+},{}_{\mathrm{DP}}\mathrm{SI}:{}_{\mathrm{UP}}\mathrm{SO})}_{\text{hyperbolic sink-to-source}} & \underbrace{(\mathrm{UP}:\mathrm{DP},\mathrm{SO})}_{\text{hyperbolic-secant source}}\end{array}\right\}$$
$$\text{for } a_{110}<0, a_{220}<0. \tag{3.16}$$

The infinite-equilibrium of $x_2^* = a_{121} = a_{221}$ is summarized as follows:

$$\begin{aligned}
&(\bar{x}_1,a_{221}) = \underbrace{(\mathrm{DP},\mathrm{SO})}_{\text{down-parabola source}} \quad \text{if } \bar{x}_1\in(-\infty,a_{211}),\\
&(\bar{x}_1,a_{221}) = \underbrace{(\mathrm{UP},\mathrm{SI})}_{\text{up-parabola sink}} \quad \text{if } \bar{x}_1\in(a_{211},a_{111}),\\
&(\bar{x}_1,a_{221}) = \underbrace{(\mathrm{DP},\mathrm{SI})}_{\text{down-parabola sink}} \quad \text{if } \bar{x}_1\in(a_{111},\infty),\\
&\text{for } a_{110}>0, a_{220}>0;
\end{aligned} \tag{3.17}$$

$$\begin{aligned}
&(\bar{x}_1,a_{221}) = \underbrace{(\mathrm{UP},\mathrm{SO})}_{\text{up-parabola source}} \quad \text{if } \bar{x}_1\in(-\infty,a_{211}),\\
&(\bar{x}_1,a_{221}) = \underbrace{(\mathrm{DP},\mathrm{SI})}_{\text{down-parabola sink}} \quad \text{if } \bar{x}_1\in(a_{211},a_{111}),\\
&(\bar{x}_1,a_{221}) = \underbrace{(\mathrm{UP},\mathrm{SI})}_{\text{up-parabola sink}} \quad \text{if } \bar{x}_1\in(a_{111},\infty),\\
&\text{for } a_{110}<0, a_{220}>0;
\end{aligned} \tag{3.18}$$

$$\begin{aligned}
&(\bar{x}_1,a_{221}) = \underbrace{(\mathrm{UP},\mathrm{SI})}_{\text{up-parabola sink}} \quad \text{if } \bar{x}_1\in(-\infty,a_{211}),\\
&(\bar{x}_1,a_{221}) = \underbrace{(\mathrm{DP},\mathrm{SO})}_{\text{down-parabola source}} \quad \text{if } \bar{x}_1\in(a_{211},a_{111}),\\
&(\bar{x}_1,a_{221}) = \underbrace{(\mathrm{UP},\mathrm{SO})}_{\text{up-parabola source}} \quad \text{if } \bar{x}_1\in(a_{111},\infty),\\
&\text{for } a_{110}>0, a_{220}<0;
\end{aligned} \tag{3.19}$$

$$\begin{aligned}
&(\bar{x}_1,a_{221}) = \underbrace{(\mathrm{DP},\mathrm{SI})}_{\text{down-parabola sink}} \quad \text{if } \bar{x}_1\in(-\infty,a_{211}),\\
&(\bar{x}_1,a_{221}) = \underbrace{(\mathrm{UP},\mathrm{SO})}_{\text{up-parabola source}} \quad \text{if } \bar{x}_1\in(a_{211},a_{111}),\\
&(\bar{x}_1,a_{221}) = \underbrace{(\mathrm{DP},\mathrm{SO})}_{\text{down-parabola source}} \quad \text{if } \bar{x}_1\in(a_{111},\infty),\\
&\text{for } a_{110}<0, a_{220}<0.
\end{aligned} \tag{3.20}$$

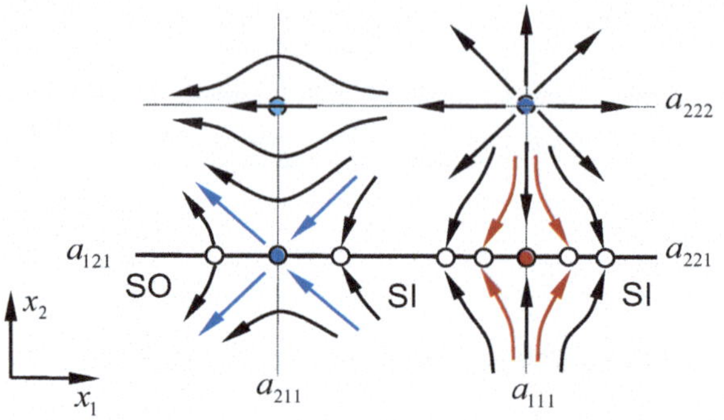

(a)

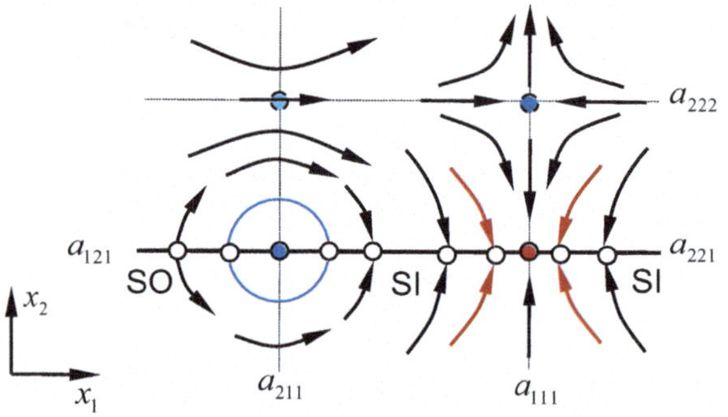

(b)

Fig. 3.4 Phase portraits ($a_{121} = a_{221}$ and $a_{211} = a_{111}$) for two-dimensional systems on the x_1-direction with $x_1^* = a_{111}, a_{211}$ and on the x_2-direction with $x_2^* = a_{121}, a_{221}, a_{222}$. The infinite-equilibriums: (**a**) ($a_{110} > 0$, $a_{220} > 0$), (**b**) ($a_{110} < 0$, $a_{220} > 0$), (**c**) ($a_{110} > 0$, $a_{220} < 0$), and (**d**) ($a_{110} < 0$, $a_{220} < 0$)

The infinite-equilibrium of $x_2^* = a_{121} = a_{221}$ for up-parabola and down-parabola source and sink is presented in Fig. 3.4. The infinite-equilibrium is for the switching bifurcation for the equilibriums for $a_{121} < a_{221}$ and $a_{221} < a_{121} < a_{222}$.

(ii) Consider a dynamical system for $a_{222} = a_{121}$ as

$$\begin{aligned} \dot{x}_1 &= a_{110}(x_1 - a_{111})(x_2 - a_{121})^2, \\ \dot{x}_2 &= a_{220}(x_1 - a_{211})(x_2 - a_{221})(x_2 - a_{222}), \end{aligned} \tag{3.21}$$

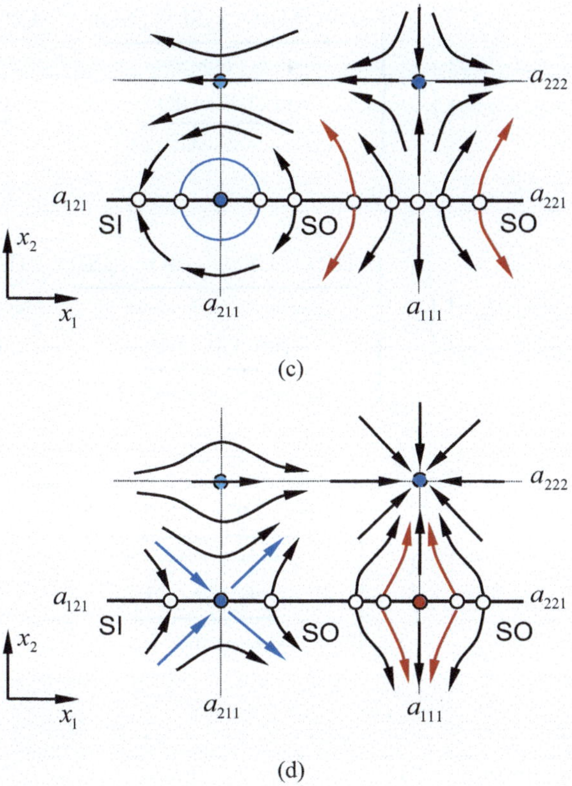

Fig. 3.4 (continued)

and the corresponding first integral manifold for $a_{221} = a_{121}$ is

$$\begin{aligned}[(x_1 - a_{111}) - (x_{10} - a_{111})] + (a_{111} - a_{211}) \ln \frac{|x_1 - a_{111}|}{|x_{10} - a_{111}|} \\ = \frac{a_{110}}{a_{220}} \left[(x_2 - x_{20}) + (a_{221} - a_{121}) \ln \frac{|x_2 - a_{221}|}{|x_{20} - a_{221}|} \right].\end{aligned} \tag{3.22}$$

The infinite-equilibriums of $x_2^* = a_{222} = a_{121}$ are

$$\left\{ \begin{matrix} (a_{121}, a_{211}) & (a_{111}, a_{222}) \\ (a_{211}, a_{221}) & (a_{111}, a_{221}) \end{matrix} \right\} = \left\{ \begin{matrix} \underbrace{(\text{UP}_- : \text{DP}_-, {}_{\text{UP}}\text{SI} : {}_{\text{DP}}\text{SO})}_{\text{circular sink-to-source}} & \underbrace{(\text{DP} : \text{UP}, \text{SO})}_{\text{hyperbolic source}} \\ \underbrace{(\text{nF}, \text{UP} : \text{DP})}_{\text{hyperbolic flow } (-)} & \underbrace{(\text{SO}, \text{SI})}_{\text{saddle}} \end{matrix} \right\}$$

for $a_{110} > 0, a_{220} > 0,$

(3.23)

$$\left\{\begin{array}{cc}(a_{121},a_{211}) & (a_{111},a_{222})\\(a_{211},a_{221}) & (a_{111},a_{221})\end{array}\right\}=\left\{\begin{array}{cc}\underbrace{(\mathrm{DP}_{+}:\mathrm{UP}_{+},{}_{\mathrm{DP}}\mathrm{SI}:{}_{\mathrm{UP}}\mathrm{SO})}_{\text{hyperbolic sink-to-source}} & \underbrace{(\mathrm{UP}:\mathrm{DP},\mathrm{SO})}_{\text{hyperbolic-secant source}}\\ \underbrace{(\mathrm{pF},\mathrm{UP}:\mathrm{DP})}_{\text{hyperbolic-secant flow }(+)} & \underbrace{(\mathrm{SI},\mathrm{SI})}_{\text{sink}}\end{array}\right\}$$

for $a_{110}<0, a_{220}>0,$

$$\tag{3.24}$$

$$\left\{\begin{array}{cc}(a_{121},a_{211}) & (a_{111},a_{222})\\(a_{211},a_{221}) & (a_{111},a_{221})\end{array}\right\}=\left\{\begin{array}{cc}\underbrace{(\mathrm{DP}_{-}:\mathrm{UP}_{-},{}_{\mathrm{DP}}\mathrm{SO}:{}_{\mathrm{UP}}\mathrm{SI})}_{\text{hyperbolic source-to-sink}} & \underbrace{(\mathrm{UP}:\mathrm{DP},\mathrm{SI})}_{\text{hyperbolic-secant sink}}\\ \underbrace{(\mathrm{nF},\mathrm{UP}:\mathrm{DP})}_{\text{hyperbolic-secant flow }(-)} & \underbrace{(\mathrm{SO},\mathrm{SO})}_{\text{source}}\end{array}\right\}$$

for $a_{110}>0, a_{220}<0,$

$$\tag{3.25}$$

$$\left\{\begin{array}{cc}(a_{121},a_{211}) & (a_{111},a_{222})\\(a_{211},a_{221}) & (a_{111},a_{221})\end{array}\right\}=\left\{\begin{array}{cc}\underbrace{(\mathrm{UP}_{+}:\mathrm{DP}_{+},{}_{\mathrm{UP}}\mathrm{SO}:{}_{\mathrm{DP}}\mathrm{SI})}_{\text{circular source-to-sink}} & \underbrace{(\mathrm{DP}:\mathrm{UP},\mathrm{SI})}_{\text{hyperbolic sink}}\\ \underbrace{(\mathrm{pF},\mathrm{UP}:\mathrm{DP})}_{\text{hyperbolic flow }(+)} & \underbrace{(\mathrm{SI},\mathrm{SO})}_{\text{saddle}}\end{array}\right\}$$

for $a_{110}<0, a_{220}<0.$

$$\tag{3.26}$$

The infinite-equilibrium of $x_2^*=a_{121}=a_{222}$ is summarized as follows:

$$\begin{array}{l}(\bar{x}_1,a_{222})=\underbrace{(\mathrm{UP},\mathrm{SI})}_{\text{up-parabola sink}}\ \text{ if } \bar{x}_1\in(-\infty,a_{211}),\\(\bar{x}_1,a_{222})=\underbrace{(\mathrm{DP},\mathrm{SO})}_{\text{down-parabola source}}\ \text{ if } \bar{x}_1\in(a_{211},a_{111}),\\(\bar{x}_1,a_{222})=\underbrace{(\mathrm{UP},\mathrm{SO})}_{\text{up-parabola source}}\ \text{ if } \bar{x}_1\in(a_{111},\infty),\\ \text{for } a_{110}>0, a_{220}>0;\end{array}\tag{3.27}$$

$$\begin{array}{l}(\bar{x}_1,a_{222})=\underbrace{(\mathrm{DP},\mathrm{SI})}_{\text{down-parabola sink}}\ \text{ if } \bar{x}_1\in(-\infty,a_{211}),\\(\bar{x}_1,a_{222})=\underbrace{(\mathrm{UP},\mathrm{SO})}_{\text{up-parabola source}}\ \text{ if } \bar{x}_1\in(a_{211},a_{111}),\\(\bar{x}_1,a_{222})=\underbrace{(\mathrm{DP},\mathrm{SO})}_{\text{down-parabola source}}\ \text{ if } \bar{x}_1\in(a_{111},\infty),\\ \text{for } a_{110}<0, a_{220}>0;\end{array}\tag{3.28}$$

$$
\begin{aligned}
(\bar{x}_1, a_{222}) &= \underbrace{(\mathrm{DP}, \mathrm{SO})}_{\text{down-parabola source}} \quad \text{if } \bar{x}_1 \in (-\infty, a_{211}),\\
(\bar{x}_1, a_{222}) &= \underbrace{(\mathrm{UP}, \mathrm{SI})}_{\text{up-parabola sink}} \quad \text{if } \bar{x}_1 \in (a_{211}, a_{111}),\\
(\bar{x}_1, a_{222}) &= \underbrace{(\mathrm{DP}, \mathrm{SI})}_{\text{down-parabola sink}} \quad \text{if } \bar{x}_1 \in (a_{111}, \infty),
\end{aligned}
\tag{3.29}
$$

for $a_{110} > 0, a_{220} < 0$;

$$
\begin{aligned}
(\bar{x}_1, a_{222}) &= \underbrace{(\mathrm{UP}, \mathrm{SO})}_{\text{up-parabola source}} \quad \text{if } \bar{x}_1 \in (-\infty, a_{211}),\\
(\bar{x}_1, a_{222}) &= \underbrace{(\mathrm{DP}, \mathrm{SI})}_{\text{down-parabola sink}} \quad \text{if } \bar{x}_1 \in (a_{211}, a_{111}),\\
(\bar{x}_1, a_{222}) &= \underbrace{(\mathrm{UP}, \mathrm{SI})}_{\text{up-parabola sink}} \quad \text{if } \bar{x}_1 \in (a_{111}, \infty),
\end{aligned}
\tag{3.30}
$$

for $a_{110} < 0, a_{220} < 0$.

The infinite-equilibrium of $x_2^* = a_{121} = a_{222}$ for up-parabola and down-parabola source and sink is presented in Fig. 3.5. The infinite-equilibrium is for the switching bifurcation for the equilibriums for $a_{121} < a_{222}$ and $a_{121} > a_{222}$.

For the two cases of $a_{211} > a_{111}$, the down-parabola and up-parabola sink and source infinite-equilibriums are discussed, and the corresponding phase portraits are illustrated. Similarly, the readers can work out the phase portraits of the infinite-equilibriums for the cases of $a_{211} < a_{111}$.

3.2.2 *Inflection-Sink and Source Infinite-Equilibriums*

Consider a dynamical system for $a_{111} = a_{211}$ as

$$
\begin{aligned}
\dot{x}_1 &= a_{110}(x_1 - a_{111})(x_2 - a_{121})^2,\\
\dot{x}_2 &= a_{220}(x_1 - a_{211})(x_2 - a_{221})(x_2 - a_{222}),
\end{aligned}
\tag{3.31}
$$

and the corresponding first integral manifold for $a_{111} = a_{211}$ is

$$
\begin{aligned}
&(x_1 - a_{111}) - (x_{10} - a_{111})\\
&= \frac{a_{110}}{a_{220}}\left[(x_2 - x_{20}) + \frac{(a_{221} - a_{121})^2}{a_{221} - a_{222}} \ln \frac{|x_2 - a_{221}|}{|x_{20} - a_{221}|}\right.\\
&\quad \left. + \frac{(a_{222} - a_{121})^2}{a_{222} - a_{221}} \ln \frac{|x_2 - a_{222}|}{|x_{20} - a_{222}|}\right].
\end{aligned}
\tag{3.32}
$$

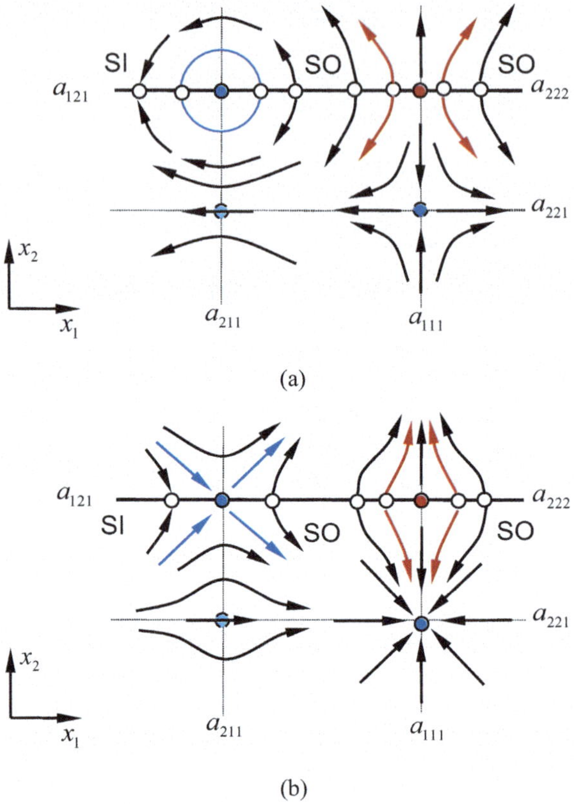

Fig. 3.5 Phase portraits ($a_{121} = a_{222}$ and $a_{211} < a_{111}$) for two-dimensional systems on the x_1-direction with $x_1^* = a_{111}, a_{211}$ and on the x_2-direction with $x_2^* = a_{121}, a_{221}, a_{222}$. The infinite-equilibriums: (**a**) ($a_{110} > 0$, $a_{220} > 0$), (**b**) ($a_{110} < 0$, $a_{220} > 0$), (**c**) ($a_{110} > 0$, $a_{220} < 0$), and (**d**) ($a_{110} < 0$, $a_{220} < 0$)

(i) The equilibriums of $x_2^* = a_{221}, a_{222}, a_{121}$ and $x_1^* = a_{111} = a_{211}$ for $a_{121} < a_{221}$ are

$$\left\{\begin{array}{c}(a_{111}, a_{222})\\(a_{111}, a_{221})\\(a_{111}, a_{121})\end{array}\right\} = \left\{\begin{array}{c}\underbrace{(\mathrm{SO}, \mathrm{DI}:\mathrm{II})}_{\text{inflection source}}\\\underbrace{(\mathrm{SO}, \mathrm{II}:\mathrm{DI})}_{\text{inflection source}}\\\underbrace{(\mathrm{SO}, \mathrm{II}:\mathrm{II})}_{\text{increasing-inflection source}}\end{array}\right\} \text{ for } a_{110} > 0, a_{220} > 0, \tag{3.33}$$

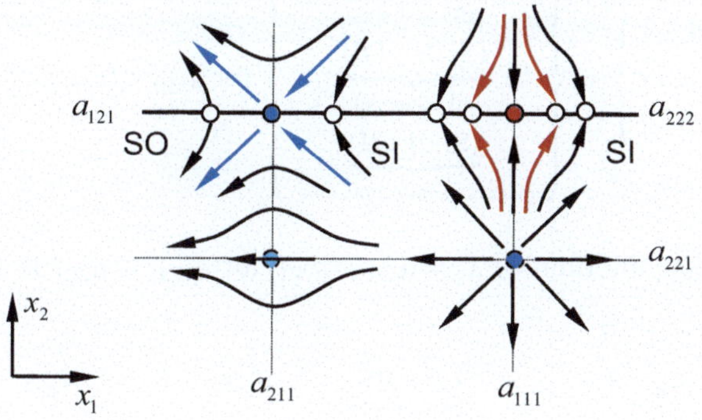

(c)

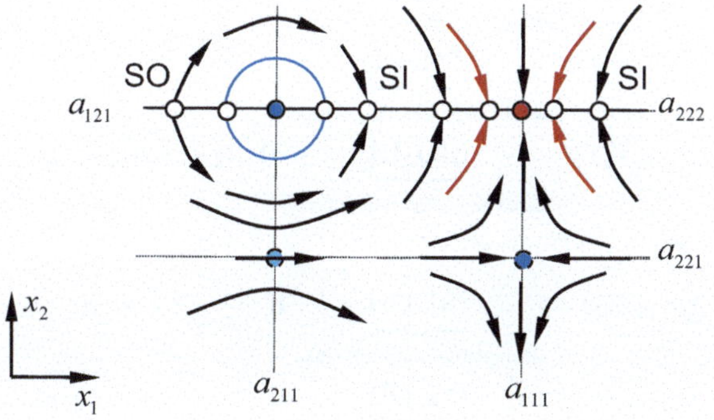

(d)

Fig. 3.5 (continued)

$$\left\{\begin{array}{c}(a_{111},a_{222})\\(a_{111},a_{221})\\(a_{111},a_{121})\end{array}\right\}=\left\{\begin{array}{c}\underbrace{(\mathrm{SI},\mathrm{II}:\mathrm{DI})}_{\text{inflection sink}}\\\underbrace{(\mathrm{SI},\mathrm{DI}:\mathrm{II})}_{\text{inflection sink}}\\\underbrace{(\mathrm{SI},\mathrm{DI}:\mathrm{DI})}_{\text{decreasing-inflection sink}}\end{array}\right\}\text{ for } a_{110}<0, a_{220}>0, \tag{3.34}$$

$$\left\{\begin{array}{c}(a_{111},a_{222})\\(a_{111},a_{221})\\(a_{111},a_{121})\end{array}\right\}=\left\{\begin{array}{c}\underbrace{(\mathrm{SO},\mathrm{II}:\mathrm{DI})}_{\text{inflection source}}\\\underbrace{(\mathrm{SO},\mathrm{DI}:\mathrm{II})}_{\text{inflection source}}\\\underbrace{(\mathrm{SO},\mathrm{DI}:\mathrm{DI})}_{\text{decreasing-inflection source}}\end{array}\right\}\text{ for } a_{110}>0, a_{220}<0, \tag{3.35}$$

$$\left\{\begin{array}{c}(a_{111},a_{222})\\(a_{111},a_{221})\\(a_{111},a_{121})\end{array}\right\}=\left\{\begin{array}{c}\underbrace{(\mathrm{SI},\mathrm{DI}:\mathrm{II})}_{\text{inflection sink}}\\\underbrace{(\mathrm{SI},\mathrm{II}:\mathrm{DI})}_{\text{inflection sink}}\\\underbrace{(\mathrm{SI},\mathrm{II}:\mathrm{II})}_{\text{increasing-inflection sink}}\end{array}\right\}\text{ for } a_{110}<0, a_{220}<0. \tag{3.36}$$

The infinite-equilibrium of $x_2^*=a_{121}=a_{221}$ for $a_{121}<a_{221}$ is summarized as follows:

$$\left.\begin{array}{ll}(a_{111},\bar{x}_2)=\underbrace{(\mathrm{SO},\mathrm{II})}_{\text{increasing-inflection source}} & \text{if } \bar{x}_2\in(-\infty,a_{121}),\\(a_{111},\bar{x}_2)=\underbrace{(\mathrm{SO},\mathrm{II})}_{\text{increasing-inflection source}} & \text{if } \bar{x}_2\in(a_{121},a_{221}),\\(a_{111},\bar{x}_2)=\underbrace{(\mathrm{SO},\mathrm{DI})}_{\text{decreasing-inflection source}} & \text{if } \bar{x}_2\in(a_{221},a_{222}),\\(a_{111},\bar{x}_2)=\underbrace{(\mathrm{SO},\mathrm{II})}_{\text{increasing-inflection source}} & \text{if } \bar{x}_2\in(a_{222},\infty),\\\text{for } a_{110}>0, a_{220}>0; &\end{array}\right. \tag{3.37}$$

$$\left.\begin{array}{ll}(a_{111},\bar{x}_2)=\underbrace{(\mathrm{SI},\mathrm{DI})}_{\text{decreasing-inflection sink}} & \text{if } \bar{x}_2\in(-\infty,a_{121}),\\(a_{111},\bar{x}_2)=\underbrace{(\mathrm{SI},\mathrm{DI})}_{\text{decreasing-inflection sink}} & \text{if } \bar{x}_2\in(a_{121},a_{221}),\\(a_{111},\bar{x}_2)=\underbrace{(\mathrm{SI},\mathrm{II})}_{\text{increasing-inflection sink}} & \text{if } \bar{x}_2\in(a_{221},a_{222}),\\(a_{111},\bar{x}_2)=\underbrace{(\mathrm{SI},\mathrm{DI})}_{\text{decreasing-inflection sink}} & \text{if } \bar{x}_2\in(a_{222},\infty),\\\text{for } a_{110}<0, a_{220}>0; &\end{array}\right. \tag{3.38}$$

$$\left.\begin{array}{ll}(a_{111},\bar{x}_2)=\underbrace{(\mathrm{SO},\mathrm{DI})}_{\text{decreasing-inflection source}} & \text{if } \bar{x}_2\in(-\infty,a_{121}),\\(a_{111},\bar{x}_2)=\underbrace{(\mathrm{SO},\mathrm{DI})}_{\text{decreasing-inflection source}} & \text{if } \bar{x}_2\in(a_{121},a_{221}),\\(a_{111},\bar{x}_2)=\underbrace{(\mathrm{SO},\mathrm{II})}_{\text{increasing-inflection source}} & \text{if } \bar{x}_2\in(a_{221},a_{222}),\\(a_{111},\bar{x}_2)=\underbrace{(\mathrm{SO},\mathrm{DI})}_{\text{decreasing-inflection source}} & \text{if } \bar{x}_2\in(a_{222},\infty),\\\text{for } a_{110}>0, a_{220}<0; &\end{array}\right. \tag{3.39}$$

$$\begin{aligned}
(a_{111}, \bar{x}_2) &= \underbrace{(\text{SI, II})}_{\text{increasing-inflection sink}} && \text{if } \bar{x}_2 \in (-\infty, a_{121}),\\
(a_{111}, \bar{x}_2) &= \underbrace{(\text{SI, II})}_{\text{increasing-inflection sink}} && \text{if } \bar{x}_2 \in (a_{121}, a_{221}),\\
(a_{111}, \bar{x}_2) &= \underbrace{(\text{SI, DI})}_{\text{decreasing-inflection sink}} && \text{if } \bar{x}_2 \in (a_{221}, a_{222}),\\
(a_{111}, \bar{x}_2) &= \underbrace{(\text{SI, II})}_{\text{increasing-inflection sink}} && \text{if } \bar{x}_2 \in (a_{222}, \infty),\\
&\text{for } a_{110} < 0, a_{220} < 0.
\end{aligned} \tag{3.40}$$

The infinite-equilibriums of $x_1^* = a_{111} = a_{211}$ with $a_{121} < a_{221}$ are for increasing-inflection and decreasing-inflection source and sink, as shown in Fig. 3.6. The infinite-equilibrium is for the switching bifurcation for the equilibriums for $a_{111} < a_{211}$ and $a_{211} < a_{111}$ with $a_{121} < a_{221}$.

(ii) The infinite-equilibriums of $x_1^* = a_{111} = a_{211}$ for $a_{222} < a_{121}$ have the same properties as for $x_1^* = a_{111} = a_{211}$ with $a_{121} < a_{221}$. The equilibrium locations of $x_2^* = a_{221}, a_{222}, a_{121}$ and $x_1^* = a_{111} = a_{211}$ are different. The corresponding phase portraits are presented in Fig. 3.7.

(iii) The equilibriums of $x_2^* = a_{221}, a_{222}, a_{121}$ and $x_1^* = a_{111} = a_{211}$ for $a_{121} \in (a_{221}, a_{222})$ are

$$\left\{\begin{array}{l}(a_{111}, a_{222})\\(a_{111}, a_{121})\\(a_{111}, a_{221})\end{array}\right\} = \left\{\begin{array}{c}\underbrace{(\text{SO, DI : II})}_{\text{inflection source}}\\ \underbrace{(\text{SO, DI : DI})}_{\text{decreasing-inflection source}}\\ \underbrace{(\text{SO, II : DI})}_{\text{inflection source}}\end{array}\right\} \text{ for } a_{110} > 0, a_{220} > 0, \tag{3.41}$$

$$\left\{\begin{array}{l}(a_{111}, a_{222})\\(a_{111}, a_{121})\\(a_{111}, a_{221})\end{array}\right\} = \left\{\begin{array}{c}\underbrace{(\text{SI, II : DI})}_{\text{inflection sink}}\\ \underbrace{(\text{SI, II : II})}_{\text{increasing-inflection sink}}\\ \underbrace{(\text{SI, DI : II})}_{\text{inflection sink}}\end{array}\right\} \text{ for } a_{110} < 0, a_{220} > 0, \tag{3.42}$$

$$\left\{\begin{array}{l}(a_{111}, a_{222})\\(a_{111}, a_{121})\\(a_{111}, a_{221})\end{array}\right\} = \left\{\begin{array}{c}\underbrace{(\text{SO, II : DI})}_{\text{inflection source}}\\ \underbrace{(\text{SO, II : II})}_{\text{increasing-inflection source}}\\ \underbrace{(\text{SO, DI : II})}_{\text{inflection source}}\end{array}\right\} \text{ for } a_{110} > 0, a_{220} < 0, \tag{3.43}$$

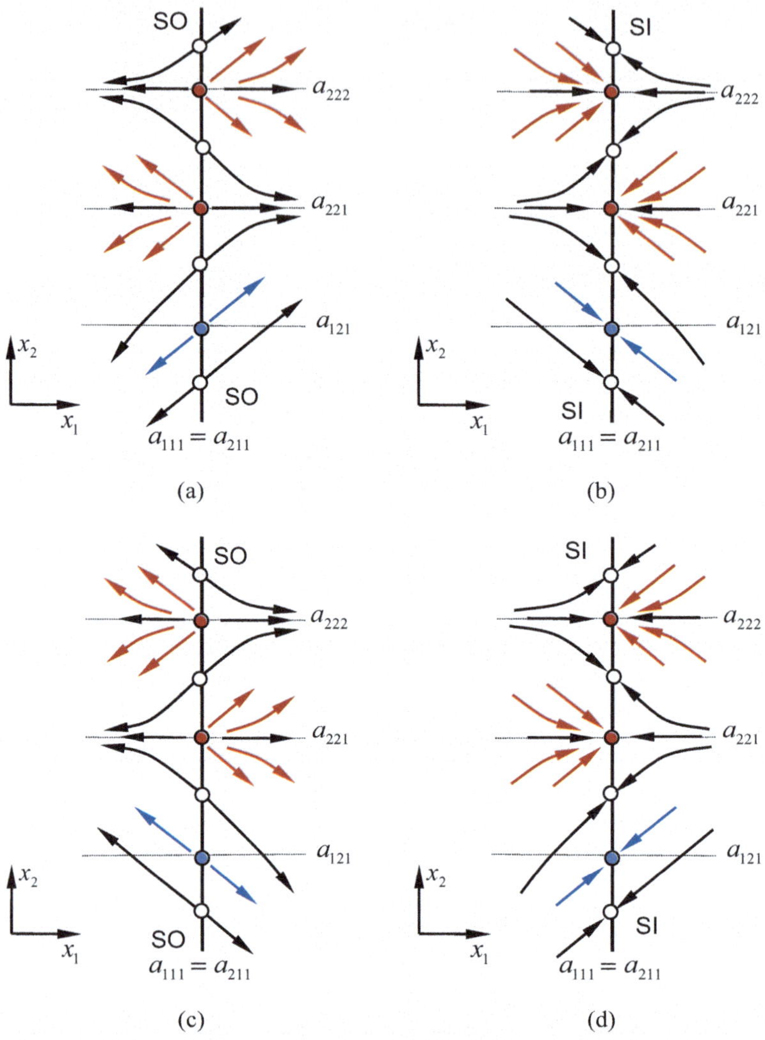

Fig. 3.6 Phase portraits ($a_{111} = a_{211}$, $a_{121} < a_{221}$) for two-dimensional systems on the x_1-direction with $x_1^* = a_{111} = a_{211}$ and on the x_2-direction with $x_2^* = a_{121}, a_{221}, a_{222}$. The equilibrium flows: (**a**) ($a_{110} > 0$, $a_{220} > 0$), (**b**) ($a_{110} < 0$, $a_{220} > 0$), (**c**) ($a_{110} > 0$, $a_{220} < 0$), and (**d**) ($a_{110} < 0$, $a_{220} < 0$)

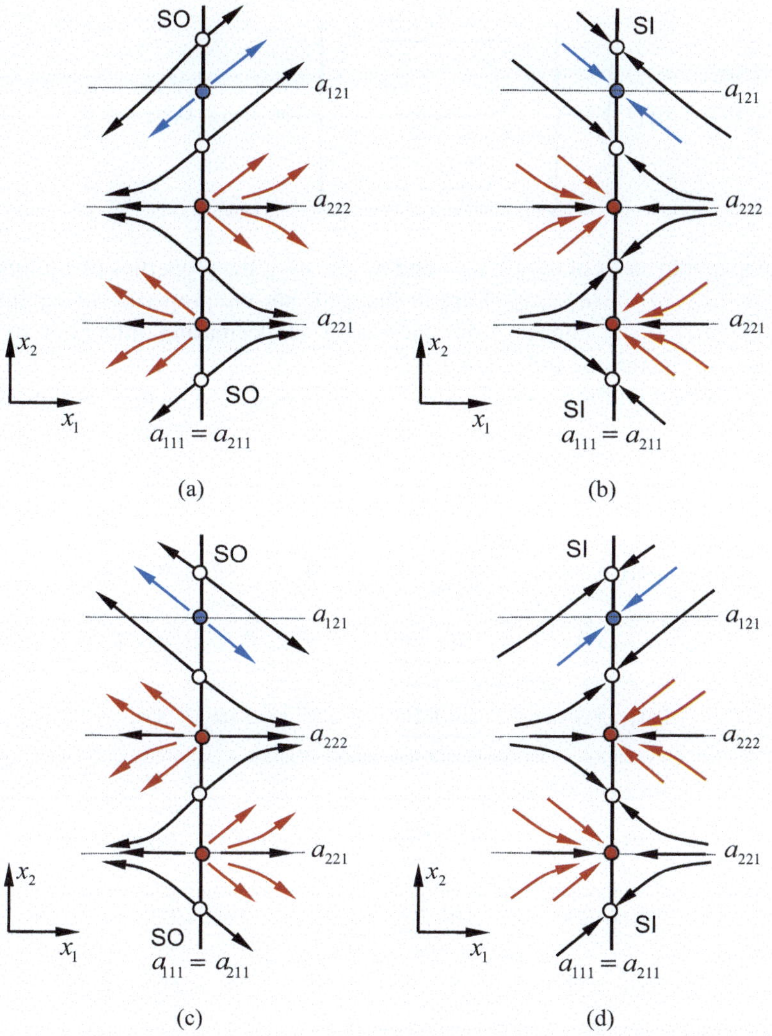

Fig. 3.7 Phase portraits ($a_{111} = a_{211}, a_{222} < a_{121}$) for two-dimensional systems on the x_1-direction with $x_1^* = a_{111} = a_{211}$ and on the x_2-direction with $x_2^* = a_{121}, a_{221}, a_{222}$. The equilibrium flows: (**a**) ($a_{110} > 0$, $a_{220} > 0$), (**b**) ($a_{110} < 0$, $a_{220} > 0$), (**c**) ($a_{110} > 0$, $a_{220} < 0$), and (**d**) ($a_{110} < 0$, $a_{220} < 0$)

$$\left\{\begin{array}{c}(a_{111},a_{222})\\(a_{111},a_{121})\\(a_{111},a_{221})\end{array}\right\}=\left\{\begin{array}{c}\underbrace{(\mathrm{SI},\mathrm{DI}:\mathrm{II})}_{\text{inflection sink}}\\\underbrace{(\mathrm{SI},\mathrm{DI}:\mathrm{DI})}_{\text{decreasing-inflection sink}}\\\underbrace{(\mathrm{SI},\mathrm{II}:\mathrm{DI})}_{\text{inflection sink}}\end{array}\right\}\text{ for }a_{110}<0,a_{220}<0. \tag{3.44}$$

Compared to cases of $a_{222} < a_{121}$ and $a_{121} < a_{221}$, the properties of equilibriums $(x_1^*, x_2^*) = (a_{111}, a_{221}), (a_{111}, a_{222})$ are unchanged, but the properties of equilibrium $(x_1^*, x_2^*) = (a_{111}, a_{121})$ change from increasing to decreasing inflection or from decreasing to increasing inflection.

The infinite-equilibrium of $x_2^* = a_{121} = a_{221}$ for $a_{121} \in (a_{221}, a_{222})$ is summarized as follows:

$$\begin{array}{ll}(a_{111},\bar{x}_2)=\underbrace{(\mathrm{SO},\mathrm{II})}_{\text{increasing-inflection source}} & \text{if } \bar{x}_2\in(-\infty,a_{221}),\\(a_{111},\bar{x}_2)=\underbrace{(\mathrm{SO},\mathrm{DI})}_{\text{decreasing-inflection source}} & \text{if } \bar{x}_2\in(a_{221},a_{121}),\\(a_{111},\bar{x}_2)=\underbrace{(\mathrm{SO},\mathrm{DI})}_{\text{decreasing-inflection source}} & \text{if } \bar{x}_2\in(a_{121},a_{222}),\\(a_{111},\bar{x}_2)=\underbrace{(\mathrm{SO},\mathrm{II})}_{\text{increasing-inflection source}} & \text{if } \bar{x}_2\in(a_{222},\infty),\\\text{for } a_{110}>0,a_{220}>0; & \end{array} \tag{3.45}$$

$$\begin{array}{ll}(a_{111},\bar{x}_2)=\underbrace{(\mathrm{SI},\mathrm{DI})}_{\text{decreasing-inflection sink}} & \text{if } \bar{x}_2\in(-\infty,a_{221}),\\(a_{111},\bar{x}_2)=\underbrace{(\mathrm{SI},\mathrm{II})}_{\text{increasing-inflection sink}} & \text{if } \bar{x}_2\in(a_{221},a_{121}),\\(a_{111},\bar{x}_2)=\underbrace{(\mathrm{SI},\mathrm{II})}_{\text{increasing-inflection sink}} & \text{if } \bar{x}_2\in(a_{121},a_{222}),\\(a_{111},\bar{x}_2)=\underbrace{(\mathrm{SI},\mathrm{DI})}_{\text{decreasing-inflection sink}} & \text{if } \bar{x}_2\in(a_{222},\infty),\\\text{for } a_{110}<0,a_{220}>0; & \end{array} \tag{3.46}$$

$$\begin{array}{ll}(a_{111},\bar{x}_2)=\underbrace{(\mathrm{SO},\mathrm{DI})}_{\text{decreasing-inflection source}} & \text{if } \bar{x}_2\in(-\infty,a_{221}),\\(a_{111},\bar{x}_2)=\underbrace{(\mathrm{SO},\mathrm{II})}_{\text{increasing-inflection source}} & \text{if } \bar{x}_2\in(a_{221},a_{121}),\\(a_{111},\bar{x}_2)=\underbrace{(\mathrm{SO},\mathrm{II})}_{\text{increasing-inflection source}} & \text{if } \bar{x}_2\in(a_{121},a_{222}),\\(a_{111},\bar{x}_2)=\underbrace{(\mathrm{SO},\mathrm{DI})}_{\text{decreasing-inflection source}} & \text{if } \bar{x}_2\in(a_{222},\infty),\\\text{for } a_{110}>0,a_{220}<0; & \end{array} \tag{3.47}$$

$$
\begin{aligned}
&(a_{111},\bar{x}_2)= \underbrace{(\mathrm{SI,II})}_{\text{increasing-inflection sink}} &&\text{if } \bar{x}_2\in(-\infty,a_{221}),\\
&(a_{111},\bar{x}_2)= \underbrace{(\mathrm{SI,DI})}_{\text{decreasing-inflection sink}} &&\text{if } \bar{x}_2\in(a_{221},a_{121}),\\
&(a_{111},\bar{x}_2)= \underbrace{(\mathrm{SI,DI})}_{\text{decreasing-inflection sink}} &&\text{if } \bar{x}_2\in(a_{121},a_{222}),\\
&(a_{111},\bar{x}_2)= \underbrace{(\mathrm{SI,II})}_{\text{increasing-inflection sink}} &&\text{if } \bar{x}_2\in(a_{222},\infty),\\
&\text{for } a_{110}<0, a_{220}<0.
\end{aligned}
\tag{3.48}
$$

The infinite-equilibriums of $x_1^*=a_{111}=a_{211}$ with $a_{121}\in(a_{221},a_{222})$ are for increasing-inflection and decreasing-inflection source and sink, as shown in Fig. 3.8. The infinite-equilibrium is for the switching bifurcation for the equilibriums for $a_{111}<a_{211}$ and $a_{211}<a_{111}$ with $a_{121}<a_{221}$. The infinite-equilibriums of $(x_1^*,x_2^*)=(a_{111},a_{121})$ are changed from increasing to decreasing inflection or from decreasing to increasing inflection.

3.2.3 Inflection Upper-Saddle and Lower-Saddle Equilibriums

Consider a dynamical system for $a_{111}=a_{211}$ and $a_{22s_1}=a_{121}$ (s_1, $s_2\in\{1,2\}$; $s_1\neq s_2$) as

$$
\begin{aligned}
\dot{x}_1&=a_{110}(x_1-a_{111})(x_2-a_{121})^2,\\
\dot{x}_2&=a_{220}(x_1-a_{211})(x_2-a_{221})(x_2-a_{222}),
\end{aligned}
\tag{3.49}
$$

and the corresponding first integral manifold for $a_{111}=a_{211}$ and $a_{22s_1}=a_{121}$ is

$$
\begin{aligned}
&(x_1-a_{111})-(x_{10}-a_{111})\\
&=\frac{a_{110}}{a_{220}}\left[(x_2-x_{20})+\frac{(a_{22s_2}-a_{121})^2}{a_{22s_2}-a_{22s_1}}\ln\frac{|x_2-a_{22s_2}|}{|x_{20}-a_{22s_2}|}\right].
\end{aligned}
\tag{3.50}
$$

(i) The equilibriums of $x_2^*=a_{221},a_{222},a_{121}$ and $x_1^*=a_{111}=a_{211}$ for $a_{121}=a_{221}$ are

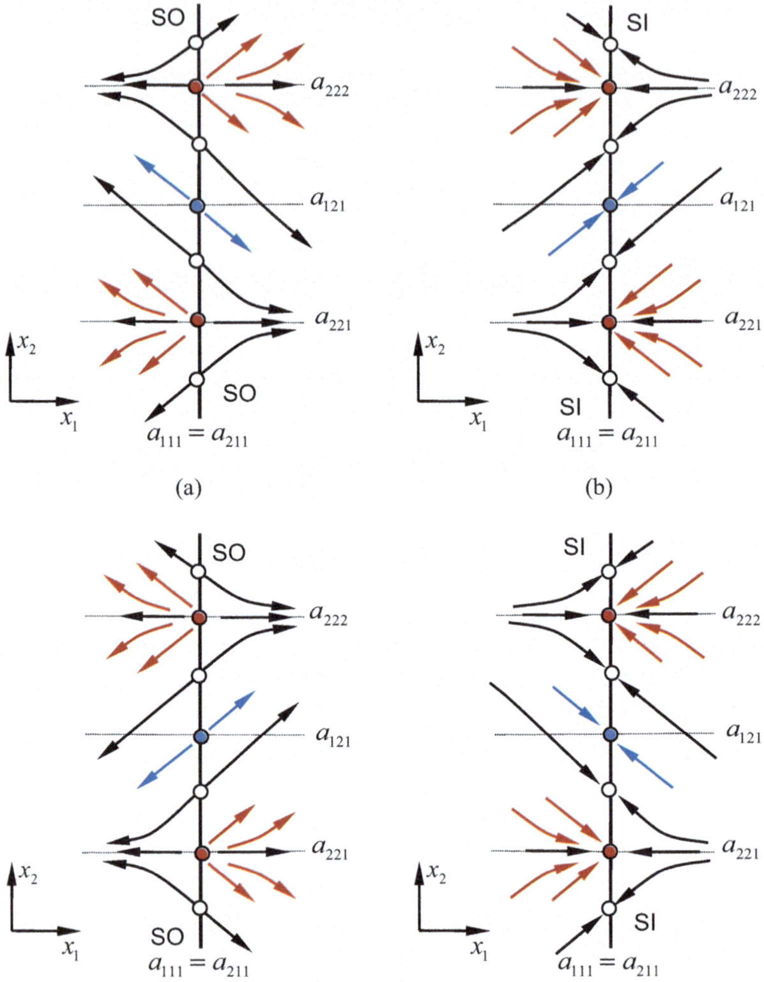

Fig. 3.8 Phase portraits ($a_{111} = a_{211}$, $a_{121} \in (a_{221}, a_{222})$) for two-dimensional systems on the x_1-direction with $x_1^* = a_{111} = a_{211}$ and on the x_2-direction with $x_2^* = a_{121}, a_{221}, a_{222}$. The infinite-equilibriums: (**a**) ($a_{110} > 0$, $a_{220} > 0$), (**b**) ($a_{110} < 0$, $a_{220} > 0$), (**c**) ($a_{110} > 0$, $a_{220} < 0$), and (**d**) ($a_{110} < 0$, $a_{220} < 0$)

$$\left\{ \begin{array}{c} (a_{111}, a_{222}) \\ (a_{111}, a_{121}) \end{array} \right\} = \left\{ \begin{array}{c} \underbrace{(\mathrm{SO}, \mathrm{DI} : \mathrm{II})}_{\text{inflection-source}} \\ \underbrace{({}_{\mathrm{II:DI}}\mathrm{SO}, {}_{\mathrm{SO:SI}}\mathrm{DP})}_{\text{down-parabola source}} \end{array} \right\} \text{ for } a_{110} > 0, a_{220} > 0, \tag{3.51}$$

$$\left\{\begin{array}{l}(a_{111},a_{222})\\(a_{111},a_{121})\end{array}\right\}=\left\{\begin{array}{c}\underbrace{(\mathrm{SI},\mathrm{II}:\mathrm{DI})}_{\text{inflection-sink}}\\\underbrace{({}_{\mathrm{DI:II}}\mathrm{SI},{}_{\mathrm{SO:SI}}\mathrm{UP})}_{\text{up-parabola sink}}\end{array}\right\}\text{for } a_{110}<0,a_{220}>0,\tag{3.52}$$

$$\left\{\begin{array}{l}(a_{111},a_{222})\\(a_{111},a_{121})\end{array}\right\}=\left\{\begin{array}{c}\underbrace{(\mathrm{SO},\mathrm{II}:\mathrm{DI})}_{\text{inflection-source}}\\\underbrace{({}_{\mathrm{DI:II}}\mathrm{SO},{}_{\mathrm{SI:SO}}\mathrm{UP})}_{\text{up-parabola source}}\end{array}\right\}\text{for } a_{110}>0,a_{220}<0,\tag{3.53}$$

$$\left\{\begin{array}{l}(a_{111},a_{222})\\(a_{111},a_{121})\end{array}\right\}=\left\{\begin{array}{c}\underbrace{(\mathrm{SI},\mathrm{DI}:\mathrm{II})}_{\text{inflection-sink}}\\\underbrace{({}_{\mathrm{II:DI}}\mathrm{SI},{}_{\mathrm{SI:SO}}\mathrm{DP})}_{\text{down-parabola sink}}\end{array}\right\}\text{for } a_{110}<0,a_{220}<0.\tag{3.54}$$

The infinite-equilibrium of $x_1^* = a_{111} = a_{211}$ for $a_{121} = a_{221}$ is summarized as follows:

$$\left.\begin{array}{ll}(a_{111},\bar{x}_2)=\underbrace{(\mathrm{SO},\mathrm{II})}_{\text{increasing-inflection source}} & \text{if } \bar{x}_2\in(-\infty,a_{221}),\\(a_{111},\bar{x}_2)=\underbrace{(\mathrm{SO},\mathrm{DI})}_{\text{decreasing-inflection source}} & \text{if } \bar{x}_2\in(a_{221},a_{222}),\\(a_{111},\bar{x}_2)=\underbrace{(\mathrm{SO},\mathrm{II})}_{\text{increasing-inflection source}} & \text{if } \bar{x}_2\in(a_{222},\infty),\\\text{for } a_{110}>0,a_{220}>0; & \end{array}\right.\tag{3.55}$$

$$\left.\begin{array}{ll}(a_{111},\bar{x}_2)=\underbrace{(\mathrm{SI},\mathrm{DI})}_{\text{decreasing-inflection sink}} & \text{if } \bar{x}_2\in(-\infty,a_{221}),\\(a_{111},\bar{x}_2)=\underbrace{(\mathrm{SI},\mathrm{II})}_{\text{increasing-inflection sink}} & \text{if } \bar{x}_2\in(a_{221},a_{222}),\\(a_{111},\bar{x}_2)=\underbrace{(\mathrm{SI},\mathrm{DI})}_{\text{decreasing-inflection sink}} & \text{if } \bar{x}_2\in(a_{222},\infty),\\\text{for } a_{110}<0,a_{220}>0; & \end{array}\right.\tag{3.56}$$

$$\left.\begin{array}{ll}(a_{111},\bar{x}_2)=\underbrace{(\mathrm{SO},\mathrm{DI})}_{\text{decreasing-inflection source}} & \text{if } \bar{x}_2\in(-\infty,a_{221}),\\(a_{111},\bar{x}_2)=\underbrace{(\mathrm{SO},\mathrm{II})}_{\text{increasing-inflection source}} & \text{if } \bar{x}_2\in(a_{221},a_{222}),\\(a_{111},\bar{x}_2)=\underbrace{(\mathrm{SO},\mathrm{DI})}_{\text{decreasing-inflection source}} & \text{if } \bar{x}_2\in(a_{222},\infty),\\\text{for } a_{110}>0,a_{220}<0; & \end{array}\right.\tag{3.57}$$

$$\left.\begin{aligned}
&(a_{111},\bar{x}_2)=\underbrace{(\mathrm{SI},\mathrm{II})}_{\text{increasing-inflection sink}} \quad \text{if } \bar{x}_2\in(-\infty,a_{221}),\\
&(a_{111},\bar{x}_2)=\underbrace{(\mathrm{SI},\mathrm{DI})}_{\text{decreasing-inflection sink}} \quad \text{if } \bar{x}_2\in(a_{221},a_{222}),\\
&(a_{111},\bar{x}_2)=\underbrace{(\mathrm{SI},\mathrm{II})}_{\text{increasing-inflection sink}} \quad \text{if } \bar{x}_2\in(a_{222},\infty),\\
&\text{for } a_{110}<0, a_{220}<0.
\end{aligned}\right\} \tag{3.58}$$

The infinite-equilibrium of $x_2^*=a_{121}=a_{221}$ for $a_{111}=a_{211}$ is summarized as follows:

$$\begin{aligned}
&(\bar{x}_1,a_{121})=\underbrace{(\mathrm{SO},\mathrm{DP})}_{\text{down-parabola source}} \quad \text{if } \bar{x}_2\in(-\infty,a_{111}),\\
&(\bar{x}_1,a_{121})=\underbrace{(\mathrm{SI},\mathrm{DP})}_{\text{down-parabola sink}} \quad \text{if } \bar{x}_2\in(a_{111},\infty),\\
&\text{for } a_{110}>0, a_{220}>0;
\end{aligned} \tag{3.59}$$

$$\begin{aligned}
&(\bar{x}_1,a_{121})=\underbrace{(\mathrm{SO},\mathrm{UP})}_{\text{up-parabola source}} \quad \text{if } \bar{x}_2\in(-\infty,a_{111}),\\
&(\bar{x}_1,a_{121})=\underbrace{(\mathrm{SI},\mathrm{UP})}_{\text{up-parabola sink}} \quad \text{if } \bar{x}_2\in(a_{111},\infty),\\
&\text{for } a_{110}<0, a_{220}>0;
\end{aligned} \tag{3.60}$$

$$\begin{aligned}
&(\bar{x}_1,a_{121})=\underbrace{(\mathrm{SI},\mathrm{UP})}_{\text{up-parabola sink}} \quad \text{if } \bar{x}_2\in(-\infty,a_{111}),\\
&(\bar{x}_1,a_{121})=\underbrace{(\mathrm{SO},\mathrm{UP})}_{\text{up-parabola source}} \quad \text{if } \bar{x}_2\in(a_{111},\infty),\\
&\text{for } a_{110}>0, a_{220}<0;
\end{aligned} \tag{3.61}$$

$$\begin{aligned}
&(\bar{x}_1,a_{121})=\underbrace{(\mathrm{SI},\mathrm{DP})}_{\text{down-parabola sink}} \quad \text{if } \bar{x}_2\in(-\infty,a_{111}),\\
&(\bar{x}_1,a_{121})=\underbrace{(\mathrm{SO},\mathrm{DP})}_{\text{down-parabola source}} \quad \text{if } \bar{x}_2\in(a_{111},\infty),\\
&\text{for } a_{110}<0, a_{220}<0.
\end{aligned} \tag{3.62}$$

The infinite-equilibriums of $x_1^*=a_{111}=a_{211}$ and $x_2^*=a_{121}=a_{221}$ are for increasing-inflection and decreasing-inflection source and sink, as shown in Fig. 3.9. The infinite-equilibrium is the switching bifurcation of the equilibriums:

- for $a_{111}<a_{211}$ and $a_{211}<a_{111}$ with $a_{121}=a_{221}$;
- for $a_{121}<a_{221}$ and $a_{221}<a_{121}$ with $a_{111}=a_{211}$;
- for $a_{111}<a_{211}$ and $a_{211}<a_{111}$ with $a_{121}<a_{221}$ and $a_{221}<a_{121}$.

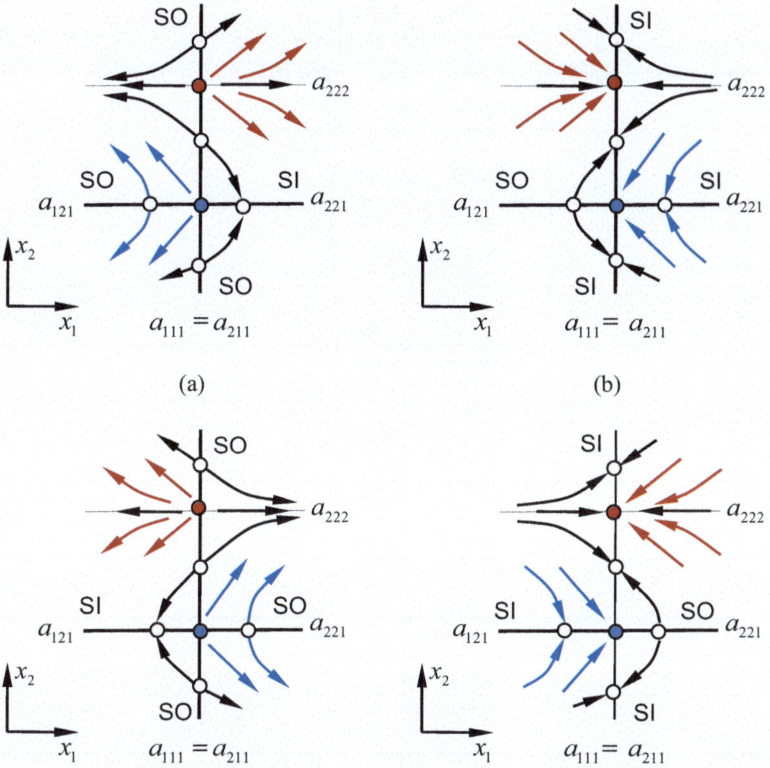

Fig. 3.9 Phase portraits ($a_{111} = a_{211}, a_{121} = a_{221}$) for two-dimensional systems on the x_1-direction with $x_1^* = a_{111} = a_{211}$ and on the x_2-direction with $x_2^* = a_{121}, a_{221}, a_{222}$. The equilibrium flows: **(a)** ($a_{110} > 0$, $a_{220} > 0$), **(b)** ($a_{110} < 0$, $a_{220} > 0$), **(c)** ($a_{110} > 0$, $a_{220} < 0$), and **(d)** ($a_{110} < 0$, $a_{220} < 0$)

(ii) Similarly, the equilibriums of $x_2^* = a_{221}, a_{121}, a_{122}$ and $x_1^* = a_{111} = a_{211}$ for $a_{122} = a_{221}$ are

$$\left\{\begin{matrix}(a_{111}, a_{121})\\(a_{111}, a_{221})\end{matrix}\right\} = \left\{\begin{matrix}\underbrace{({}_{\text{DI:II}}\text{SO}, {}_{\text{SI:SO}}\text{UP})}_{\text{up-parabola source}}\\\underbrace{(\text{SO}, \text{II} : \text{DI})}_{\text{inflection-source}}\end{matrix}\right\} \text{ for } a_{110} > 0, a_{220} > 0, \tag{3.63}$$

$$\left\{\begin{array}{l}(a_{111},a_{121})\\(a_{111},a_{221})\end{array}\right\}=\left\{\begin{array}{c}\underbrace{({}_{\text{II:DI}}\text{SI},{}_{\text{SO:SI}}\text{DP})}_{\text{down-parabola sink}}\\\underbrace{(\text{SI},\text{DI}:\text{II})}_{\text{inflection-sink}}\end{array}\right\}\text{for } a_{110}<0, a_{220}>0, \tag{3.64}$$

$$\left\{\begin{array}{l}(a_{111},a_{121})\\(a_{111},a_{221})\end{array}\right\}=\left\{\begin{array}{c}\underbrace{({}_{\text{II:DI}}\text{SO},{}_{\text{SI:SO}}\text{DP})}_{\text{down-parabola source}}\\\underbrace{(\text{SO},\text{DI}:\text{II})}_{\text{inflection-source}}\end{array}\right\}\text{for } a_{110}>0, a_{220}<0, \tag{3.65}$$

$$\left\{\begin{array}{l}(a_{111},a_{121})\\(a_{111},a_{221})\end{array}\right\}=\left\{\begin{array}{c}\underbrace{({}_{\text{DI:II}}\text{SI},{}_{\text{SO:SI}}\text{UP})}_{\text{up-parabola sink}}\\\underbrace{(\text{SI},\text{II}:\text{DI})}_{\text{inflection-sink}}\end{array}\right\}\text{for } a_{110}<0, a_{220}<0. \tag{3.66}$$

The infinite-equilibrium of $x_1^* = a_{111} = a_{211}$ for $a_{121} = a_{222}$ is summarized as follows:

$$\begin{aligned}&(a_{111},\bar{x}_2)=\underbrace{(\text{SO},\text{II})}_{\text{increasing-inflection source}} &&\text{if } \bar{x}_2\in(-\infty,a_{221}),\\&(a_{111},\bar{x}_2)=\underbrace{(\text{SO},\text{DI})}_{\text{decreasing-inflection source}} &&\text{if } \bar{x}_2\in(a_{221},a_{222}),\\&(a_{111},\bar{x}_2)=\underbrace{(\text{SO},\text{II})}_{\text{increasing-inflection source}} &&\text{if } \bar{x}_2\in(a_{222},\infty),\\&\text{for } a_{110}>0, a_{220}>0;\end{aligned} \tag{3.67}$$

$$\begin{aligned}&(a_{111},\bar{x}_2)=\underbrace{(\text{SI},\text{DI})}_{\text{decreasing-inflection sink}} &&\text{if } \bar{x}_2\in(-\infty,a_{221}),\\&(a_{111},\bar{x}_2)=\underbrace{(\text{SI},\text{II})}_{\text{increasing-inflection sink}} &&\text{if } \bar{x}_2\in(a_{221},a_{222}),\\&(a_{111},\bar{x}_2)=\underbrace{(\text{SI},\text{DI})}_{\text{decreasing-inflection sink}} &&\text{if } \bar{x}_2\in(a_{222},\infty),\\&\text{for } a_{110}<0, a_{220}>0;\end{aligned} \tag{3.68}$$

$$\begin{aligned}&(a_{111},\bar{x}_2)=\underbrace{(\text{SO},\text{DI})}_{\text{decreasing-inflection source}} &&\text{if } \bar{x}_2\in(-\infty,a_{221}),\\&(a_{111},\bar{x}_2)=\underbrace{(\text{SO},\text{II})}_{\text{increasing-inflection source}} &&\text{if } \bar{x}_2\in(a_{221},a_{222}),\\&(a_{111},\bar{x}_2)=\underbrace{(\text{SO},\text{DI})}_{\text{decreasing-inflection source}} &&\text{if } \bar{x}_2\in(a_{222},\infty),\\&\text{for } a_{110}>0, a_{220}<0;\end{aligned} \tag{3.69}$$

$$
\begin{aligned}
&(a_{111}, \bar{x}_2) = \underbrace{(\text{SI, II})}_{\text{increasing-inflection sink}} \quad \text{if } \bar{x}_2 \in (-\infty, a_{221}),\\
&(a_{111}, \bar{x}_2) = \underbrace{(\text{SI, DI})}_{\text{decreasing-inflection sink}} \quad \text{if } \bar{x}_2 \in (a_{221}, a_{222}),\\
&(a_{111}, \bar{x}_2) = \underbrace{(\text{SI, II})}_{\text{increasing-inflection sink}} \quad \text{if } \bar{x}_2 \in (a_{222}, \infty),\\
&\text{for } a_{110} < 0, a_{220} < 0.
\end{aligned} \tag{3.70}
$$

The infinite-equilibrium of $x_1^* = a_{111} = a_{211}$ for $a_{121} = a_{222}$ is identical to the infinite-equilibriums of $x_1^* = a_{111} = a_{211}$ for $a_{121} = a_{221}$, but the global phase portraits are different.

The infinite-equilibrium of $x_2^* = a_{121} = a_{222}$ for $a_{111} = a_{211}$ is summarized as follows:

$$
\begin{aligned}
&(\bar{x}_1, a_{121}) = \underbrace{(\text{SI, UP})}_{\text{up-parabola sink}} \quad \text{if } \bar{x}_1 \in (-\infty, a_{111}),\\
&(\bar{x}_1, a_{121}) = \underbrace{(\text{SO, UP})}_{\text{up-parabola source}} \quad \text{if } \bar{x}_1 \in (a_{111}, \infty),\\
&\text{for } a_{110} > 0, a_{220} > 0;
\end{aligned} \tag{3.71}
$$

$$
\begin{aligned}
&(\bar{x}_1, a_{121}) = \underbrace{(\text{SI, DP})}_{\text{down-parabola sink}} \quad \text{if } \bar{x}_2 \in (-\infty, a_{111}),\\
&(\bar{x}_1, a_{121}) = \underbrace{(\text{SO, DP})}_{\text{down-parabola source}} \quad \text{if } \bar{x}_2 \in (a_{111}, \infty),\\
&\text{for } a_{110} < 0, a_{220} > 0;
\end{aligned} \tag{3.72}
$$

$$
\begin{aligned}
&(\bar{x}_1, a_{121}) = \underbrace{(\text{SO, DP})}_{\text{down-parabola source}} \quad \text{if } \bar{x}_2 \in (-\infty, a_{111}),\\
&(\bar{x}_1, a_{121}) = \underbrace{(\text{SI, DP})}_{\text{down-parabola sink}} \quad \text{if } \bar{x}_2 \in (a_{111}, \infty),\\
&\text{for } a_{110} > 0, a_{220} < 0;
\end{aligned} \tag{3.73}
$$

$$
\begin{aligned}
&(\bar{x}_1, a_{121}) = \underbrace{(\text{SO, UP})}_{\text{up-parabola source}} \quad \text{if } \bar{x}_1 \in (-\infty, a_{111}),\\
&(\bar{x}_1, a_{121}) = \underbrace{(\text{SI, UP})}_{\text{up-parabola sink}} \quad \text{if } \bar{x}_1 \in (a_{111}, \infty),\\
&\text{for } a_{110} < 0, a_{220} < 0.
\end{aligned} \tag{3.74}
$$

The infinite-equilibriums of $x_1^* = a_{111} = a_{211}$ and $x_2^* = a_{121} = a_{222}$ are for increasing-inflection and decreasing-inflection source and sink, as shown in Fig. 3.10. The infinite-equilibrium is the switching bifurcation of the equilibriums:

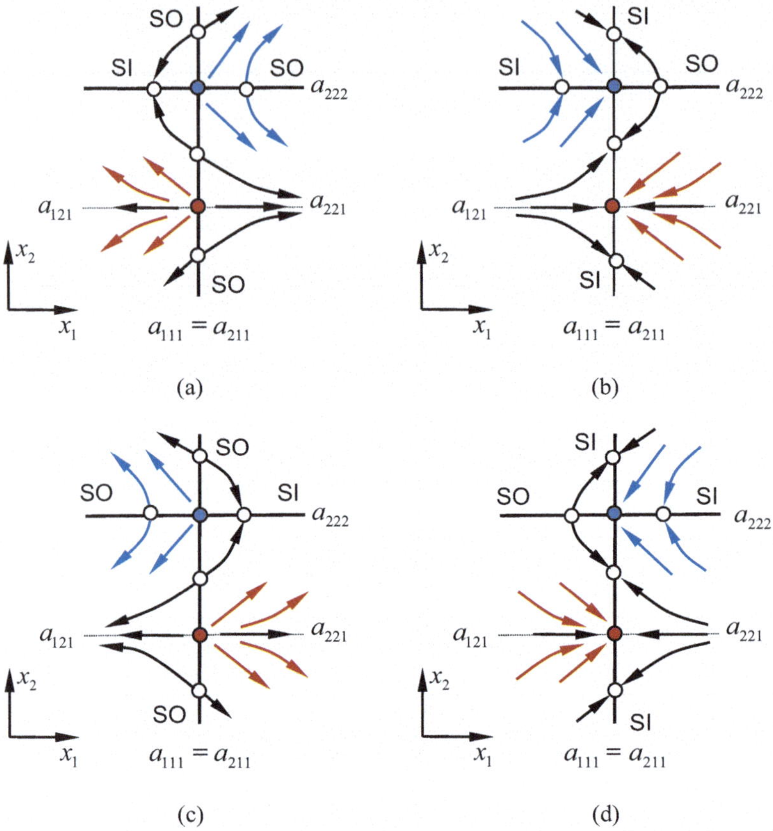

Fig. 3.10 Phase portraits ($a_{111} = a_{211}$, $a_{121} = a_{222}$) for two-dimensional systems on the x_1-direction with $x_1^* = a_{111} = a_{211}$ and on the x_2-direction with $x_2^* = a_{121}, a_{221}, a_{222}$. The equilibrium flows: (**a**) ($a_{110} > 0$, $a_{220} > 0$), (**b**) ($a_{110} < 0$, $a_{220} > 0$), (**c**) ($a_{110} > 0$, $a_{220} < 0$), and (**d**) ($a_{110} < 0$, $a_{220} < 0$)

- for $a_{111} < a_{211}$ and $a_{211} < a_{111}$ with $a_{121} = a_{222}$;
- for $a_{121} < a_{222}$ and $a_{222} < a_{121}$ with $a_{111} = a_{211}$;
- for $a_{111} < a_{211}$ and $a_{211} < a_{111}$ with $a_{121} < a_{222}$ and $a_{222} < a_{121}$.

Chapter 4
Saddle-Source (Sink) with Hyperbolic Flow Singularity

In this chapter, saddle-source (sink) flows and hyperbolic flow singularity with hyperbolic flows, positive or negative saddles, and center are discussed, and the corresponding networks are presented for saddle-source (sink) and hyperbolic flow singularity with simple flows and equilibriums. The corresponding switching bifurcations are discussed through the infinite-equilibrium, which includes up-down and down-up infinite-equilibriums, inflection-source (sink) infinite-equilibriums, and two infinite-equilibriums of inflection-source (sink) and up-down (down-up)-saddles.

4.1 Saddle-Source (Sink) with Hyperbolic Flow Singularity

Consider a dynamical system as

$$\begin{aligned}\dot{x}_1 &= a_{110}(x_1 - a_{111})(x_2 - a_{121})(x_2 - a_{122}),\\ \dot{x}_2 &= a_{220}(x_1 - a_{211})(x_2 - a_{221})^2,\end{aligned} \tag{4.1}$$

and the corresponding first integral manifold is

$$\begin{aligned}&[(x_1 - a_{111}) - (x_{10} - a_{111})] + (a_{111} - a_{211})\ln\frac{|x_1 - a_{111}|}{|x_{10} - a_{111}|}\\ &= \frac{a_{110}}{a_{220}}\Big[(x_2 - x_{20}) + (2a_{221} - a_{121} - a_{122})\ln\frac{|x_2 - a_{221}|}{|x_{20} - a_{221}|}\\ &\quad - (a_{221} - a_{121})(a_{221} - a_{122})(\frac{1}{x_2 - a_{221}} - \frac{1}{x_{20} - a_{221}})\Big].\end{aligned} \tag{4.2}$$

(i) For $a_{122} < a_{221}$ and $a_{211} < a_{111}$, the equilibrium matrices of $x_1^* = a_{111}, a_{211}$ with $x_2^* = a_{121}, a_{122}, a_{221}$ have the following properties, as shown in Fig. 4.1.

A. C. J. Luo, *Two-dimensional Two-product Cubic Systems, Vol. I*,
https://doi.org/10.1007/978-3-031-48487-2_4

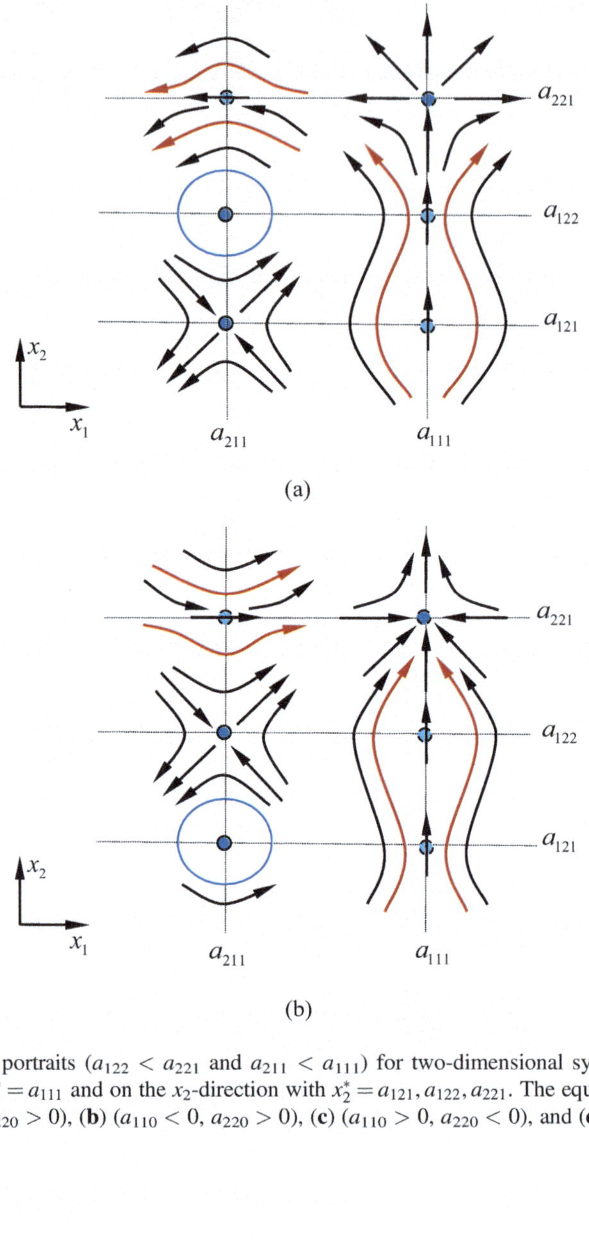

Fig. 4.1 Phase portraits ($a_{122} < a_{221}$ and $a_{211} < a_{111}$) for two-dimensional systems on the x_1-direction with $x_1^* = a_{111}$ and on the x_2-direction with $x_2^* = a_{121}, a_{122}, a_{221}$. The equilibrium flows: (**a**) ($a_{110} > 0$, $a_{220} > 0$), (**b**) ($a_{110} < 0$, $a_{220} > 0$), (**c**) ($a_{110} > 0$, $a_{220} < 0$), and (**d**) ($a_{110} < 0$, $a_{220} < 0$)

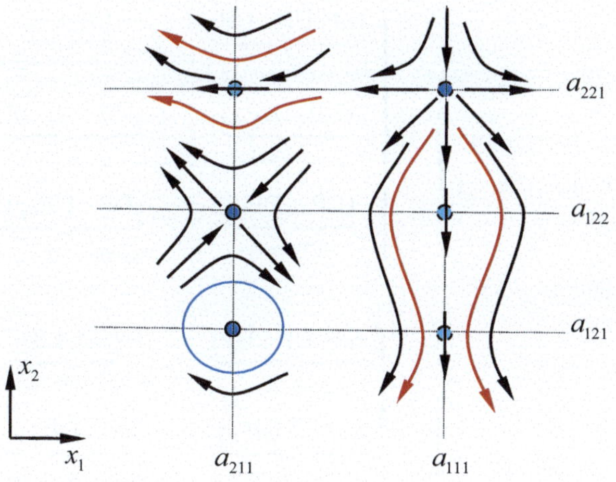

(c)

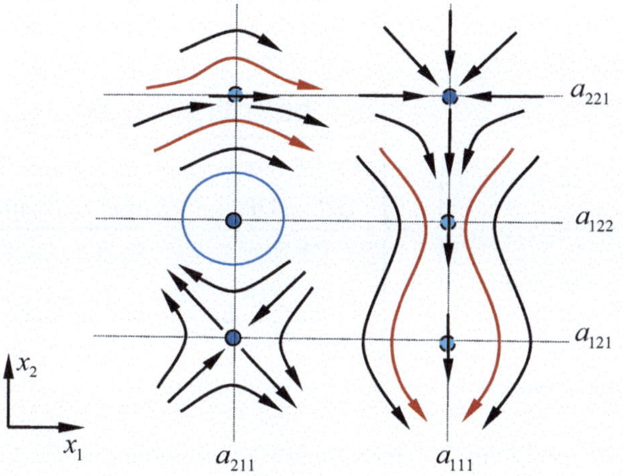

(d)

Fig. 4.1 (continued)

$$\left\{\begin{matrix} (a_{221},a_{211}) & (a_{111},a_{122}) \\ (a_{122},a_{211}) & (a_{111},a_{221}) \\ (a_{121},a_{211}) & (a_{111},a_{121}) \end{matrix}\right\} = \left\{\begin{matrix} \underbrace{(\text{nF},\text{DP}:\text{DP})}_{\text{HB-to-HS flow }(-)} & \underbrace{(\text{SO},\text{US})}_{\text{upper-saddle-source}} \\ \underbrace{(\text{DP}_{+},\text{DP}_{-})}_{\text{CCW center}} & \underbrace{(\text{DP}:\text{UP},\text{pF})}_{\text{hyperbolic flow }(+)} \\ \underbrace{(\text{UP}_{+},\text{UP}_{+})}_{\text{positive saddle}} & \underbrace{(\text{UP}:\text{DP},\text{pF})}_{\text{hyperbolic-secant flow }(+)} \end{matrix}\right\} \quad (4.3)$$

for $a_{110} > 0$ and $a_{220} > 0$;

$$\left\{\begin{array}{cc}(a_{221},a_{211}) & (a_{111},a_{122})\\ (a_{122},a_{211}) & (a_{111},a_{221})\\ (a_{121},a_{211}) & (a_{111},a_{121})\end{array}\right\}=\left\{\begin{array}{cc}\underbrace{(\mathrm{pF},\mathrm{UP}:\mathrm{UP})}_{\text{HS-to-HB flow }(+)} & \underbrace{(\mathrm{SI},\mathrm{US})}_{\text{upper-saddle-sink}}\\ \underbrace{(\mathrm{UP}_{+},\mathrm{UP}_{+})}_{\text{positive saddle}} & \underbrace{(\mathrm{UP}:\mathrm{DP},\mathrm{pF})}_{\text{hyperbolic-secant flow }(+)}\\ \underbrace{(\mathrm{DP}_{+},\mathrm{DP}_{-})}_{\text{CCW center}} & \underbrace{(\mathrm{DP}:\mathrm{UP},\mathrm{pF})}_{\text{hyperbolic flow }(+)}\end{array}\right\} \tag{4.4}$$

for $a_{110}<0$ and $a_{220}>0$;

$$\left\{\begin{array}{cc}(a_{221},a_{211}) & (a_{111},a_{122})\\ (a_{122},a_{211}) & (a_{111},a_{221})\\ (a_{121},a_{211}) & (a_{111},a_{121})\end{array}\right\}=\left\{\begin{array}{cc}\underbrace{(\mathrm{nF},\mathrm{UP}:\mathrm{UP})}_{\text{HS-to-HB flow }(-)} & \underbrace{(\mathrm{SO},\mathrm{LS})}_{\text{lower-saddle-source}}\\ \underbrace{(\mathrm{UP}_{-},\mathrm{UP}_{-})}_{\text{negative saddle}} & \underbrace{(\mathrm{UP}:\mathrm{DP},\mathrm{nF})}_{\text{hyperbolic-secant flow }(-)}\\ \underbrace{(\mathrm{DP}_{-},\mathrm{DP}_{+})}_{\text{CW center}} & \underbrace{(\mathrm{DP}:\mathrm{UP},\mathrm{nF})}_{\text{hyperbolic flow }(-)}\end{array}\right\} \tag{4.5}$$

for $a_{110}>0$ and $a_{220}<0$;

$$\left\{\begin{array}{cc}(a_{221},a_{211}) & (a_{111},a_{122})\\ (a_{122},a_{211}) & (a_{111},a_{221})\\ (a_{121},a_{211}) & (a_{111},a_{121})\end{array}\right\}=\left\{\begin{array}{cc}\underbrace{(\mathrm{pF},\mathrm{DP}:\mathrm{DP})}_{\text{HB-to-HS flow }(+)} & \underbrace{(\mathrm{SI},\mathrm{LS})}_{\text{lower-saddle-sink}}\\ \underbrace{(\mathrm{DP}_{-},\mathrm{DP}_{+})}_{\text{CW center}} & \underbrace{(\mathrm{DP}:\mathrm{UP},\mathrm{nF})}_{\text{hyperbolic flow }(-)}\\ \underbrace{(\mathrm{UP}_{-},\mathrm{UP}_{-})}_{\text{negative saddle}} & \underbrace{(\mathrm{UP}:\mathrm{DP},\mathrm{nF})}_{\text{hyperbolic-secant flow }(-)}\end{array}\right\} \tag{4.6}$$

for $a_{110}<0$ and $a_{220}<0$.

The hyperbolic and hyperbolic-secant flows with upper-saddle sink and source, lower-saddle-sink and source are presented. The upper- and lower-saddle-sink and saddle-source are the appearing bifurcations for saddle and sink and for saddle and source, respectively. All the equilibriums on two lines of $x_1^* = a_{111}, a_{211}$ form a matrix of equilibriums. On the line of $x_1^* = a_{111}$, the saddle-sink and saddle-source are on the top of the series of equilibriums. The hyperbolic and hyperbolic-secant flows are connected. On the line of $x_1^* = a_{211}$, the hyperbolic-to-hyperbolic-secant flow (HB-to-HS flow) or the hyperbolic-secant-to-hyperbolic flow (HS-to-HB flow) exists for the appearing bifurcations of the two separated hyperbolic and hyperbolic-secant flows. The clockwise and counterclockwise centers with positive and negative saddles, respectively, as simple equilibriums are also on the line of $x_1^* = a_{211}$.

(ii) For $a_{221} \in (a_{121}, a_{122})$ and $a_{211} < a_{111}$, the equilibriums of $x_1^* = a_{111}, a_{211}$ with $x_2^* = a_{121}, a_{122}, a_{221}$ have the following properties:

$$\left\{\begin{array}{ll}(a_{122}, a_{211}) & (a_{111}, a_{122})\\(a_{221}, a_{211}) & (a_{111}, a_{221})\\(a_{121}, a_{211}) & (a_{111}, a_{121})\end{array}\right\} = \left\{\begin{array}{ll}\underbrace{(\mathrm{DP}_+, \mathrm{DP}_-)}_{\text{CCW center}} & \underbrace{(\mathrm{DP}:\mathrm{UP}, \mathrm{pF})}_{\text{hyperbolic flow }(+)}\\\underbrace{(\mathrm{pF}, \mathrm{UP}:\mathrm{UP})}_{\text{HS-to-HB flow }(+)} & \underbrace{(\mathrm{SI}, \mathrm{US})}_{\text{upper-saddle-source}}\\\underbrace{(\mathrm{UP}_+, \mathrm{UP}_+)}_{\text{positive saddle}} & \underbrace{(\mathrm{UP}:\mathrm{DP}, \mathrm{pF})}_{\text{hyperbolic-secant flow }(+)}\end{array}\right\} \tag{4.7}$$

for $a_{110} > 0$ and $a_{220} > 0$;

$$\left\{\begin{array}{ll}(a_{122}, a_{211}) & (a_{111}, a_{122})\\(a_{221}, a_{211}) & (a_{111}, a_{221})\\(a_{121}, a_{211}) & (a_{111}, a_{121})\end{array}\right\} = \left\{\begin{array}{ll}\underbrace{(\mathrm{UP}_+, \mathrm{UP}_+)}_{\text{positive saddle}} & \underbrace{(\mathrm{UP}:\mathrm{DP}, \mathrm{pF})}_{\text{hyperbolic-secant flow }(+)}\\\underbrace{(\mathrm{nF}, \mathrm{DP}:\mathrm{DP})}_{\text{HB-to-HS flow }(-)} & \underbrace{(\mathrm{SO}, \mathrm{US})}_{\text{upper-saddle-source}}\\\underbrace{(\mathrm{DP}_+, \mathrm{DP}_-)}_{\text{CCW center}} & \underbrace{(\mathrm{DP}:\mathrm{UP}, \mathrm{pF})}_{\text{hyperbolic flow }(+)}\end{array}\right\} \tag{4.8}$$

for $a_{110} < 0$ and $a_{220} > 0$;

$$\left\{\begin{array}{ll}(a_{122}, a_{211}) & (a_{111}, a_{122})\\(a_{221}, a_{211}) & (a_{111}, a_{221})\\(a_{121}, a_{211}) & (a_{111}, a_{121})\end{array}\right\} = \left\{\begin{array}{ll}\underbrace{(\mathrm{UP}_-, \mathrm{UP}_-)}_{\text{negative saddle}} & \underbrace{(\mathrm{UP}:\mathrm{DP}, \mathrm{nF})}_{\text{hyperbolic-secant flow }(-)}\\\underbrace{(\mathrm{pF}, \mathrm{DP}:\mathrm{DP})}_{\text{HB-to-HS flow }(+)} & \underbrace{(\mathrm{SI}, \mathrm{LS})}_{\text{lower-saddle-sink}}\\\underbrace{(\mathrm{DP}_-, \mathrm{DP}_+)}_{\text{CW center}} & \underbrace{(\mathrm{DP}:\mathrm{UP}, \mathrm{nF})}_{\text{hyperbolic flow }(-)}\end{array}\right\} \tag{4.9}$$

for $a_{110} > 0$ and $a_{220} < 0$;

$$\left\{\begin{array}{ll}(a_{122}, a_{211}) & (a_{111}, a_{122})\\(a_{221}, a_{211}) & (a_{111}, a_{221})\\(a_{121}, a_{211}) & (a_{111}, a_{121})\end{array}\right\} = \left\{\begin{array}{ll}\underbrace{(\mathrm{DP}_-, \mathrm{DP}_+)}_{\text{CW center}} & \underbrace{(\mathrm{DP}:\mathrm{UP}, \mathrm{nF})}_{\text{hyperbolic flow }(-)}\\\underbrace{(\mathrm{nF}, \mathrm{UP}:\mathrm{UP})}_{\text{HS-to-HB flow }(-)} & \underbrace{(\mathrm{SO}, \mathrm{LS})}_{\text{lower-saddle-source}}\\\underbrace{(\mathrm{UP}_-, \mathrm{UP}_-)}_{\text{negaitve saddle}} & \underbrace{(\mathrm{UP}:\mathrm{DP}, \mathrm{nF})}_{\text{hyperbolic-secant flow }(-)}\end{array}\right\} \tag{4.10}$$

for $a_{110} < 0$ and $a_{220} < 0$,

as shown in Fig. 4.2.

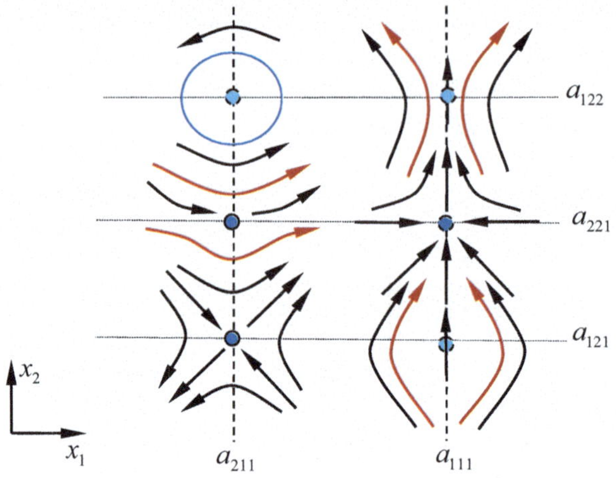

(a)

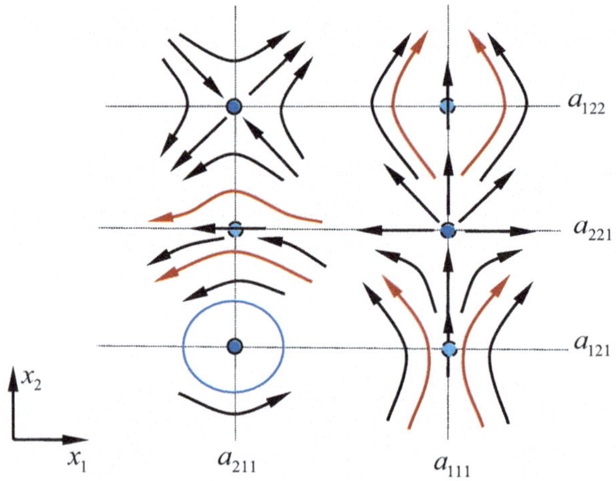

(b)

Fig. 4.2 Phase portraits ($a_{121} < a_{221}$ and $a_{211} < a_{111}$) for two-dimensional systems on the x_1-direction with $x_1^* = a_{111}$ and on the x_2-direction with $x_2^* = a_{121}, a_{122}, a_{221}, a_{222}$. The equilibrium flows: **(a)** ($a_{110} > 0$, $a_{220} > 0$), **(b)** ($a_{110} < 0$, $a_{220} > 0$), **(c)** ($a_{110} > 0$, $a_{220} < 0$), and **(d)** ($a_{110} < 0$, $a_{220} < 0$)

The saddle-source and sink equilibriums on the line of $x_1^* = a_{111}$ lie between hyperbolic and hyperbolic-secant flows. The hyperbolic and hyperbolic-secant flows are separated by the saddle-source or saddle-sinks. The hyperbolic and hyperbolic-secant flow for $a_{221} \in (a_{121}, a_{122})$ and $a_{211} < a_{111}$ are identical to the case of $a_{221} > a_{122}$ and $a_{211} < a_{111}$. Only the saddle-source and saddle-sink are switched.

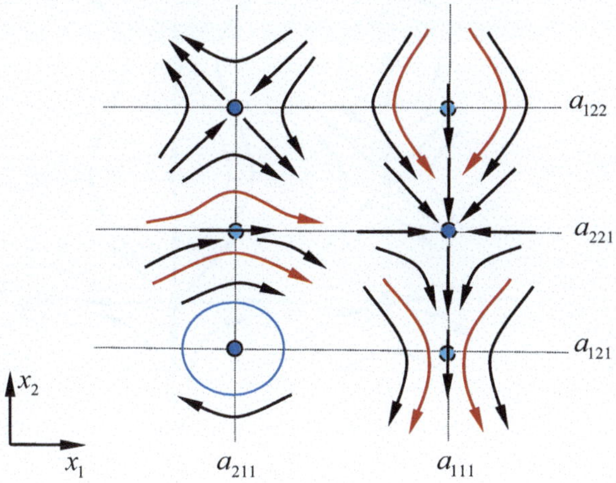

(c)

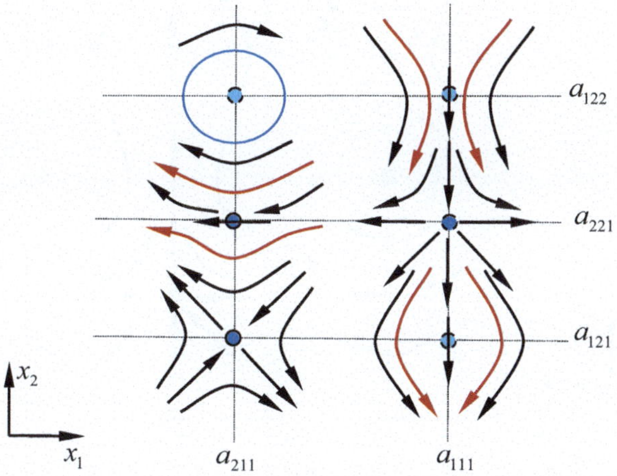

(d)

Fig. 4.2 (continued)

The hyperbolic-to-hyperbolic-secant flow or the hyperbolic-secant-to-hyperbolic flow on the line of $x_1^* = a_{211}$ are between the center and saddle. The saddle and centers for $a_{221} \in (a_{121}, a_{122})$ are identical to the case of $a_{221} > a_{122}$ and $a_{211} < a_{111}$. Only the hyperbolic-to-hyperbolic-secant flow and the hyperbolic-to-hyperbolic-secant flow are switched.

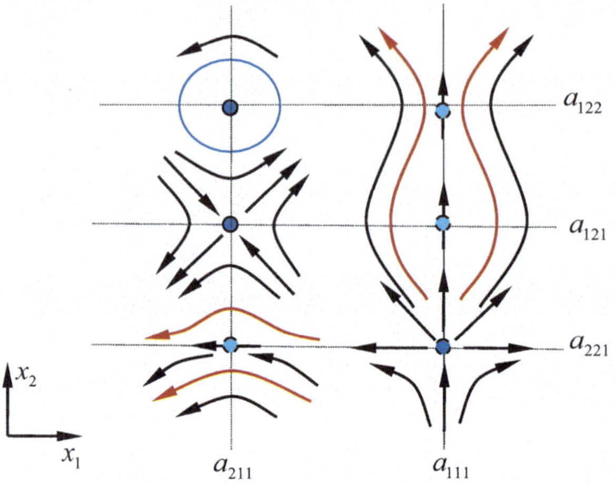

(a)

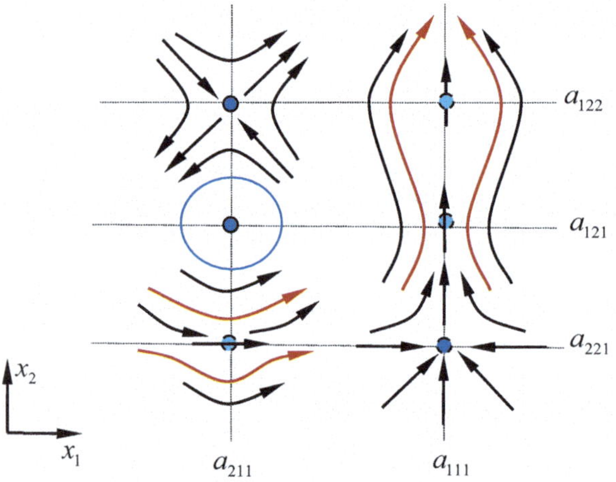

(b)

Fig. 4.3 Phase portraits ($a_{121} < a_{221}$ and $a_{211} < a_{111}$) for two-dimensional systems on the x_1-direction with $x_1^* = a_{111}$ and on the x_2-direction with $x_2^* = a_{121}, a_{122}, a_{221}, a_{222}$. The equilibrium flows: (**a**) ($a_{110} > 0$, $a_{220} > 0$), (**b**) ($a_{110} < 0$, $a_{220} > 0$), (**c**) ($a_{110} > 0$, $a_{220} < 0$), and (**d**) ($a_{110} < 0$, $a_{220} < 0$)

(iii) For $a_{221} < a_{121}$ and $a_{211} < a_{111}$, the equilibriums of $x_1^* = a_{111}, a_{211}$ with $x_2^* = a_{121}, a_{122}, a_{221}$ have the following properties as in Eqs. (4.3)–(4.6), as shown in Fig. 4.3. The dynamical behaviors of equilibriums for $a_{221} < a_{121}$ and $a_{211} < a_{111}$ are identical to the case of $a_{122} < a_{221}$ and $a_{211} < a_{111}$. Just the locations of equilibriums are different. The saddle-sink and saddle-source are at the bottom of

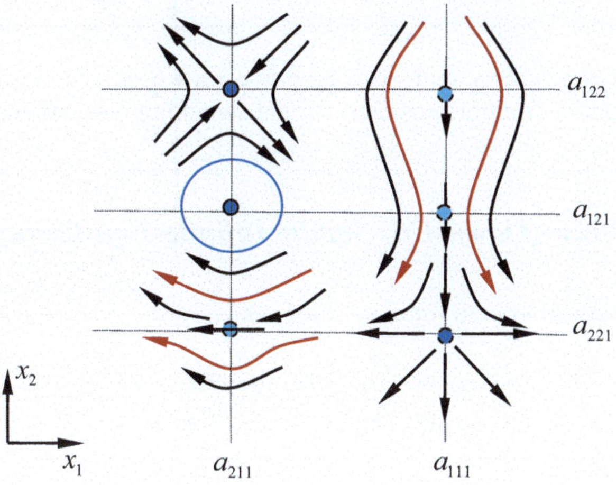

(c)

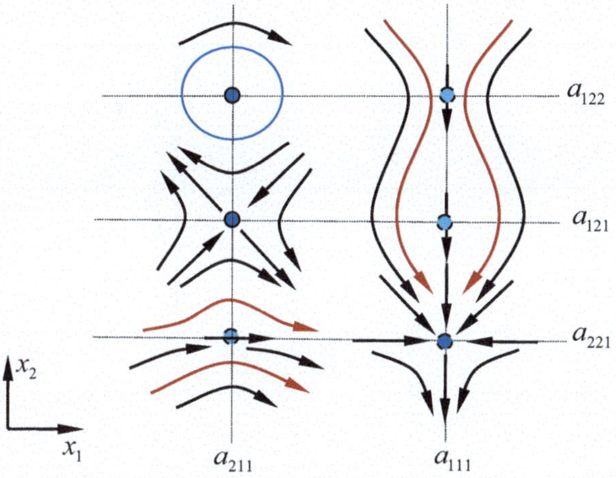

(d)

Fig. 4.3 (continued)

the series of equilibriums. The hyperbolic and hyperbolic-secant flows are connected.

Similarly, the readers can discuss the dynamical behaviors of equilibriums for three cases of $a_{221} < a_{121}$, $a_{221} \in (a_{121}, a_{122})$ and $a_{221} > a_{122}$ with $a_{211} > a_{111}$. The corresponding phase portraits for the equilibrium matrix can be sketched.

4.2 Infinite-Equilibriums and Switching Bifurcations

In this section, the infinite-equilibriums and switching dynamics are discussed for such a cubic dynamical system. The up-down (down-up)-saddle and inflection-source (sink) infinite-equilibriums exist for the switching bifurcations.

4.2.1 Up-Down (Down-Up)-Saddle Infinite-Equilibriums

Consider a dynamical system for $a_{121} = a_{221}$ as

$$\begin{aligned} \dot{x}_1 &= a_{110}(x_1 - a_{111})(x_2 - a_{121})(x_2 - a_{122}), \\ \dot{x}_2 &= a_{220}(x_1 - a_{211})(x_2 - a_{221})^2. \end{aligned} \tag{4.11}$$

(i) For $a_{121} = a_{221}$ and $a_{211} < a_{111}$, the corresponding first integral manifold is

$$\begin{aligned} &[(x_1 - a_{111}) - (x_{10} - a_{111})] + (a_{111} - a_{211}) \ln \frac{|x_1 - a_{111}|}{|x_{10} - a_{111}|} \\ &= \frac{a_{110}}{a_{220}} \left[(x_2 - x_{20}) + (a_{221} - a_{122}) \ln \frac{|x_2 - a_{221}|}{|x_{20} - a_{221}|} \right]. \end{aligned} \tag{4.12}$$

The equilibriums of $x_2^* = a_{121}, a_{122}, a_{221}$ with $x_1^* = a_{111}, a_{211}$ for $a_{121} = a_{221}$ have the following properties:

$$\left\{ \begin{matrix} (a_{122}, a_{211}) & (a_{111}, a_{122}) \\ (a_{121}, a_{211}) & (a_{111}, a_{221}) \end{matrix} \right\} = \left\{ \begin{matrix} \underbrace{(\text{DP}_+ : \text{DP}_-)}_{\text{CCW center}} & \underbrace{(\text{DP} : \text{UP}, \text{pF})}_{\text{hyperbolic flow } (+)} \\ \underbrace{(\text{DP} : \text{UP}, \text{LS} : \text{US})}_{\text{hyperbolic lower-to-upper saddle}} & \underbrace{(\text{UD} : \text{DU}, \text{US})}_{\text{source-to-sink upper-saddle}} \end{matrix} \right\}$$

for $a_{110} > 0$ and $a_{220} > 0$;

(4.13)

$$\left\{ \begin{matrix} (a_{122}, a_{211}) & (a_{111}, a_{122}) \\ (a_{121}, a_{211}) & (a_{111}, a_{221}) \end{matrix} \right\} = \left\{ \begin{matrix} \underbrace{(\text{UP}_+ : \text{UP}_+)}_{\text{positive saddle}} & \underbrace{(\text{UP} : \text{DP}, \text{pF})}_{\text{hyperbolic-secant flow } (+)} \\ \underbrace{(\text{UP} : \text{DP}, \text{US} : \text{LS})}_{\text{hyperbolic-secant upper-to-lower saddle}} & \underbrace{(\text{DU} : \text{UD}, \text{US})}_{\text{sink-to-source upper-saddle}} \end{matrix} \right\}$$

for $a_{110} < 0$ and $a_{220} > 0$;

(4.14)

$$\left\{\begin{array}{cc}(a_{122},a_{211}) & (a_{111},a_{122})\\(a_{121},a_{211}) & (a_{111},a_{221})\end{array}\right\}=\left\{\begin{array}{cc}\underbrace{(\mathrm{UP}_{-}:\mathrm{UP}_{-})}_{\text{negative saddle}} & \underbrace{(\mathrm{UP}:\mathrm{DP},\mathrm{nF})}_{\text{hyperbolic-secant flow }(-)}\\ \underbrace{(\mathrm{UP}:\mathrm{DP},\mathrm{LS}:\mathrm{US})}_{\text{hyperbolic-secant lower-to-upper saddle}} & \underbrace{(\mathrm{DU}:\mathrm{UD},\mathrm{LS})}_{\text{sink-to-source lower-saddle}}\end{array}\right\}$$

for $a_{110}>0$ and $a_{220}<0$;

(4.15)

$$\left\{\begin{array}{cc}(a_{122},a_{211}) & (a_{111},a_{122})\\(a_{121},a_{211}) & (a_{111},a_{221})\end{array}\right\}=\left\{\begin{array}{cc}\underbrace{(\mathrm{DP}_{-}:\mathrm{DP}_{+})}_{\text{CW center}} & \underbrace{(\mathrm{DP}:\mathrm{UP},\mathrm{nF})}_{\text{hyperbolic flow }(-)}\\ \underbrace{(\mathrm{DP}:\mathrm{UP},\mathrm{US}:\mathrm{LS})}_{\text{hyperbolic upper-to-lower saddle}} & \underbrace{(\mathrm{UD}:\mathrm{DU},\mathrm{LS})}_{\text{source-to-sink lower-saddle}}\end{array}\right\}$$

for $a_{110}<0$ and $a_{220}<0$.

(4.16)

The infinite-equilibriums of $x_2^*=a_{121}=a_{221}$ with $a_{211}<a_{111}$ are

$$\begin{aligned}(\overline{x}_1,a_{121})&=\underbrace{(\mathrm{UD},\mathrm{LS})}_{\text{up-down lower-saddle}}\quad\text{if }\overline{x}\in(-\infty,a_{211}),\\(\overline{x}_1,a_{121})&=\underbrace{(\mathrm{DU},\mathrm{US})}_{\text{down-up upper-saddle}}\quad\text{if }\overline{x}\in(a_{211},a_{111})\\(\overline{x}_1,a_{121})&=\underbrace{(\mathrm{UD},\mathrm{US})}_{\text{up-down upper-saddle}}\quad\text{if }\overline{x}\in(a_{111},\infty),\end{aligned}\tag{4.17}$$

for $a_{110}>0$ and $a_{220}>0$;

$$\begin{aligned}(\overline{x}_1,a_{121})&=\underbrace{(\mathrm{DU},\mathrm{LS})}_{\text{down-up lower-saddle}}\quad\text{if }\overline{x}\in(-\infty,a_{211}),\\(\overline{x}_1,a_{121})&=\underbrace{(\mathrm{UD},\mathrm{US})}_{\text{up-down upper-saddle}}\quad\text{if }\overline{x}\in(a_{211},a_{111}),\\(\overline{x}_1,a_{121})&=\underbrace{(\mathrm{DU},\mathrm{US})}_{\text{down-up upper-saddle}}\quad\text{if }\overline{x}\in(a_{111},\infty),\end{aligned}\tag{4.18}$$

for $a_{110}<0$ and $a_{220}>0$;

$$\begin{aligned}(\overline{x}_1,a_{121})&=\underbrace{(\mathrm{DU},\mathrm{US})}_{\text{down-up upper-saddle}}\quad\text{if }\overline{x}\in(-\infty,a_{211}),\\(\overline{x}_1,a_{121})&=\underbrace{(\mathrm{UD},\mathrm{LS})}_{\text{up-down lower-saddle}}\quad\text{if }\overline{x}\in(a_{211},a_{111}),\\(\overline{x}_1,a_{121})&=\underbrace{(\mathrm{DU},\mathrm{LS})}_{\text{down-up lower-saddle}}\quad\text{if }\overline{x}\in(a_{111},\infty),\end{aligned}\tag{4.19}$$

for $a_{110}>0$ and $a_{220}<0$;

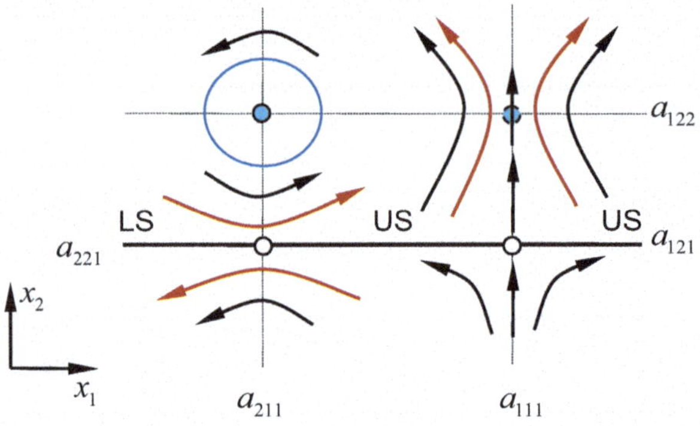

(a)

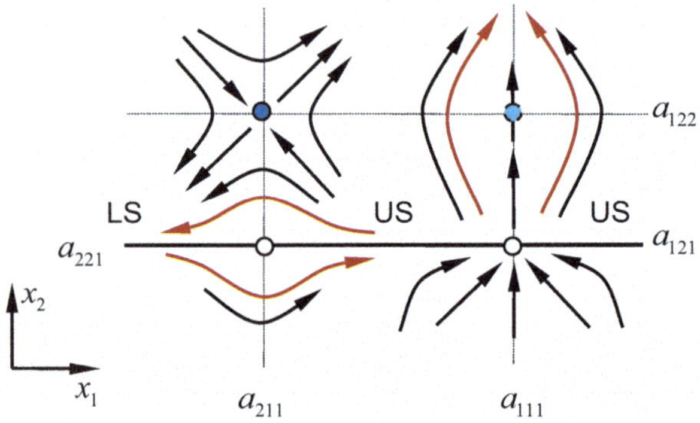

(b)

Fig. 4.4 Phase portraits ($a_{121} = a_{221}$ and $a_{211} < a_{111}$) for two-dimensional systems on the x_1-direction with $x_1^* = a_{111}$ and on the x_2-direction with $x_2^* = a_{121}, a_{122}, a_{221}, a_{222}$. The equilibrium flows: (**a**) ($a_{110} > 0$, $a_{220} > 0$), (**b**) ($a_{110} < 0$, $a_{220} > 0$), (**c**) ($a_{110} > 0$, $a_{220} < 0$), and (**d**) ($a_{110} < 0$, $a_{220} < 0$)

$$
\begin{aligned}
(\overline{x}_1, a_{121}) &= \underbrace{(\mathrm{UD}, \mathrm{US})}_{\text{up-down upper-saddle}} \quad \text{if } \overline{x} \in (-\infty, a_{211}),\\
(\overline{x}_1, a_{121}) &= \underbrace{(\mathrm{DU}, \mathrm{LS})}_{\text{down-up lower-saddle}} \quad \text{if } \overline{x} \in (a_{211}, a_{111})\\
(\overline{x}_1, a_{121}) &= \underbrace{(\mathrm{UD}, \mathrm{LS})}_{\text{up-down lower-saddle}} \quad \text{if } \overline{x} \in (a_{111}, \infty),\\
&\text{for } a_{110} < 0 \text{ and } a_{220} < 0.
\end{aligned}
\tag{5.20}
$$

The above infinite-equilibriums are shown in Fig. 4.4. The hyperbolic flow or hyperbolic-secant flows are on the upper sides of the up-down and down-up upper-saddle and lower-saddle infinite-equilibrium. The infinite-equilibriums are for the

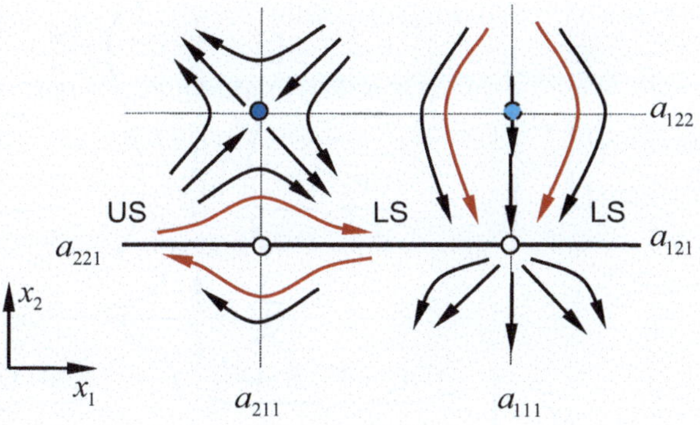

(c)

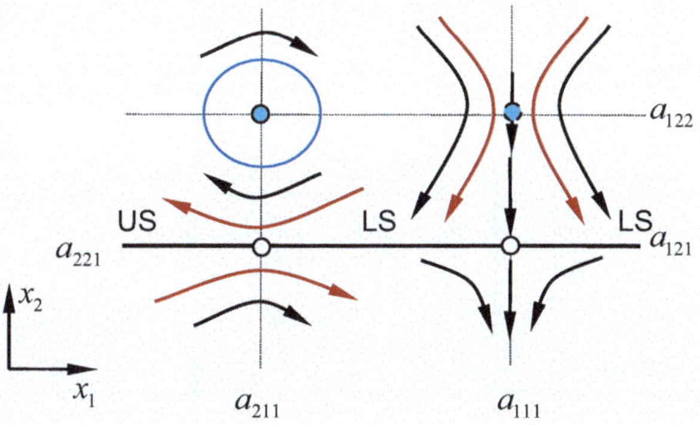

(d)

Fig. 4.4 (continued)

switching bifurcations of the saddle-sink and saddle-source with hyperbolic and hyperbolic-secant flows. The hyperbolic lower-to-upper saddle and upper-to-lower saddle infinite-equilibriums are for switching bifurcations of the positive and negative saddle with the hyperbolic-to-hyperbolic-secant flow and hyperbolic-secant-to-hyperbolic flows.

(ii) For $a_{221} = a_{122}$ and $a_{211} < a_{111}$, the corresponding first integral manifold is

$$\begin{aligned} &[(x_1 - a_{111}) - (x_{10} - a_{111})] + (a_{111} - a_{211}) \ln \frac{|x_1 - a_{111}|}{|x_{10} - a_{111}|} \\ &= \frac{a_{110}}{a_{220}} \left[(x_2 - x_{20}) + (a_{221} - a_{121}) \ln \frac{|x_2 - a_{221}|}{|x_{20} - a_{221}|} \right]. \end{aligned} \tag{5.21}$$

The equilibriums of $x_2^* = a_{121}, a_{122}, a_{221}$ with $x_1^* = a_{111}$ for $a_{122} = a_{221}$ are

$$\left\{\begin{array}{cc}(a_{122},a_{211}) & (a_{111},a_{122})\\ (a_{121},a_{211}) & (a_{111},a_{221})\end{array}\right\} = \left\{\begin{array}{cc}\underbrace{(\mathrm{UP}:\mathrm{DP},\mathrm{LS}:\mathrm{US})}_{\text{hyperbolic-secant lower-to-upper saddle}} & \underbrace{(\mathrm{DU}:\mathrm{UD},\mathrm{US})}_{\text{down-up-to-up-down upper-saddle}}\\ \underbrace{(\mathrm{UP}_+:\mathrm{UP}_+)}_{\text{positive saddle}} & \underbrace{(\mathrm{UP}:\mathrm{DP},\mathrm{pF})}_{\text{hyperbolic-secant flow}(+)}\end{array}\right\}$$

for $a_{110} > 0$ and $a_{220} > 0$; (4.22)

$$\left\{\begin{array}{cc}(a_{122},a_{211}) & (a_{111},a_{122})\\ (a_{121},a_{211}) & (a_{111},a_{221})\end{array}\right\} = \left\{\begin{array}{cc}\underbrace{(\mathrm{DP}:\mathrm{UP},\mathrm{LS}:\mathrm{US})}_{\text{hyperbolic lower-to-upper saddle}} & \underbrace{(\mathrm{UD}:\mathrm{DU},\mathrm{US})}_{\text{up-down-to-down-up upper-saddle}}\\ \underbrace{(\mathrm{DP}_+:\mathrm{DP}_-)}_{\text{CCW center}} & \underbrace{(\mathrm{DP}:\mathrm{UP},\mathrm{pF})}_{\text{hyperbolic flow }(+)}\end{array}\right\}$$

for $a_{110} < 0$ and $a_{220} > 0$; (4.23)

$$\left\{\begin{array}{cc}(a_{122},a_{211}) & (a_{111},a_{122})\\ (a_{121},a_{211}) & (a_{111},a_{221})\end{array}\right\} = \left\{\begin{array}{cc}\underbrace{(\mathrm{DP}:\mathrm{UP},\mathrm{US}:\mathrm{LS})}_{\text{hyperbolic upper-to-lower saddle}} & \underbrace{(\mathrm{UD}:\mathrm{DU},\mathrm{LS})}_{\text{up-down-to-down-up lower-saddle}}\\ \underbrace{(\mathrm{DP}_-:\mathrm{DP}_+)}_{\text{CW center}} & \underbrace{(\mathrm{DP}:\mathrm{UP},\mathrm{nF})}_{\text{hyperbolic flow }(-)}\end{array}\right\}$$

for $a_{110} > 0$ and $a_{220} < 0$; (4.24)

$$\left\{\begin{array}{cc}(a_{122},a_{211}) & (a_{111},a_{122})\\ (a_{121},a_{211}) & (a_{111},a_{221})\end{array}\right\} = \left\{\begin{array}{cc}\underbrace{(\mathrm{UP}:\mathrm{DP},\mathrm{US}:\mathrm{LS})}_{\text{hyperbolic-secant upper-to-lower saddle}} & \underbrace{(\mathrm{DU}:\mathrm{UD},\mathrm{LS})}_{\text{down-up-to-up-down lower-saddle}}\\ \underbrace{(\mathrm{UP}_-:\mathrm{UP}_-)}_{\text{negative saddle}} & \underbrace{(\mathrm{UP}:\mathrm{DP},\mathrm{nF})}_{\text{hyperbolic-secant flow }(-)}\end{array}\right\}$$

for $a_{110} < 0$ and $a_{220} < 0$. (4.25)

The infinite-equilibriums of $x_2^* = a_{122} = a_{221}$ for $a_{211} < a_{111}$ are

$$\begin{aligned}(\bar{x}_1, a_{122}) &= \underbrace{(\mathrm{UD},\mathrm{LS})}_{\text{up-down lower-saddle}} \quad \text{if } \bar{x} \in (-\infty, a_{211}),\\ (\bar{x}_1, a_{122}) &= \underbrace{(\mathrm{DU},\mathrm{US})}_{\text{down-up upper-saddle}} \quad \text{if } \bar{x} \in (a_{211}, a_{111})\\ (\bar{x}_1, a_{122}) &= \underbrace{(\mathrm{UD},\mathrm{US})}_{\text{up-down upper-saddle}} \quad \text{if } \bar{x} \in (a_{111}, \infty),\end{aligned} \tag{4.26}$$

for $a_{110} > 0$ and $a_{220} > 0$.

$$
\begin{aligned}
(\bar{x}_1, a_{122}) &= \underbrace{(\mathrm{DU}, \mathrm{LS})}_{\text{down-up lower-saddle}} \quad \text{if } \bar{x} \in (-\infty, a_{211}),\\
(\bar{x}_1, a_{122}) &= \underbrace{(\mathrm{UD}, \mathrm{US})}_{\text{up-down upper-saddle}} \quad \text{if } \bar{x} \in (a_{211}, a_{111}),\\
(\bar{x}_1, a_{122}) &= \underbrace{(\mathrm{DU}, \mathrm{US})}_{\text{down-up upper-saddle}} \quad \text{if } \bar{x} \in (a_{111}, \infty),\\
&\text{for } a_{110} < 0 \text{ and } a_{220} > 0;
\end{aligned} \tag{4.27}
$$

$$
\begin{aligned}
(\bar{x}_1, a_{122}) &= \underbrace{(\mathrm{DU}, \mathrm{US})}_{\text{down-up upper-saddle}} \quad \text{if } \bar{x} \in (-\infty, a_{211}),\\
(\bar{x}_1, a_{122}) &= \underbrace{(\mathrm{UD}, \mathrm{LS})}_{\text{up-down lower-saddle}} \quad \text{if } \bar{x} \in (a_{211}, a_{111})\\
(\bar{x}_1, a_{122}) &= \underbrace{(\mathrm{DU}, \mathrm{LS})}_{\text{down-up lower-saddle}} \quad \text{if } \bar{x} \in (a_{111}, \infty),\\
&\text{for } a_{110} > 0 \text{ and } a_{220} < 0;
\end{aligned} \tag{4.28}
$$

$$
\begin{aligned}
(\bar{x}_1, a_{122}) &= \underbrace{(\mathrm{UD}, \mathrm{US})}_{\text{up-down upper-saddle}} \quad \text{if } \bar{x} \in (-\infty, a_{211}),\\
(\bar{x}_1, a_{122}) &= \underbrace{(\mathrm{DU}, \mathrm{LS})}_{\text{down-up lower-saddle}} \quad \text{if } \bar{x} \in (a_{211}, a_{111})\\
(\bar{x}_1, a_{122}) &= \underbrace{(\mathrm{UD}, \mathrm{LS})}_{\text{up-down lower-saddle}} \quad \text{if } \bar{x} \in (a_{111}, \infty),\\
&\text{for } a_{110} < 0 \text{ and } a_{220} < 0.
\end{aligned} \tag{4.29}
$$

The corresponding infinite-equilibriums are sketched in Fig. 4.5. The hyperbolic flow or hyperbolic-secant flows are on the lower sides of the down-up and up-down upper-saddle and lower-saddle infinite-equilibriums. The infinite-equilibriums are for the switching bifurcations of the saddle-sink and saddle-source with hyperbolic and hyperbolic-secant flows. The hyperbolic lower-to-upper saddle and upper-to-lower saddle infinite-equilibriums are switching bifurcations for the positive and negative saddle with the hyperbolic-to-hyperbolic-secant flow and hyperbolic-secant-to-hyperbolic flows, respectively.

The readers can complete the analysis for the properties of equilibriums for $a_{221} = a_{121}$ and $a_{221} = a_{122}$ with $a_{211} > a_{111}$. The corresponding infinite-equilibriums and switching bifurcations can be discussed. The phase portraits can be sketched.

4.2.2 *Inflection-Sink and Source Infinite-Equilibriums*

Consider a dynamical system for $a_{111} = a_{211}$ as

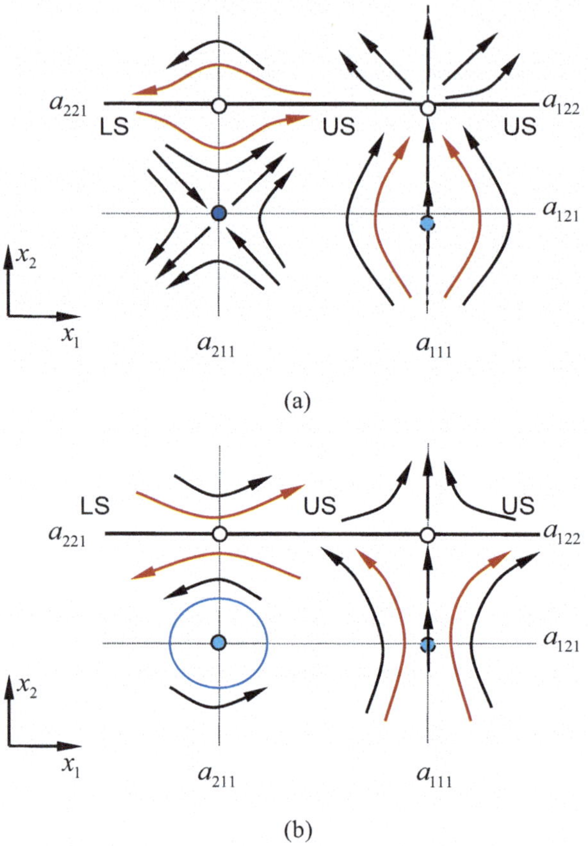

Fig. 4.5 Phase portraits ($a_{122} = a_{221}$ and $a_{211} < a_{111}$) for two-dimensional systems on the x_1-direction with $x_1^* = a_{111}$ and on the x_2-direction with $x_2^* = a_{121}, a_{122}, a_{221}, a_{222}$. The equilibrium flows: (**a**) ($a_{110} > 0$, $a_{220} > 0$), (**b**) ($a_{110} < 0$, $a_{220} > 0$), (**c**) ($a_{110} > 0$, $a_{220} < 0$), and (**d**) ($a_{110} < 0$, $a_{220} < 0$)

$$\begin{aligned} \dot{x}_1 &= a_{110}(x_1 - a_{111})(x_2 - a_{121})(x_2 - a_{122}), \\ \dot{x}_2 &= a_{220}(x_1 - a_{211})(x_2 - a_{221})^2, \end{aligned} \tag{4.30}$$

and the corresponding first integral manifold for $a_{111} = a_{211}$ is

$$\begin{aligned} &[(x_1 - a_{111}) - (x_{10} - a_{111})] \\ &= \frac{a_{110}}{a_{220}} \Bigg[(x_2 - x_{20}) + (2a_{221} - a_{121} - a_{122}) \ln \frac{|x_2 - a_{221}|}{|x_{20} - a_{221}|} \\ &\quad - (a_{221} - a_{121})(a_{221} - a_{122})\Big(\frac{1}{x_2 - a_{221}} - \frac{1}{x_{20} - a_{221}}\Big) \Bigg]. \end{aligned} \tag{4.31}$$

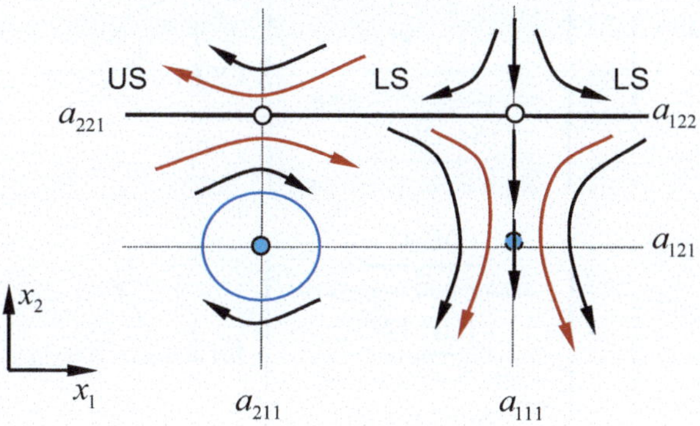

(c)

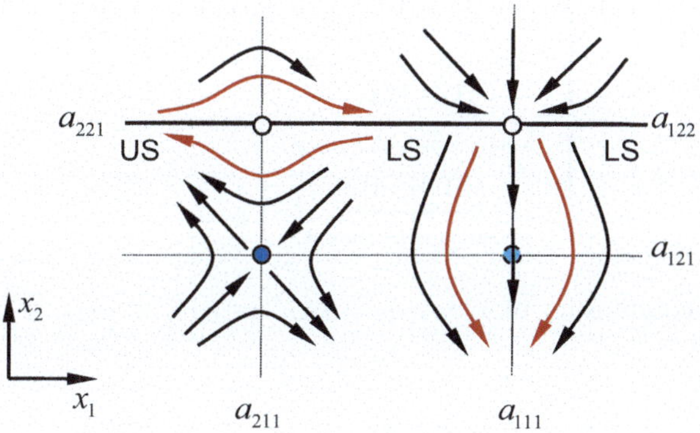

(d)

Fig. 4.5 (continued)

(i) The equilibriums of $x_2^* = a_{121}, a_{122}, a_{221}$ and $x_1^* = a_{111} = a_{211}$ for $a_{221} < a_{121}$ are

$$\begin{Bmatrix} (a_{122}, a_{211}) \\ (a_{121}, a_{211}) \\ (a_{111}, a_{221}) \end{Bmatrix} = \begin{Bmatrix} \underbrace{(\mathrm{UP}, \mathrm{US})}_{\text{up-parabola upper-saddle}} \\ \underbrace{(\mathrm{DP}, \mathrm{LS})}_{\text{down-parabola lower-saddle}} \\ \underbrace{(\mathrm{SO}, \mathrm{II} : \mathrm{II})}_{\text{increasing-inflection source}} \end{Bmatrix} \text{ for } a_{110} > 0, a_{220} > 0, \tag{4.32}$$

$$\left\{\begin{array}{l}(a_{122}, a_{211})\\(a_{121}, a_{211})\\(a_{111}, a_{221})\end{array}\right\} = \left\{\begin{array}{c}\underbrace{(\mathrm{DP}, \mathrm{LS})}_{\text{down-parabola lower-saddle}}\\\underbrace{(\mathrm{UP}, \mathrm{US})}_{\text{up-parabola upper-saddle}}\\\underbrace{(\mathrm{SI}, \mathrm{DI}:\mathrm{DI})}_{\text{decreasing-inflection sink}}\end{array}\right\} \text{ for } a_{110} < 0, a_{220} > 0, \tag{4.33}$$

$$\left\{\begin{array}{l}(a_{122}, a_{211})\\(a_{121}, a_{211})\\(a_{111}, a_{221})\end{array}\right\} = \left\{\begin{array}{c}\underbrace{(\mathrm{DP}, \mathrm{US})}_{\text{down-parabola upper-saddle}}\\\underbrace{(\mathrm{UP}, \mathrm{LS})}_{\text{up-parabola lower-saddle}}\\\underbrace{(\mathrm{SO}, \mathrm{DI}:\mathrm{DI})}_{\text{decreasing-inflection source}}\end{array}\right\} \text{ for } a_{110} > 0, a_{220} < 0, \tag{4.34}$$

$$\left\{\begin{array}{l}(a_{122}, a_{211})\\(a_{221}, a_{211})\\(a_{111}, a_{221})\end{array}\right\} = \left\{\begin{array}{c}\underbrace{(\mathrm{UP}, \mathrm{LS})}_{\text{up-parabola lower-saddle}}\\\underbrace{(\mathrm{DP}, \mathrm{US})}_{\text{down-parabola upper-saddle}}\\\underbrace{(\mathrm{SI}, \mathrm{II}:\mathrm{II})}_{\text{increasing-inflection sink}}\end{array}\right\} \text{ for } a_{110} < 0, a_{220} < 0. \tag{4.35}$$

The infinite-equilibrium of $x_1^* = a_{111} = a_{211}$ for $a_{211} < a_{121}$ is summarized as follows:

$$\begin{aligned}
(a_{111}, \bar{x}_2) &= \underbrace{(\mathrm{SO}, \mathrm{II})}_{\text{increasing-inflection source}} && \text{if } \bar{x}_2 \in (-\infty, a_{221}),\\
(a_{111}, \bar{x}_2) &= \underbrace{(\mathrm{SO}, \mathrm{II})}_{\text{increasing-inflection source}} && \text{if } \bar{x}_2 \in (a_{221}, a_{121}),\\
(a_{111}, \bar{x}_2) &= \underbrace{(\mathrm{SI}, \mathrm{DI})}_{\text{decreasing-inflection sink}} && \text{if } \bar{x}_2 \in (a_{121}, a_{122}),\\
(a_{111}, \bar{x}_2) &= \underbrace{(\mathrm{SO}, \mathrm{II})}_{\text{increasing-inflection source}} && \text{if } \bar{x}_2 \in (a_{122}, \infty),\\
&\text{for } a_{110} > 0, a_{220} > 0;
\end{aligned} \tag{4.36}$$

$$\begin{aligned}
(a_{111}, \bar{x}_2) &= \underbrace{(\mathrm{SI}, \mathrm{DI})}_{\text{decreasing-inflection sink}} && \text{if } \bar{x}_2 \in (-\infty, a_{221}),\\
(a_{111}, \bar{x}_2) &= \underbrace{(\mathrm{SI}, \mathrm{DI})}_{\text{decreasing-inflection sink}} && \text{if } \bar{x}_2 \in (a_{221}, a_{121}),\\
(a_{111}, \bar{x}_2) &= \underbrace{(\mathrm{SO}, \mathrm{II})}_{\text{increasing-inflection source}} && \text{if } \bar{x}_2 \in (a_{121}, a_{122}),\\
(a_{111}, \bar{x}_2) &= \underbrace{(\mathrm{SI}, \mathrm{DI})}_{\text{decreasing-inflection sink}} && \text{if } \bar{x}_2 \in (a_{122}, \infty),\\
&\text{for } a_{110} < 0, a_{220} > 0;
\end{aligned} \tag{4.37}$$

$$\begin{aligned}
&(a_{111},\bar{x}_2) = \underbrace{(\text{SO, DI})}_{\text{decreasing-inflection source}} && \text{if } \bar{x}_2 \in (-\infty, a_{221}),\\
&(a_{111},\bar{x}_2) = \underbrace{(\text{SO, DI})}_{\text{decreasing-inflection source}} && \text{if } \bar{x}_2 \in (a_{221}, a_{121}),\\
&(a_{111},\bar{x}_2) = \underbrace{(\text{SI, II})}_{\text{increasing-inflection sink}} && \text{if } \bar{x}_2 \in (a_{121}, a_{122}),\\
&(a_{111},\bar{x}_2) = \underbrace{(\text{SO, DI})}_{\text{decreasing-inflection source}} && \text{if } \bar{x}_2 \in (a_{122}, \infty),\\
&\text{for } a_{110} > 0, a_{220} < 0;
\end{aligned} \tag{4.38}$$

$$\begin{aligned}
&(a_{111},\bar{x}_2) = \underbrace{(\text{SI, II})}_{\text{increasing-inflection sink}} && \text{if } \bar{x}_2 \in (-\infty, a_{221}),\\
&(a_{111},\bar{x}_2) = \underbrace{(\text{SO, DI})}_{\text{decreasing-inflection source}} && \text{if } \bar{x}_2 \in (a_{221}, a_{121}),\\
&(a_{111},\bar{x}_2) = \underbrace{(\text{SI, II})}_{\text{increasing-inflection sink}} && \text{if } \bar{x}_2 \in (a_{121}, a_{122}),\\
&(a_{111},\bar{x}_2) = \underbrace{(\text{SI, II})}_{\text{increasing-inflection sink}} && \text{if } \bar{x}_2 \in (a_{122}, \infty),\\
&\text{for } a_{110} < 0, a_{220} < 0.
\end{aligned} \tag{4.39}$$

The infinite-equilibrium of $x_1^* = a_{111} = a_{211}$ with $a_{122} < a_{221}$ are for increasing-inflection and decreasing-inflection source and sink, as shown in Fig. 4.6. The infinite-equilibrium is for the switching bifurcation for the equilibriums for $a_{111} < a_{211}$ and $a_{211} < a_{111}$ with $a_{122} < a_{221}$.

(ii) The infinite-equilibriums of $x_1^* = a_{111} = a_{211}$ for $a_{221} < a_{121}$ have the same properties as for $x_1^* = a_{111} = a_{211}$ with $a_{121} < a_{221}$. The equilibrium locations of $x_2^* = a_{121}, a_{122}, a_{221}$ and $x_1^* = a_{111} = a_{211}$ are different. The corresponding phase portraits are presented in Fig. 4.7.

(iii) The equilibriums of $x_2^* = a_{121}, a_{122}, a_{221}$ and $x_1^* = a_{111} = a_{211}$ for $a_{221} \in (a_{121}, a_{122})$ are

$$\left\{\begin{array}{c}(a_{122}, a_{211})\\(a_{111}, a_{221})\\(a_{121}, a_{211})\end{array}\right\} = \left\{\begin{array}{c}\underbrace{(\text{UP, US})}_{\text{up-parabola upper-saddle}}\\ \underbrace{(\text{SI, DI : DI})}_{\text{decreasing-inflection sink}}\\ \underbrace{(\text{DP, LS})}_{\text{down-parabola lower-saddle}}\end{array}\right\} \text{ for } a_{110} > 0, a_{220} > 0, \tag{4.40}$$

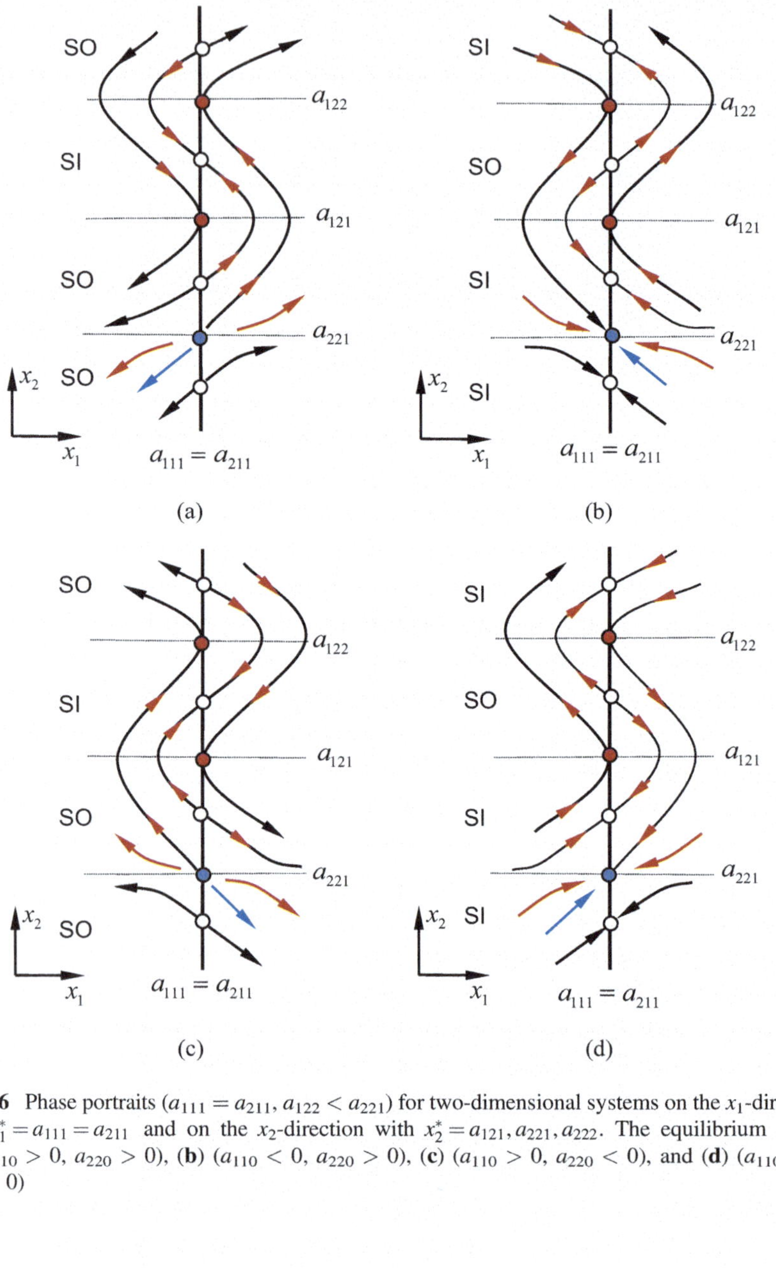

Fig. 4.6 Phase portraits ($a_{111} = a_{211}, a_{122} < a_{221}$) for two-dimensional systems on the x_1-direction with $x_1^* = a_{111} = a_{211}$ and on the x_2-direction with $x_2^* = a_{121}, a_{221}, a_{222}$. The equilibrium flows: **(a)** ($a_{110} > 0$, $a_{220} > 0$), **(b)** ($a_{110} < 0$, $a_{220} > 0$), **(c)** ($a_{110} > 0$, $a_{220} < 0$), and **(d)** ($a_{110} < 0$, $a_{220} < 0$)

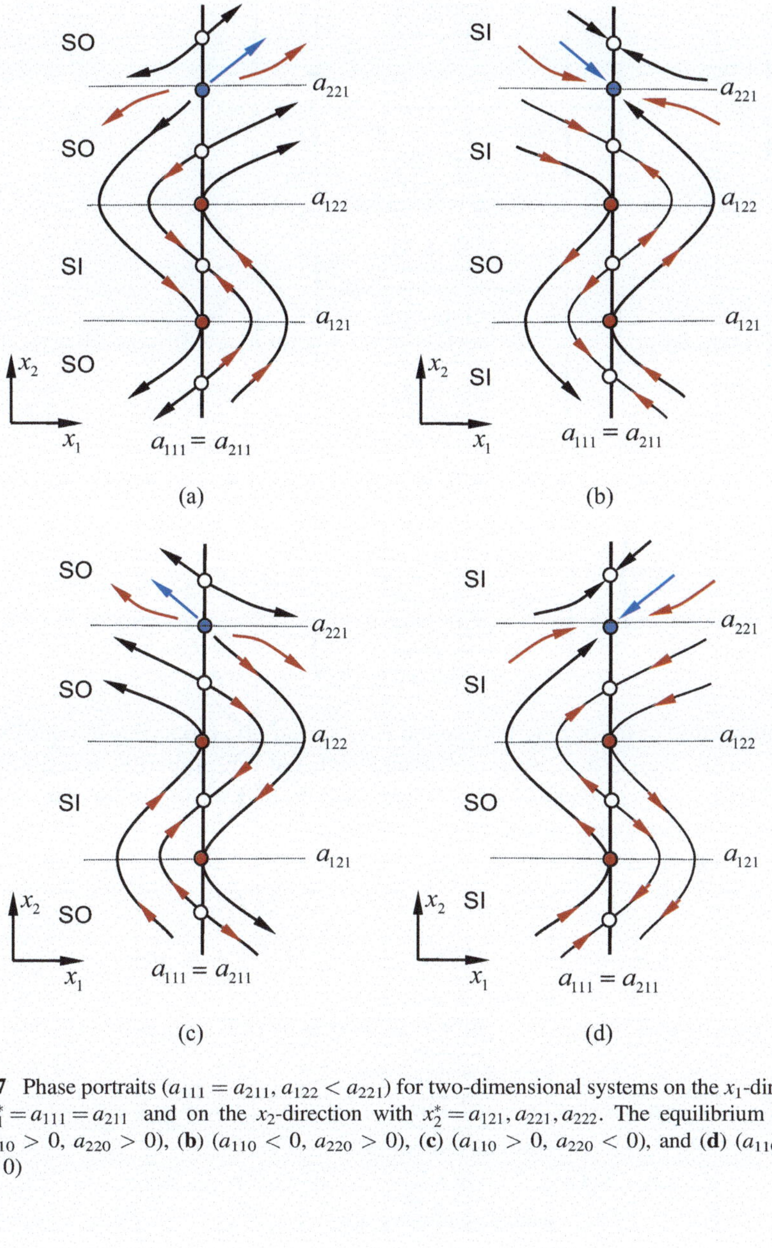

Fig. 4.7 Phase portraits ($a_{111} = a_{211}$, $a_{122} < a_{221}$) for two-dimensional systems on the x_1-direction with $x_1^* = a_{111} = a_{211}$ and on the x_2-direction with $x_2^* = a_{121}, a_{221}, a_{222}$. The equilibrium flows: (**a**) ($a_{110} > 0$, $a_{220} > 0$), (**b**) ($a_{110} < 0$, $a_{220} > 0$), (**c**) ($a_{110} > 0$, $a_{220} < 0$), and (**d**) ($a_{110} < 0$, $a_{220} < 0$)

$$\left\{\begin{array}{l}(a_{122},a_{211})\\(a_{111},a_{221})\\(a_{121},a_{211})\end{array}\right\}=\left\{\begin{array}{c}\underbrace{(\mathrm{DP},\mathrm{LS})}_{\text{down-parabola lower-saddle}}\\\underbrace{(\mathrm{SO},\mathrm{II}:\mathrm{II})}_{\text{increasing-inflection source}}\\\underbrace{(\mathrm{UP},\mathrm{US})}_{\text{up-parabola upper-saddle}}\end{array}\right\}\text{ for }a_{110}<0,a_{220}>0,\tag{4.41}$$

$$\left\{\begin{array}{l}(a_{122},a_{211})\\(a_{111},a_{221})\\(a_{121},a_{211})\end{array}\right\}=\left\{\begin{array}{c}\underbrace{(\mathrm{DP},\mathrm{US})}_{\text{down-parabola upper-saddle}}\\\underbrace{(\mathrm{SI},\mathrm{II}:\mathrm{II})}_{\text{increasing-inflection sink}}\\\underbrace{(\mathrm{UP},\mathrm{LS})}_{\text{up-parabola lower-saddle}}\end{array}\right\}\text{ for }a_{110}>0,a_{220}<0,\tag{4.42}$$

$$\left\{\begin{array}{l}(a_{122},a_{211})\\(a_{111},a_{221})\\(a_{121},a_{211})\end{array}\right\}=\left\{\begin{array}{c}\underbrace{(\mathrm{UP},\mathrm{LS})}_{\text{up-parabola lower-saddle}}\\\underbrace{(\mathrm{SO},\mathrm{DI}:\mathrm{DI})}_{\text{decreasing-inflection source}}\\\underbrace{(\mathrm{DP},\mathrm{US})}_{\text{down-parabola upper-saddle}}\end{array}\right\}\text{ for }a_{110}<0,a_{220}<0.\tag{4.43}$$

Compared to cases of $a_{221} < a_{122}$ and $a_{221} < a_{121}$, the properties of equilibriums $(x_1^*, x_2^*) = (a_{111}, a_{121}), (a_{111}, a_{122})$ are unchanged, but the properties of equilibrium $(x_1^*, x_2^*) = (a_{111}, a_{221})$ change from increasing to decreasing-inflection or from decreasing to increasing-inflection.

The infinite-equilibrium of $x_2^* = a_{121} = a_{221}$ for $a_{121} \in (a_{221}, a_{222})$ is summarized as follows:

$$\begin{aligned}&(a_{111},\bar{x}_2)=\underbrace{(\mathrm{SO},\mathrm{II})}_{\text{increasing-inflection source}}\quad\text{if }\bar{x}_2\in(-\infty,a_{121}),\\&(a_{111},\bar{x}_2)=\underbrace{(\mathrm{SI},\mathrm{DI})}_{\text{decreasing-inflection sink}}\quad\text{if }\bar{x}_2\in(a_{121},a_{221}),\\&(a_{111},\bar{x}_2)=\underbrace{(\mathrm{SI},\mathrm{DI})}_{\text{decreasing-inflection sink}}\quad\text{if }\bar{x}_2\in(a_{221},a_{122}),\\&(a_{111},\bar{x}_2)=\underbrace{(\mathrm{SO},\mathrm{II})}_{\text{increasing-inflection source}}\quad\text{if }\bar{x}_2\in(a_{122},\infty),\\&\text{for }a_{110}>0,a_{220}>0;\end{aligned}\tag{4.44}$$

$$
\begin{aligned}
&(a_{111}, \bar{x}_2) = \underbrace{(\text{SI}, \text{DI})}_{\text{decreasing-inflection sink}} &&\text{if } \bar{x}_2 \in (-\infty, a_{121}),\\
&(a_{111}, \bar{x}_2) = \underbrace{(\text{SO}, \text{II})}_{\text{increasing-inflection source}} &&\text{if } \bar{x}_2 \in (a_{121}, a_{221}),\\
&(a_{111}, \bar{x}_2) = \underbrace{(\text{SO}, \text{II})}_{\text{increasing-inflection source}} &&\text{if } \bar{x}_2 \in (a_{221}, a_{122}),\\
&(a_{111}, \bar{x}_2) = \underbrace{(\text{SI}, \text{DI})}_{\text{decreasing-inflection sink}} &&\text{if } \bar{x}_2 \in (a_{122}, \infty),\\
&\text{for } a_{110} < 0, a_{220} > 0;
\end{aligned} \tag{4.45}
$$

$$
\begin{aligned}
&(a_{111}, \bar{x}_2) = \underbrace{(\text{SO}, \text{DI})}_{\text{decreasing-inflection source}} &&\text{if } \bar{x}_2 \in (-\infty, a_{121}),\\
&(a_{111}, \bar{x}_2) = \underbrace{(\text{SI}, \text{II})}_{\text{increasing-inflection sink}} &&\text{if } \bar{x}_2 \in (a_{121}, a_{221}),\\
&(a_{111}, \bar{x}_2) = \underbrace{(\text{SI}, \text{II})}_{\text{increasing-inflection sink}} &&\text{if } \bar{x}_2 \in (a_{221}, a_{122}),\\
&(a_{111}, \bar{x}_2) = \underbrace{(\text{SO}, \text{DI})}_{\text{decreasing-inflection source}} &&\text{if } \bar{x}_2 \in (a_{122}, \infty),\\
&\text{for } a_{110} > 0, a_{220} < 0;
\end{aligned} \tag{4.46}
$$

$$
\begin{aligned}
&(a_{111}, \bar{x}_2) = \underbrace{(\text{SI}, \text{II})}_{\text{increasing-inflection sink}} &&\text{if } \bar{x}_2 \in (-\infty, a_{121}),\\
&(a_{111}, \bar{x}_2) = \underbrace{(\text{SO}, \text{DI})}_{\text{decreasing-inflection source}} &&\text{if } \bar{x}_2 \in (a_{121}, a_{221}),\\
&(a_{111}, \bar{x}_2) = \underbrace{(\text{SO}, \text{DI})}_{\text{decreasing-inflection source}} &&\text{if } \bar{x}_2 \in (a_{221}, a_{122}),\\
&(a_{111}, \bar{x}_2) = \underbrace{(\text{SI}, \text{II})}_{\text{increasing-inflection sink}} &&\text{if } \bar{x}_2 \in (a_{122}, \infty),\\
&\text{for } a_{110} < 0, a_{220} < 0.
\end{aligned} \tag{4.47}
$$

The infinite-equilibriums of $x_1^* = a_{111} = a_{211}$ with $a_{121} \in (a_{221}, a_{222})$ are for increasing-inflection and decreasing-inflection source and sink, as shown in Fig. 4.8. The infinite-equilibrium is for the switching bifurcation for the equilibriums for $a_{111} < a_{211}$ and $a_{211} < a_{111}$ with $a_{121} \in (a_{221}, a_{222})$. The infinite-equilibriums of $(x_1^*, x_2^*) = (a_{111}, a_{221})$ are changed from the increasing to decreasing-inflection or from decreasing to increasing-inflection.

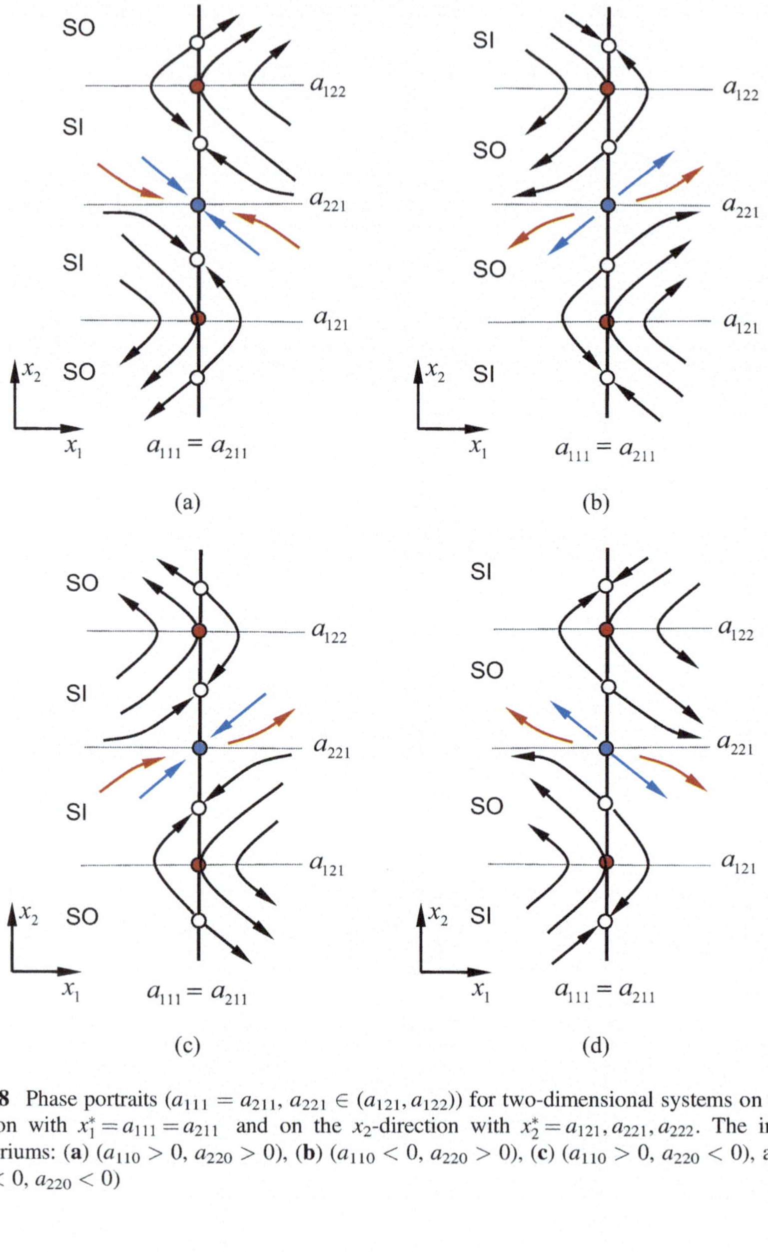

Fig. 4.8 Phase portraits ($a_{111} = a_{211}$, $a_{221} \in (a_{121}, a_{122})$) for two-dimensional systems on the x_1-direction with $x_1^* = a_{111} = a_{211}$ and on the x_2-direction with $x_2^* = a_{121}, a_{221}, a_{222}$. The infinite-equilibriums: (**a**) ($a_{110} > 0$, $a_{220} > 0$), (**b**) ($a_{110} < 0$, $a_{220} > 0$), (**c**) ($a_{110} > 0$, $a_{220} < 0$), and (**d**) ($a_{110} < 0$, $a_{220} < 0$)

4.2.3 *Inflection-Source (Sink) and Up-Down (Down-Up)-Saddle*

Consider a dynamical system for $a_{111} = a_{211}$ and $a_{221} = a_{121}, a_{122}$ as

$$\begin{aligned}\dot{x}_1 &= a_{110}(x_1 - a_{111})(x_2 - a_{121})(x_2 - a_{122}),\\ \dot{x}_2 &= a_{220}(x_1 - a_{211})(x_2 - a_{221})^2,\end{aligned} \tag{4.48}$$

and the corresponding first integral manifold for $a_{111} = a_{211}$ and $a_{221} = a_{12s_1}$ $(s_1, s_2 \in \{1,2\}, s_1 \neq s_2)$ is

$$[(x_1 - a_{111}) - (x_{10} - a_{111})] = \frac{a_{110}}{a_{220}}\left[(x_2 - x_{20}) + (a_{221} - a_{12s_1})\ln\frac{|x_2 - a_{22s_2}|}{|x_{20} - a_{22s_2}|}\right]. \tag{4.49}$$

(i) The equilibriums of $x_2^* = a_{221}, a_{222}, a_{121}$ and $x_1^* = a_{111} = a_{211}$ for $a_{121} = a_{221}$ are

$$\left\{\begin{matrix}(a_{122}, a_{211})\\(a_{111}, a_{121})\end{matrix}\right\} = \left\{\begin{matrix}\underbrace{(\text{UP}, \text{US})}_{\text{up-parabola upper-saddle}}\\ \underbrace{({}_{\text{LS:US}}\text{UD}, {}_{\text{II}}\text{SO:}{}_{\text{DI}}\text{SI})}_{\text{up-down inflection source-to-sink}}\end{matrix}\right\} \text{ for } a_{110} > 0, a_{220} > 0, \tag{4.50}$$

$$\left\{\begin{matrix}(a_{122}, a_{211})\\(a_{111}, a_{121})\end{matrix}\right\} = \left\{\begin{matrix}\underbrace{(\text{DP}, \text{LS})}_{\text{down-parabola lower-saddle}}\\ \underbrace{({}_{\text{LS:US}}\text{DU}, {}_{\text{DI}}\text{SI:}{}_{\text{II}}\text{SO})}_{\text{down-up inflection sink-to-source}}\end{matrix}\right\} \text{ for } a_{110} < 0, a_{220} > 0, \tag{4.51}$$

$$\left\{\begin{matrix}(a_{122}, a_{211})\\(a_{111}, a_{121})\end{matrix}\right\} = \left\{\begin{matrix}\underbrace{(\text{DP}, \text{US})}_{\text{down-parabola upper-saddle}}\\ \underbrace{({}_{\text{US:LS}}\text{DU}, {}_{\text{DI}}\text{SO:}{}_{\text{II}}\text{SI})}_{\text{down-up inflection source-to-sink}}\end{matrix}\right\} \text{ for } a_{110} > 0, a_{220} < 0, \tag{4.52}$$

$$\left\{\begin{matrix}(a_{122}, a_{211})\\(a_{111}, a_{121})\end{matrix}\right\} = \left\{\begin{matrix}\underbrace{(\text{UP}, \text{LS})}_{\text{up-parabola lower-saddle}}\\ \underbrace{({}_{\text{US:LS}}\text{UD}, {}_{\text{II}}\text{SI:}{}_{\text{DI}}\text{SO})}_{\text{up-down inflection sink-to-source}}\end{matrix}\right\} \text{ for } a_{110} < 0, a_{220} < 0. \tag{4.53}$$

The infinite-equilibrium of $x_1^* = a_{111} = a_{211}$ for $a_{121} = a_{221}$ is summarized as follows:

$$
\begin{aligned}
&(a_{111},\bar{x}_2)=\underbrace{(\text{SO, II})}_{\text{increasing-inflection source}} \quad \text{if } \bar{x}_2\in(-\infty,a_{121}),\\
&(a_{111},\bar{x}_2)=\underbrace{(\text{SI, DI})}_{\text{decreasing-inflection sink}} \quad \text{if } \bar{x}_2\in(a_{121},a_{122}),\\
&(a_{111},\bar{x}_2)=\underbrace{(\text{SO, II})}_{\text{increasing-inflection source}} \quad \text{if } \bar{x}_2\in(a_{122},\infty);\\
&\text{for } a_{110}>0, a_{220}>0;
\end{aligned}
\tag{4.54}
$$

$$
\begin{aligned}
&(a_{111},\bar{x}_2)=\underbrace{(\text{SI, DI})}_{\text{decreasing-inflection sink}} \quad \text{if } \bar{x}_2\in(-\infty,a_{121}),\\
&(a_{111},\bar{x}_2)=\underbrace{(\text{SO, II})}_{\text{increasing-inflection source}} \quad \text{if } \bar{x}_2\in(a_{121},a_{122}),\\
&(a_{111},\bar{x}_2)=\underbrace{(\text{SI, DI})}_{\text{decreasing-inflection sink}} \quad \text{if } \bar{x}_2\in(a_{122},\infty);\\
&\text{for } a_{110}<0, a_{220}>0;
\end{aligned}
\tag{4.55}
$$

$$
\begin{aligned}
&(a_{111},\bar{x}_2)=\underbrace{(\text{SO, DI})}_{\text{decreasing-inflection source}} \quad \text{if } \bar{x}_2\in(-\infty,a_{121}),\\
&(a_{111},\bar{x}_2)=\underbrace{(\text{SI, II})}_{\text{increasing-inflection sink}} \quad \text{if } \bar{x}_2\in(a_{121},a_{122}),\\
&(a_{111},\bar{x}_2)=\underbrace{(\text{SO, DI})}_{\text{decreasing-inflection source}} \quad \text{if } \bar{x}_2\in(a_{122},\infty);\\
&\text{for } a_{110}>0, a_{220}<0;
\end{aligned}
\tag{4.56}
$$

$$
\begin{aligned}
&(a_{111},\bar{x}_2)=\underbrace{(\text{SI, II})}_{\text{increasing-inflection sink}} \quad \text{if } \bar{x}_2\in(-\infty,a_{121}),\\
&(a_{111},\bar{x}_2)=\underbrace{(\text{SO, DI})}_{\text{decreasing-inflection source}} \quad \text{if } \bar{x}_2\in(a_{121},a_{122}),\\
&(a_{111},\bar{x}_2)=\underbrace{(\text{SI, II})}_{\text{increasing-inflection sink}} \quad \text{if } \bar{x}_2\in(a_{122},\infty);\\
&\text{for } a_{110}<0, a_{220}<0.
\end{aligned}
\tag{4.57}
$$

The infinite-equilibrium of $x_2^*=a_{121}=a_{221}$ for $a_{111}=a_{211}$ is summarized as follows:

$$
\begin{aligned}
&(\bar{x}_1,a_{221})=\underbrace{(\text{UD, LS})}_{\text{up-down lower-saddle}} \quad \text{if } \bar{x}_1\in(-\infty,a_{111}),\\
&(\bar{x}_1,a_{221})=\underbrace{(\text{UD, US})}_{\text{up-down upper-saddle}} \quad \text{if } \bar{x}_1\in(a_{111},\infty);\\
&\text{for } a_{110}>0, a_{220}>0;
\end{aligned}
\tag{4.58}
$$

$$(\bar{x}_1, a_{221}) = \underbrace{(\text{DU}, \text{LS})}_{\text{down-up lower-saddle}} \quad \text{if } \bar{x}_1 \in (-\infty, a_{111}),$$

$$(\bar{x}_1, a_{221}) = \underbrace{(\text{DU}, \text{US})}_{\text{down-up upper-saddle}} \quad \text{if } \bar{x}_1 \in (a_{111}, \infty); \tag{4.59}$$

for $a_{110} < 0, a_{220} > 0$;

$$(\bar{x}_1, a_{221}) = \underbrace{(\text{DU}, \text{US})}_{\text{down-up upper-saddle}} \quad \text{if } \bar{x}_1 \in (-\infty, a_{111}),$$

$$(\bar{x}_1, a_{221}) = \underbrace{(\text{DU}, \text{LS})}_{\text{down-up lower-saddle}} \quad \text{if } \bar{x}_1 \in (a_{111}, \infty); \tag{4.60}$$

for $a_{110} > 0, a_{220} < 0$;

$$(\bar{x}_1, a_{221}) = \underbrace{(\text{UD}, \text{US})}_{\text{up-down upper-saddle}} \quad \text{if } \bar{x}_1 \in (-\infty, a_{111}),$$

$$(\bar{x}_1, a_{221}) = \underbrace{(\text{UD}, \text{LS})}_{\text{up-down lower-saddle}} \quad \text{if } \bar{x}_1 \in (a_{111}, \infty); \tag{4.61}$$

for $a_{110} < 0, a_{220} < 0$.

The infinite-equilibriums of $x_1^* = a_{111} = a_{211}$ and $x_2^* = a_{121} = a_{221}$ are for increasing-inflection and decreasing-inflection source and sink, as shown in Fig. 4.9. The infinite-equilibrium is the switching bifurcation of equilibriums:

- for $a_{111} < a_{211}$ and $a_{211} < a_{111}$ with $a_{121} = a_{221}$;
- for $a_{121} < a_{221}$ and $a_{221} < a_{121}$ with $a_{111} = a_{211}$;
- for $a_{111} < a_{211}$ and $a_{211} < a_{111}$ with $a_{121} < a_{221}$ and $a_{221} < a_{121}$.

(ii) Similarly, the equilibriums of $x_2^* = a_{221}, a_{222}, a_{121}$ and $x_1^* = a_{111} = a_{211}$ for $a_{221} = a_{122}$ are

$$\left\{ \begin{array}{l} (a_{111}, a_{122}) \\ (a_{121}, a_{211}) \end{array} \right\} = \left\{ \begin{array}{c} \underbrace{({}_{\text{LS:US}}\text{DU}, {}_{\text{DI}}\text{SI:}{}_{\text{II}}\text{SO})}_{\text{down-up inflection sink-to-source}} \\ \underbrace{(\text{DP}, \text{LS})}_{\text{down-parabola lower-saddle}} \end{array} \right\} \text{for } a_{110} > 0, a_{220} > 0, \tag{4.62}$$

$$\left\{ \begin{array}{l} (a_{111}, a_{122}) \\ (a_{121}, a_{211}) \end{array} \right\} = \left\{ \begin{array}{c} \underbrace{({}_{\text{LS:US}}\text{UD}, {}_{\text{II}}\text{SO:}{}_{\text{DI}}\text{SI})}_{\text{up-down inflection source-to-sink}} \\ \underbrace{(\text{UP}, \text{US})}_{\text{up-parabola upper-saddle}} \end{array} \right\} \text{for } a_{110} < 0, a_{220} > 0, \tag{4.63}$$

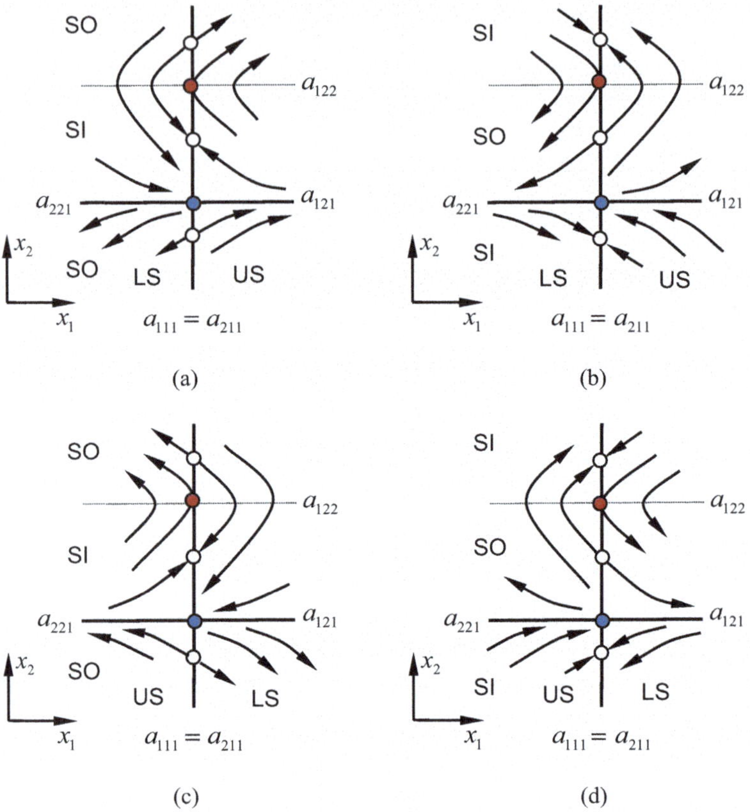

Fig. 4.9 Phase portraits ($a_{111} = a_{211}, a_{121} = a_{221}$) for two-dimensional systems on the x_1-direction with $x_1^* = a_{111} = a_{211}$ and on the x_2-direction with $x_2^* = a_{121}, a_{221}, a_{222}$. The equilibrium flows: (**a**) ($a_{110} > 0, a_{220} > 0$), (**b**) ($a_{110} < 0, a_{220} > 0$), (**c**) ($a_{110} > 0, a_{220} < 0$), and (**d**) ($a_{110} < 0, a_{220} < 0$)

$$\left\{\begin{array}{l}(a_{111}, a_{122}) \\ (a_{121}, a_{211})\end{array}\right\} = \left\{\begin{array}{c}\underbrace{({}_{\text{US:LS}}\text{UD}, {}_{\text{II}}\text{SI:}{}_{\text{DI}}\text{SO})}_{\text{up-down inflection sink-to-source}} \\ \underbrace{(\text{UP}, \text{LS})}_{\text{up-parabola lower-saddle}}\end{array}\right\} \text{for } a_{110} > 0, a_{220} < 0, \quad (4.64)$$

$$\left\{\begin{array}{l}(a_{111}, a_{122}) \\ (a_{121}, a_{211})\end{array}\right\} = \left\{\begin{array}{c}\underbrace{({}_{\text{US:LS}}\text{DU}, {}_{\text{DI}}\text{SO:}{}_{\text{II}}\text{SI})}_{\text{down-up inflection source-to-sink}} \\ \underbrace{(\text{DP}, \text{US})}_{\text{down-parabola upper-saddle}}\end{array}\right\} \text{for } a_{110} < 0, a_{220} < 0. \quad (4.65)$$

The infinite-equilibrium of $x_1^* = a_{111} = a_{211}$ for $a_{121} = a_{221}$ is summarized as follows:

$$
\left.
\begin{aligned}
&(a_{111}, \bar{x}_2) = \underbrace{(\mathrm{SO}, \mathrm{II})}_{\text{increasing-inflection source}} && \text{if } \bar{x}_2 \in (-\infty, a_{121}),\\
&(a_{111}, \bar{x}_2) = \underbrace{(\mathrm{SI}, \mathrm{DI})}_{\text{decreasing-inflection sink}} && \text{if } \bar{x}_2 \in (a_{121}, a_{122}),\\
&(a_{111}, \bar{x}_2) = \underbrace{(\mathrm{SO}, \mathrm{II})}_{\text{increasing-inflection source}} && \text{if } \bar{x}_2 \in (a_{122}, \infty);\\
&\text{for } a_{110} > 0, a_{220} > 0;
\end{aligned}
\right\} \tag{4.66}
$$

$$
\left.
\begin{aligned}
&(a_{111}, \bar{x}_2) = \underbrace{(\mathrm{SI}, \mathrm{DI})}_{\text{decreasing-inflection sink}} && \text{if } \bar{x}_2 \in (-\infty, a_{121}),\\
&(a_{111}, \bar{x}_2) = \underbrace{(\mathrm{SO}, \mathrm{II})}_{\text{increasing-inflection source}} && \text{if } \bar{x}_2 \in (a_{121}, a_{122}),\\
&(a_{111}, \bar{x}_2) = \underbrace{(\mathrm{SI}, \mathrm{DI})}_{\text{decreasing-inflection sink}} && \text{if } \bar{x}_2 \in (a_{122}, \infty);\\
&\text{for } a_{110} < 0, a_{220} > 0;
\end{aligned}
\right\} \tag{4.67}
$$

$$
\left.
\begin{aligned}
&(a_{111}, \bar{x}_2) = \underbrace{(\mathrm{SO}, \mathrm{DI})}_{\text{decreasing-inflection source}} && \text{if } \bar{x}_2 \in (-\infty, a_{121}),\\
&(a_{111}, \bar{x}_2) = \underbrace{(\mathrm{SI}, \mathrm{II})}_{\text{increasing-inflection sink}} && \text{if } \bar{x}_2 \in (a_{121}, a_{122}),\\
&(a_{111}, \bar{x}_2) = \underbrace{(\mathrm{SO}, \mathrm{DI})}_{\text{decreasing-inflection source}} && \text{if } \bar{x}_2 \in (a_{122}, \infty);\\
&\text{for } a_{110} > 0, a_{220} < 0;
\end{aligned}
\right\} \tag{4.68}
$$

$$
\left.
\begin{aligned}
&(a_{111}, \bar{x}_2) = \underbrace{(\mathrm{SI}, \mathrm{II})}_{\text{increasing-inflection sink}} && \text{if } \bar{x}_2 \in (-\infty, a_{121}),\\
&(a_{111}, \bar{x}_2) = \underbrace{(\mathrm{SO}, \mathrm{DI})}_{\text{decreasing-inflection source}} && \text{if } \bar{x}_2 \in (a_{121}, a_{122}),\\
&(a_{111}, \bar{x}_2) = \underbrace{(\mathrm{SI}, \mathrm{II})}_{\text{increasing-inflection sink}} && \text{if } \bar{x}_2 \in (a_{122}, \infty);\\
&\text{for } a_{110} < 0, a_{220} < 0.
\end{aligned}
\right\} \tag{4.69}
$$

The infinite-equilibrium of $x_2^* = a_{121} = a_{221}$ for $a_{111} = a_{211}$ is summarized as follows:

$$
\begin{aligned}
&(\bar{x}_1, a_{221}) = \underbrace{(\mathrm{DU}, \mathrm{LS})}_{\text{down-up lower-saddle}} \quad \text{if } \bar{x}_1 \in (-\infty, a_{111}),\\
&(\bar{x}_1, a_{221}) = \underbrace{(\mathrm{DU}, \mathrm{US})}_{\text{down-up upper-saddle}} \quad \text{if } \bar{x}_1 \in (a_{111}, \infty);\\
&\text{for } a_{110} > 0, a_{220} > 0;
\end{aligned}
\tag{4.70}
$$

$$
\begin{aligned}
&(\bar{x}_1, a_{221}) = \underbrace{(\mathrm{UD}, \mathrm{LS})}_{\text{up-down lower-saddle}} \quad \text{if } \bar{x}_1 \in (-\infty, a_{111}),\\
&(\bar{x}_1, a_{221}) = \underbrace{(\mathrm{UD}, \mathrm{US})}_{\text{up-down upper-saddle}} \quad \text{if } \bar{x}_1 \in (a_{111}, \infty);\\
&\text{for } a_{110} < 0, a_{220} > 0;
\end{aligned}
\tag{4.71}
$$

$$
\begin{aligned}
&(\bar{x}_1, a_{221}) = \underbrace{(\mathrm{UD}, \mathrm{US})}_{\text{up-down upper-saddle}} \quad \text{if } \bar{x}_1 \in (-\infty, a_{111}),\\
&(\bar{x}_1, a_{221}) = \underbrace{(\mathrm{UD}, \mathrm{LS})}_{\text{up-down lower-saddle}} \quad \text{if } \bar{x}_1 \in (a_{111}, \infty);\\
&\text{for } a_{110} > 0, a_{220} < 0;
\end{aligned}
\tag{4.72}
$$

$$
\begin{aligned}
&(\bar{x}_1, a_{221}) = \underbrace{(\mathrm{DU}, \mathrm{US})}_{\text{down-up upper-saddle}} \quad \text{if } \bar{x}_1 \in (-\infty, a_{111}),\\
&(\bar{x}_1, a_{221}) = \underbrace{(\mathrm{DU}, \mathrm{LS})}_{\text{down-up lower-saddle}} \quad \text{if } \bar{x}_1 \in (a_{111}, \infty);\\
&\text{for } a_{110} < 0, a_{220} < 0.
\end{aligned}
\tag{4.73}
$$

The infinite-equilibriums of $x_1^* = a_{111} = a_{211}$ and $x_2^* = a_{122} = a_{221}$ are for increasing-inflection and decreasing inflection source and sink, as shown in Fig. 4.10. The infinite-equilibrium is the switching bifurcation of equilibriums:

- for $a_{111} < a_{211}$ and $a_{211} < a_{111}$ with $a_{122} = a_{221}$;
- for $a_{122} < a_{221}$ and $a_{221} < a_{122}$ with $a_{111} = a_{211}$;
- for $a_{111} < a_{211}$ and $a_{211} < a_{111}$ with $a_{122} < a_{221}$ and $a_{221} < a_{122}$.

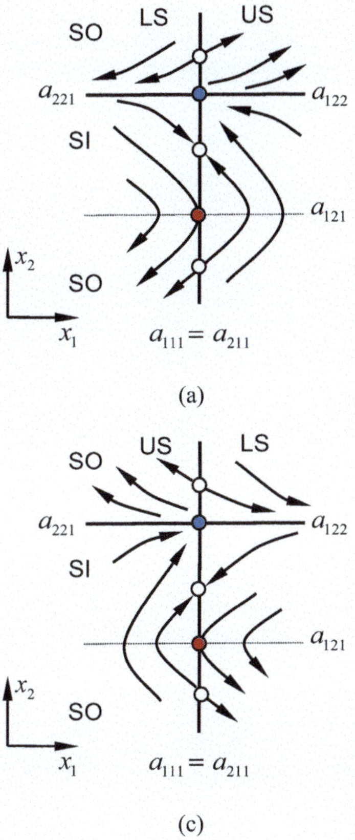

Fig. 4.10 Phase portraits ($a_{111} = a_{211}$, $a_{221} = a_{122}$) for two-dimensional systems on the x_1-direction with $x_1^* = a_{111} = a_{211}$ and on the x_2-direction with $x_2^* = a_{121}, a_{221}, a_{222}$. The equilibrium flows: **(a)** ($a_{110} > 0$, $a_{220} > 0$), **(b)** ($a_{110} < 0$, $a_{220} > 0$), **(c)** ($a_{110} > 0$, $a_{220} < 0$), and **(d)** ($a_{110} < 0$, $a_{220} < 0$)

Chapter 5
Equilibrium Matrices with Hyperbolic Flows

In this chapter, equilibrium matrices of saddle, center, saddle, source, and sink with hyperbolic flows are discussed, and the corresponding switching bifurcations are discussed. The inflection source (sink) infinite-equilibriums exist in the vertical and horizontal directions. The parabola-saddles are for a hyperbolic-secant flow with a paralleled saddle and for a hyperbolic-secant flow with a paralleled center. The inflection sources (sinks) are for a hyperbolic-secant flow with a connected source (sink) and for a hyperbolic flow with a connected saddle.

5.1 Equilibrium Matrices with Hyperbolic Flows

Consider a dynamical system as

$$\begin{aligned}\dot{x}_1 &= a_{110}(x_1 - a_{111})(x_2 - a_{121})(x_2 - a_{122}),\\ \dot{x}_2 &= a_{220}(x_2 - a_{111})(x_2 - a_{221})(x_2 - a_{222}),\end{aligned} \tag{5.1}$$

and the corresponding first integral manifold is

$$\begin{aligned}&[(x_1 - a_{111}) - (x_{10} - a_{111})] + (a_{111} - a_{211})\ln\frac{|x_1 - a_{111}|}{|x_{10} - a_{111}|}\\ &= \frac{a_{110}}{a_{220}}\left[(x_2 - x_{20}) + \frac{(a_{221} - a_{121})(a_{221} - a_{122})}{a_{221} - a_{222}}\ln\frac{|x_2 - a_{221}|}{|x_{20} - a_{221}|}\right.\\ &\quad\left. + \frac{(a_{222} - a_{121})(a_{222} - a_{121})}{a_{222} - a_{221}}\ln\frac{|x_2 - a_{222}|}{|x_{20} - a_{222}|}\right].\end{aligned} \tag{5.2}$$

(i) For the case of $a_{122} < a_{221}$ and $a_{211} < a_{111}$, the equilibrium matrices of $x_1^* = a_{111}, a_{211}$ with $x_2^* = a_{121}, a_{122}, a_{221}, a_{222}$ have the following properties, as shown in Fig. 5.1.

A. C. J. Luo, *Two-dimensional Two-product Cubic Systems, Vol. I*,
https://doi.org/10.1007/978-3-031-48487-2_5

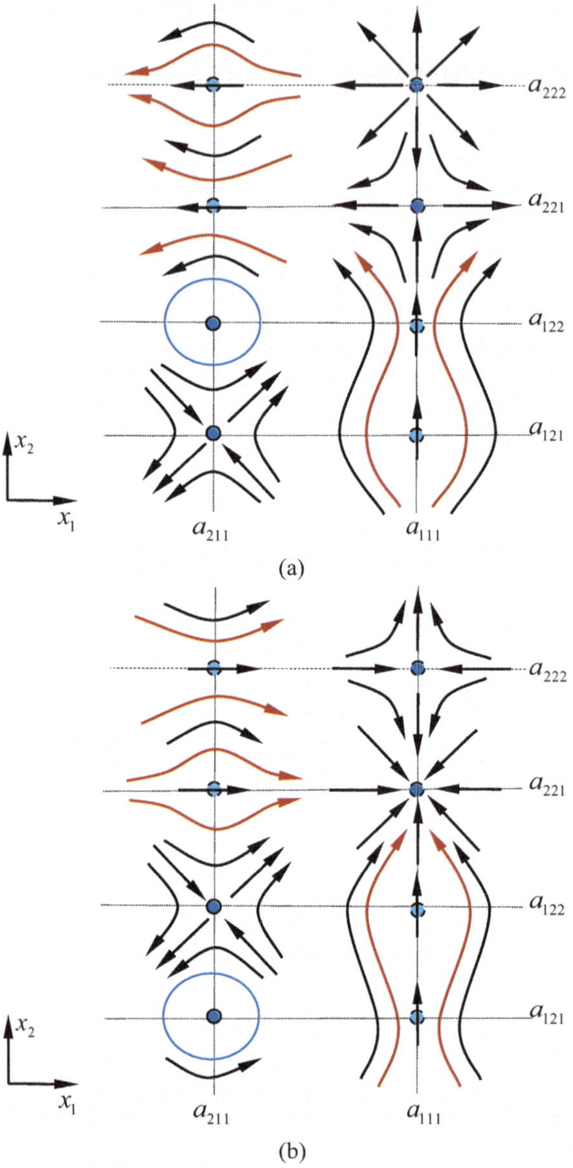

Fig. 5.1 Phase portraits ($a_{122} < a_{221}$ and $a_{211} < a_{111}$) for two-dimensional systems on the x_1-direction with $x_1^* = a_{111}, a_{211}$ and on the x_2-direction with $x_2^* = a_{121}, a_{122}, a_{221}, a_{222}$. (**a**) ($a_{110} > 0$, $a_{220} > 0$), (**b**) ($a_{110} < 0$, $a_{220} > 0$), (**c**) ($a_{110} > 0$, $a_{220} < 0$), and (**d**) ($a_{110} < 0$, $a_{220} < 0$)

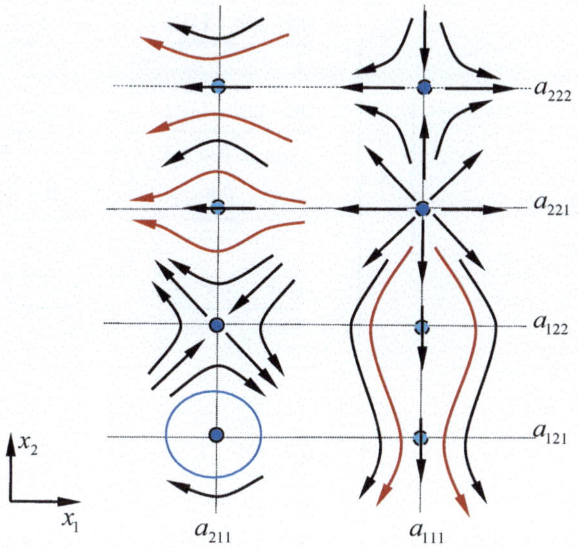

(c)

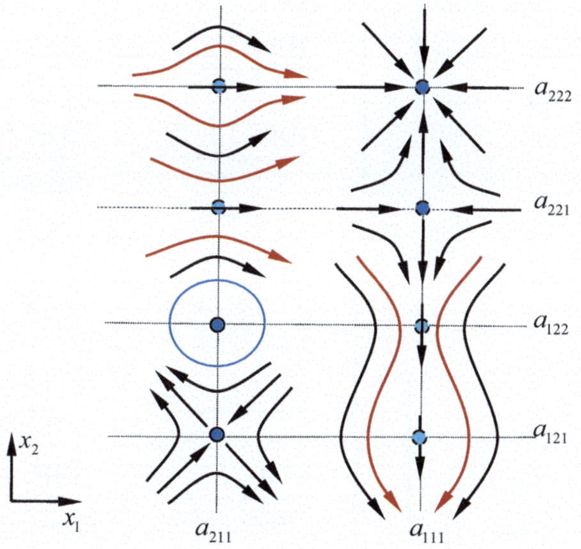

(d)

Fig. 5.1 (continued)

$$\left\{\begin{array}{ll}(a_{211},a_{222}) & (a_{111},a_{222})\\ (a_{211},a_{221}) & (a_{111},a_{221})\\ (a_{122},a_{211}) & (a_{111},a_{122})\\ (a_{121},a_{211}) & (a_{111},a_{121})\end{array}\right\} = \left\{\begin{array}{ll}\underbrace{(\mathrm{nF},\mathrm{UP}:\mathrm{DP})}_{\text{hyperbolic-secant flow }(-)} & \underbrace{(\mathrm{SO},\mathrm{SO})}_{\text{source}}\\ \underbrace{(\mathrm{nF},\mathrm{DP}:\mathrm{UP})}_{\text{hyperbolic flow }(-)} & \underbrace{(\mathrm{SO},\mathrm{SI})}_{\text{saddle}}\\ \underbrace{(\mathrm{DP}_+,\mathrm{DP}_-)}_{\text{CCW center}} & \underbrace{(\mathrm{DP}:\mathrm{UP},\mathrm{pF})}_{\text{hyperbolic flow }(+)}\\ \underbrace{(\mathrm{UP}_+,\mathrm{UP}_+)}_{\text{positive saddle}} & \underbrace{(\mathrm{UP}:\mathrm{DP},\mathrm{pF})}_{\text{hyperbolic-secant flow }(+)}\end{array}\right\}$$

for $a_{110} > 0$ and $a_{220} > 0$;

(5.3)

$$\left\{\begin{array}{ll}(a_{211},a_{222}) & (a_{111},a_{222})\\ (a_{211},a_{221}) & (a_{111},a_{221})\\ (a_{122},a_{211}) & (a_{111},a_{122})\\ (a_{121},a_{211}) & (a_{111},a_{121})\end{array}\right\} = \left\{\begin{array}{ll}\underbrace{(\mathrm{pF},\mathrm{DP}:\mathrm{UP})}_{\text{hyperbolic flow }(+)} & \underbrace{(\mathrm{SI},\mathrm{SO})}_{\text{saddle}}\\ \underbrace{(\mathrm{pF},\mathrm{UP}:\mathrm{DP})}_{\text{hyperbolic-secant flow }(+)} & \underbrace{(\mathrm{SI},\mathrm{SI})}_{\text{sink}}\\ \underbrace{(\mathrm{UP}_+,\mathrm{UP}_+)}_{\text{positive saddle}} & \underbrace{(\mathrm{UP}:\mathrm{DP},\mathrm{pF})}_{\text{hyperbolic-secant flow }(+)}\\ \underbrace{(\mathrm{DP}_+,\mathrm{DP}_-)}_{\text{CCW center}} & \underbrace{(\mathrm{DP}:\mathrm{UP},\mathrm{pF})}_{\text{hyperbolic flow }(+)}\end{array}\right\}$$

for $a_{110} < 0$ and $a_{220} > 0$;

(5.4)

$$\left\{\begin{array}{ll}(a_{211},a_{222}) & (a_{111},a_{222})\\ (a_{211},a_{221}) & (a_{111},a_{221})\\ (a_{122},a_{211}) & (a_{111},a_{122})\\ (a_{121},a_{211}) & (a_{111},a_{121})\end{array}\right\} = \left\{\begin{array}{ll}\underbrace{(\mathrm{nF},\mathrm{DP}:\mathrm{UP})}_{\text{hyperbolic flow }(-)} & \underbrace{(\mathrm{SO},\mathrm{SI})}_{\text{saddle}}\\ \underbrace{(\mathrm{nF},\mathrm{UP}:\mathrm{DP})}_{\text{hyperbolic-secant flow }(-)} & \underbrace{(\mathrm{SO},\mathrm{SO})}_{\text{source}}\\ \underbrace{(\mathrm{UP}_-,\mathrm{UP}_-)}_{\text{negative saddle}} & \underbrace{(\mathrm{UP}:\mathrm{DP},\mathrm{nF})}_{\text{hyperbolic-secant flow }(-)}\\ \underbrace{(\mathrm{DP}_-,\mathrm{DP}_+)}_{\text{CW center}} & \underbrace{(\mathrm{DP}:\mathrm{UP},\mathrm{nF})}_{\text{hyperbolic flow }(-)}\end{array}\right\}$$

for $a_{110} > 0$ and $a_{220} < 0$;

(5.5)

$$\begin{Bmatrix} (a_{211},a_{222}) & (a_{111},a_{222}) \\ (a_{211},a_{221}) & (a_{111},a_{221}) \\ (a_{122},a_{211}) & (a_{111},a_{122}) \\ (a_{121},a_{211}) & (a_{111},a_{121}) \end{Bmatrix} = \begin{Bmatrix} \underbrace{(\text{pF}, \text{UP}:\text{DP})}_{\text{hyperbolic-secant flow }(+)} & \underbrace{(\text{SI}, \text{SI})}_{\text{sink}} \\ \underbrace{(\text{pF}, \text{DP}:\text{UP})}_{\text{hyperbolic flow }(+)} & \underbrace{(\text{SI}, \text{SO})}_{\text{saddle}} \\ \underbrace{(\text{DP}_{-}, \text{DP}_{+})}_{\text{CW center}} & \underbrace{(\text{DP}:\text{UP}, \text{nF})}_{\text{hyperbolic flow }(-)} \\ \underbrace{(\text{UP}_{-}, \text{UP}_{-})}_{\text{negative saddle}} & \underbrace{(\text{UP}:\text{DP}, \text{nF})}_{\text{hyperbolic-secant flow }(-)} \end{Bmatrix}$$

for $a_{110} < 0$ and $a_{220} < 0$.

(5.6)

The saddle and source or sink with hyperbolic and hyperbolic-secant flows on the line of $x_1^* = a_{111}$ exist in the equilibrium matrix. The hyperbolic and hyperbolic-secant flows with directrix are connected together on the bottom. The hyperbolic and hyperbolic-secant flows with center and positive saddle on the line of $x_1^* = a_{211}$ exist in the equilibrium matrix, and hyperbolic and hyperbolic-secant flows are separated with two directrices.

(ii) For $a_{221} \in (a_{121}, a_{122})$ and $a_{122} < a_{222}$ with $a_{211} < a_{111}$, the equilibriums of $x_1^* = a_{111}, a_{211}$ with $x_2^* = a_{121}, a_{122}, a_{221}, a_{222}$ have the following properties, as shown in Fig. 5.2.

$$\begin{Bmatrix} (a_{211},a_{222}) & (a_{111},a_{222}) \\ (a_{122},a_{211}) & (a_{111},a_{122}) \\ (a_{211},a_{221}) & (a_{111},a_{221}) \\ (a_{121},a_{211}) & (a_{111},a_{121}) \end{Bmatrix} = \begin{Bmatrix} \underbrace{(\text{nF}, \text{UP}:\text{DP})}_{\text{hyperbolic-secant flow }(-)} & \underbrace{(\text{SO}, \text{SO})}_{\text{source}} \\ \underbrace{(\text{UP}_{-}, \text{UP}_{-})}_{\text{negative saddle}} & \underbrace{(\text{UP}:\text{DP}, \text{nF})}_{\text{hyperbolic-secant flow }(-)} \\ \underbrace{(\text{pF}, \text{UP}:\text{DP})}_{\text{hyperbolic-secant flow }(+)} & \underbrace{(\text{SI}, \text{SI})}_{\text{sink}} \\ \underbrace{(\text{UP}_{+}, \text{UP}_{+})}_{\text{positive saddle}} & \underbrace{(\text{UP}:\text{DP}, \text{pF})}_{\text{hyperbolic-secant flow }(+)} \end{Bmatrix}$$

for $a_{110} > 0$ and $a_{220} > 0$;

(5.7)

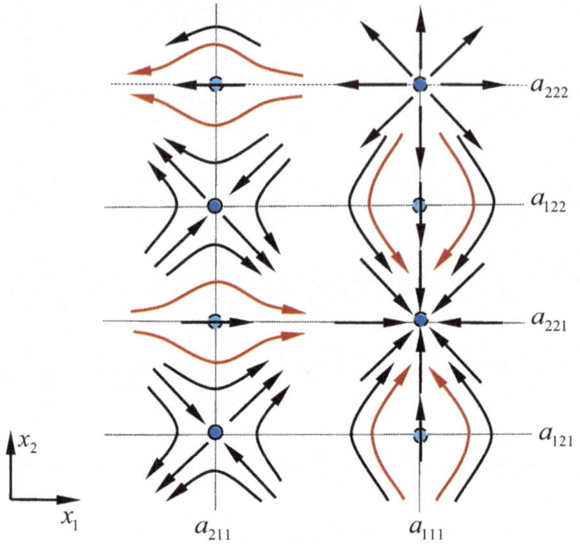

(a)

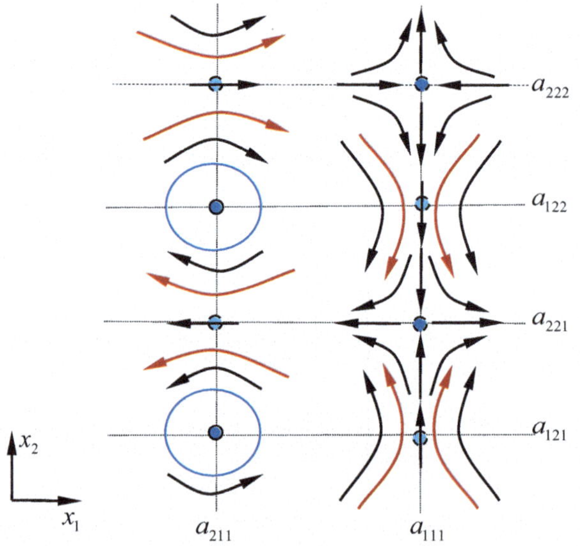

(b)

Fig. 5.2 Phase portraits ($a_{221} \in (a_{121}, a_{122})$ and $a_{211} < a_{111}$) for two-dimensional systems on the x_1-direction with $x_1^* = a_{111}, a_{211}$ and on the x_2-direction with $x_2^* = a_{121}, a_{122}, a_{221}, a_{222}$. **(a)** ($a_{110} > 0$, $a_{220} > 0$), **(b)** ($a_{110} < 0$, $a_{220} > 0$), **(c)** ($a_{110} > 0$, $a_{220} < 0$), and **(d)** ($a_{110} < 0$, $a_{220} < 0$)

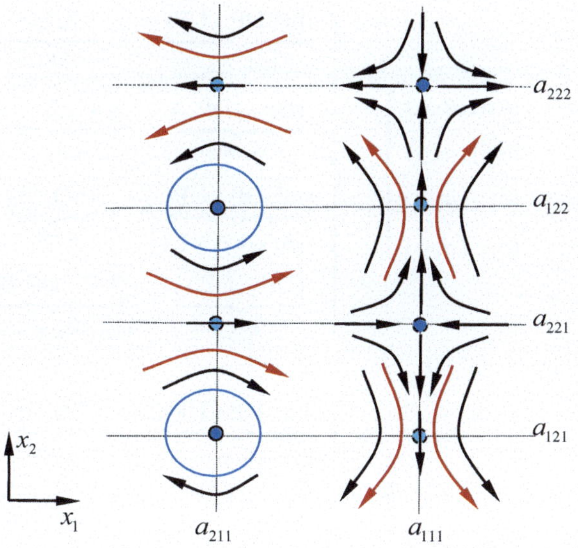

(c)

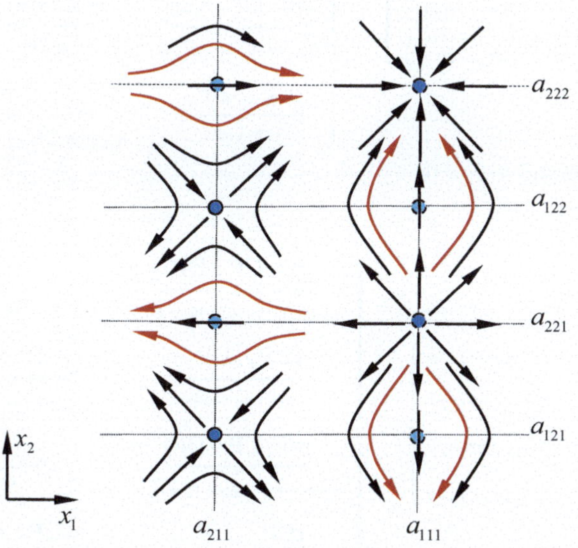

(d)

Fig. 5.2 (continued)

$$
\left\{\begin{array}{cc}
(a_{211},a_{222}) & (a_{111},a_{222})\\
(a_{122},a_{211}) & (a_{111},a_{122})\\
(a_{211},a_{221}) & (a_{111},a_{221})\\
(a_{121},a_{211}) & (a_{111},a_{121})
\end{array}\right\}
=\left\{\begin{array}{cc}
\underbrace{(\mathrm{pF},\mathrm{DP}:\mathrm{UP})}_{\text{hyperbolic flow }(+)} & \underbrace{(\mathrm{SI},\mathrm{SO})}_{\text{saddle}}\\
\underbrace{(\mathrm{DP}_{-},\mathrm{DP}_{+})}_{\text{CW center}} & \underbrace{(\mathrm{DP}:\mathrm{UP},\mathrm{nF})}_{\text{hyperbolic flow }(-)}\\
\underbrace{(\mathrm{nF},\mathrm{DP}:\mathrm{UP})}_{\text{hyperbolic flow }(-)} & \underbrace{(\mathrm{SO},\mathrm{SI})}_{\text{saddle}}\\
\underbrace{(\mathrm{DP}_{+},\mathrm{DP}_{-})}_{\text{CCW center}} & \underbrace{(\mathrm{DP}:\mathrm{UP},\mathrm{pF})}_{\text{hyperbolic flow }(+)}
\end{array}\right\}
\tag{5.8}
$$

for $a_{110}<0$ and $a_{220}>0$;

$$
\left\{\begin{array}{cc}
(a_{211},a_{222}) & (a_{111},a_{222})\\
(a_{122},a_{211}) & (a_{111},a_{122})\\
(a_{211},a_{221}) & (a_{111},a_{221})\\
(a_{121},a_{211}) & (a_{111},a_{121})
\end{array}\right\}
=\left\{\begin{array}{cc}
\underbrace{(\mathrm{nF},\mathrm{DP}:\mathrm{UP})}_{\text{hyperbolic flow }(-)} & \underbrace{(\mathrm{SO},\mathrm{SI})}_{\text{saddle}}\\
\underbrace{(\mathrm{DP}_{+},\mathrm{DP}_{-})}_{\text{CCW center}} & \underbrace{(\mathrm{DP}:\mathrm{UP},\mathrm{pF})}_{\text{hyperbolic flow }(+)}\\
\underbrace{(\mathrm{pF},\mathrm{DP}:\mathrm{UP})}_{\text{hyperbolic flow }(+)} & \underbrace{(\mathrm{SI},\mathrm{SO})}_{\text{saddle}}\\
\underbrace{(\mathrm{DP}_{-},\mathrm{DP}_{+})}_{\text{CW center}} & \underbrace{(\mathrm{DP}:\mathrm{UP},\mathrm{nF})}_{\text{hyperbolic flow }(-)}
\end{array}\right\}
\tag{5.9}
$$

for $a_{110}>0$ and $a_{220}<0$;

$$
\left\{\begin{array}{cc}
(a_{211},a_{222}) & (a_{111},a_{222})\\
(a_{122},a_{211}) & (a_{111},a_{122})\\
(a_{211},a_{221}) & (a_{111},a_{221})\\
(a_{121},a_{211}) & (a_{111},a_{121})
\end{array}\right\}
=\left\{\begin{array}{cc}
\underbrace{(\mathrm{pF},\mathrm{UP}:\mathrm{DP})}_{\text{hyperbolic-secant flow }(+)} & \underbrace{(\mathrm{SI},\mathrm{SI})}_{\text{sink}}\\
\underbrace{(\mathrm{UP}_{+},\mathrm{UP}_{+})}_{\text{positive saddle}} & \underbrace{(\mathrm{UP}:\mathrm{DP},\mathrm{pF})}_{\text{hyperbolic-secant flow }(+)}\\
\underbrace{(\mathrm{nF},\mathrm{UP}:\mathrm{DP})}_{\text{hyperbolic-secant flow }(-)} & \underbrace{(\mathrm{SO},\mathrm{SO})}_{\text{source}}\\
\underbrace{(\mathrm{UP}_{-},\mathrm{UP}_{-})}_{\text{negative saddle}} & \underbrace{(\mathrm{UP}:\mathrm{DP},\mathrm{nF})}_{\text{hyperbolic-secant flow }(-)}
\end{array}\right\}
$$

for $a_{110}<0$ and $a_{220}<0$.

(5.10)

The saddle and source or sink with hyperbolic and hyperbolic-secant flows exist in the line of $x_1^* = a_{111}$. The hyperbolic and hyperbolic-secant flows with directrix are separated through saddle or source and sink. The source and sink are separated by a hyperbolic-secant flow. The (SI,SO)-saddle and (SO,SI)-saddle are separated by a hyperbolic flow. The hyperbolic and hyperbolic-secant flows with centers and saddles on the line of $x_1^* = a_{211}$ exist in the equilibrium matrix. The positive or negative saddles are separated by the two hyperbolic-secant flows, and the counter-clockwise and clockwise centers are separated by the two hyperbolic flows.

(iii) For $a_{121}, a_{122} \in (a_{221}, a_{222})$ and $a_{211} < a_{111}$, the equilibriums of $x_1^* = a_{111}, a_{211}$ with $x_2^* = a_{121}, a_{122},\ a_{221}, a_{222}$ have the following properties, as shown in Fig. 5.3.

$$\begin{Bmatrix} (a_{211}, a_{222}) & (a_{111}, a_{222}) \\ (a_{122}, a_{211}) & (a_{111}, a_{122}) \\ (a_{121}, a_{211}) & (a_{111}, a_{121}) \\ (a_{211}, a_{221}) & (a_{111}, a_{221}) \end{Bmatrix} = \begin{Bmatrix} \underbrace{(\text{nF, UP : DP})}_{\text{hyperbolic-secant flow } (-)} & \underbrace{(\text{SO, SO})}_{\text{source}} \\ \underbrace{(\text{UP}_-, \text{UP}_-)}_{\text{negative saddle}} & \underbrace{(\text{UP : DP, nF})}_{\text{hyperbolic-secant flow } (-)} \\ \underbrace{(\text{DP}_-, \text{DP}_+)}_{\text{CW center}} & \underbrace{(\text{DP : UP, nF})}_{\text{hyperbolic flow } (-)} \\ \underbrace{(\text{nF, DP : UP})}_{\text{hyperbolic flow } (-)} & \underbrace{(\text{SI, SO})}_{\text{saddle}} \end{Bmatrix}$$

for $a_{110} > 0$ and $a_{220} > 0$;

(5.11)

$$\begin{Bmatrix} (a_{211}, a_{222}) & (a_{111}, a_{222}) \\ (a_{122}, a_{211}) & (a_{111}, a_{122}) \\ (a_{121}, a_{211}) & (a_{111}, a_{121}) \\ (a_{211}, a_{221}) & (a_{111}, a_{221}) \end{Bmatrix} = \begin{Bmatrix} \underbrace{(\text{pF, DP : UP})}_{\text{hyperbolic flow } (+)} & \underbrace{(\text{SI, SO})}_{\text{saddle}} \\ \underbrace{(\text{DP}_-, \text{DP}_+)}_{\text{CW center}} & \underbrace{(\text{DP : UP, nF})}_{\text{hyperbolic flow } (-)} \\ \underbrace{(\text{UP}_-, \text{UP}_-)}_{\text{negative saddle}} & \underbrace{(\text{UP : DP, nF})}_{\text{hyperbolic-secant flow } (-)} \\ \underbrace{(\text{pF, UP : DP})}_{\text{hyperbolic-secant flow } (+)} & \underbrace{(\text{SI, SI})}_{\text{sink}} \end{Bmatrix}$$

for $a_{110} < 0$ and $a_{220} > 0$;

(5.12)

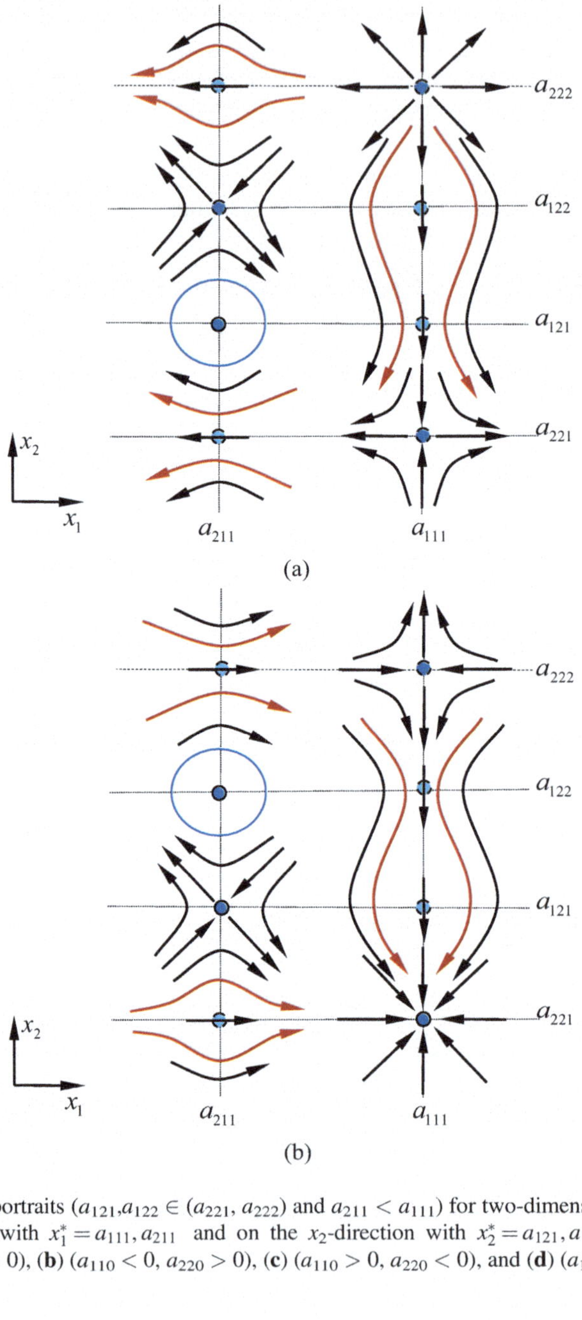

Fig. 5.3 Phase portraits (a_{121},$a_{122} \in (a_{221}, a_{222})$ and $a_{211} < a_{111}$) for two-dimensional systems on the x_1-direction with $x_1^* = a_{111}, a_{211}$ and on the x_2-direction with $x_2^* = a_{121}, a_{122}, a_{221}, a_{222}$. (**a**) ($a_{110} > 0$, $a_{220} > 0$), (**b**) ($a_{110} < 0$, $a_{220} > 0$), (**c**) ($a_{110} > 0$, $a_{220} < 0$), and (**d**) ($a_{110} < 0$, $a_{220} < 0$)

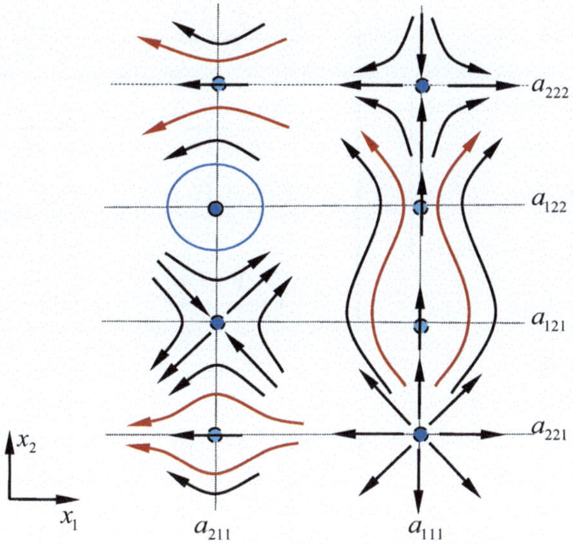

(c)

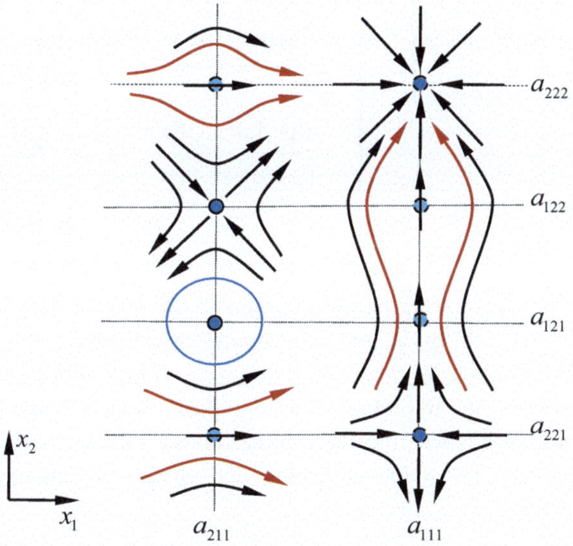

(d)

Fig. 5.3 (continued)

$$\left\{\begin{array}{ll}(a_{211},a_{222}) & (a_{111},a_{222})\\ (a_{122},a_{211}) & (a_{111},a_{122})\\ (a_{121},a_{211}) & (a_{111},a_{121})\\ (a_{211},a_{221}) & (a_{111},a_{221})\end{array}\right\} = \left\{\begin{array}{ll}\underbrace{(\text{nF},\text{DP}:\text{UP})}_{\text{hyperbolic flow }(-)} & \underbrace{(\text{SO},\text{SI})}_{\text{saddle}}\\ \underbrace{(\text{DP}_{+},\text{DP}_{-})}_{\text{CCW center}} & \underbrace{(\text{DP}:\text{UP},\text{pF})}_{\text{hyperbolic flow }(+)}\\ \underbrace{(\text{UP}_{+},\text{UP}_{+})}_{\text{positive saddle}} & \underbrace{(\text{UP}:\text{DP},\text{pF})}_{\text{hyperbolic-secant flow }(+)}\\ \underbrace{(\text{nF},\text{UP}:\text{DP})}_{\text{hyperbolic-secant flow }(-)} & \underbrace{(\text{SO},\text{SO})}_{\text{source}}\end{array}\right\}$$

for $a_{110} > 0$ and $a_{220} < 0$;

$$(5.13)$$

$$\left\{\begin{array}{ll}(a_{211},a_{222}) & (a_{111},a_{222})\\ (a_{122},a_{211}) & (a_{111},a_{122})\\ (a_{121},a_{211}) & (a_{111},a_{121})\\ (a_{211},a_{221}) & (a_{111},a_{221})\end{array}\right\} = \left\{\begin{array}{ll}\underbrace{(\text{pF},\text{UP}:\text{DP})}_{\text{hyperbolic-secant flow }(+)} & \underbrace{(\text{SI},\text{SI})}_{\text{sink}}\\ \underbrace{(\text{UP}_{+},\text{UP}_{+})}_{\text{positive saddle}} & \underbrace{(\text{UP}:\text{DP},\text{pF})}_{\text{hyperbolic-secant flow }(+)}\\ \underbrace{(\text{DP}_{+},\text{DP}_{-})}_{\text{CCW center}} & \underbrace{(\text{DP}:\text{UP},\text{pF})}_{\text{hyperbolic flow }(+)}\\ \underbrace{(\text{nF},\text{DP}:\text{UP})}_{\text{hyperbolic flow }(-)} & \underbrace{(\text{SI},\text{SO})}_{\text{saddle}}\end{array}\right\}$$

for $a_{110} < 0$ and $a_{220} < 0$.

$$(5.14)$$

The saddle and source or sink with hyperbolic and hyperbolic-secant flows exist on the line of $x_1^* = a_{111}$ in the equilibrium matrix. The hyperbolic and hyperbolic-secant flows with directrix are connected together. Source and saddle or sink and saddle are separated by the hyperbolic and hyperbolic-secant flows. The hyperbolic and hyperbolic-secant flows with centers and saddles on the line of $x_1^* = a_{211}$ exist in the equilibrium matrix. The saddle and center are separated by the hyperbolic-secant and hyperbolic flows.

(iv) For $a_{121} \in (a_{221}, a_{222})$ and $a_{222} < a_{122}$, the equilibriums of $x_1^* = a_{111}, a_{211}$ with $x_2^* = a_{121}, a_{122}, a_{221}, a_{222}$ have the following properties, as shown in Fig. 5.4.

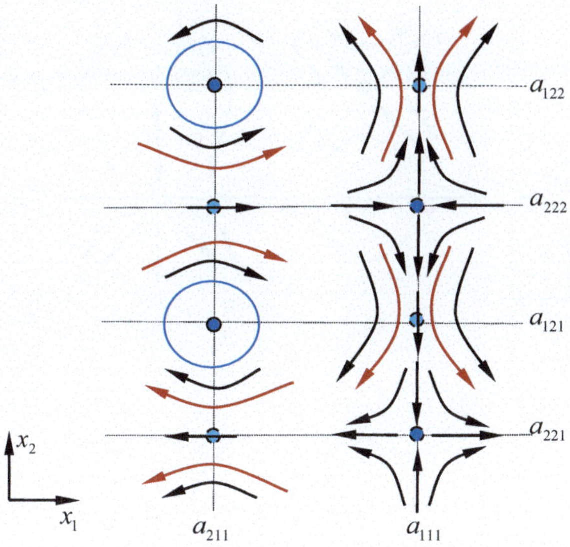

(a)

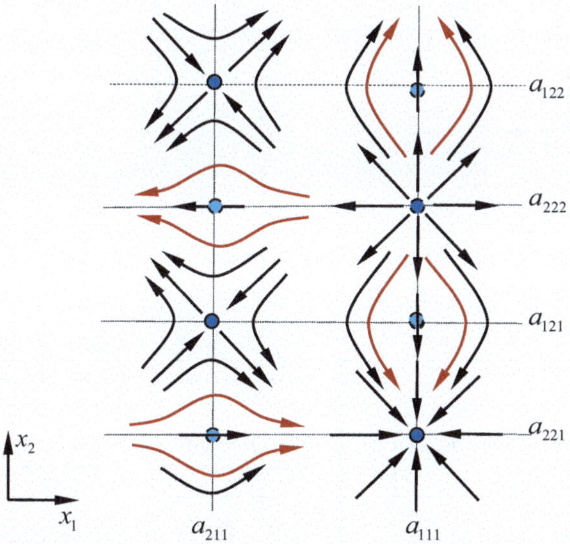

(b)

Fig. 5.4 Phase portraits ($a_{121} \in (a_{221}, a_{222})$, $a_{222} < a_{122}$) for two-dimensional systems on the x_1 direction with $x_1^* = a_{111}, a_{211}$ and on the x_2-direction with $x_2^* = a_{121}, a_{122}, a_{221}, a_{222}$. (**a**) ($a_{110} > 0$, $a_{220} > 0$), (**b**) ($a_{110} < 0$, $a_{220} > 0$), (**c**) ($a_{110} > 0$, $a_{220} < 0$), and (**d**) ($a_{110} < 0$, $a_{220} < 0$)

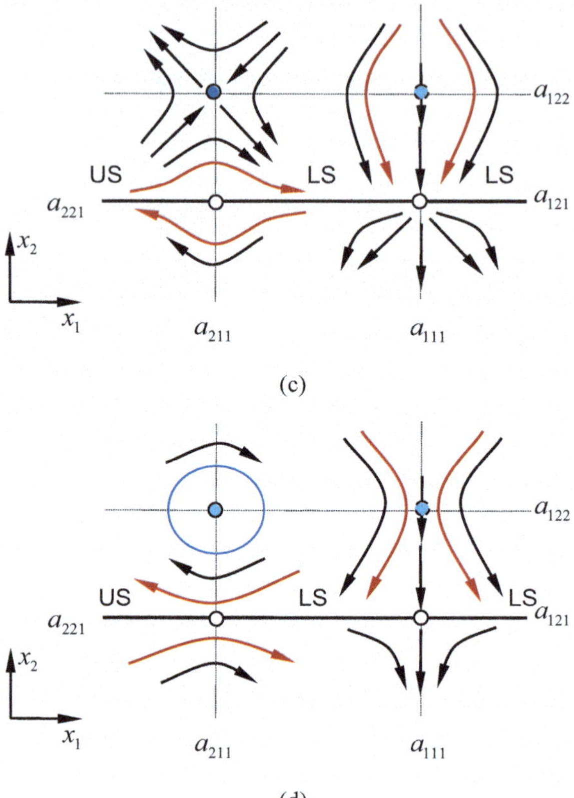

Fig. 5.4 (continued)

$$\left\{\begin{array}{cc}(a_{122},a_{211}) & (a_{111},a_{122})\\(a_{211},a_{222}) & (a_{111},a_{222})\\(a_{121},a_{211}) & (a_{111},a_{121})\\(a_{211},a_{221}) & (a_{111},a_{221})\end{array}\right\}=\left\{\begin{array}{cc}\underbrace{(\mathrm{DP}_+,\mathrm{DP}_-)}_{\text{CCW center}} & \underbrace{(\mathrm{DP}:\mathrm{UP},\mathrm{pF})}_{\text{hyperbolic flow }(+)}\\\underbrace{(\mathrm{pF},\mathrm{DP}:\mathrm{UP})}_{\text{hyperbolic flow }(+)} & \underbrace{(\mathrm{SI},\mathrm{SO})}_{\text{saddle}}\\\underbrace{(\mathrm{DP}_-,\mathrm{DP}_+)}_{\text{CW center}} & \underbrace{(\mathrm{DP}:\mathrm{UP},\mathrm{nF})}_{\text{hyperbolic flow }(-)}\\\underbrace{(\mathrm{nF},\mathrm{DP}:\mathrm{UP})}_{\text{hyperbolic flow }(-)} & \underbrace{(\mathrm{SO},\mathrm{SI})}_{\text{saddle}}\end{array}\right\} \quad (5.15)$$

for $a_{110} > 0$ and $a_{220} > 0$;

$$\left\{\begin{array}{cc} (a_{122},a_{211}) & (a_{111},a_{122}) \\ (a_{211},a_{222}) & (a_{111},a_{222}) \\ (a_{121},a_{211}) & (a_{111},a_{121}) \\ (a_{211},a_{221}) & (a_{111},a_{221}) \end{array}\right\} = \left\{\begin{array}{cc} \underbrace{(\mathrm{UP}_{+},\mathrm{UP}_{+})}_{\text{positive saddle}} & \underbrace{(\mathrm{UP}:\mathrm{DP},\mathrm{pF})}_{\text{hyperbolic-secant flow }(+)} \\ \underbrace{(\mathrm{nF},\mathrm{UP}:\mathrm{DP})}_{\text{hyperbolic-secant flow }(-)} & \underbrace{(\mathrm{SO},\mathrm{SO})}_{\text{source}} \\ \underbrace{(\mathrm{UP}_{-},\mathrm{UP}_{-})}_{\text{negative saddle}} & \underbrace{(\mathrm{UP}:\mathrm{DP},\mathrm{nF})}_{\text{hyperbolic-secant flow }(-)} \\ \underbrace{(\mathrm{pF},\mathrm{UP}:\mathrm{DP})}_{\text{hyperbolic-secant flow }(+)} & \underbrace{(\mathrm{SI},\mathrm{SI})}_{\text{sink}} \end{array}\right\}$$

for $a_{110} < 0$ and $a_{220} > 0$; (5.16)

$$\left\{\begin{array}{cc} (a_{122},a_{211}) & (a_{111},a_{122}) \\ (a_{211},a_{222}) & (a_{111},a_{222}) \\ (a_{121},a_{211}) & (a_{111},a_{121}) \\ (a_{211},a_{221}) & (a_{111},a_{221}) \end{array}\right\} = \left\{\begin{array}{cc} \underbrace{(\mathrm{UP}_{-},\mathrm{UP}_{-})}_{\text{negative saddle}} & \underbrace{(\mathrm{DP}:\mathrm{UP},\mathrm{nF})}_{\text{hyperbolic-secant flow }(-)} \\ \underbrace{(\mathrm{pF},\mathrm{UP}:\mathrm{DP})}_{\text{hyperbolic-secant flow }(+)} & \underbrace{(\mathrm{SI},\mathrm{SI})}_{\text{sink}} \\ \underbrace{(\mathrm{UP}_{+},\mathrm{UP}_{+})}_{\text{positive saddle}} & \underbrace{(\mathrm{UP}:\mathrm{DP},\mathrm{pF})}_{\text{hyperbolic-secant flow }(+)} \\ \underbrace{(\mathrm{nF},\mathrm{UP}:\mathrm{DP})}_{\text{hyperbolic-secant flow }(-)} & \underbrace{(\mathrm{SO},\mathrm{SO})}_{\text{source}} \end{array}\right\}$$

for $a_{110} > 0$ and $a_{220} < 0$; (5.17)

$$\left\{\begin{array}{cc} (a_{122},a_{211}) & (a_{111},a_{122}) \\ (a_{211},a_{222}) & (a_{111},a_{222}) \\ (a_{121},a_{211}) & (a_{111},a_{121}) \\ (a_{211},a_{221}) & (a_{111},a_{221}) \end{array}\right\} = \left\{\begin{array}{cc} \underbrace{(\mathrm{DP}_{-},\mathrm{DP}_{+})}_{\text{CW center}} & \underbrace{(\mathrm{DP}:\mathrm{UP},\mathrm{nF})}_{\text{hyperbolic flow }(-)} \\ \underbrace{(\mathrm{nF},\mathrm{DP}:\mathrm{UP})}_{\text{hyperbolic flow }(-)} & \underbrace{(\mathrm{SO},\mathrm{SI})}_{\text{saddle}} \\ \underbrace{(\mathrm{DP}_{+},\mathrm{DP}_{-})}_{\text{CCW center}} & \underbrace{(\mathrm{DP}:\mathrm{UP},\mathrm{pF})}_{\text{hyperbolic flow }(+)} \\ \underbrace{(\mathrm{pF},\mathrm{DP}:\mathrm{UP})}_{\text{hyperbolic flow }(+)} & \underbrace{(\mathrm{SI},\mathrm{SO})}_{\text{saddle}} \end{array}\right\} \tag{5.18}$$

for $a_{110} < 0$ and $a_{220} < 0$.

The saddle and source or sink with hyperbolic and hyperbolic-secant flows exist on the line of $x_1^* = a_{111}$ in the equilibrium matrix. The hyperbolic and hyperbolic-secant flows with directrix are separated through saddles or source and sink. The source and sink are separated by a hyperbolic-secant flow. The (SI,SO)-saddle and (SO,SI)-saddle are separated by a hyperbolic flow. The hyperbolic and hyperbolic-secant flows with centers and saddles on the line of $x_1^* = a_{211}$ exist in the equilibrium matrix. The positive or negative saddles are separated by the two hyperbolic-secant flows, and the counterclockwise and clockwise centers are separated by the two hyperbolic flows.

(v) For the case of $a_{222} < a_{121}$ and $a_{211} < a_{111}$, the equilibrium series of $x_1^* = a_{111}, a_{211}$ with $x_2^* = a_{121}, a_{122}, a_{221}, a_{222}$ have the properties as in Eqs. (5.3)–(5.6), as shown in Fig. 5.5. The saddle and source or sink with hyperbolic and hyperbolic-secant flows exist on the line of $x_1^* = a_{111}$ in the equilibrium matrix. The hyperbolic and hyperbolic-secant flows with directrix are connected together on the top. The hyperbolic and hyperbolic-secant flows with centers and saddles on the line of $x_1^* = a_{211}$ exist in the equilibrium matrix. The saddle and center are separated by the hyperbolic-secant and hyperbolic flows.

The equilibrium matrices for the above five cases with $a_{211} > a_{111}$ can be similarly completed, which can be done by readers.

5.2 Inflection-Source and Sink Infinite-Equilibriums

In this section, inflection-source and sink infinite-equilibriums are discussed. The parabola-saddles are the switching bifurcations for the hyperbolic flows with paralleled center and saddles. The inflection-sources (sinks) are the switching bifurcations for the hyperbolic flows with connected saddles, source, and sink.

5.2.1 Infinite-Equilibriums in the Horizontal Direction

A. Consider a dynamical system as

$$
\begin{aligned}
\dot{x}_1 &= a_{110}(x_1 - a_{111})(x_2 - a_{121})(x_2 - a_{122}),\\
\dot{x}_2 &= a_{220}(x_1 - a_{211})(x_2 - a_{221})(x_2 - a_{222}),
\end{aligned}
\tag{5.19}
$$

and the corresponding first integral manifold for $a_{12s_1} = a_{22l_1}$ ($s_1, s_2, l_1, l_2 \in \{1,2\}$, $s_1 \neq s_2$, $l_1 \neq l_2$) and $a_{111} \neq a_{211}$ is

$$
\begin{aligned}
&[(x_1 - a_{111}) - (x_{10} - a_{111})] + (a_{111} - a_{211}) \ln \frac{|x_1 - a_{111}|}{|x_{10} - a_{111}|}\\
&= \frac{a_{110}}{a_{220}} \left[(x_2 - x_{20}) + (a_{22l_2} - a_{12s_2}) \ln \frac{|x_2 - a_{22l_2}|}{|x_{20} - a_{22l_2}|} \right].
\end{aligned}
\tag{5.20}
$$

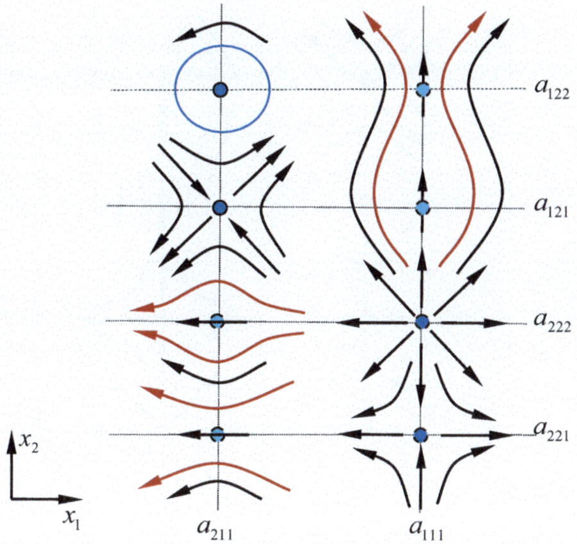

(a)

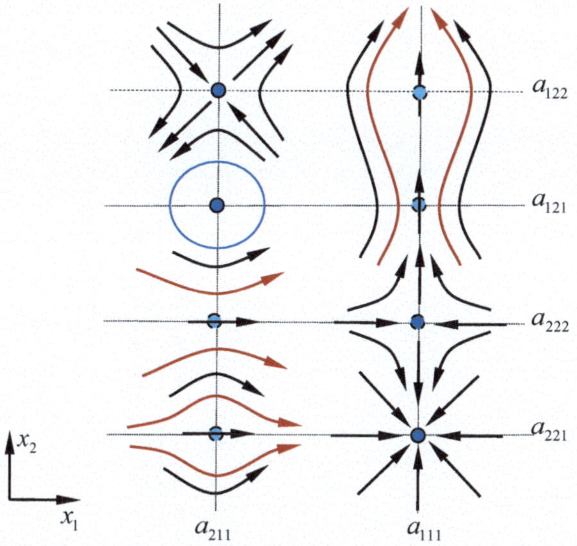

(b)

Fig. 5.5 Phase portraits ($a_{222} < a_{121}$ and $a_{211} < a_{111}$) for two-dimensional systems on the x_1-direction with $x_1^* = a_{111}, a_{211}$ and on the x_2-direction with $x_2^* = a_{121}, a_{122}, a_{221}, a_{222}$. (**a**) ($a_{110} > 0, a_{220} > 0$), (**b**) ($a_{110} < 0, a_{220} > 0$), (**c**) ($a_{110} > 0, a_{220} < 0$), and (**d**) ($a_{110} < 0, a_{220} < 0$)

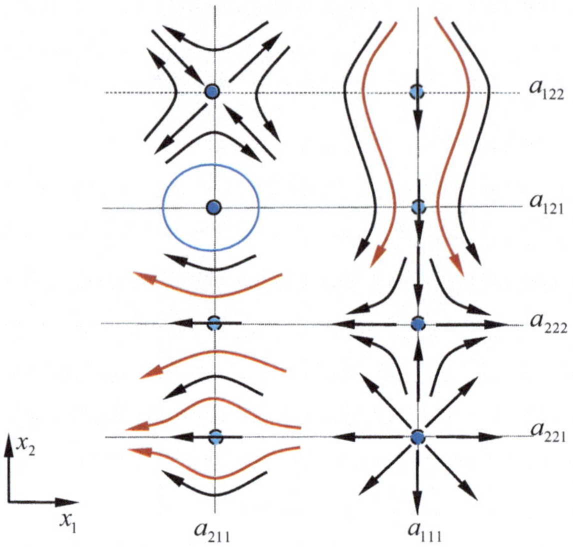

(c)

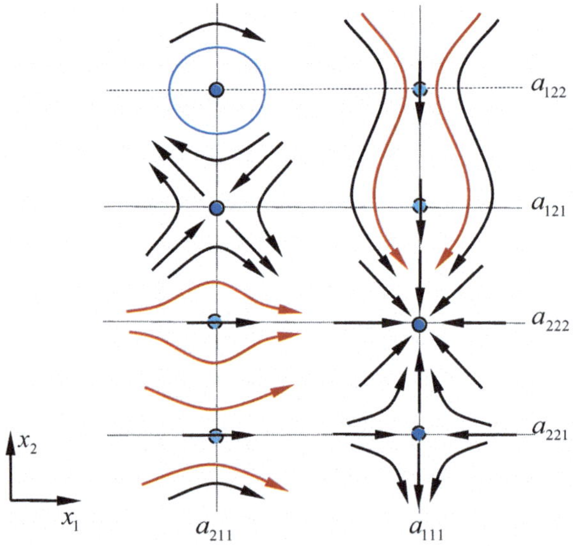

(d)

Fig. 5.5 (continued)

(i) The equilibrium matrices of $x_2^* = a_{121}, a_{122}, a_{221}, a_{222}$ with $x_1^* = a_{111}, a_{211}$ for $a_{221} = a_{121}$ with $a_{111} > a_{211}$ are as follows:

$$\left\{\begin{matrix} (a_{211}, a_{222}) & (a_{111}, a_{222}) \\ (a_{122}, a_{211}) & (a_{111}, a_{122}) \\ (a_{121}, a_{211}) & (a_{111}, a_{221}) \end{matrix}\right\} = \left\{\begin{matrix} \underbrace{(\text{nF}, \text{UP} : \text{DP})}_{\text{hyperbolic-secant flow } (-)} & \underbrace{(\text{SO}, \text{SO})}_{\text{source}} \\ \underbrace{(\text{UP}_-, \text{UP}_-)}_{\text{negative saddle}} & \underbrace{(\text{UP} : \text{DP}, \text{nF})}_{\text{hyperbolic-secant flow } (-)} \\ \underbrace{(\text{LS}, \text{DP})}_{\text{down-parabola lower-saddle}} & \underbrace{(\text{DI} : \text{II}, \text{SI})}_{\text{inflection-sink}} \end{matrix}\right\}$$

for $a_{110} > 0, a_{220} > 0,$

(5.21)

$$\left\{\begin{matrix} (a_{211}, a_{222}) & (a_{111}, a_{222}) \\ (a_{122}, a_{211}) & (a_{111}, a_{122}) \\ (a_{121}, a_{211}) & (a_{111}, a_{221}) \end{matrix}\right\} = \left\{\begin{matrix} \underbrace{(\text{pF}, \text{DP} : \text{UP})}_{\text{hyperbolic flow } (+)} & \underbrace{(\text{SI}, \text{SO})}_{\text{saddle}} \\ \underbrace{(\text{DP}_-, \text{DP}_+)}_{\text{CW center}} & \underbrace{(\text{DP} : \text{UP}, \text{nF})}_{\text{hyperbolic flow } (-)} \\ \underbrace{(\text{LS}, \text{UP})}_{\text{up-parabola lower-saddle}} & \underbrace{(\text{II} : \text{DI}, \text{SI})}_{\text{inflection-sink}} \end{matrix}\right\} \quad (5.22)$$

for $a_{110} < 0, a_{220} > 0,$

$$\left\{\begin{matrix} (a_{211}, a_{222}) & (a_{111}, a_{222}) \\ (a_{122}, a_{211}) & (a_{111}, a_{122}) \\ (a_{121}, a_{211}) & (a_{111}, a_{221}) \end{matrix}\right\} = \left\{\begin{matrix} \underbrace{(\text{nF}, \text{DP} : \text{UP})}_{\text{hyperbolic flow } (-)} & \underbrace{(\text{SO}, \text{SI})}_{\text{saddle}} \\ \underbrace{(\text{DP}_+, \text{DP}_-)}_{\text{CCW center}} & \underbrace{(\text{DP} : \text{UP}, \text{pF})}_{\text{hyperbolic flow } (+)} \\ \underbrace{(\text{US}, \text{UP})}_{\text{up-parabola upper-saddle}} & \underbrace{(\text{II} : \text{DI}, \text{SO})}_{\text{inflection-source}} \end{matrix}\right\} \quad (5.23)$$

for $a_{110} > 0, a_{220} < 0,$

$$\left\{\begin{matrix} (a_{211}, a_{222}) & (a_{111}, a_{222}) \\ (a_{122}, a_{211}) & (a_{111}, a_{122}) \\ (a_{121}, a_{211}) & (a_{111}, a_{221}) \end{matrix}\right\} = \left\{\begin{matrix} \underbrace{(\text{pF}, \text{UP} : \text{DP})}_{\text{hyperbolic-secant flow } (+)} & \underbrace{(\text{SI}, \text{SI})}_{\text{sink}} \\ \underbrace{(\text{UP}_+, \text{UP}_+)}_{\text{positive saddle}} & \underbrace{(\text{UP} : \text{DP}, \text{pF})}_{\text{hyperbolic-secant flow } (+)} \\ \underbrace{(\text{US}, \text{DP})}_{\text{down-parabola upper-saddle}} & \underbrace{(\text{DI} : \text{II}, \text{SO})}_{\text{inflection-source}} \end{matrix}\right\}$$

for $a_{110} < 0, a_{220} < 0.$

(5.24)

The infinite-equilibrium of $x_2^* = a_{121} = a_{221}$ is summarized as follows:

$$\left.\begin{array}{ll}(\bar{x}_1, a_{221}) = \underbrace{(\text{II, SO})}_{\text{increasing-infelction source}} & \text{if } \bar{x}_1 \in (-\infty, a_{211}),\\ (\bar{x}_1, a_{221}) = \underbrace{(\text{DI, SI})}_{\text{decreasing-infelction sink}} & \text{if } \bar{x}_1 \in (a_{211}, a_{111}),\\ (\bar{x}_1, a_{221}) = \underbrace{(\text{II, SI})}_{\text{increasing-inflection sink}} & \text{if } \bar{x}_1 \in (a_{111}, \infty),\\ \text{for } a_{110} > 0, a_{220} > 0; & \end{array}\right\} \tag{5.25}$$

$$\left.\begin{array}{ll}(\bar{x}_1, a_{221}) = \underbrace{(\text{DI, SO})}_{\text{decreasing-infelction source}} & \text{if } \bar{x}_1 \in (-\infty, a_{211}),\\ (\bar{x}_1, a_{221}) = \underbrace{(\text{II, SI})}_{\text{increasing-inflection sink}} & \text{if } \bar{x}_1 \in (a_{211}, a_{111}),\\ (\bar{x}_1, a_{221}) = \underbrace{(\text{DI, SI})}_{\text{decreasing-infelction sink}} & \text{if } \bar{x}_1 \in (a_{111}, \infty),\\ \text{for } a_{110} < 0, a_{220} > 0; & \end{array}\right\} \tag{5.26}$$

$$\left.\begin{array}{ll}(\bar{x}_1, a_{221}) = \underbrace{(\text{DI, SI})}_{\text{decreasing-infelction sink}} & \text{if } \bar{x}_1 \in (-\infty, a_{211}),\\ (\bar{x}_1, a_{221}) = \underbrace{(\text{II, SO})}_{\text{increasing-infelction source}} & \text{if } \bar{x}_1 \in (a_{211}, a_{111}),\\ (\bar{x}_1, a_{221}) = \underbrace{(\text{DI, SO})}_{\text{decreasing-inflection source}} & \text{if } \bar{x}_1 \in (a_{111}, \infty),\\ \text{for } a_{110} > 0, a_{220} < 0; & \end{array}\right\} \tag{5.27}$$

$$\left.\begin{array}{ll}(\bar{x}_1, a_{221}) = \underbrace{(\text{II, SI})}_{\text{increasing-infelction sink}} & \text{if } \bar{x}_1 \in (-\infty, a_{211}),\\ (\bar{x}_1, a_{221}) = \underbrace{(\text{DI, SO})}_{\text{decreasing-infelction source}} & \text{if } \bar{x}_1 \in (a_{211}, a_{111}),\\ (\bar{x}_1, a_{221}) = \underbrace{(\text{II, SO})}_{\text{increasing-inflection source}} & \text{if } \bar{x}_1 \in (a_{111}, \infty),\\ \text{for } a_{110} < 0, a_{220} < 0. & \end{array}\right\} \tag{5.28}$$

The inflection-source and sink infinite-equilibriums of $x_2^* = a_{121} = a_{221}$ with $a_{122} \neq a_{222}$ and $a_{111} > a_{211}$ are the switching bifurcations of sink, source, and saddles with hyperbolic flows for $a_{221} \in (a_{121}, a_{122})$ and $a_{221} < a_{121}$ with $a_{122} < a_{222}$ and $a_{111} > a_{211}$, as shown in Fig. 5.6. In addition, the saddle and hyperbolic flows exist, or a hyperbolic-secant flow with sink or source exists. The inflection sink and source infinite-equilibriums are at the bottom of the equilibrium matrix.

(ii) The equilibrium matrices of $x_2^* = a_{121}, a_{122}, a_{221}, a_{222}$ with $x_1^* = a_{111}, a_{211}$ for $a_{221} = a_{121}$ with $a_{111} > a_{211}$ are as follows:

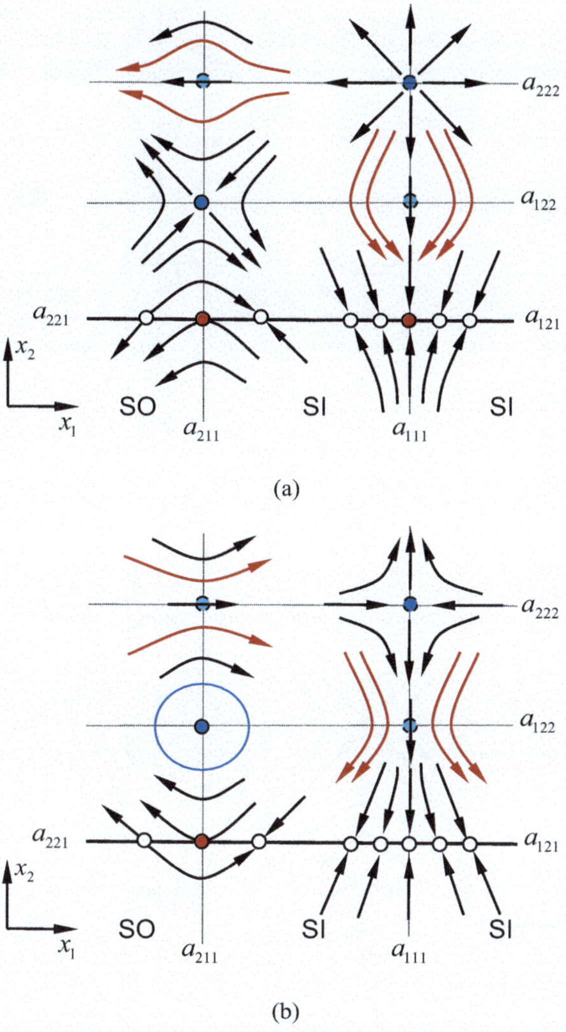

Fig. 5.6 Phase portraits ($a_{121} = a_{221}$) for two-dimensional systems on the x_1-direction with $x_1^* = a_{111}$ and on the x_2-direction with $x_2^* = a_{121}, a_{122}, a_{221}, a_{222}$. The equilibrium flows: (**a**) ($a_{110} > 0$, $a_{220} > 0$), (**b**) ($a_{110} < 0$, $a_{220} > 0$), (**c**) ($a_{110} > 0$, $a_{220} < 0$), and (**d**) ($a_{110} < 0$, $a_{220} < 0$)

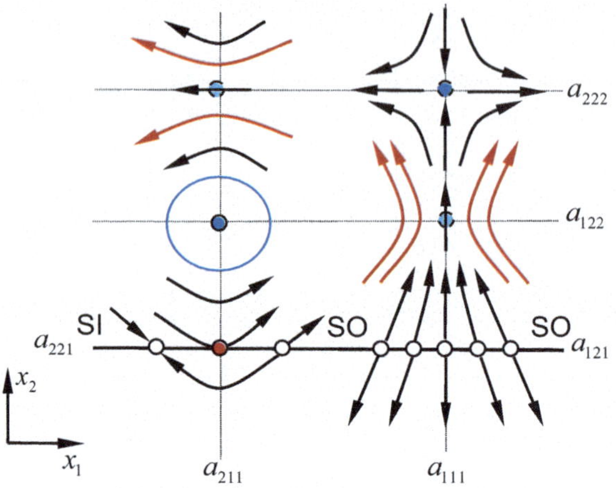

(c)

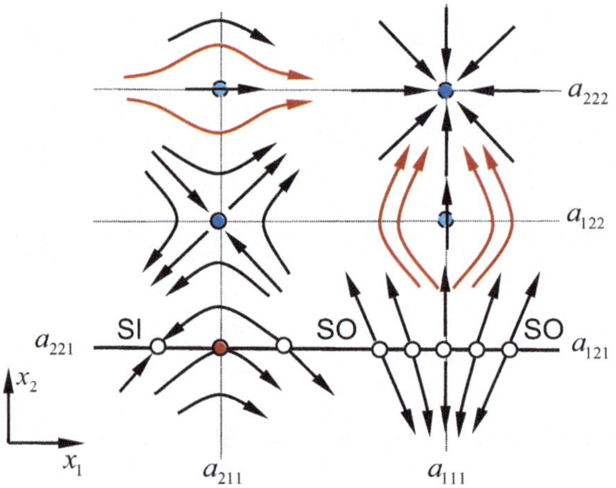

(d)

Fig. 5.6 (continued)

$$\left\{\begin{matrix} (a_{122}, a_{211}) & (a_{111}, a_{122}) \\ (a_{211}, a_{222}) & (a_{111}, a_{222}) \\ (a_{121}, a_{211}) & (a_{111}, a_{221}) \end{matrix}\right\} = \left\{\begin{matrix} \underbrace{(\text{DP}_+, \text{DP}_-)}_{\text{CCW center}} & \underbrace{(\text{DP}:\text{UP}, \text{pF})}_{\text{hyperbolic flow }(+)} \\ \underbrace{(\text{pF}, \text{DP}:\text{UP})}_{\text{hyperbolic-secant flow }(+)} & \underbrace{(\text{SI}, \text{SO})}_{\text{saddle}} \\ \underbrace{(\text{LS}, \text{DP})}_{\text{down-parabola lower-saddle}} & \underbrace{(\text{DI}:\text{II}, \text{SI})}_{\text{inflection-sink}} \end{matrix}\right\}$$

for $a_{110} > 0, a_{220} > 0,$

$$(5.29)$$

$$\begin{Bmatrix} (a_{122},a_{211}) & (a_{111},a_{122}) \\ (a_{211},a_{222}) & (a_{111},a_{222}) \\ (a_{121},a_{211}) & (a_{111},a_{221}) \end{Bmatrix} = \begin{Bmatrix} \underbrace{(\mathrm{UP}_+,\mathrm{UP}_+)}_{\text{positive saddle}} & \underbrace{(\mathrm{UP}:\mathrm{DP},\mathrm{pF})}_{\text{hyperbolic-secant flow }(+)} \\ \underbrace{(\mathrm{nF},\mathrm{UP}:\mathrm{DP})}_{\text{hyperbolic-secant flow }(-)} & \underbrace{(\mathrm{SO},\mathrm{SO})}_{\text{source}} \\ \underbrace{(\mathrm{LS},\mathrm{UP})}_{\text{up-parabola lower-saddle}} & \underbrace{(\mathrm{II}:\mathrm{DI},\mathrm{SI})}_{\text{inflection-sink}} \end{Bmatrix}$$

for $a_{110} < 0, a_{220} > 0,$

(5.30)

$$\begin{Bmatrix} (a_{122},a_{211}) & (a_{111},a_{122}) \\ (a_{211},a_{222}) & (a_{111},a_{222}) \\ (a_{121},a_{211}) & (a_{111},a_{221}) \end{Bmatrix} = \begin{Bmatrix} \underbrace{(\mathrm{UP}_-,\mathrm{UP}_-)}_{\text{negative saddle}} & \underbrace{(\mathrm{UP}:\mathrm{DP},\mathrm{nF})}_{\text{hyperbolic-secant flow }(-)} \\ \underbrace{(\mathrm{pF},\mathrm{UP}:\mathrm{DP})}_{\text{hyperbolic-secant flow }(+)} & \underbrace{(\mathrm{SI},\mathrm{SI})}_{\text{sink}} \\ \underbrace{(\mathrm{US},\mathrm{UP})}_{\text{up-parabola upper-saddle}} & \underbrace{(\mathrm{II}:\mathrm{DI},\mathrm{SO})}_{\text{inflection-source}} \end{Bmatrix}$$

for $a_{110} > 0, a_{220} < 0,$

(5.31)

$$\begin{Bmatrix} (a_{122},a_{211}) & (a_{111},a_{122}) \\ (a_{211},a_{222}) & (a_{111},a_{222}) \\ (a_{121},a_{211}) & (a_{111},a_{221}) \end{Bmatrix} = \begin{Bmatrix} \underbrace{(\mathrm{DP}_-,\mathrm{DP}_+)}_{\text{CW center}} & \underbrace{(\mathrm{DP}:\mathrm{UP},\mathrm{nF})}_{\text{hyperbolic flow }(-)} \\ \underbrace{(\mathrm{nF},\mathrm{DP}:\mathrm{UP})}_{\text{hyperbolic flow }(-)} & \underbrace{(\mathrm{SO},\mathrm{SI})}_{\text{saddle}} \\ \underbrace{(\mathrm{US},\mathrm{DP})}_{\text{down-parabola upper-saddle}} & \underbrace{(\mathrm{DI}:\mathrm{II},\mathrm{SO})}_{\text{inflection-source}} \end{Bmatrix}$$

for $a_{110} < 0, a_{220} < 0.$

(5.32)

The infinite-equilibrium of $x_2^* = a_{121} = a_{221}$ is summarized as in Eqs. (5.25)–(5.28). The inflection-source and sink infinite-equilibriums of $x_2^* = a_{121} = a_{221}$ with $a_{122} \neq a_{222}$ and $a_{111} > a_{211}$ are the switching bifurcations of sink, source, and saddles with hyperbolic flows for $a_{221} \in (a_{121}, a_{122})$ and $a_{221} < a_{121}$ with $a_{222} < a_{122}$ and $a_{111} > a_{211}$, as shown in Fig. 5.7. In addition, the saddle and hyperbolic flows exist, or a hyperbolic-secant flow with sink or source exists. The inflection-sink and source infinite-equilibriums are at the bottom of the equilibrium matrix.

(iii) The equilibrium matrices of $x_2^* = a_{121}, a_{122}, a_{221}, a_{222}$ with $x_1^* = a_{111}$ for $a_{221} = a_{122}$ with $a_{111} \neq a_{211}$ are as follows:

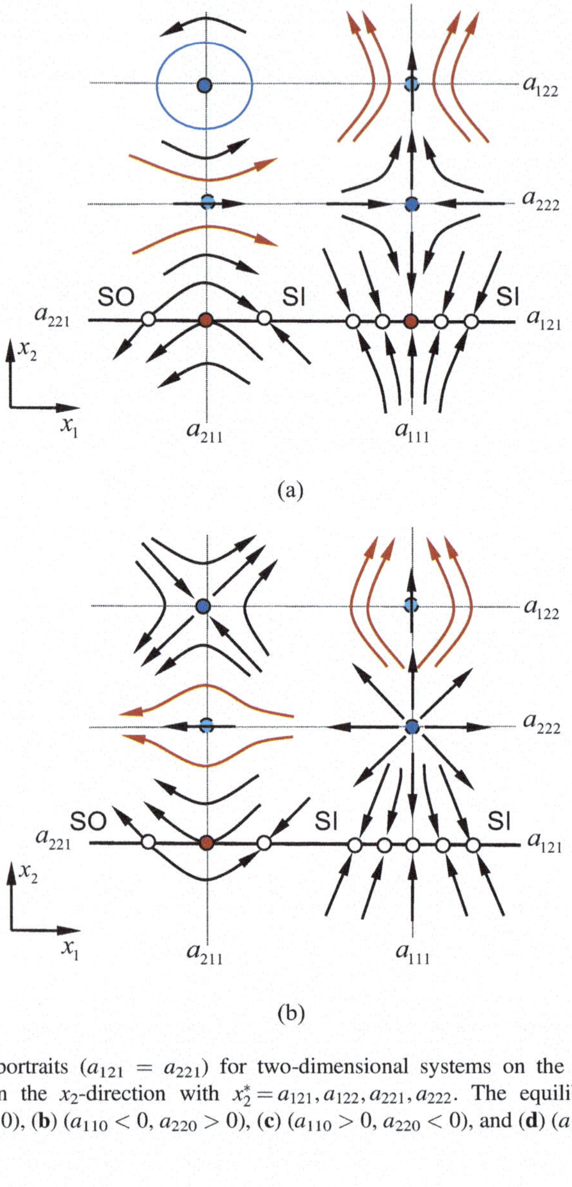

(b)

Fig. 5.7 Phase portraits ($a_{121} = a_{221}$) for two-dimensional systems on the x_1-direction with $x_1^* = a_{111}$ and on the x_2-direction with $x_2^* = a_{121}, a_{122}, a_{221}, a_{222}$. The equilibrium flows: (**a**) ($a_{110} > 0$, $a_{220} > 0$), (**b**) ($a_{110} < 0$, $a_{220} > 0$), (**c**) ($a_{110} > 0$, $a_{220} < 0$), and (**d**) ($a_{110} < 0$, $a_{220} < 0$)

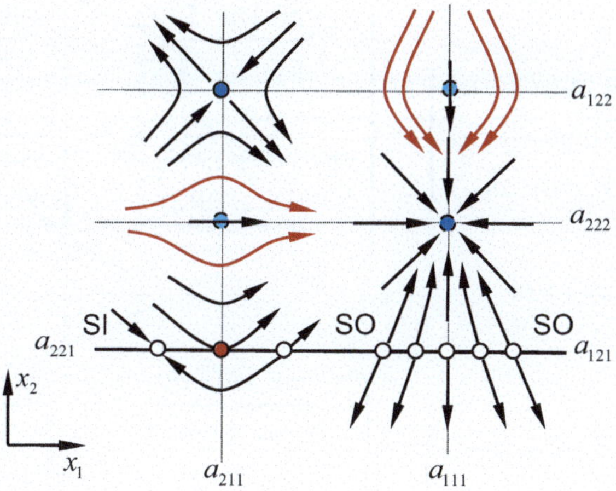

(c)

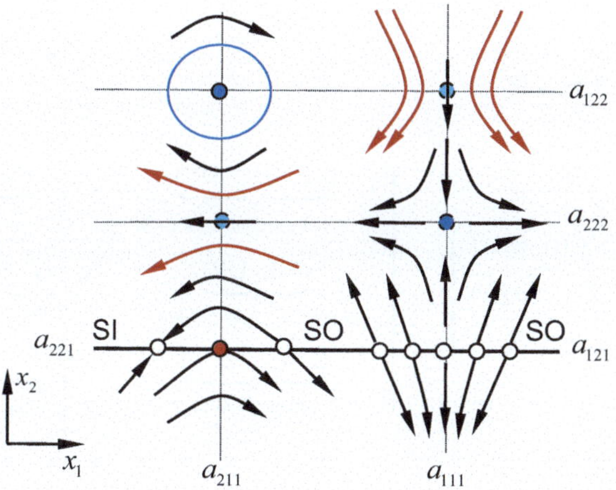

(d)

Fig. 5.7 (continued)

$$
\left\{
\begin{matrix}
(a_{211}, a_{222}) & (a_{111}, a_{222}) \\
(a_{122}, a_{211}) & (a_{111}, a_{221}) \\
(a_{211}, a_{121}) & (a_{111}, a_{121})
\end{matrix}
\right\}
=
\left\{
\begin{matrix}
\underbrace{(\text{nF}, \text{UP} : \text{DP})}_{\text{hyperbolic-secant flow } (-)} & \underbrace{(\text{SO}, \text{SO})}_{\text{source}} \\
\underbrace{(\text{LS}, \text{UP})}_{\text{up-parabola lower-saddle}} & \underbrace{(\text{II} : \text{DI}, \text{SI})}_{\text{inflection-sink}} \\
\underbrace{(\text{UP}_+, \text{UP}_+)}_{\text{positive saddle}} & \underbrace{(\text{UP} : \text{DP}, \text{pF})}_{\text{hyperbolic-secant flow } (+)}
\end{matrix}
\right\}
$$

for $a_{110} > 0, a_{220} > 0,$

(5.33)

$$\left\{\begin{array}{cc}(a_{211},a_{222}) & (a_{111},a_{222})\\ (a_{122},a_{211}) & (a_{111},a_{221})\\ (a_{211},a_{121}) & (a_{111},a_{121})\end{array}\right\}=\left\{\begin{array}{cc}\underbrace{(\text{pF},\text{DP}:\text{UP})}_{\text{hyperbolic flow }(+)} & \underbrace{(\text{SI},\text{SO})}_{\text{saddle}}\\ \underbrace{(\text{LS},\text{DP})}_{\text{down-parabola lower-saddle}} & \underbrace{(\text{DI}:\text{II},\text{SI})}_{\text{inflection-sink}}\\ \underbrace{(\text{DP}_{+},\text{DP}_{-})}_{\text{CCW center}} & \underbrace{(\text{DP}:\text{UP},\text{pF})}_{\text{hyperbolic flow }(+)}\end{array}\right\}$$

for $a_{110}<0, a_{220}>0$,

(5.34)

$$\left\{\begin{array}{cc}(a_{211},a_{222}) & (a_{111},a_{222})\\ (a_{122},a_{211}) & (a_{111},a_{221})\\ (a_{211},a_{121}) & (a_{111},a_{121})\end{array}\right\}=\left\{\begin{array}{cc}\underbrace{(\text{nF},\text{DP}:\text{UP})}_{\text{hyperbolic flow }(-)} & \underbrace{(\text{SO},\text{SI})}_{\text{saddle}}\\ \underbrace{(\text{US},\text{DP})}_{\text{down-parabola upper-saddle}} & \underbrace{(\text{DI}:\text{II},\text{SO})}_{\text{inflection-source}}\\ \underbrace{(\text{DP}_{-},\text{DP}_{+})}_{\text{CW center}} & \underbrace{(\text{DP}:\text{UP},\text{nF})}_{\text{hyperbolic flow }(-)}\end{array}\right\}$$

for $a_{110}>0, a_{220}<0$,

(5.35)

$$\left\{\begin{array}{cc}(a_{211},a_{222}) & (a_{111},a_{222})\\ (a_{122},a_{211}) & (a_{111},a_{221})\\ (a_{211},a_{121}) & (a_{111},a_{121})\end{array}\right\}=\left\{\begin{array}{cc}\underbrace{(\text{pF},\text{UP}:\text{DP})}_{\text{hyperbolic-secant flow }(+)} & \underbrace{(\text{SI},\text{SI})}_{\text{sink}}\\ \underbrace{(\text{US},\text{UP})}_{\text{up-parabola upper-saddle}} & \underbrace{(\text{II}:\text{DI},\text{SO})}_{\text{inflection-source}}\\ \underbrace{(\text{UP}_{-},\text{UP}_{-})}_{\text{negative saddle}} & \underbrace{(\text{UP}:\text{DP},\text{nF})}_{\text{hyperbolic-secant flow }(-)}\end{array}\right\}$$

for $a_{110}<0, a_{220}<0$.

(5.36)

The infinite-equilibrium of $x_2^*=a_{122}=a_{221}$ is summarized as follows:

$$\begin{aligned}(\bar{x}_1,a_{221}) &= \underbrace{(\text{DI},\text{SO})}_{\text{decreasing-inflection source}} \quad \text{if } \bar{x}_1\in(-\infty,a_{211}),\\ (\bar{x}_1,a_{221}) &= \underbrace{(\text{II},\text{SI})}_{\text{increasing-infelction sink}} \quad \text{if } \bar{x}_1\in(a_{211},a_{111}),\\ (\bar{x}_1,a_{221}) &= \underbrace{(\text{DI},\text{SI})}_{\text{decreasing-inflection sink}} \quad \text{if } \bar{x}_1\in(a_{111},\infty),\end{aligned} \tag{5.37}$$

for $a_{110}>0, a_{220}>0$;

$$
\left.
\begin{aligned}
&(\bar{x}_1, a_{221}) = \underbrace{(\text{II, SO})}_{\text{increasing-inflection source}} && \text{if } \bar{x}_1 \in (-\infty, a_{211}),\\
&(\bar{x}_1, a_{221}) = \underbrace{(\text{DI, SI})}_{\text{decreasing-inflection sink}} && \text{if } \bar{x}_1 \in (a_{211}, a_{111}),\\
&(\bar{x}_1, a_{221}) = \underbrace{(\text{II, SI})}_{\text{increasing-infelction sink}} && \text{if } \bar{x}_1 \in (a_{111}, \infty),
\end{aligned}
\right.
\tag{5.38}
$$

for $a_{110} < 0, a_{220} > 0$;

$$
\left.
\begin{aligned}
&(\bar{x}_1, a_{221}) = \underbrace{(\text{II, SI})}_{\text{increasing-inflection sink}} && \text{if } \bar{x}_1 \in (-\infty, a_{211}),\\
&(\bar{x}_1, a_{221}) = \underbrace{(\text{DI, SO})}_{\text{decreasing-infelction source}} && \text{if } \bar{x}_1 \in (a_{211}, a_{111}),\\
&(\bar{x}_1, a_{221}) = \underbrace{(\text{II, SO})}_{\text{increasing-inflection source}} && \text{if } \bar{x}_1 \in (a_{111}, \infty),
\end{aligned}
\right.
\tag{5.39}
$$

for $a_{110} > 0, a_{220} < 0$;

$$
\left.
\begin{aligned}
&(\bar{x}_1, a_{221}) = \underbrace{(\text{DI, SI})}_{\text{decreasing-inflection sink}} && \text{if } \bar{x}_1 \in (-\infty, a_{211}),\\
&(\bar{x}_1, a_{221}) = \underbrace{(\text{II, SO})}_{\text{increasing-inflection source}} && \text{if } \bar{x}_1 \in (a_{211}, a_{111}),\\
&(\bar{x}_1, a_{221}) = \underbrace{(\text{DI, SO})}_{\text{decreasing-infelction source}} && \text{if } \bar{x}_1 \in (a_{111}, \infty),
\end{aligned}
\right.
\tag{5.40}
$$

for $a_{110} < 0, a_{220} < 0$.

The inflection-source and sink infinite-equilibriums of $x_2^* = a_{122} = a_{221}$ with $a_{111} > a_{211}$ are the switching bifurcations of sink, source, and saddles with hyperbolic flows for $a_{221} \in (a_{121}, a_{122})$ and $a_{122} < a_{221}$ with $a_{111} > a_{211}$, as shown in Fig. 5.8. In addition, on the line of $x_1^* = a_{111}$, the saddle and hyperbolic flows exist or a hyperbolic-secant flow with sink or source exists. The infinite-equilibrium is in the middle between saddle and hyperbolic flow or between the hyperbolic-secant flow and sink or source for $x_1^* = a_{111}$. On the line of $x_1^* = a_{211}$, the saddle and hyperbolic-secant flow exist or a hyperbolic flow and center exist.

(iv) The equilibrium matrices of $x_2^* = a_{121}, a_{122}, a_{221}, a_{222}$ with $x_1^* = a_{111}$ for $a_{121} = a_{222}$ with $a_{111} > a_{211}$ are

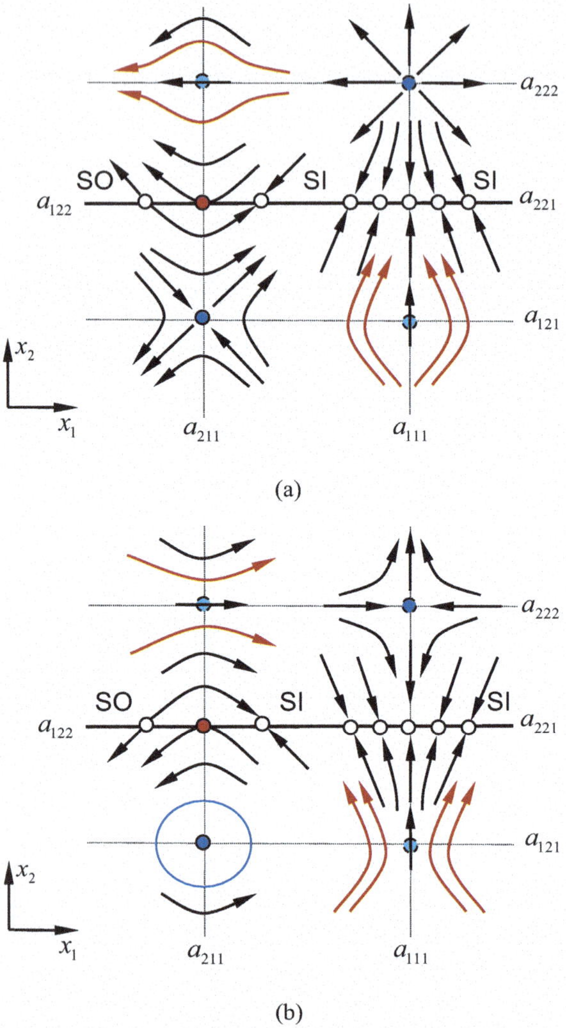

Fig. 5.8 Phase portraits ($a_{122} = a_{221}$) for two-dimensional systems on the x_1-direction with $x_1^* = a_{111}$ and on the x_2-direction with $x_2^* = a_{121}, a_{122}, a_{221}, a_{222}$. The equilibrium flows: (**a**) ($a_{110} > 0, a_{220} > 0$), (**b**) ($a_{110} < 0, a_{220} > 0$), (**c**) ($a_{110} > 0, a_{220} < 0$), and (**d**) ($a_{110} < 0, a_{220} < 0$)

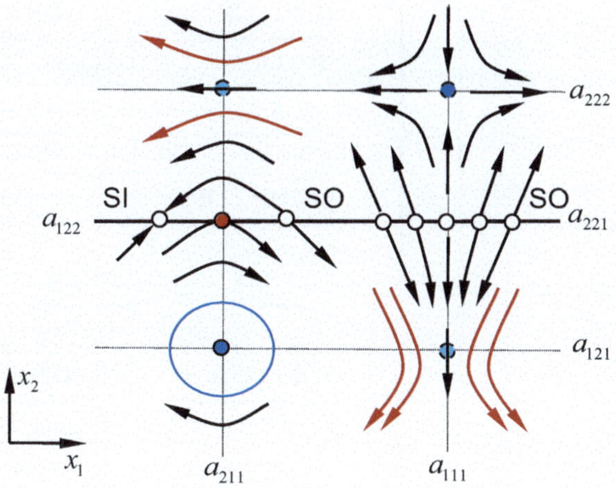

(c)

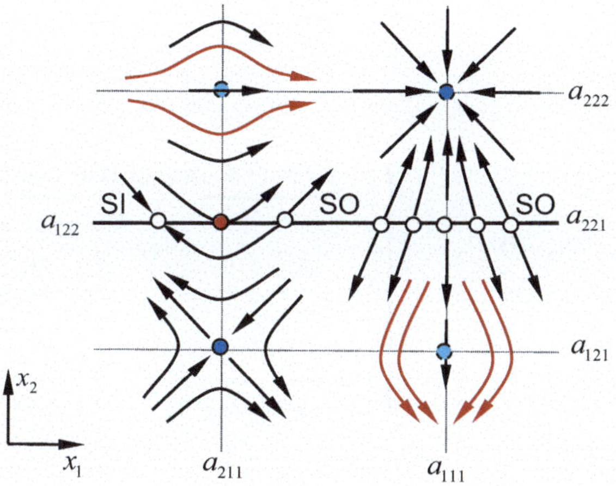

(d)

Fig. 5.8 (continued)

$$\left\{\begin{matrix} (a_{211}, a_{122}) & (a_{111}, a_{122}) \\ (a_{121}, a_{211}) & (a_{111}, a_{222}) \\ (a_{211}, a_{221}) & (a_{111}, a_{221}) \end{matrix}\right\} = \left\{\begin{matrix} \underbrace{(\mathrm{DP}_{+}, \mathrm{DP}_{-})}_{\text{CCW center}} & \underbrace{(\mathrm{DP}:\mathrm{UP}, \mathrm{pF})}_{\text{hyperbolic flow }(+)} \\ \underbrace{(\mathrm{US}, \mathrm{UP})}_{\text{up-parabola upper-saddle}} & \underbrace{(\mathrm{II}:\mathrm{DI}, \mathrm{SO})}_{\text{inflection-source}} \\ \underbrace{(\mathrm{nF}, \mathrm{DP}:\mathrm{UP})}_{\text{hyperbolic flow }(-)} & \underbrace{(\mathrm{SO}, \mathrm{SI})}_{\text{saddle}} \end{matrix}\right\} \tag{5.41}$$

for $a_{110} > 0, a_{220} > 0$,

$$\left\{\begin{array}{ll}(a_{211},a_{122}) & (a_{111},a_{122})\\(a_{121},a_{211}) & (a_{111},a_{222})\\(a_{211},a_{221}) & (a_{111},a_{221})\end{array}\right\}=\left\{\begin{array}{ll}\underbrace{(\mathrm{UP}_+,\mathrm{UP}_+)}_{\text{positive saddle}} & \underbrace{(\mathrm{UP}:\mathrm{DP},\mathrm{pF})}_{\text{hyperbolic-secant flow }(+)}\\\underbrace{(\mathrm{US},\mathrm{DP})}_{\text{down-parabola upper-saddle}} & \underbrace{(\mathrm{DI}:\mathrm{II},\mathrm{SO})}_{\text{inflection-source}}\\\underbrace{(\mathrm{pF},\mathrm{UP}:\mathrm{DP})}_{\text{hyperbolic-secant flow }(+)} & \underbrace{(\mathrm{SI},\mathrm{SI})}_{\text{sink}}\end{array}\right\}$$

for $a_{110}<0, a_{220}>0$,

(5.42)

$$\left\{\begin{array}{ll}(a_{211},a_{122}) & (a_{111},a_{122})\\(a_{121},a_{211}) & (a_{111},a_{222})\\(a_{211},a_{221}) & (a_{111},a_{221})\end{array}\right\}=\left\{\begin{array}{ll}\underbrace{(\mathrm{UP}_-,\mathrm{UP}_-)}_{\text{negative saddle}} & \underbrace{(\mathrm{UP}:\mathrm{DP},\mathrm{nF})}_{\text{hyperbolic-secant flow }(-)}\\\underbrace{(\mathrm{LS},\mathrm{DP})}_{\text{down-parabola lower-saddle}} & \underbrace{(\mathrm{DI}:\mathrm{II},\mathrm{SI})}_{\text{inflection-sink}}\\\underbrace{(\mathrm{nF},\mathrm{UP}:\mathrm{DP})}_{\text{hyperbolic-secant flow }(-)} & \underbrace{(\mathrm{SO},\mathrm{SO})}_{\text{source}}\end{array}\right\}$$

for $a_{110}>0, a_{220}<0$,

(5.43)

$$\left\{\begin{array}{ll}(a_{211},a_{122}) & (a_{111},a_{122})\\(a_{121},a_{211}) & (a_{111},a_{222})\\(a_{211},a_{221}) & (a_{111},a_{221})\end{array}\right\}=\left\{\begin{array}{ll}\underbrace{(\mathrm{DP}_-,\mathrm{DP}_+)}_{\text{CW center}} & \underbrace{(\mathrm{DP}:\mathrm{UP},\mathrm{nF})}_{\text{hyperbolic flow }(-)}\\\underbrace{(\mathrm{LS},\mathrm{UP})}_{\text{up-parabola lower-saddle}} & \underbrace{(\mathrm{II}:\mathrm{DI},\mathrm{SI})}_{\text{inflection-sink}}\\\underbrace{(\mathrm{pF},\mathrm{DP}:\mathrm{UP})}_{\text{hyperbolic flow }(+)} & \underbrace{(\mathrm{SI},\mathrm{SO})}_{\text{saddle}}\end{array}\right\} \quad (5.44)$$

for $a_{110}<0, a_{220}<0$.

The infinite-equilibrium of $x_2^*=a_{122}=a_{222}$ is summarized as follows:

$$\begin{array}{ll}(\bar{x}_1,a_{222})=\underbrace{(\mathrm{DI},\mathrm{SI})}_{\text{decreasing-inflection sink}} & \text{if } \bar{x}_1\in(-\infty,a_{211}),\\(\bar{x}_1,a_{222})=\underbrace{(\mathrm{II},\mathrm{SO})}_{\text{increasing-inflection source}} & \text{if } \bar{x}_1\in(a_{211},a_{111}),\\(\bar{x}_1,a_{222})=\underbrace{(\mathrm{DI},\mathrm{SO})}_{\text{decreasing-infelction source}} & \text{if } \bar{x}_1\in(a_{111},\infty),\end{array} \quad (5.45)$$

for $a_{110}>0, a_{220}>0$;

$$\left.\begin{aligned}
(\bar{x}_1, a_{222}) &= \underbrace{(\mathrm{II}, \mathrm{SI})}_{\text{increasing-inflection sink}} && \text{if } \bar{x}_1 \in (-\infty, a_{211}),\\
(\bar{x}_1, a_{222}) &= \underbrace{(\mathrm{DI}, \mathrm{SO})}_{\text{decreasing-infelction source}} && \text{if } \bar{x}_1 \in (a_{211}, a_{111}),\\
(\bar{x}_1, a_{222}) &= \underbrace{(\mathrm{II}, \mathrm{SO})}_{\text{increasing-inflection source}} && \text{if } \bar{x}_1 \in (a_{111}, \infty),
\end{aligned}\right\} \tag{5.46}$$

for $a_{110} < 0, a_{220} > 0$;

$$\left.\begin{aligned}
(\bar{x}_1, a_{222}) &= \underbrace{(\mathrm{II}, \mathrm{SO})}_{\text{increasing-inflection source}} && \text{if } \bar{x}_1 \in (-\infty, a_{211}),\\
(\bar{x}_1, a_{222}) &= \underbrace{(\mathrm{DI}, \mathrm{SI})}_{\text{decreasing-infelction sink}} && \text{if } \bar{x}_1 \in (a_{211}, a_{111}),\\
(\bar{x}_1, a_{222}) &= \underbrace{(\mathrm{II}, \mathrm{SI})}_{\text{increasing-inflection sink}} && \text{if } \bar{x}_1 \in (a_{111}, \infty),
\end{aligned}\right\} \tag{5.47}$$

for $a_{110} > 0, a_{220} < 0$;

$$\left.\begin{aligned}
(\bar{x}_1, a_{222}) &= \underbrace{(\mathrm{DI}, \mathrm{SO})}_{\text{decreasing-inflection source}} && \text{if } \bar{x}_1 \in (-\infty, a_{211}),\\
(\bar{x}_1, a_{222}) &= \underbrace{(\mathrm{II}, \mathrm{SI})}_{\text{increasing-inflection sink}} && \text{if } \bar{x}_1 \in (a_{211}, a_{111}),\\
(\bar{x}_1, a_{222}) &= \underbrace{(\mathrm{DI}, \mathrm{SI})}_{\text{decreasing-infelction sink}} && \text{if } \bar{x}_1 \in (a_{111}, \infty),
\end{aligned}\right\} \tag{5.48}$$

for $a_{110} < 0, a_{220} < 0$.

The inflection sink and source infinite-equilibriums of $x_2^* = a_{121} = a_{222}$ with $a_{111} > a_{211}$ are the switching bifurcations for the equilibriums for $a_{121} \in (a_{221}, a_{222})$ and $a_{121} > a_{222}$ with $a_{111} > a_{211}$, as shown in Fig. 5.9. In addition, the saddle and hyperbolic flows exist, or a hyperbolic-secant flow with sink or source exist. The infinite-equilibrium is also in the middle between saddle and hyperbolic flow or between the hyperbolic-secant flow and sink or source.

(v) The equilibrium series of $x_2^* = a_{121}, a_{122}, a_{221}, a_{222}$ with $x_1^* = a_{111}$ for $a_{122} = a_{222}$ are

$$\left\{\begin{matrix} (a_{122}, a_{211}) & (a_{111}, a_{222}) \\ (a_{211}, a_{121}) & (a_{111}, a_{121}) \\ (a_{211}, a_{221}) & (a_{111}, a_{221}) \end{matrix}\right\} = \left\{\begin{matrix} \underbrace{(\mathrm{US}, \mathrm{DP})}_{\text{down-parabola upper-saddle}} & \underbrace{(\mathrm{DI}:\mathrm{II}, \mathrm{SO})}_{\text{inflection-source}} \\ \underbrace{(\mathrm{DP}_-, \mathrm{DP}_+)}_{\text{CW center}} & \underbrace{(\mathrm{DP}:\mathrm{UP}, \mathrm{nF})}_{\text{hyperbolic flow (–)}} \\ \underbrace{(\mathrm{nF}, \mathrm{DP}:\mathrm{UP})}_{\text{hyperbolic flow (–)}} & \underbrace{(\mathrm{SO}, \mathrm{SI})}_{\text{saddle}} \end{matrix}\right\}$$

for $a_{110} > 0, a_{220} > 0$,

(5.49)

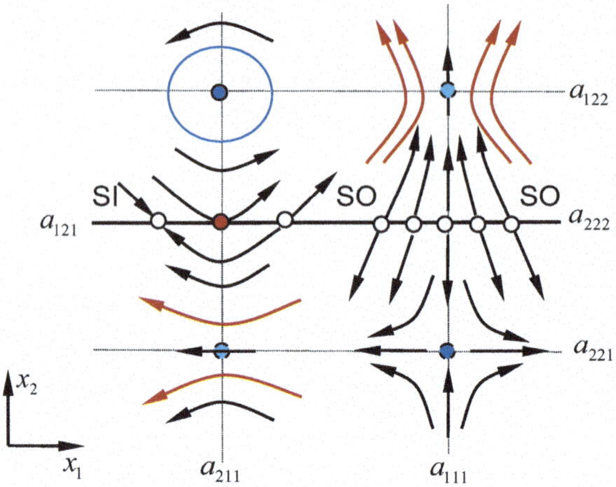

(a)

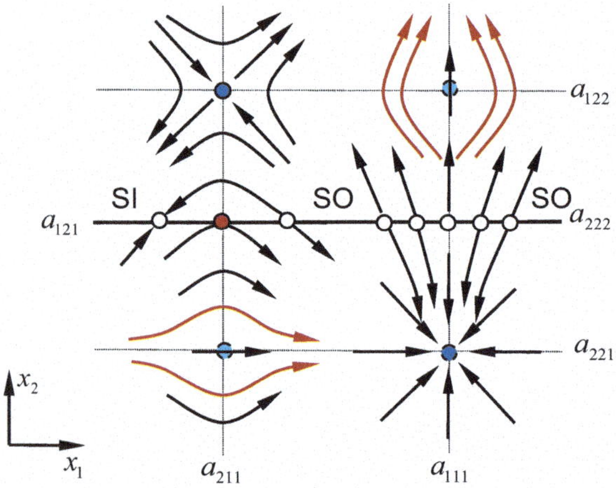

(b)

Fig. 5.9 Phase portraits ($a_{121} = a_{222}$) for two-dimensional systems on the x_1-direction with $x_1^* = a_{111}$ and on the x_2-direction with $x_2^* = a_{121}, a_{221}, a_{222}$. The equilibrium matrix with infinite-equilibrium: (**a**) ($a_{110} > 0$, $a_{220} > 0$), (**b**) ($a_{110} < 0$, $a_{220} > 0$), (**c**) ($a_{110} > 0$, $a_{220} < 0$), and (**d**) ($a_{110} < 0$, $a_{220} < 0$)

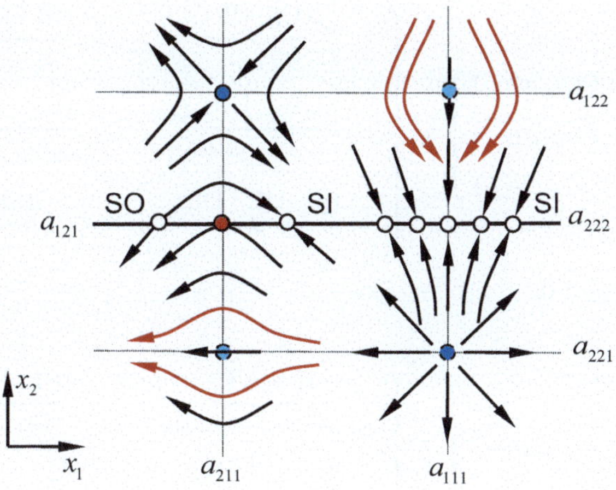

(c)

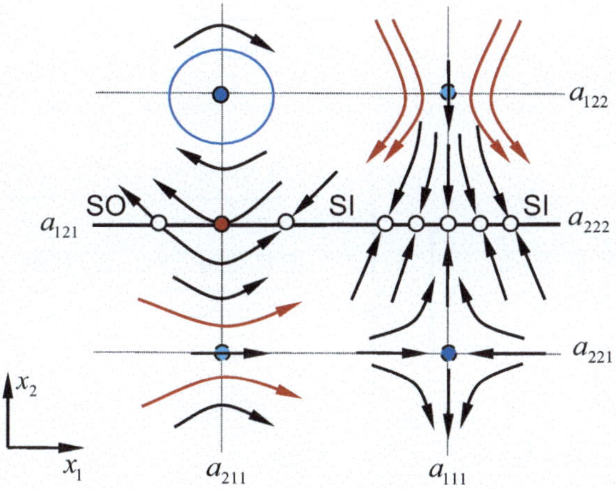

(d)

Fig. 5.9 (continued)

$$\left\{\begin{matrix}(a_{211},a_{122}) & (a_{111},a_{222})\\ (a_{211},a_{121}) & (a_{111},a_{121})\\ (a_{211},a_{221}) & (a_{111},a_{221})\end{matrix}\right\} = \left\{\begin{matrix}\underbrace{(\text{US},\text{UP})}_{\text{up-parabola upper-saddle}} & \underbrace{(\text{II}:\text{DI},\text{SO})}_{\text{inflection-source}}\\ \underbrace{(\text{UP}_-,\text{UP}_-)}_{\text{negative saddle}} & \underbrace{(\text{UP}:\text{DP},\text{nF})}_{\text{hyperbolic-secant flow }(-)}\\ \underbrace{(\text{pF},\text{UP}:\text{DP})}_{\text{hyperbolic-secant flow }(+)} & \underbrace{(\text{SI},\text{SI})}_{\text{sink}}\end{matrix}\right\}$$

for $a_{110} < 0, a_{220} > 0,$ (5.50)

$$\left\{\begin{matrix}(a_{122},a_{211}) & (a_{111},a_{222})\\ (a_{211},a_{121}) & (a_{111},a_{121})\\ (a_{211},a_{221}) & (a_{111},a_{221})\end{matrix}\right\} = \left\{\begin{matrix}\underbrace{(\text{LS},\text{UP})}_{\text{up-parabola lower-saddle}} & \underbrace{(\text{II}:\text{DI},\text{SI})}_{\text{inflection-sink}}\\ \underbrace{(\text{UP}_+,\text{UP}_+)}_{\text{positive saddle}} & \underbrace{(\text{UP}:\text{DP},\text{pF})}_{\text{hyperbolic-secant flow }(+)}\\ \underbrace{(\text{nF},\text{UP}:\text{DP})}_{\text{hyperbolic-secant flow }(-)} & \underbrace{(\text{SO},\text{SO})}_{\text{source}}\end{matrix}\right\}$$

for $a_{110} > 0, a_{220} < 0,$ (5.51)

$$\left\{\begin{matrix}(a_{122},a_{211}) & (a_{111},a_{222})\\ (a_{211},a_{121}) & (a_{111},a_{121})\\ (a_{211},a_{221}) & (a_{111},a_{221})\end{matrix}\right\} = \left\{\begin{matrix}\underbrace{(\text{LS},\text{DP})}_{\text{down-parabola lower-saddle}} & \underbrace{(\text{DI}:\text{II},\text{SI})}_{\text{inflection-sink}}\\ \underbrace{(\text{DP}_+,\text{DP}_-)}_{\text{CCW center}} & \underbrace{(\text{DP}:\text{UP},\text{pF})}_{\text{hyperbolic flow }(+)}\\ \underbrace{(\text{pF},\text{DP}:\text{UP})}_{\text{hyperbolic flow }(+)} & \underbrace{(\text{SI},\text{SO})}_{\text{saddle}}\end{matrix}\right\}$$

for $a_{110} < 0, a_{220} < 0.$ (5.52)

The infinite-equilibrium of $x_2^* = a_{122} = a_{222}$ is summarized as follows:

$$\begin{aligned}&(\bar{x}_1, a_{222}) = \underbrace{(\text{II},\text{SI})}_{\text{increasing-infelction sink}} \quad \text{if } \bar{x}_1 \in (-\infty, a_{211})\\ &(\bar{x}_1, a_{222}) = \underbrace{(\text{DI},\text{SO})}_{\text{decreasing-infelction source}} \quad \text{if } \bar{x}_1 \in (a_{211}, a_{111}),\\ &(\bar{x}_1, a_{222}) = \underbrace{(\text{II},\text{SO})}_{\text{increasing-inflection source}} \quad \text{if } \bar{x}_1 \in (a_{111}, \infty),\end{aligned} \tag{5.53}$$

for $a_{110} > 0, a_{220} > 0;$

$$
\begin{aligned}
&(\bar{x}_1, a_{222}) = \underbrace{(\mathrm{DI}, \mathrm{SI})}_{\text{decreasing-infelction sink}} \quad \text{if } \bar{x}_1 \in (-\infty, a_{211})\\
&(\bar{x}_1, a_{222}) = \underbrace{(\mathrm{II}, \mathrm{SO})}_{\text{increasing-inflection source}} \quad \text{if } \bar{x}_1 \in (a_{211}, a_{111}),\\
&(\bar{x}_1, a_{222}) = \underbrace{(\mathrm{DI}, \mathrm{SO})}_{\text{decreasing-infelction source}} \quad \text{if } \bar{x}_1 \in (a_{111}, \infty),\\
&\text{for } a_{110} < 0, a_{220} > 0;
\end{aligned}
\tag{5.54}
$$

$$
\begin{aligned}
&(\bar{x}_1, a_{222}) = \underbrace{(\mathrm{DI}, \mathrm{SO})}_{\text{decreasing-infelction source}} \quad \text{if } \bar{x}_1 \in (-\infty, a_{211})\\
&(\bar{x}_1, a_{222}) = \underbrace{(\mathrm{II}, \mathrm{SI})}_{\text{increasing-infelction sink}} \quad \text{if } \bar{x}_1 \in (a_{211}, a_{111}),\\
&(\bar{x}_1, a_{222}) = \underbrace{(\mathrm{DI}, \mathrm{SI})}_{\text{decreasing-inflection sink}} \quad \text{if } \bar{x}_1 \in (a_{111}, \infty),\\
&\text{for } a_{110} > 0, a_{220} < 0;
\end{aligned}
\tag{5.55}
$$

$$
\begin{aligned}
&(\bar{x}_1, a_{222}) = \underbrace{(\mathrm{II}, \mathrm{SO})}_{\text{increasing-infelction source}} \quad \text{if } \bar{x}_1 \in (-\infty, a_{211})\\
&(\bar{x}_1, a_{222}) = \underbrace{(\mathrm{DI}, \mathrm{SI})}_{\text{decreasing-inflection sink}} \quad \text{if } \bar{x}_1 \in (a_{211}, a_{111}),\\
&(\bar{x}_1, a_{222}) = \underbrace{(\mathrm{II}, \mathrm{SI})}_{\text{increasing-infelction sink}} \quad \text{if } \bar{x}_1 \in (a_{111}, \infty),\\
&\text{for } a_{110} < 0, a_{220} < 0.
\end{aligned}
\tag{5.56}
$$

The inflection-source and sink infinite-equilibriums of $x_2^* = a_{122} = a_{222}$ with $a_{121} \neq a_{221}$ and $a_{111} > a_{211}$ are the switching bifurcations for sink, source, and saddles for $a_{222} \in (a_{121}, a_{122})$ and $a_{122} < a_{222}$ with $a_{111} > a_{211}$, as presented in Fig. 5.10. On the line of $x_1^* = a_{111}$, the saddle and hyperbolic flows exist, or a hyperbolic-secant flow with sink or source exists. On the line of $x_1^* = a_{211}$, the saddle and hyperbolic-secant flows exist, or a hyperbolic flow with centers exists.

(vi) The equilibrium series of $x_2^* = a_{121}, a_{122}, a_{221}, a_{222}$ with $x_1^* = a_{111}$ for $a_{122} = a_{222}$ are

$$
\left\{
\begin{matrix}
(a_{122}, a_{211}) & (a_{111}, a_{222})\\
(a_{211}, a_{221}) & (a_{111}, a_{221})\\
(a_{121}, a_{211}) & (a_{111}, a_{121})
\end{matrix}
\right\}
=
\left\{
\begin{matrix}
\underbrace{(\mathrm{US}, \mathrm{DP})}_{\text{down-parabola upper-saddle}} & \underbrace{(\mathrm{DI} : \mathrm{II}, \mathrm{SO})}_{\text{inflection-source}}\\
\underbrace{(\mathrm{pF}, \mathrm{UP} : \mathrm{DP})}_{\text{hyperbolic flow } (+)} & \underbrace{(\mathrm{SI}, \mathrm{SI})}_{\text{sink}}\\
\underbrace{(\mathrm{UP}_+, \mathrm{UP}_+)}_{\text{positive saddle}} & \underbrace{(\mathrm{UP} : \mathrm{PP}, \mathrm{pF})}_{\text{hyperbolic-secant flow } (+)}
\end{matrix}
\right\}
$$

$$
\text{for } a_{110} > 0, a_{220} > 0, \tag{5.57}
$$

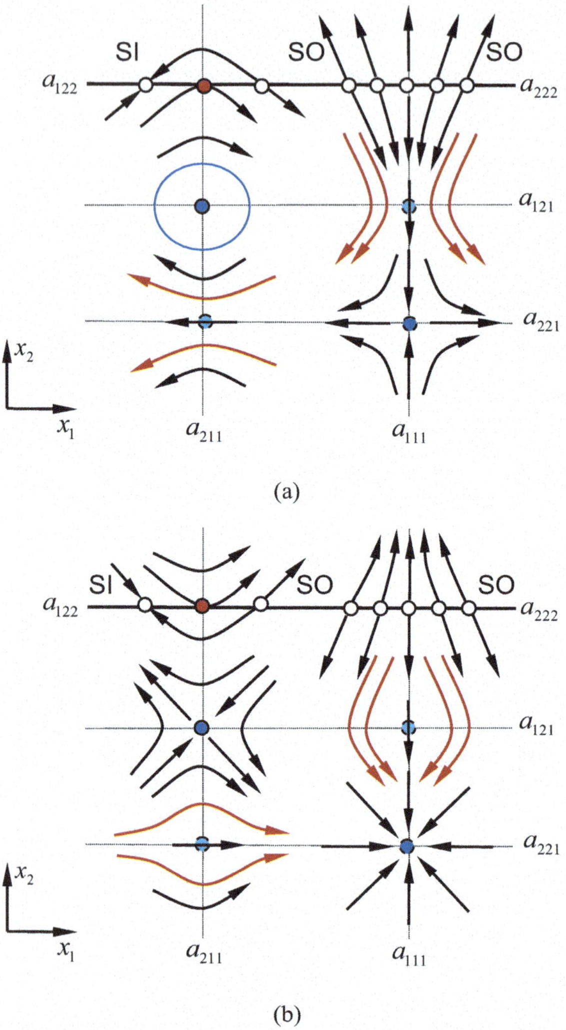

Fig. 5.10 Phase portraits ($a_{122} = a_{222}$, $a_{221} < a_{121}$, $a_{211} < a_{111}$) for two-dimensional systems on the x_1-direction with $x_1^* = a_{111}, a_{211}$ and on the x_2-direction with $x_2^* = a_{121}, a_{122}, a_{221}, a_{222}$. The equilibrium flows: (**a**) ($a_{110} > 0$, $a_{220} > 0$), (**b**) ($a_{110} < 0$, $a_{220} > 0$), (**c**) ($a_{110} > 0$, $a_{220} < 0$), and (**d**) ($a_{110} < 0$, $a_{220} < 0$)

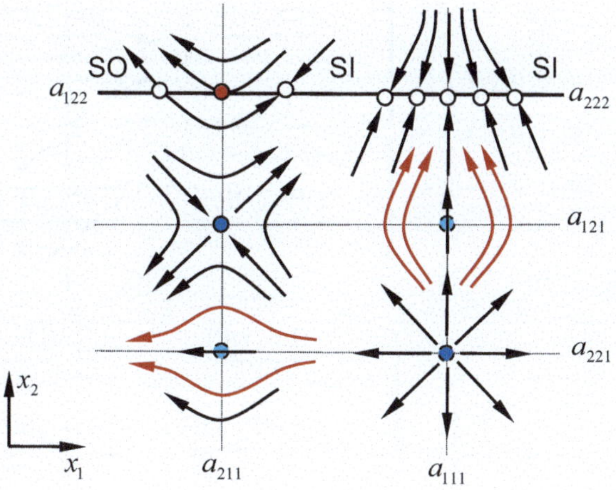

(c)

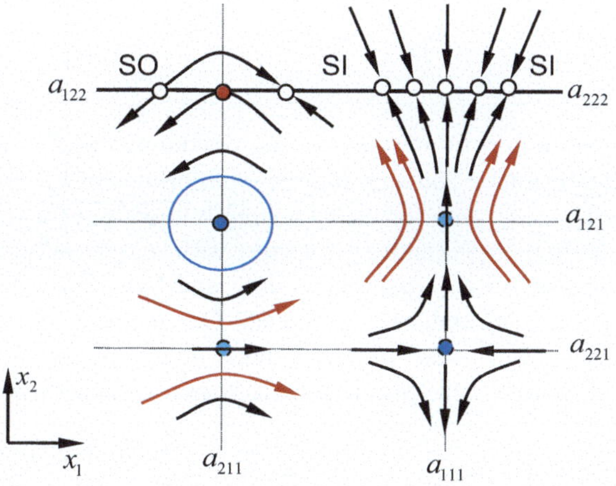

(d)

Fig. 5.10 (continued)

$$\left\{\begin{matrix}(a_{122},a_{211}) & (a_{111},a_{222})\\(a_{211},a_{221}) & (a_{111},a_{221})\\(a_{121},a_{211}) & (a_{111},a_{121})\end{matrix}\right\}=\left\{\begin{matrix}\underbrace{(\mathrm{US},\mathrm{UP})}_{\text{up-parabola upper-saddle}} & \underbrace{(\mathrm{II}:\mathrm{DI},\mathrm{SO})}_{\text{inflection-source}}\\\underbrace{(\mathrm{nF},\mathrm{DP}:\mathrm{UP})}_{\text{hyperbolic flow }(-)} & \underbrace{(\mathrm{SO},\mathrm{SI})}_{\text{saddle}}\\\underbrace{(\mathrm{DP}_{+},\mathrm{DP}_{-})}_{\text{CCW center}} & \underbrace{(\mathrm{DP}:\mathrm{UP},\mathrm{pF})}_{\text{hyperbolic flow }(+)}\end{matrix}\right\} \tag{5.58}$$

for $a_{110} < 0, a_{220} > 0$,

$$\left\{\begin{matrix} (a_{122}, a_{211}) & (a_{111}, a_{222}) \\ (a_{211}, a_{221}) & (a_{111}, a_{221}) \\ (a_{121}, a_{211}) & (a_{111}, a_{121}) \end{matrix}\right\} = \left\{\begin{matrix} \underbrace{(\mathrm{LS}, \mathrm{UP})}_{\text{up-parabola lower-saddle}} & \underbrace{(\mathrm{II} : \mathrm{DI}, \mathrm{SI})}_{\text{inflection-sink}} \\ \underbrace{(\mathrm{pF}, \mathrm{DP} : \mathrm{UP})}_{\text{hyperbolic flow } (+)} & \underbrace{(\mathrm{SI}, \mathrm{SO})}_{\text{saddle}} \\ \underbrace{(\mathrm{DP}_-, \mathrm{DP}_+)}_{\text{CW center}} & \underbrace{(\mathrm{DP} : \mathrm{UP}, \mathrm{nF})}_{\text{hyperbolic flow } (-)} \end{matrix}\right\} \tag{5.59}$$

for $a_{110} > 0, a_{220} < 0$,

$$\left\{\begin{matrix} (a_{122}, a_{211}) & (a_{111}, a_{222}) \\ (a_{211}, a_{221}) & (a_{111}, a_{221}) \\ (a_{121}, a_{211}) & (a_{111}, a_{121}) \end{matrix}\right\} = \left\{\begin{matrix} \underbrace{(\mathrm{LS}, \mathrm{DP})}_{\text{down-parabola lower-saddle}} & \underbrace{(\mathrm{DI} : \mathrm{II}, \mathrm{SI})}_{\text{inflection-sink}} \\ \underbrace{(\mathrm{nF}, \mathrm{UP} : \mathrm{DP})}_{\text{hyperbolic-secant flow } (-)} & \underbrace{(\mathrm{SO}, \mathrm{SO})}_{\text{source}} \\ \underbrace{(\mathrm{UP}_-, \mathrm{UP}_-)}_{\text{negative saddle}} & \underbrace{(\mathrm{UP} : \mathrm{PP}, \mathrm{nF})}_{\text{hyperbolic-secant flow } (-)} \end{matrix}\right\}$$

for $a_{110} < 0, a_{220} < 0$.

(5.60)

The infinite-equilibrium of $x_2^* = a_{122} = a_{222}$ is summarized as in Eqs. (5.53)–(5.56). The inflection source and sink infinite-equilibriums of $x_2^* = a_{122} = a_{222}$ with $a_{121} \neq a_{221}$ and $a_{111} > a_{211}$ are for the switching bifurcations for sink, source, and saddles for $a_{122} \in (a_{221}, a_{222})$ and $a_{222} < a_{122}$ with $a_{111} > a_{211}$, as presented in Fig. 5.11. On the line of $x_1^* = a_{111}$, the saddle and hyperbolic flows exist, or a hyperbolic-secant flow with sink or source exists. On the line of $x_1^* = a_{211}$, the saddle and hyperbolic-secant flows exist, or a hyperbolic flow with centers exists.

The readers can complete the phase portraits for six cases with $a_{111} < a_{211}$ as an exercise.

B. Consider a dynamical system as

$$\begin{aligned} \dot{x}_1 &= a_{110}(x_1 - a_{111})(x_2 - a_{121})(x_2 - a_{122}), \\ \dot{x}_2 &= a_{220}(x_1 - a_{211})(x_2 - a_{221})(x_2 - a_{222}), \end{aligned} \tag{5.61}$$

and the corresponding first integral manifold for $a_{121} = a_{221}$ and $a_{122} = a_{222}$ with $a_{111} \neq a_{211}$ is

$$(x_1 - x_{10}) + (a_{111} - a_{211}) \ln \frac{|x_1 - a_{111}|}{|x_{10} - a_{111}|} = \frac{a_{110}}{a_{220}}(x_2 - x_{20}). \tag{5.62}$$

The equilibrium matrices of $x_2^* = a_{121}, a_{122}, a_{221}, a_{222}$ with $x_1^* = a_{111}, a_{211}$ for $a_{121} = a_{221}$ and $a_{122} = a_{222}$ with $a_{111} \neq a_{211}$ are

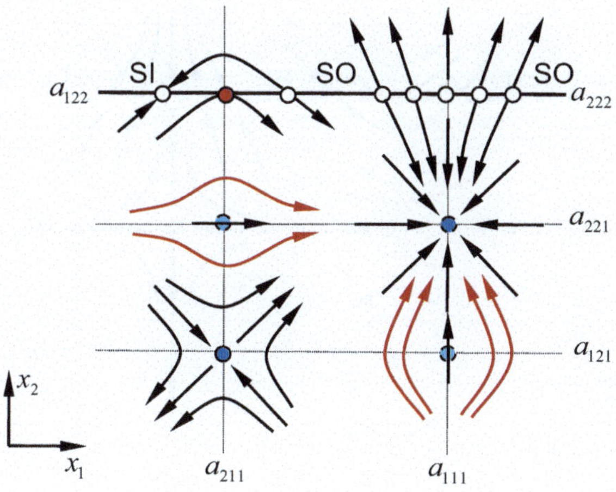

(a)

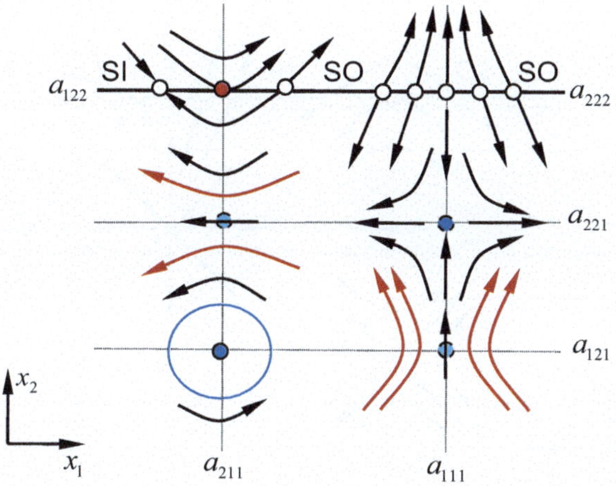

(b)

Fig. 5.11 Phase portraits ($a_{122} = a_{222}$, $a_{121} < a_{221}$, $a_{211} < a_{111}$) for two-dimensional systems on the x_1-direction with $x_1^* = a_{111}, a_{211}$ and on the x_2-direction with $x_2^* = a_{121}, a_{122}, a_{221}, a_{222}$. The equilibrium flows: (**a**) ($a_{110} > 0$, $a_{220} > 0$), (**b**) ($a_{110} < 0$, $a_{220} > 0$), (**c**) ($a_{110} > 0$, $a_{220} < 0$), and (**d**) ($a_{110} < 0$, $a_{220} < 0$)

$$\begin{Bmatrix} (a_{122}, a_{211}) & (a_{111}, a_{222}) \\ (a_{121}, a_{211}) & (a_{111}, a_{221}) \end{Bmatrix} = \begin{Bmatrix} \underbrace{(\text{US}, \text{DP})}_{\text{down-parabola upper-saddle}} & \underbrace{(\text{DI}:\text{II}, \text{SO})}_{\text{inflection-source}} \\ \underbrace{(\text{LS}, \text{DP})}_{\text{down-parabola lower-saddle}} & \underbrace{(\text{DI}:\text{II}, \text{SI})}_{\text{inflection-sink}} \end{Bmatrix} \tag{5.63}$$

for $a_{110} > 0, a_{220} > 0$,

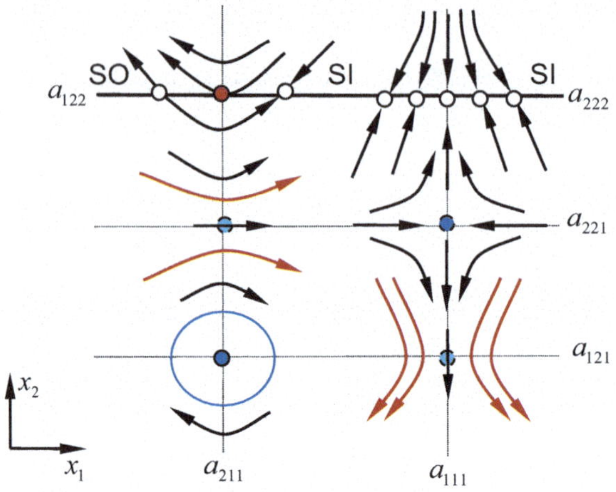

(c)

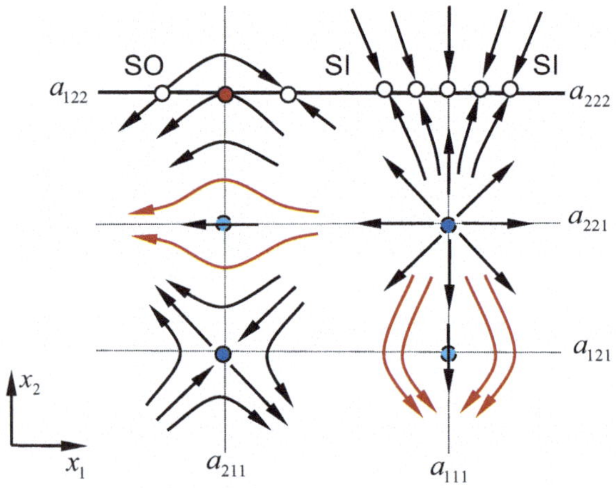

(d)

Fig. 5.11 (continued)

$$\left\{ \begin{array}{cc} (a_{122}, a_{211}) & (a_{111}, a_{222}) \\ (a_{121}, a_{211}) & (a_{111}, a_{221}) \end{array} \right\} = \left\{ \begin{array}{cc} \underbrace{(\text{US}, \text{UP})}_{\text{up-parabola upper-saddle}} & \underbrace{(\text{II} : \text{DI}, \text{SO})}_{\text{inflection-source}} \\ \underbrace{(\text{LS}, \text{UP})}_{\text{up-parabola lower-saddle}} & \underbrace{(\text{II} : \text{DI}, \text{SI})}_{\text{inflection-sink}} \end{array} \right\} \tag{5.64}$$

for $a_{110} < 0, a_{220} > 0,$

$$\left\{\begin{matrix} (a_{122},a_{211}) & (a_{111},a_{222}) \\ (a_{121},a_{211}) & (a_{111},a_{221}) \end{matrix}\right\} = \left\{\begin{matrix} \underbrace{(\text{LS},\text{UP})}_{\text{up-parabola lower-saddle}} & \underbrace{(\text{II}:\text{DI},\text{SI})}_{\text{inflection-sink}} \\ \underbrace{(\text{US},\text{UP})}_{\text{up-parabola upper-saddle}} & \underbrace{(\text{II}:\text{DI},\text{SO})}_{\text{inflection-source}} \end{matrix}\right\} \tag{5.24}$$

for $a_{110} > 0, a_{220} < 0$,

$$\left\{\begin{matrix} (a_{122},a_{211}) & (a_{111},a_{222}) \\ (a_{121},a_{211}) & (a_{111},a_{221}) \end{matrix}\right\} = \left\{\begin{matrix} \underbrace{(\text{LS},\text{DP})}_{\text{down-parabola lower-saddle}} & \underbrace{(\text{DI}:\text{II},\text{SI})}_{\text{inflection-sink}} \\ \underbrace{(\text{US},\text{DP})}_{\text{down-parabola upper-saddle}} & \underbrace{(\text{DI}:\text{II},\text{SO})}_{\text{inflection-source}} \end{matrix}\right\} \tag{5.65}$$

for $a_{110} < 0, a_{220} < 0$,

and the infinite-equilibrium of $x_2^* = a_{122} = a_{222}$ with $a_{111} \neq a_{211}$ is summarized as follows:

$$\begin{aligned} (\bar{x}_1, a_{222}) &= \underbrace{(\text{II},\text{SI})}_{\text{increasing-inflection sink}} && \text{if } \bar{x}_1 \in (-\infty, a_{211}), \\ (\bar{x}_1, a_{222}) &= \underbrace{(\text{DI},\text{SO})}_{\text{decreasing-infelction source}} && \text{if } \bar{x}_1 \in (a_{211}, a_{111}), \\ (\bar{x}_1, a_{222}) &= \underbrace{(\text{II},\text{SO})}_{\text{increasing-inflection source}} && \text{if } \bar{x}_1 \in (a_{111}, \infty), \end{aligned} \tag{5.66}$$

for $a_{110} > 0, a_{220} > 0$;

$$\begin{aligned} (\bar{x}_1, a_{222}) &= \underbrace{(\text{DI},\text{SI})}_{\text{decreasing-inflection sink}} && \text{if } \bar{x}_1 \in (-\infty, a_{211}), \\ (\bar{x}_1, a_{222}) &= \underbrace{(\text{II},\text{SO})}_{\text{increasing-infelction source}} && \text{if } \bar{x}_1 \in (a_{211}, a_{111}), \\ (\bar{x}_1, a_{222}) &= \underbrace{(\text{DI},\text{SO})}_{\text{decreasing-inflection source}} && \text{if } \bar{x}_1 \in (a_{111}, \infty), \end{aligned} \tag{5.67}$$

for $a_{110} < 0, a_{220} > 0$.

$$\begin{aligned} (\bar{x}_1, a_{222}) &= \underbrace{(\text{DI},\text{SO})}_{\text{decreasing-infelction source}} && \text{if } \bar{x}_1 \in (-\infty, a_{211}) \\ (\bar{x}_1, a_{222}) &= \underbrace{(\text{II},\text{SI})}_{\text{increasing-infelction sink}} && \text{if } \bar{x}_1 \in (a_{211}, a_{111}), \\ (\bar{x}_1, a_{222}) &= \underbrace{(\text{DI},\text{SI})}_{\text{decreasing-inflection sink}} && \text{if } \bar{x}_1 \in (a_{111}, \infty), \end{aligned} \tag{5.68}$$

for $a_{110} > 0, a_{220} < 0$;

$$\begin{aligned}
(\bar{x}_1, a_{222}) &= \underbrace{(\text{II}, \text{SO})}_{\text{increasing-infelction source}} \quad \text{if } \bar{x}_1 \in (-\infty, a_{211})\\
(\bar{x}_1, a_{222}) &= \underbrace{(\text{DI}, \text{SI})}_{\text{decreasing-infelction sink}} \quad \text{if } \bar{x}_1 \in (a_{211}, a_{111}),\\
(\bar{x}_1, a_{222}) &= \underbrace{(\text{II}, \text{SI})}_{\text{increasing-inflection sink}} \quad \text{if } \bar{x}_1 \in (a_{111}, \infty),\\
&\text{for } a_{110} < 0, a_{220} < 0.
\end{aligned} \tag{5.69}$$

The infinite-equilibrium of $x_2^* = a_{121} = a_{221}$ with $a_{111} \neq a_{211}$ is summarized as follows:

$$\begin{aligned}
(\bar{x}_1, a_{221}) &= \underbrace{(\text{II}, \text{SO})}_{\text{increasing-inflection source}} \quad \text{if } \bar{x}_1 \in (-\infty, a_{211}),\\
(\bar{x}_1, a_{221}) &= \underbrace{(\text{DI}, \text{SI})}_{\text{decreasing-infelction sink}} \quad \text{if } \bar{x}_1 \in (a_{211}, a_{111}),\\
(\bar{x}_1, a_{221}) &= \underbrace{(\text{II}, \text{SI})}_{\text{increasing-inflection sink}} \quad \text{if } \bar{x}_1 \in (a_{211}, \infty),\\
&\text{for } a_{110} > 0, a_{220} > 0,
\end{aligned} \tag{5.70}$$

$$\begin{aligned}
(\bar{x}_1, a_{221}) &= \underbrace{(\text{DI}, \text{SO})}_{\text{decreasing-inflection source}} \quad \text{if } \bar{x}_1 \in (-\infty, a_{211}),\\
(\bar{x}_1, a_{221}) &= \underbrace{(\text{II}, \text{SI})}_{\text{increasing-infelction sink}} \quad \text{if } \bar{x}_1 \in (a_{211}, a_{111}),\\
(\bar{x}_1, a_{221}) &= \underbrace{(\text{DI}, \text{SI})}_{\text{decreasing-inflection sink}} \quad \text{if } \bar{x}_1 \in (a_{111}, \infty),\\
&\text{for } a_{110} < 0, a_{220} > 0;
\end{aligned} \tag{5.71}$$

$$\begin{aligned}
(\bar{x}_1, a_{221}) &= \underbrace{(\text{DI}, \text{SI})}_{\text{decreasing-inflection sink}} \quad \text{if } \bar{x}_1 \in (-\infty, a_{211}),\\
(\bar{x}_1, a_{221}) &= \underbrace{(\text{II}, \text{SO})}_{\text{increasing-infelction source}} \quad \text{if } \bar{x}_1 \in (a_{211}, a_{111}),\\
(\bar{x}_1, a_{221}) &= \underbrace{(\text{DI}, \text{SO})}_{\text{decreasing-inflection source}} \quad \text{if } \bar{x}_1 \in (a_{111}, \infty),\\
&\text{for } a_{110} > 0, a_{220} < 0,
\end{aligned} \tag{5.72}$$

$$\begin{aligned}
(\bar{x}_1, a_{221}) &= \underbrace{(\text{II}, \text{SI})}_{\text{increasing-inflection sink}} \quad \text{if } \bar{x}_1 \in (-\infty, a_{211}),\\
(\bar{x}_1, a_{221}) &= \underbrace{(\text{DI}, \text{SO})}_{\text{decreasing-infelction source}} \quad \text{if } \bar{x}_1 \in (a_{211}, a_{111}),\\
(\bar{x}_1, a_{221}) &= \underbrace{(\text{II}, \text{SO})}_{\text{increasing-inflection source}} \quad \text{if } \bar{x}_1 \in (a_{111}, \infty),\\
&\text{for } a_{110} < 0, a_{220} < 0.
\end{aligned} \tag{5.73}$$

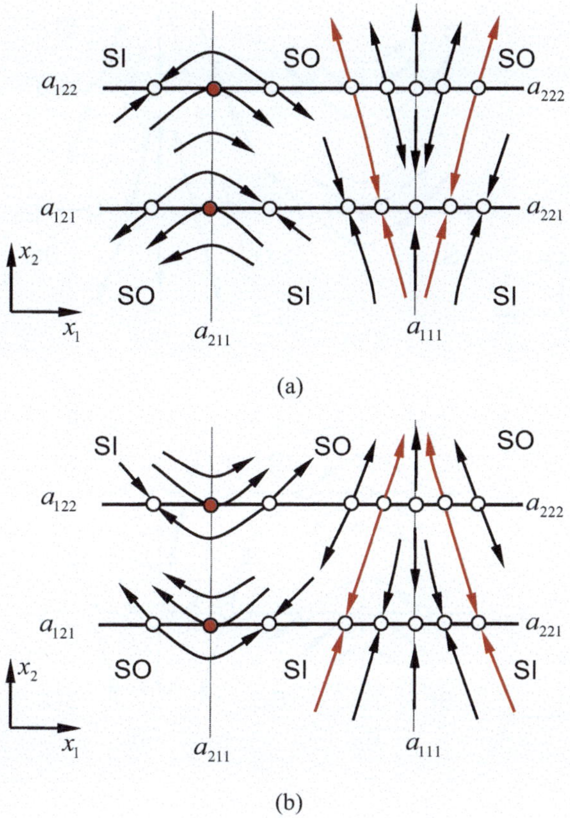

Fig. 5.12 Phase portraits ($a_{121} = a_{221}$ and $a_{122} = a_{222}$) for two-dimensional systems on the x_1-direction with $x_1^* = a_{111}$ and on the x_2-direction with $x_2^* = a_{121}, a_{221}, a_{222}$. The infinite-equilibriums: **(a)** ($a_{110} > 0$, $a_{220} > 0$), **(b)** ($a_{110} < 0$, $a_{220} > 0$), **(c)** ($a_{110} > 0$, $a_{220} < 0$), and **(d)** ($a_{110} < 0$, $a_{220} < 0$)

The double inflection-sink and source infinite-equilibriums of $x_2^* = a_{121} = a_{221}$ and $x_2^* = a_{122} = a_{222}$ are presented in Fig. 5.12.

Similarly, the readers can complete the phase portraits for $a_{111} < a_{211}$ as an exercise.

5.2.2 *Infinite-Equilibriums on the Vertical Direction*

Consider a dynamical system for $a_{111} = a_{211}$ as

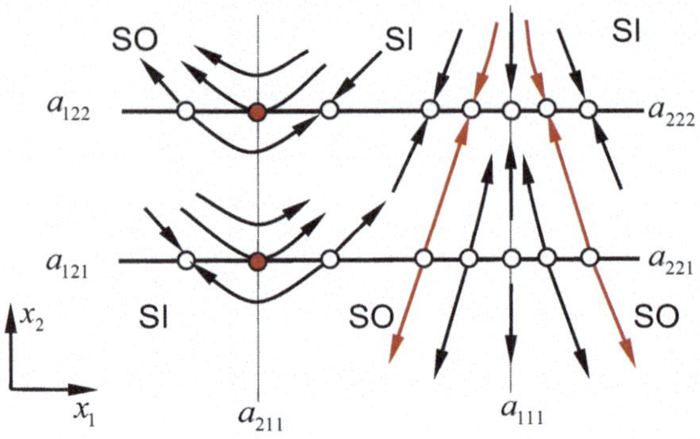

(c)

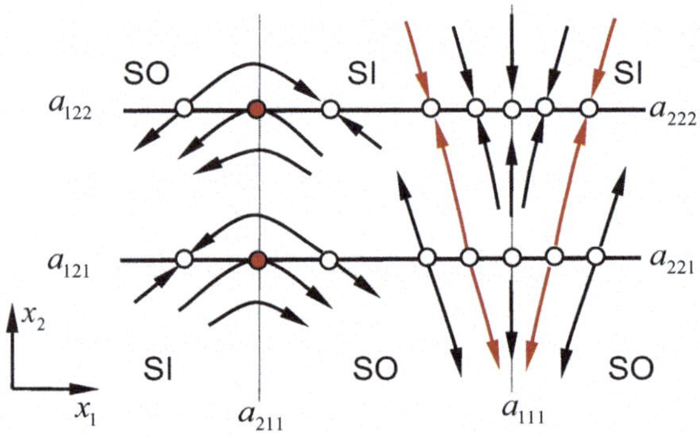

(d)

Fig. 5.12 (continued)

$$\begin{aligned}\dot{x}_1 &= a_{110}(x_1 - a_{111})(x_2 - a_{121})(x_2 - a_{122}),\\ \dot{x}_2 &= a_{220}(x_1 - a_{211})(x_2 - a_{221})(x_2 - a_{222}),\end{aligned} \tag{5.74}$$

and the corresponding first integral manifold for $a_{111} = a_{211}$ is

$$\begin{aligned}x_1 - x_{10} = \frac{a_{110}}{a_{220}} \Bigg[(x_2 - x_{20}) &+ \frac{(a_{221} - a_{121})(a_{221} - a_{122})}{a_{221} - a_{222}} \ln \frac{|x_2 - a_{221}|}{|x_{20} - a_{221}|} \\ &+ \frac{(a_{222} - a_{121})(a_{222} - a_{121})}{a_{222} - a_{221}} \ln \frac{|x_2 - a_{222}|}{|x_{20} - a_{222}|} \Bigg].\end{aligned} \tag{4.75}$$

(i) For the case of $a_{122} < a_{221}$, the equilibrium series of $x_1^* = a_{111} = a_{211}$ with $x_2^* = a_{121}, a_{122}, a_{221}, a_{222}$ have the following properties, as shown in Fig. 5.13.

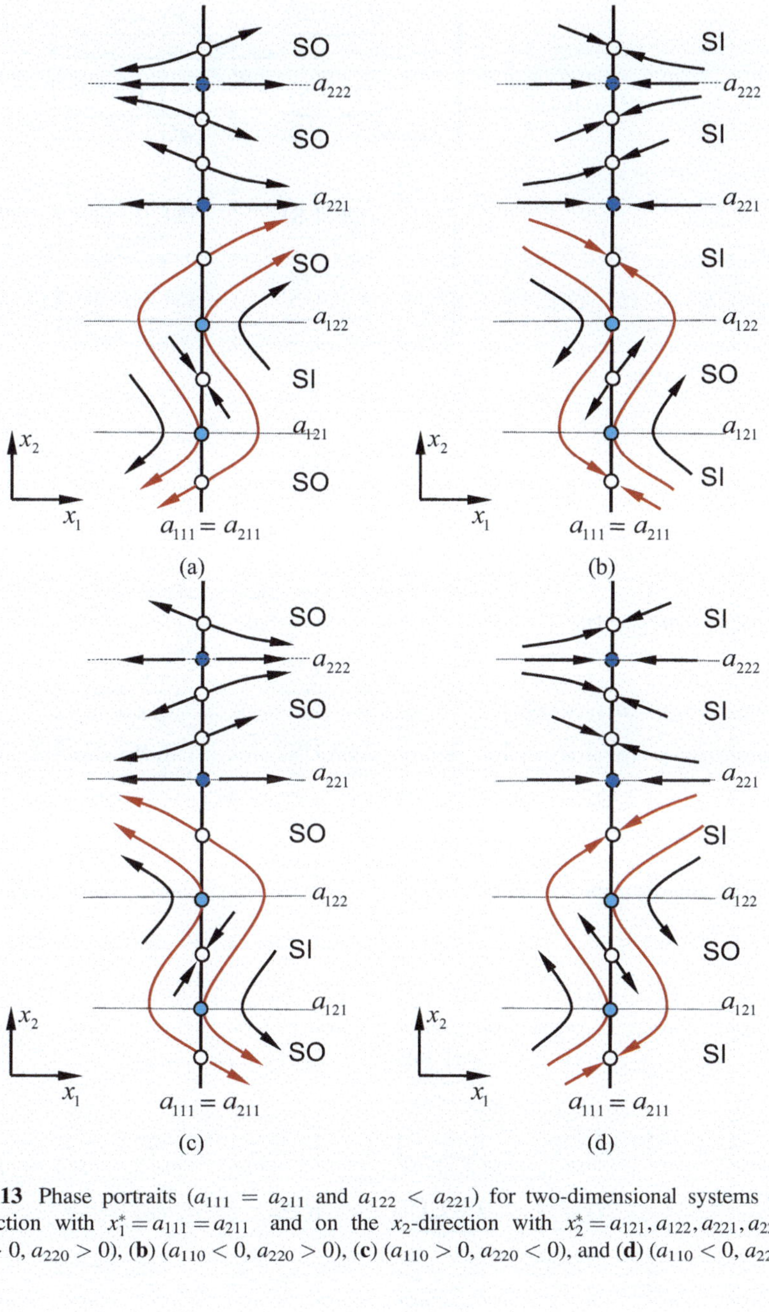

Fig. 5.13 Phase portraits ($a_{111} = a_{211}$ and $a_{122} < a_{221}$) for two-dimensional systems on the x_1-direction with $x_1^* = a_{111} = a_{211}$ and on the x_2-direction with $x_2^* = a_{121}, a_{122}, a_{221}, a_{222}$. **(a)** ($a_{110} > 0$, $a_{220} > 0$), **(b)** ($a_{110} < 0$, $a_{220} > 0$), **(c)** ($a_{110} > 0$, $a_{220} < 0$), and **(d)** ($a_{110} < 0$, $a_{220} < 0$)

$$\left\{\begin{array}{cc}(a_{122},a_{211}) & (a_{111},a_{222})\\ (a_{121},a_{211}) & (a_{111},a_{221})\end{array}\right\} = \left\{\begin{array}{cc}\underbrace{(\mathrm{UP},\mathrm{US})}_{\text{up-parabola upper-saddle}} & \underbrace{(\mathrm{SO},\mathrm{DI}:\mathrm{II})}_{\text{inflection-source}}\\ \underbrace{(\mathrm{DP},\mathrm{LS})}_{\text{down-parabola lower-saddle}} & \underbrace{(\mathrm{SO},\mathrm{II}:\mathrm{DI})}_{\text{inflection-source}}\end{array}\right\} \tag{5.76}$$

for $a_{110} > 0, a_{220} > 0$,

$$\left\{\begin{array}{cc}(a_{122},a_{211}) & (a_{111},a_{222})\\ (a_{121},a_{211}) & (a_{111},a_{221})\end{array}\right\} = \left\{\begin{array}{cc}\underbrace{(\mathrm{DP},\mathrm{LS})}_{\text{down-parabola lower-saddle}} & \underbrace{(\mathrm{SI},\mathrm{II}:\mathrm{DI})}_{\text{inflection-sink}}\\ \underbrace{(\mathrm{UP},\mathrm{US})}_{\text{up-parabola upper-saddle}} & \underbrace{(\mathrm{SI},\mathrm{DI}:\mathrm{II})}_{\text{inflection-sink}}\end{array}\right\} \tag{5.77}$$

for $a_{110} < 0, a_{220} > 0$,

$$\left\{\begin{array}{cc}(a_{122},a_{211}) & (a_{111},a_{222})\\ (a_{121},a_{211}) & (a_{111},a_{221})\end{array}\right\} = \left\{\begin{array}{cc}\underbrace{(\mathrm{DP},\mathrm{US})}_{\text{down-parabola upper-saddle}} & \underbrace{(\mathrm{SO},\mathrm{II}:\mathrm{DI})}_{\text{inflection-source}}\\ \underbrace{(\mathrm{UP},\mathrm{LS})}_{\text{up-parabola lower-saddle}} & \underbrace{(\mathrm{SO},\mathrm{DI}:\mathrm{II})}_{\text{inflection-source}}\end{array}\right\} \tag{5.78}$$

for $a_{110} > 0, a_{220} < 0$,

$$\left\{\begin{array}{cc}(a_{122},a_{211}) & (a_{111},a_{222})\\ (a_{121},a_{211}) & (a_{111},a_{221})\end{array}\right\} = \left\{\begin{array}{cc}\underbrace{(\mathrm{UP},\mathrm{LS})}_{\text{up-parabola lower-saddle}} & \underbrace{(\mathrm{SI},\mathrm{DI}:\mathrm{II})}_{\text{inflection-sink}}\\ \underbrace{(\mathrm{DP},\mathrm{US})}_{\text{down-parabola upper-saddle}} & \underbrace{(\mathrm{SI},\mathrm{II}:\mathrm{DI})}_{\text{inflection-sink}}\end{array}\right\} \tag{5.79}$$

for $a_{110} < 0, a_{220} < 0$,

and the infinite-equilibrium of $x_1^* = a_{111} = a_{211}$ with $a_{111} \neq a_{211}$ is summarized as follows:

$$\begin{array}{lll}(a_{111},\bar{x}_2) = & \underbrace{(\mathrm{SO},\mathrm{II})}_{\text{increasing-inflection source}} & \text{if } \bar{x}_2 \in (-\infty, a_{121}),\\ (a_{111},\bar{x}_2) = & \underbrace{(\mathrm{SI},\mathrm{DI})}_{\text{decreasing-inflection source}} & \text{if } \bar{x}_2 \in (a_{121}, a_{122}),\\ (a_{111},\bar{x}_2) = & \underbrace{(\mathrm{SO},\mathrm{II})}_{\text{increasing-inflection source}} & \text{if } \bar{x}_2 \in (a_{122}, a_{221}),\\ (a_{111},\bar{x}_2) = & \underbrace{(\mathrm{SO},\mathrm{DI})}_{\text{increasing-inflection source}} & \text{if } \bar{x}_2 \in (a_{221}, a_{222}),\\ (a_{111},\bar{x}_2) = & \underbrace{(\mathrm{SO},\mathrm{II})}_{\text{increasing-inflection source}} & \text{if } \bar{x}_2 \in (a_{222}, \infty),\end{array} \tag{5.80}$$

for $a_{110} > 0, a_{220} > 0$;

$$
\begin{aligned}
&(a_{111},\bar{x}_2) = \underbrace{(\mathrm{SI},\mathrm{DI})}_{\text{decreasing-inflection sink}} \quad \text{if } \bar{x}_2 \in (-\infty, a_{121}),\\
&(a_{111},\bar{x}_2) = \underbrace{(\mathrm{SO},\mathrm{II})}_{\text{increasing-inflection source}} \quad \text{if } \bar{x}_2 \in (a_{121}, a_{122}),\\
&(a_{111},\bar{x}_2) = \underbrace{(\mathrm{SI},\mathrm{DI})}_{\text{decreasing-inflection sink}} \quad \text{if } \bar{x}_2 \in (a_{122}, a_{221}),\\
&(a_{111},\bar{x}_2) = \underbrace{(\mathrm{SI},\mathrm{II})}_{\text{increasing-inflection sink}} \quad \text{if } \bar{x}_2 \in (a_{221}, a_{222}),\\
&(a_{111},\bar{x}_2) = \underbrace{(\mathrm{SI},\mathrm{DI})}_{\text{decreasing-inflection sink}} \quad \text{if } \bar{x}_2 \in (a_{222}, \infty),\\
&\text{for } a_{110} < 0, a_{220} > 0;
\end{aligned}
\tag{5.81}
$$

$$
\begin{aligned}
&(a_{111},\bar{x}_2) = \underbrace{(\mathrm{SO},\mathrm{DI})}_{\text{decreasing-inflection source}} \quad \text{if } \bar{x}_2 \in (-\infty, a_{121}),\\
&(a_{111},\bar{x}_2) = \underbrace{(\mathrm{SI},\mathrm{II})}_{\text{increasing-inflection sink}} \quad \text{if } \bar{x}_2 \in (a_{121}, a_{122}),\\
&(a_{111},\bar{x}_2) = \underbrace{(\mathrm{SO},\mathrm{DI})}_{\text{decreasing-inflection source}} \quad \text{if } \bar{x}_2 \in (a_{122}, a_{221}),\\
&(a_{111},\bar{x}_2) = \underbrace{(\mathrm{SO},\mathrm{II})}_{\text{increasing-inflection source}} \quad \text{if } \bar{x}_2 \in (a_{221}, a_{222}),\\
&(a_{111},\bar{x}_2) = \underbrace{(\mathrm{SO},\mathrm{DI})}_{\text{decreasing-inflection source}} \quad \text{if } \bar{x}_2 \in (a_{222}, \infty),\\
&\text{for } a_{110} > 0, a_{220} < 0;
\end{aligned}
\tag{5.82}
$$

$$
\begin{aligned}
&(a_{111},\bar{x}_2) = \underbrace{(\mathrm{SI},\mathrm{II})}_{\text{increasing-inflection sink}} \quad \text{if } \bar{x}_2 \in (-\infty, a_{121}),\\
&(a_{111},\bar{x}_2) = \underbrace{(\mathrm{SO},\mathrm{DI})}_{\text{decreasing-inflection source}} \quad \text{if } \bar{x}_2 \in (a_{121}, a_{122}),\\
&(a_{111},\bar{x}_2) = \underbrace{(\mathrm{SI},\mathrm{II})}_{\text{increasing-inflection sink}} \quad \text{if } \bar{x}_2 \in (a_{122}, a_{221}),\\
&(a_{111},\bar{x}_2) = \underbrace{(\mathrm{SI},\mathrm{DI})}_{\text{decreasing-inflection sink}} \quad \text{if } \bar{x}_2 \in (a_{221}, a_{222}),\\
&(a_{111},\bar{x}_2) = \underbrace{(\mathrm{SI},\mathrm{II})}_{\text{increasing-inflection sink}} \quad \text{if } \bar{x}_2 \in (a_{222}, \infty),\\
&\text{for } a_{110} < 0, a_{220} < 0.
\end{aligned}
\tag{5.83}
$$

The infinite-equilibriums of $x_1^* = a_{111} = a_{211}$ are in the vertical direction. The inflection-sources (sinks) on the top and the parabola-saddles on the bottom are given, which are the switching bifurcations of $a_{111} > a_{211}$ and $a_{111} < a_{211}$ with $a_{122} < a_{221}$.

(ii) For $a_{221} \in (a_{121}, a_{122})$ and $a_{122} < a_{222}$, the infinite-equilibriums of $x_1^* = a_{111} = a_{211}$ with $x_2^* = a_{121}, a_{221}, a_{222}$ have the following properties, as shown in Fig. 5.14.

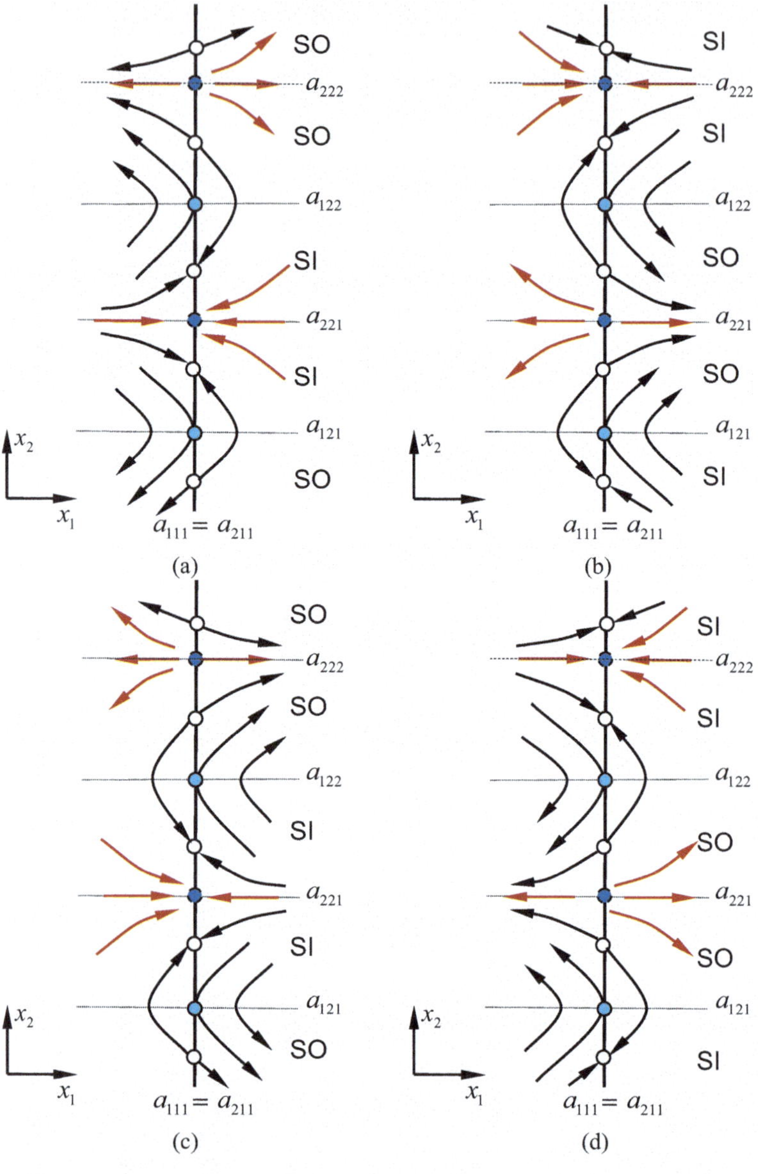

Fig. 5.14 Phase portraits ($a_{221} \in (a_{121}, a_{122})$, $a_{122} < a_{222}$) for two-dimensional systems on the x_1-direction with $x_1^* = a_{111} = a_{211}$ and on the x_2-direction with $x_2^* = a_{121}, a_{122}, a_{221}, a_{222}$. (**a**) ($a_{110} > 0$, $a_{220} > 0$), (**b**) ($a_{110} < 0$, $a_{220} > 0$), (**c**) ($a_{110} > 0$, $a_{220} < 0$), and (**d**) ($a_{110} < 0$, $a_{220} < 0$)

$$\left\{\begin{array}{ll}(a_{111},a_{221}) & (a_{111},a_{222})\\ (a_{121},a_{211}) & (a_{122},a_{211})\end{array}\right\}=\left\{\begin{array}{cc}\underbrace{(\mathrm{SI},\mathrm{DI}:\mathrm{II})}_{\text{inflection-sink}} & \underbrace{(\mathrm{SO},\mathrm{DI}:\mathrm{II})}_{\text{inflection-source}}\\ \underbrace{(\mathrm{DP},\mathrm{LS})}_{\text{down-parabola lower-saddle}} & \underbrace{(\mathrm{DP},\mathrm{US})}_{\text{down-parabola upper-saddle}}\end{array}\right\}$$

for $a_{110}>0, a_{220}>0$, (5.84)

$$\left\{\begin{array}{ll}(a_{111},a_{221}) & (a_{111},a_{222})\\ (a_{121},a_{211}) & (a_{122},a_{211})\end{array}\right\}=\left\{\begin{array}{cc}\underbrace{(\mathrm{SO},\mathrm{II}:\mathrm{DI})}_{\text{inflection-source}} & \underbrace{(\mathrm{SI},\mathrm{II}:\mathrm{DI})}_{\text{inflection-sink}}\\ \underbrace{(\mathrm{UP},\mathrm{US})}_{\text{up-parabola upper-saddle}} & \underbrace{(\mathrm{UP},\mathrm{LS})}_{\text{up-parabola lower-saddle}}\end{array}\right\}$$

for $a_{110}<0, a_{220}>0$, (5.85)

$$\left\{\begin{array}{ll}(a_{111},a_{221}) & (a_{111},a_{222})\\ (a_{121},a_{211}) & (a_{122},a_{211})\end{array}\right\}=\left\{\begin{array}{cc}\underbrace{(\mathrm{SI},\mathrm{II}:\mathrm{DI})}_{\text{inflection-sink}} & \underbrace{(\mathrm{SO},\mathrm{II}:\mathrm{DI})}_{\text{inflection-source}}\\ \underbrace{(\mathrm{UP},\mathrm{LS})}_{\text{up-parabola lower-saddle}} & \underbrace{(\mathrm{UP},\mathrm{US})}_{\text{up-parabola upper-saddle}}\end{array}\right\}$$

for $a_{110}>0, a_{220}<0$, (5.86)

$$\left\{\begin{array}{ll}(a_{111},a_{221}) & (a_{111},a_{222})\\ (a_{121},a_{211}) & (a_{122},a_{211})\end{array}\right\}=\left\{\begin{array}{cc}\underbrace{(\mathrm{SO},\mathrm{DI}:\mathrm{II})}_{\text{inflection-source}} & \underbrace{(\mathrm{SI},\mathrm{DI}:\mathrm{II})}_{\text{inflection-sink}}\\ \underbrace{(\mathrm{DP},\mathrm{US})}_{\text{down-parabola upper-saddle}} & \underbrace{(\mathrm{DP},\mathrm{LS})}_{\text{down-parabola lower-saddle}}\end{array}\right\}$$

for $a_{110}<0, a_{220}<0$, (5.87)

and the infinite-equilibrium of $x_1^*=a_{111}=a_{211}$ with $a_{111}\neq a_{211}$ is summarized as follows:

$$\begin{array}{ll}(a_{111},\bar{x}_2)=\underbrace{(\mathrm{SO},\mathrm{II})}_{\text{increasing-inflection source}} & \text{if } \bar{x}_2\in(-\infty,a_{121}),\\ (a_{111},\bar{x}_2)=\underbrace{(\mathrm{SI},\mathrm{DI})}_{\text{decreasing-inflection sink}} & \text{if } \bar{x}_2\in(a_{121},a_{221}),\\ (a_{111},\bar{x}_2)=\underbrace{(\mathrm{SI},\mathrm{II})}_{\text{increasing-inflection sink}} & \text{if } \bar{x}_2\in(a_{221},a_{122}),\\ (a_{111},\bar{x}_2)=\underbrace{(\mathrm{SO},\mathrm{DI})}_{\text{decreasing-inflection source}} & \text{if } \bar{x}_2\in(a_{122},a_{222}),\\ (a_{111},\bar{x}_2)=\underbrace{(\mathrm{SO},\mathrm{II})}_{\text{increasing-inflection source}} & \text{if } \bar{x}_2\in(a_{222},\infty),\\ \text{for } a_{110}>0, a_{220}>0; & \end{array} \tag{5.88}$$

$$\begin{aligned}
&(a_{111},\bar{x}_2)=\underbrace{(\mathrm{SI},\mathrm{DI})}_{\text{decreasing-inflection sink}} && \text{if } \bar{x}_2\in(-\infty,a_{121}),\\
&(a_{111},\bar{x}_2)=\underbrace{(\mathrm{SO},\mathrm{II})}_{\text{increasing-inflection source}} && \text{if } \bar{x}_2\in(a_{121},a_{221}),\\
&(a_{111},\bar{x}_2)=\underbrace{(\mathrm{SO},\mathrm{DI})}_{\text{decreasing-inflection source}} && \text{if } \bar{x}_2\in(a_{221},a_{122}),\\
&(a_{111},\bar{x}_2)=\underbrace{(\mathrm{SI},\mathrm{II})}_{\text{increasing-inflection sink}} && \text{if } \bar{x}_2\in(a_{122},a_{222}),\\
&(a_{111},\bar{x}_2)=\underbrace{(\mathrm{SI},\mathrm{DI})}_{\text{decreasing-inflection sink}} && \text{if } \bar{x}_2\in(a_{222},\infty),\\
&\text{for } a_{110}<0, a_{220}>0;
\end{aligned} \tag{5.89}$$

$$\begin{aligned}
&(a_{111},\bar{x}_2)=\underbrace{(\mathrm{SO},\mathrm{DI})}_{\text{decreasing-inflection source}} && \text{if } \bar{x}_2\in(-\infty,a_{121}),\\
&(a_{111},\bar{x}_2)=\underbrace{(\mathrm{SI},\mathrm{II})}_{\text{increasing-inflection sink}} && \text{if } \bar{x}_2\in(a_{121},a_{221}),\\
&(a_{111},\bar{x}_2)=\underbrace{(\mathrm{SI},\mathrm{DI})}_{\text{decreasing-inflection sink}} && \text{if } \bar{x}_2\in(a_{221},a_{122}),\\
&(a_{111},\bar{x}_2)=\underbrace{(\mathrm{SO},\mathrm{II})}_{\text{increasing-inflection source}} && \text{if } \bar{x}_2\in(a_{122},a_{222}),\\
&(a_{111},\bar{x}_2)=\underbrace{(\mathrm{SO},\mathrm{DI})}_{\text{decreasing-inflection sink}} && \text{if } \bar{x}_2\in(a_{222},\infty),\\
&\text{for } a_{110}>0, a_{220}<0;
\end{aligned} \tag{5.90}$$

$$\begin{aligned}
&(a_{111},\bar{x}_2)=\underbrace{(\mathrm{SI},\mathrm{II})}_{\text{increasing-inflection sink}} && \text{if } \bar{x}_2\in(-\infty,a_{121}),\\
&(a_{111},\bar{x}_2)=\underbrace{(\mathrm{SO},\mathrm{DI})}_{\text{decreasing-inflection source}} && \text{if } \bar{x}_2\in(a_{121},a_{221}),\\
&(a_{111},\bar{x}_2)=\underbrace{(\mathrm{SO},\mathrm{II})}_{\text{increasing-inflection source}} && \text{if } \bar{x}_2\in(a_{221},a_{122}),\\
&(a_{111},\bar{x}_2)=\underbrace{(\mathrm{SI},\mathrm{DI})}_{\text{decreasing-inflection sink}} && \text{if } \bar{x}_2\in(a_{122},a_{222}),\\
&(a_{111},\bar{x}_2)=\underbrace{(\mathrm{SI},\mathrm{II})}_{\text{increasing-inflection sink}} && \text{if } \bar{x}_2\in(a_{222},\infty),\\
&\text{for } a_{110}<0, a_{220}<0.
\end{aligned} \tag{5.91}$$

The infinite-equilibriums of $x_1^*=a_{111}=a_{211}$ are in the vertical direction. The inflection-sources (sinks) on the top and the parabola-saddles on the bottom are given, which are the switching bifurcations of $a_{111}>a_{211}$ and $a_{111}<a_{211}$ with $a_{221}\in(a_{121},a_{122})$ and $a_{122}<a_{222}$.

(iii) For a_{121}, $a_{122}\in(a_{221},a_{222})$, the infinite-equilibriums of $x_1^*=a_{111}=a_{211}$ with $x_2^*=a_{121},a_{122}$, a_{221}, a_{222} have the following properties, as shown in Fig. 5.15.

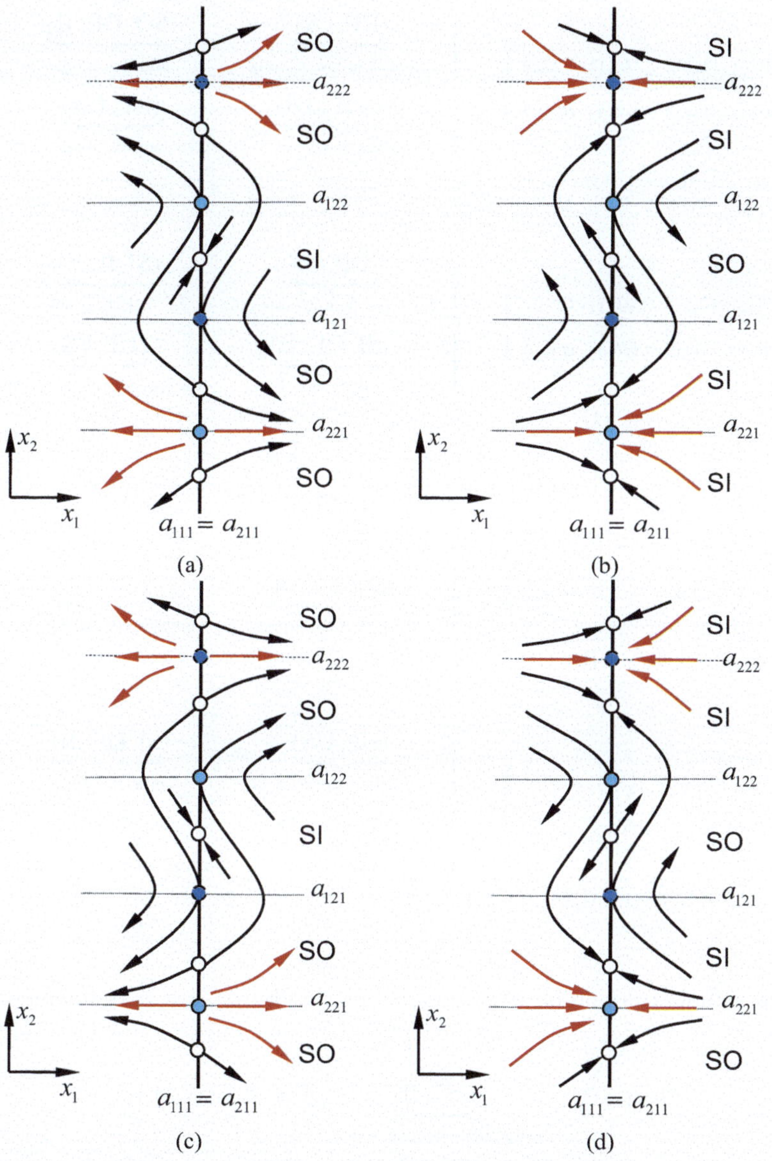

Fig. 5.15 Phase portraits ($a_{121}, a_{122} \in (a_{221}, a_{222})$) for two-dimensional systems on the x_1-direction with $x_1^* = a_{111} = a_{211}$ and on the x_2-direction with $x_2^* = a_{121}, a_{122}, a_{221}, a_{222}$. **(a)** ($a_{110} > 0$, $a_{220} > 0$), **(b)** ($a_{110} < 0$, $a_{220} > 0$), **(c)** ($a_{110} > 0$, $a_{220} < 0$), and **(d)** ($a_{110} < 0$, $a_{220} < 0$)

$$\left\{\begin{array}{cc}(a_{121},a_{211}) & (a_{111},a_{222})\\(a_{111},a_{221}) & (a_{122},a_{211})\end{array}\right\}=\left\{\begin{array}{cc}\underbrace{(\mathrm{UP},\mathrm{LS})}_{\text{up-parabola lower-saddle}} & \underbrace{(\mathrm{SO},\mathrm{DI}:\mathrm{II})}_{\text{inflection-source}}\\\underbrace{(\mathrm{SO},\mathrm{II}:\mathrm{DI})}_{\text{inflection-source}} & \underbrace{(\mathrm{DP},\mathrm{US})}_{\text{down-parabola upper-saddle}}\end{array}\right\}$$

for $a_{110}>0, a_{220}>0,$

(5.92)

$$\left\{\begin{array}{cc}(a_{121},a_{211}) & (a_{111},a_{222})\\(a_{111},a_{221}) & (a_{122},a_{211})\end{array}\right\}=\left\{\begin{array}{cc}\underbrace{(\mathrm{DP},\mathrm{US})}_{\text{down-parabola upper-saddle}} & \underbrace{(\mathrm{SI},\mathrm{II}:\mathrm{DI})}_{\text{inflection-sink}}\\\underbrace{(\mathrm{SI},\mathrm{DI}:\mathrm{II})}_{\text{inflection-sink}} & \underbrace{(\mathrm{UP},\mathrm{LS})}_{\text{up-parabola lower-saddle}}\end{array}\right\}$$

for $a_{110}<0, a_{220}>0,$

(5.93)

$$\left\{\begin{array}{cc}(a_{121},a_{211}) & (a_{111},a_{222})\\(a_{111},a_{221}) & (a_{122},a_{211})\end{array}\right\}=\left\{\begin{array}{cc}\underbrace{(\mathrm{DP},\mathrm{LS})}_{\text{down-parabola lower-saddle}} & \underbrace{(\mathrm{SO},\mathrm{II}:\mathrm{DI})}_{\text{inflection-source}}\\\underbrace{(\mathrm{SO},\mathrm{DI}:\mathrm{II})}_{\text{inflection-source}} & \underbrace{(\mathrm{UP},\mathrm{US})}_{\text{up-parabola upper-saddle}}\end{array}\right\}$$

for $a_{110}>0, a_{220}<0,$

(5.94)

$$\left\{\begin{array}{cc}(a_{121},a_{211}) & (a_{111},a_{222})\\(a_{111},a_{221}) & (a_{122},a_{211})\end{array}\right\}=\left\{\begin{array}{cc}\underbrace{(\mathrm{UP},\mathrm{US})}_{\text{up-parabola upper-saddle}} & \underbrace{(\mathrm{SI},\mathrm{DI}:\mathrm{II})}_{\text{inflection-sink}}\\\underbrace{(\mathrm{SI},\mathrm{II}:\mathrm{DI})}_{\text{inflection-sink}} & \underbrace{(\mathrm{DP},\mathrm{LS})}_{\text{down-parabola lower-saddle}}\end{array}\right\}$$

for $a_{110}<0, a_{220}<0,$

(5.95)

and the infinite-equilibrium of $x_1^*=a_{111}=a_{211}$ with $a_{111}\neq a_{211}$ is summarized as follows:

$$\begin{aligned}
&(a_{111},\bar{x}_2)=\underbrace{(\mathrm{SO},\mathrm{II})}_{\text{increasing-inflection source}} \text{ if } \bar{x}_2\in(-\infty,a_{221}),\\
&(a_{111},\bar{x}_2)=\underbrace{(\mathrm{SO},\mathrm{DI})}_{\text{decreasing-inflection source}} \text{ if } \bar{x}_2\in(a_{221},a_{121}),\\
&(a_{111},\bar{x}_2)=\underbrace{(\mathrm{SI},\mathrm{II})}_{\text{increasing-inflection sink}} \text{ if } \bar{x}_2\in(a_{121},a_{122}),\\
&(a_{111},\bar{x}_2)=\underbrace{(\mathrm{SO},\mathrm{DI})}_{\text{decreasing-inflection source}} \text{ if } \bar{x}_2\in(a_{122},a_{222}),\\
&(a_{111},\bar{x}_2)=\underbrace{(\mathrm{SO},\mathrm{II})}_{\text{increasing-inflection source}} \text{ if } \bar{x}_2\in(a_{222},\infty),\\
&\text{for } a_{110}>0, a_{220}>0;
\end{aligned} \tag{5.96}$$

$$
\begin{aligned}
(a_{111},\bar{x}_2) &= \underbrace{(\mathrm{SI},\mathrm{DI})}_{\text{decreasing-inflection sink}} && \text{if } \bar{x}_2 \in (-\infty, a_{221}),\\
(a_{111},\bar{x}_2) &= \underbrace{(\mathrm{SI},\mathrm{II})}_{\text{increasing-inflection sink}} && \text{if } \bar{x}_2 \in (a_{221}, a_{121}),\\
(a_{111},\bar{x}_2) &= \underbrace{(\mathrm{SO},\mathrm{DI})}_{\text{decreasing-inflection source}} && \text{if } \bar{x}_2 \in (a_{121}, a_{122}),\\
(a_{111},\bar{x}_2) &= \underbrace{(\mathrm{SI},\mathrm{II})}_{\text{increasing-inflection sink}} && \text{if } \bar{x}_2 \in (a_{122}, a_{222}),\\
(a_{111},\bar{x}_2) &= \underbrace{(\mathrm{SI},\mathrm{DI})}_{\text{decreasing-inflection sink}} && \text{if } \bar{x}_2 \in (a_{222}, \infty),\\
&\text{for } a_{110} < 0, a_{220} > 0;
\end{aligned}
\tag{5.97}
$$

$$
\begin{aligned}
(a_{111},\bar{x}_2) &= \underbrace{(\mathrm{SO},\mathrm{DI})}_{\text{decreasing-inflection source}} && \text{if } \bar{x}_2 \in (-\infty, a_{221}),\\
(a_{111},\bar{x}_2) &= \underbrace{(\mathrm{SO},\mathrm{II})}_{\text{increasing-inflection source}} && \text{if } \bar{x}_2 \in (a_{221}, a_{121}),\\
(a_{111},\bar{x}_2) &= \underbrace{(\mathrm{SI},\mathrm{DI})}_{\text{decreasing-inflection sink}} && \text{if } \bar{x}_2 \in (a_{121}, a_{122}),\\
(a_{111},\bar{x}_2) &= \underbrace{(\mathrm{SO},\mathrm{II})}_{\text{increasing-inflection source}} && \text{if } \bar{x}_2 \in (a_{122}, a_{222}),\\
(a_{111},\bar{x}_2) &= \underbrace{(\mathrm{SO},\mathrm{DI})}_{\text{decreasing-inflection sink}} && \text{if } \bar{x}_2 \in (a_{222}, \infty),\\
&\text{for } a_{110} > 0, a_{220} < 0;
\end{aligned}
\tag{5.98}
$$

$$
\begin{aligned}
(a_{111},\bar{x}_2) &= \underbrace{(\mathrm{SI},\mathrm{II})}_{\text{increasing-inflection sink}} && \text{if } \bar{x}_2 \in (-\infty, a_{221}),\\
(a_{111},\bar{x}_2) &= \underbrace{(\mathrm{SI},\mathrm{DI})}_{\text{decreasing-inflection source}} && \text{if } \bar{x}_2 \in (a_{221}, a_{121}),\\
(a_{111},\bar{x}_2) &= \underbrace{(\mathrm{SO},\mathrm{II})}_{\text{increasing-inflection source}} && \text{if } \bar{x}_2 \in (a_{121}, a_{122}),\\
(a_{111},\bar{x}_2) &= \underbrace{(\mathrm{SI},\mathrm{DI})}_{\text{decreasing-inflection sink}} && \text{if } \bar{x}_2 \in (a_{122}, a_{222}),\\
(a_{111},\bar{x}_2) &= \underbrace{(\mathrm{SI},\mathrm{II})}_{\text{increasing-inflection sink}} && \text{if } \bar{x}_2 \in (a_{222}, \infty),\\
&\text{for } a_{110} < 0, a_{220} < 0.
\end{aligned}
\tag{5.99}
$$

The infinite-equilibriums of $x_1^* = a_{111} = a_{211}$ are in the vertical direction. The inflection-sources (sinks) on the top and the parabola-saddles on the bottom are given, which are the switching bifurcations of $a_{111} > a_{211}$ and $a_{111} < a_{211}$ with a_{121}, $a_{122} \in (a_{221}, a_{222})$.

(iv) For $a_{121} \in (a_{221}, a_{222})$ and $a_{222} < a_{122}$, the equilibriums of $x_1^* = a_{111}$ with $x_2^* = a_{121}, a_{221}, a_{222}$ have the following properties, as shown in Fig. 5.16.

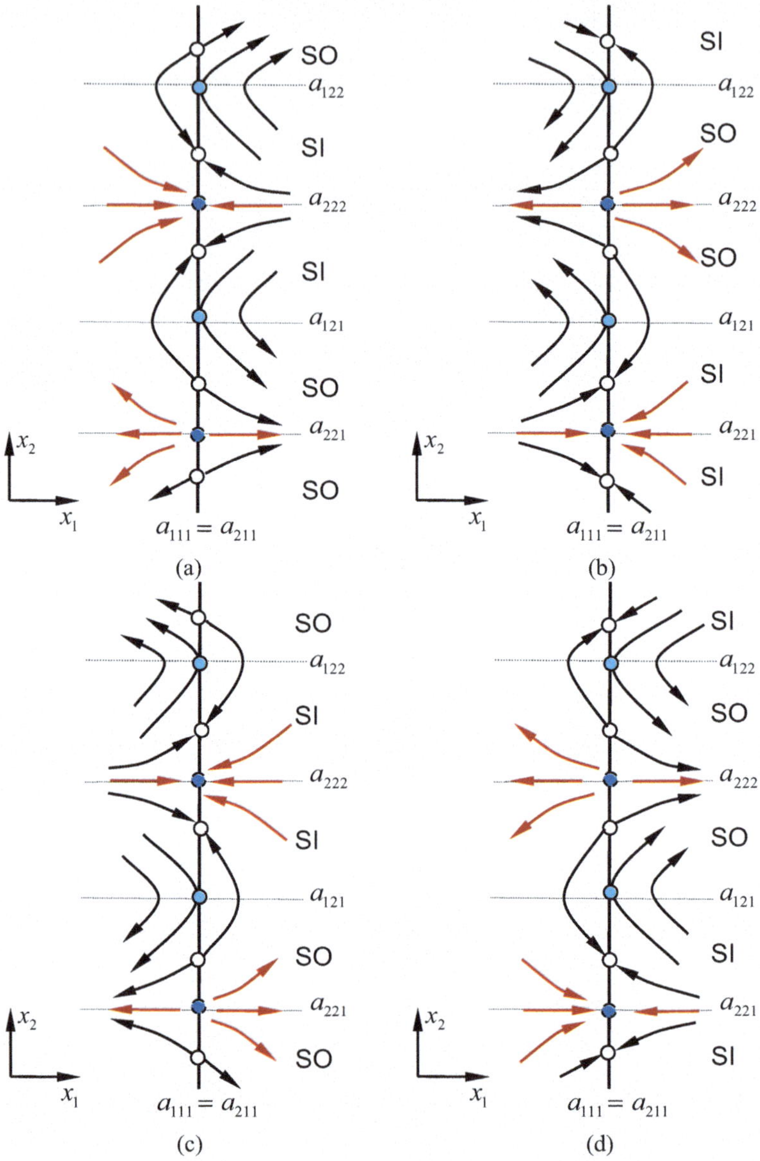

Fig. 5.16 Phase portraits ($a_{121} \in (a_{221}, a_{222})$, $a_{222} < a_{122}$) for two-dimensional systems on the x_1-direction with $x_1^* = a_{111} = a_{211}$ and on the x_2-direction with $x_2^* = a_{121}, a_{122}, a_{221}, a_{222}$. **(a)** ($a_{110} > 0$, $a_{220} > 0$), **(b)** ($a_{110} < 0$, $a_{220} > 0$), **(c)** ($a_{110} > 0$, $a_{220} < 0$), and **(d)** ($a_{110} < 0$, $a_{220} < 0$)

$$\left\{\begin{array}{ll}(a_{121},a_{211}) & (a_{122},a_{211})\\ (a_{111},a_{221}) & (a_{111},a_{222})\end{array}\right\}=\left\{\begin{array}{cc}\underbrace{(\mathrm{UP},\mathrm{LS})}_{\text{up-parabola lower-saddle}} & \underbrace{(\mathrm{UP},\mathrm{US})}_{\text{up-parabola upper-saddle}}\\ \underbrace{(\mathrm{SO},\mathrm{II}:\mathrm{DI})}_{\text{inflection-sink}} & \underbrace{(\mathrm{SI},\mathrm{II}:\mathrm{DI})}_{\text{inflection-sink}}\end{array}\right\}$$
$$\text{for } a_{110}>0, a_{220}>0, \tag{5.100}$$

$$\left\{\begin{array}{ll}(a_{121},a_{211}) & (a_{122},a_{211})\\ (a_{111},a_{221}) & (a_{111},a_{222})\end{array}\right\}=\left\{\begin{array}{cc}\underbrace{(\mathrm{DP},\mathrm{US})}_{\text{down-parabola upper-saddle}} & \underbrace{(\mathrm{DP},\mathrm{LS})}_{\text{down-parabola lower-saddle}}\\ \underbrace{(\mathrm{SI},\mathrm{II}:\mathrm{DI})}_{\text{inflection-sink}} & \underbrace{(\mathrm{SO},\mathrm{DI}:\mathrm{II})}_{\text{inflection-source}}\end{array}\right\}$$
$$\text{for } a_{110}<0, a_{220}>0, \tag{5.101}$$

$$\left\{\begin{array}{ll}(a_{121},a_{211}) & (a_{122},a_{211})\\ (a_{111},a_{221}) & (a_{111},a_{222})\end{array}\right\}=\left\{\begin{array}{cc}\underbrace{(\mathrm{DP},\mathrm{LS})}_{\text{down-parabola lower-saddle}} & \underbrace{(\mathrm{DP},\mathrm{US})}_{\text{down-parabola upper-saddle}}\\ \underbrace{(\mathrm{SO},\mathrm{DI}:\mathrm{II})}_{\text{inflection-source}} & \underbrace{(\mathrm{SI},\mathrm{DI}:\mathrm{II})}_{\text{inflection-sink}}\end{array}\right\}$$
$$\text{for } a_{110}>0, a_{220}<0, \tag{5.102}$$

$$\left\{\begin{array}{ll}(a_{121},a_{211}) & (a_{122},a_{211})\\ (a_{111},a_{221}) & (a_{111},a_{222})\end{array}\right\}=\left\{\begin{array}{cc}\underbrace{(\mathrm{UP},\mathrm{US})}_{\text{up-parabola upper-saddle}} & \underbrace{(\mathrm{UP},\mathrm{LS})}_{\text{down-parabola lower-saddle}}\\ \underbrace{(\mathrm{SI},\mathrm{II}:\mathrm{DI})}_{\text{inflection-sink}} & \underbrace{(\mathrm{SO},\mathrm{II}:\mathrm{DI})}_{\text{inflection-source}}\end{array}\right\}$$
$$\text{for } a_{110}<0, a_{220}<0, \tag{5.103}$$

and the infinite-equilibrium of $x_1^* = a_{111} = a_{211}$ with $a_{111} \neq a_{211}$ is summarized as follows:

$$\begin{array}{lll}(a_{111},\bar{x}_2)= & \underbrace{(\mathrm{SO},\mathrm{II})}_{\text{increasing-inflection source}} & \text{if } \bar{x}_2\in(-\infty,a_{221}),\\ (a_{111},\bar{x}_2)= & \underbrace{(\mathrm{SO},\mathrm{DI})}_{\text{decreasing-inflection source}} & \text{if } \bar{x}_2\in(a_{221},a_{121}),\\ (a_{111},\bar{x}_2)= & \underbrace{(\mathrm{SI},\mathrm{II})}_{\text{increasing-inflection sink}} & \text{if } \bar{x}_2\in(a_{121},a_{222}),\\ (a_{111},\bar{x}_2)= & \underbrace{(\mathrm{SI},\mathrm{DI})}_{\text{decreasing-inflection sink}} & \text{if } \bar{x}_2\in(a_{222},a_{122}),\\ (a_{111},\bar{x}_2)= & \underbrace{(\mathrm{SO},\mathrm{II})}_{\text{increasing-inflection source}} & \text{if } \bar{x}_2\in(a_{122},\infty),\end{array} \tag{5.104}$$
$$\text{for } a_{110}>0, a_{220}>0;$$

$$
\begin{aligned}
&(a_{111},\bar{x}_2) = \underbrace{(\mathrm{SI},\mathrm{DI})}_{\text{decreasing-inflection sink}} &&\text{if } \bar{x}_2 \in (-\infty, a_{221}),\\
&(a_{111},\bar{x}_2) = \underbrace{(\mathrm{SI},\mathrm{II})}_{\text{increasing-inflection sink}} &&\text{if } \bar{x}_2 \in (a_{221}, a_{121}),\\
&(a_{111},\bar{x}_2) = \underbrace{(\mathrm{SO},\mathrm{DI})}_{\text{decreasing-inflection source}} &&\text{if } \bar{x}_2 \in (a_{121}, a_{222}),\\
&(a_{111},\bar{x}_2) = \underbrace{(\mathrm{SO},\mathrm{II})}_{\text{increasing-inflection source}} &&\text{if } \bar{x}_2 \in (a_{222}, a_{122}),\\
&(a_{111},\bar{x}_2) = \underbrace{(\mathrm{SI},\mathrm{DI})}_{\text{decreasing-inflection sink}} &&\text{if } \bar{x}_2 \in (a_{122}, \infty),\\
&\text{for } a_{110} < 0, a_{220} > 0;
\end{aligned}
\tag{5.105}
$$

$$
\begin{aligned}
&(a_{111},\bar{x}_2) = \underbrace{(\mathrm{SO},\mathrm{DI})}_{\text{decreasing-inflection source}} &&\text{if } \bar{x}_2 \in (-\infty, a_{221}),\\
&(a_{111},\bar{x}_2) = \underbrace{(\mathrm{SO},\mathrm{II})}_{\text{increasing-inflection source}} &&\text{if } \bar{x}_2 \in (a_{221}, a_{121}),\\
&(a_{111},\bar{x}_2) = \underbrace{(\mathrm{SI},\mathrm{DI})}_{\text{decreasing-inflection sink}} &&\text{if } \bar{x}_2 \in (a_{121}, a_{222}),\\
&(a_{111},\bar{x}_2) = \underbrace{(\mathrm{SI},\mathrm{II})}_{\text{increasing-inflection sink}} &&\text{if } \bar{x}_2 \in (a_{222}, a_{122}),\\
&(a_{111},\bar{x}_2) = \underbrace{(\mathrm{SO},\mathrm{DI})}_{\text{decreasing-inflection sink}} &&\text{if } \bar{x}_2 \in (a_{122}, \infty),\\
&\text{for } a_{110} > 0, a_{220} < 0;
\end{aligned}
\tag{5.106}
$$

$$
\begin{aligned}
&(a_{111},\bar{x}_2) = \underbrace{(\mathrm{SI},\mathrm{II})}_{\text{increasing-inflection sink}} &&\text{if } \bar{x}_2 \in (-\infty, a_{221}),\\
&(a_{111},\bar{x}_2) = \underbrace{(\mathrm{SI},\mathrm{DI})}_{\text{decreasing-inflection source}} &&\text{if } \bar{x}_2 \in (a_{221}, a_{121}),\\
&(a_{111},\bar{x}_2) = \underbrace{(\mathrm{SO},\mathrm{II})}_{\text{increasing-inflection source}} &&\text{if } \bar{x}_2 \in (a_{121}, a_{222}),\\
&(a_{111},\bar{x}_2) = \underbrace{(\mathrm{SO},\mathrm{DI})}_{\text{decreasing-inflection source}} &&\text{if } \bar{x}_2 \in (a_{222}, a_{122}),\\
&(a_{111},\bar{x}_2) = \underbrace{(\mathrm{SI},\mathrm{II})}_{\text{increasing-inflection sink}} &&\text{if } \bar{x}_2 \in (a_{122}, \infty),\\
&\text{for } a_{110} < 0, a_{220} < 0.
\end{aligned}
\tag{5.107}
$$

The infinite-equilibriums of $x_1^* = a_{111} = a_{211}$ are in the vertical direction. The inflection-sources (sinks) on the top and the parabola-saddles on the bottom are given, which are the switching bifurcations of $a_{111} > a_{211}$ and $a_{111} < a_{211}$ with $a_{121} \in (a_{221}, a_{222})$ with $a_{222} < a_{122}$.

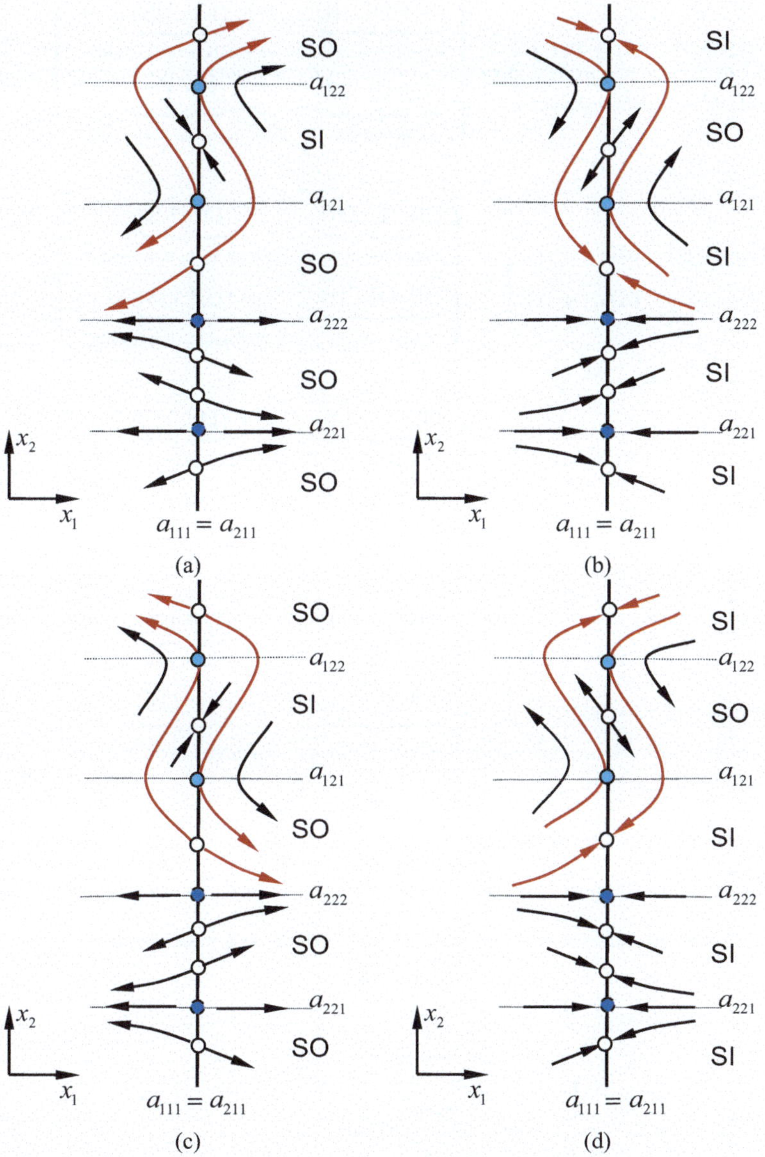

Fig. 5.17 Phase portraits ($a_{111} = a_{211}$ and $a_{222} < a_{121}$) for two-dimensional systems on the x_1-direction with $x_1^* = a_{111} = a_{211}$ and on the x_2-direction with $x_2^* = a_{121}, a_{122}, a_{221}, a_{222}$. **(a)** ($a_{110} > 0$, $a_{220} > 0$), **(b)** ($a_{110} < 0$, $a_{220} > 0$), **(c)** ($a_{110} > 0$, $a_{220} < 0$), and **(d)** ($a_{110} < 0$, $a_{220} < 0$)

(v) For the case $a_{222} < a_{121}$, the equilibrium series of $x_1^* = a_{111} = a_{211}$ with $x_2^* = a_{121}, a_{122}, a_{221}, a_{222}$ have the properties as in Eqs. (5.76)–(5.79), as shown in Fig. 5.17.

$$\left\{\begin{matrix}(a_{111},a_{222}) & (a_{122},a_{211})\\ (a_{111},a_{221}) & (a_{121},a_{211})\end{matrix}\right\}=\left\{\begin{matrix}\underbrace{(\text{SO},\text{DI}:\text{II})}_{\text{inflection-source}} & \underbrace{(\text{UP},\text{US})}_{\text{up-parabola upper-saddle}}\\ \underbrace{(\text{SO},\text{II}:\text{DI})}_{\text{inflection-sink}} & \underbrace{(\text{DP},\text{LS})}_{\text{down-parabola lower-saddle}}\end{matrix}\right\} \quad (5.108)$$

for $a_{110}>0, a_{220}>0$,

$$\left\{\begin{matrix}(a_{111},a_{222}) & (a_{122},a_{211})\\ (a_{111},a_{221}) & (a_{121},a_{211})\end{matrix}\right\}=\left\{\begin{matrix}\underbrace{(\text{SI},\text{II}:\text{DI})}_{\text{inflection-sink}} & \underbrace{(\text{DP},\text{LS})}_{\text{down-parabola lower-saddle}}\\ \underbrace{(\text{SI},\text{DI}:\text{II})}_{\text{inflection-sink}} & \underbrace{(\text{UP},\text{US})}_{\text{down-parabola upper-saddle}}\end{matrix}\right\} \quad (5.109)$$

for $a_{110}<0, a_{220}>0$,

$$\left\{\begin{matrix}(a_{121},a_{211}) & (a_{122},a_{211})\\ (a_{111},a_{221}) & (a_{111},a_{222})\end{matrix}\right\}=\left\{\begin{matrix}\underbrace{(\text{SO},\text{II}:\text{DI})}_{\text{inflection-source}} & \underbrace{(\text{DP},\text{US})}_{\text{down-parabola upper-saddle}}\\ \underbrace{(\text{SO},\text{DI}:\text{II})}_{\text{inflection-source}} & \underbrace{(\text{DP},\text{LS})}_{\text{down-parabola lower-saddle}}\end{matrix}\right\} \quad (5.110)$$

for $a_{110}>0, a_{220}<0$,

$$\left\{\begin{matrix}(a_{121},a_{211}) & (a_{122},a_{211})\\ (a_{111},a_{221}) & (a_{111},a_{222})\end{matrix}\right\}=\left\{\begin{matrix}\underbrace{(\text{SI},\text{DI}:\text{II})}_{\text{inflection-sink}} & \underbrace{(\text{UP},\text{LS})}_{\text{down-parabola lower-saddle}}\\ \underbrace{(\text{SI},\text{II}:\text{DI})}_{\text{inflection-sink}} & \underbrace{(\text{UP},\text{US})}_{\text{up-parabola upper-saddle}}\end{matrix}\right\} \quad (5.111)$$

for $a_{110}<0, a_{220}<0$,

and the infinite-equilibrium of $x_1^*=a_{111}=a_{211}$ with $a_{111}\neq a_{211}$ is summarized as follows:

$$\begin{aligned}
(a_{111},\bar{x}_2)&=\underbrace{(\text{SO},\text{II})}_{\text{increasing-inflection source}} && \text{if } \bar{x}_2\in(-\infty,a_{221}),\\
(a_{111},\bar{x}_2)&=\underbrace{(\text{SO},\text{DI})}_{\text{decreasing-inflection source}} && \text{if } \bar{x}_2\in(a_{221},a_{121}),\\
(a_{111},\bar{x}_2)&=\underbrace{(\text{SO},\text{II})}_{\text{increasing-inflection source}} && \text{if } \bar{x}_2\in(a_{121},a_{222}),\\
(a_{111},\bar{x}_2)&=\underbrace{(\text{SI},\text{DI})}_{\text{decreasing-inflection sink}} && \text{if } \bar{x}_2\in(a_{222},a_{122}),\\
(a_{111},\bar{x}_2)&=\underbrace{(\text{SO},\text{II})}_{\text{increasing-inflection source}} && \text{if } \bar{x}_2\in(a_{122},\infty),
\end{aligned} \quad (5.112)$$

for $a_{110}>0, a_{220}>0$;

$$
\begin{aligned}
&(a_{111},\bar{x}_2)=\underbrace{(\text{SI},\text{DI})}_{\text{decreasing-inflection sink}} && \text{if } \bar{x}_2\in(-\infty,a_{221}),\\
&(a_{111},\bar{x}_2)=\underbrace{(\text{SI},\text{II})}_{\text{increasing-inflection sink}} && \text{if } \bar{x}_2\in(a_{221},a_{121}),\\
&(a_{111},\bar{x}_2)=\underbrace{(\text{SI},\text{DI})}_{\text{decreasing-inflection sink}} && \text{if } \bar{x}_2\in(a_{121},a_{222}),\\
&(a_{111},\bar{x}_2)=\underbrace{(\text{SO},\text{II})}_{\text{increasing-inflection source}} && \text{if } \bar{x}_2\in(a_{222},a_{122}),\\
&(a_{111},\bar{x}_2)=\underbrace{(\text{SI},\text{DI})}_{\text{decreasing-inflection sink}} && \text{if } \bar{x}_2\in(a_{122},\infty),\\
&\text{for } a_{110}<0, a_{220}>0;
\end{aligned}
\tag{5.113}
$$

$$
\begin{aligned}
&(a_{111},\bar{x}_2)=\underbrace{(\text{SO},\text{DI})}_{\text{decreasing-inflection source}} && \text{if } \bar{x}_2\in(-\infty,a_{221}),\\
&(a_{111},\bar{x}_2)=\underbrace{(\text{SO},\text{II})}_{\text{increasing-inflection source}} && \text{if } \bar{x}_2\in(a_{221},a_{121}),\\
&(a_{111},\bar{x}_2)=\underbrace{(\text{SO},\text{DI})}_{\text{decreasing-inflection source}} && \text{if } \bar{x}_2\in(a_{121},a_{222}),\\
&(a_{111},\bar{x}_2)=\underbrace{(\text{SI},\text{II})}_{\text{increasing-inflection sink}} && \text{if } \bar{x}_2\in(a_{222},a_{122}),\\
&(a_{111},\bar{x}_2)=\underbrace{(\text{SO},\text{DI})}_{\text{decreasing-inflection source}} && \text{if } \bar{x}_2\in(a_{122},\infty),\\
&\text{for } a_{110}>0, a_{220}<0;
\end{aligned}
\tag{5.114}
$$

$$
\begin{aligned}
&(a_{111},\bar{x}_2)=\underbrace{(\text{SI},\text{II})}_{\text{increasing-inflection sink}} && \text{if } \bar{x}_2\in(-\infty,a_{221}),\\
&(a_{111},\bar{x}_2)=\underbrace{(\text{SI},\text{DI})}_{\text{decreasing-inflection sink}} && \text{if } \bar{x}_2\in(a_{221},a_{121}),\\
&(a_{111},\bar{x}_2)=\underbrace{(\text{SI},\text{II})}_{\text{increasing-inflection sink}} && \text{if } \bar{x}_2\in(a_{121},a_{222}),\\
&(a_{111},\bar{x}_2)=\underbrace{(\text{SO},\text{DI})}_{\text{decreasing-inflection source}} && \text{if } \bar{x}_2\in(a_{222},a_{122}),\\
&(a_{111},\bar{x}_2)=\underbrace{(\text{SI},\text{II})}_{\text{increasing-inflection sink}} && \text{if } \bar{x}_2\in(a_{122},\infty),\\
&\text{for } a_{110}<0, a_{220}<0.
\end{aligned}
\tag{5.115}
$$

The infinite-equilibriums of $x_1^*=a_{111}=a_{211}$ are in the vertical direction. The inflection sources (sinks) on the top and the parabola-saddles on the bottom are given, which are the switching bifurcations of $a_{111}>a_{211}$ and $a_{111}<a_{211}$ with $a_{222}<a_{121}$.

5.2.3 *Infinite-Equilibriums on Two Directions*

A. Consider a dynamical system as

$$\begin{aligned}\dot{x}_1 &= a_{110}(x_1 - a_{111})(x_2 - a_{121})(x_2 - a_{122}),\\ \dot{x}_2 &= a_{220}(x_1 - a_{211})(x_2 - a_{221})(x_2 - a_{222}),\end{aligned} \tag{5.116}$$

and the corresponding first integral manifold for $a_{12s_1} = a_{22l_1}$ (s_1, s_2, l_1, $l_2 \in \{1,2\}$, $s_1 \neq s_2$, $l_1 \neq l_2$) and $a_{111} = a_{211}$ is

$$x_1 - a_{111} = \frac{a_{110}}{a_{220}}\left[(x_2 - x_{20}) + (a_{22l_2} - a_{12s_2})\ln\frac{|x_2 - a_{22l_2}|}{|x_{20} - a_{22l_2}|}\right]. \tag{5.117}$$

(i) The equilibrium series of $x_2^* = a_{121}, a_{122}, a_{221}, a_{222}$ with $x_1^* = a_{111}, a_{211}$ for $a_{221} = a_{121}$ with $a_{111} = a_{211}$ are

$$\left\{\begin{matrix}(a_{111}, a_{222})\\(a_{122}, a_{211})\\(a_{111}, a_{221})\end{matrix}\right\} = \left\{\begin{matrix}\underbrace{(\text{SO}, \text{DI} : \text{II})}_{\text{inflection-source}}\\ \underbrace{(\text{DP}, \text{US})}_{\text{down-parabola upper-saddle}}\\ \underbrace{({}_{\text{SO:SI}}\text{II}, {}_{\text{SO:SI}}\text{II})}_{\text{increasing-inflection saddle}}\end{matrix}\right\} \text{ for } a_{110} > 0, a_{220} > 0, \tag{5.118}$$

$$\left\{\begin{matrix}(a_{111}, a_{222})\\(a_{122}, a_{211})\\(a_{111}, a_{221})\end{matrix}\right\} = \left\{\begin{matrix}\underbrace{(\text{SI}, \text{II} : \text{DI})}_{\text{inflection-sink}}\\ \underbrace{(\text{UP}, \text{LS})}_{\text{up-parabola lower-saddle}}\\ \underbrace{({}_{\text{SO:SI}}\text{DI}, {}_{\text{SI:SO}}\text{DI})}_{\text{increasing-inflection saddle}}\end{matrix}\right\} \text{ for } a_{110} < 0, a_{220} > 0, \tag{5.119}$$

$$\left\{\begin{matrix}(a_{111}, a_{222})\\(a_{122}, a_{211})\\(a_{111}, a_{221})\end{matrix}\right\} = \left\{\begin{matrix}\underbrace{(\text{SO}, \text{II} : \text{DI})}_{\text{inflection-sink}}\\ \underbrace{(\text{UP}, \text{US})}_{\text{up-parabola upper-saddle}}\\ \underbrace{({}_{\text{SI:SO}}\text{DI}, {}_{\text{SO:SI}}\text{DI})}_{\text{increasing-inflection saddle}}\end{matrix}\right\} \text{ for } a_{110} > 0, a_{220} < 0, \tag{5.120}$$

$$\left\{\begin{array}{l}(a_{111},a_{222})\\(a_{122},a_{211})\\(a_{111},a_{221})\end{array}\right\}=\left\{\begin{array}{c}\underbrace{(\text{SI},\text{DI}:\text{II})}_{\text{inflection-sink}}\\\underbrace{(\text{DP},\text{LS})}_{\text{down-parabola lower-saddle}}\\\underbrace{({}_{\text{SI:SO}}\text{II},{}_{\text{SI:SO}}\text{II})}_{\text{increasing-inflection saddle}}\end{array}\right\}\text{ for } a_{110}<0, a_{220}<0. \tag{5.121}$$

The infinite-equilibrium of $x_2^* = a_{121} = a_{221}$ is summarized as follows:

$$\begin{aligned}&(\bar{x}_1,a_{221})=\underbrace{(\text{II},\text{SO})}_{\text{increasing-inflection source}}\quad\text{if } \bar{x}_1\in(-\infty,a_{111}),\\&(\bar{x}_1,a_{221})=\underbrace{(\text{II},\text{SI})}_{\text{increasing-inflection sink}}\quad\text{if } \bar{x}_1\in(a_{111},\infty),\\&\text{for } a_{110}>0, a_{220}>0;\end{aligned} \tag{5.122}$$

$$\begin{aligned}&(\bar{x}_1,a_{221})=\underbrace{(\text{DI},\text{SO})}_{\text{decreasing-inflection source}}\quad\text{if } \bar{x}_1\in(-\infty,a_{111}),\\&(\bar{x}_1,a_{221})=\underbrace{(\text{DI},\text{SI})}_{\text{decreasing-inflection sink}}\quad\text{if } \bar{x}_1\in(a_{111},\infty),\\&\text{for } a_{110}<0, a_{220}>0;\end{aligned} \tag{5.123}$$

$$\begin{aligned}&(\bar{x}_1,a_{221})=\underbrace{(\text{DI},\text{SI})}_{\text{decreasing-inflection sink}}\quad\text{if } \bar{x}_1\in(-\infty,a_{111}),\\&(\bar{x}_1,a_{221})=\underbrace{(\text{DI},\text{SO})}_{\text{decreasing-inflection source}}\quad\text{if } \bar{x}_1\in(a_{111},\infty),\\&\text{for } a_{110}>0, a_{220}<0;\end{aligned} \tag{5.124}$$

$$\begin{aligned}&(\bar{x}_1,a_{221})=\underbrace{(\text{II},\text{SI})}_{\text{increasing-inflection sink}}\quad\text{if } \bar{x}_1\in(-\infty,a_{111}),\\&(\bar{x}_1,a_{221})=\underbrace{(\text{II},\text{SO})}_{\text{increasing-inflection source}}\quad\text{if } \bar{x}_1\in(a_{111},\infty),\\&\text{for } a_{110}<0, a_{220}<0.\end{aligned} \tag{5.125}$$

The infinite-equilibrium of $x_1^* = a_{111} = a_{211}$ is summarized as follows:

$$
\begin{aligned}
&(a_{111},\bar{x}_2) = \underbrace{(\text{SO},\text{II})}_{\text{increasing-inflection source}} \quad \text{if } \bar{x}_2 \in (-\infty, a_{121}),\\
&(a_{111},\bar{x}_2) = \underbrace{(\text{SI},\text{II})}_{\text{decreasing-inflection sink}} \quad \text{if } \bar{x}_2 \in (a_{121}, a_{221}),\\
&(a_{111},\bar{x}_2) = \underbrace{(\text{SO},\text{DI})}_{\text{decreasing-inflection source}} \quad \text{if } \bar{x}_2 \in (a_{221}, a_{222}),\\
&(a_{111},\bar{x}_2) = \underbrace{(\text{SO},\text{II})}_{\text{increasing-inflection source}} \quad \text{if } \bar{x}_2 \in (a_{222}, \infty),\\
&\text{for } a_{110} > 0, a_{220} > 0;
\end{aligned}
\tag{5.126}
$$

$$
\begin{aligned}
&(a_{111},\bar{x}_2) = \underbrace{(\text{SI},\text{DI})}_{\text{decreasing-inflection sink}} \quad \text{if } \bar{x}_2 \in (-\infty, a_{121}),\\
&(a_{111},\bar{x}_2) = \underbrace{(\text{SO},\text{DI})}_{\text{decreasing-inflection source}} \quad \text{if } \bar{x}_2 \in (a_{121}, a_{221}),\\
&(a_{111},\bar{x}_2) = \underbrace{(\text{SI},\text{II})}_{\text{increasing-inflection sink}} \quad \text{if } \bar{x}_2 \in (a_{221}, a_{222}),\\
&(a_{111},\bar{x}_2) = \underbrace{(\text{SI},\text{DI})}_{\text{decreasing-inflection sink}} \quad \text{if } \bar{x}_2 \in (a_{222}, \infty),\\
&\text{for } a_{110} < 0, a_{220} > 0;
\end{aligned}
\tag{5.127}
$$

$$
\begin{aligned}
&(a_{111},\bar{x}_2) = \underbrace{(\text{SO},\text{DI})}_{\text{decreasing-inflection source}} \quad \text{if } \bar{x}_2 \in (-\infty, a_{121}),\\
&(a_{111},\bar{x}_2) = \underbrace{(\text{SI},\text{DI})}_{\text{decreasing-inflection sink}} \quad \text{if } \bar{x}_2 \in (a_{121}, a_{221}),\\
&(a_{111},\bar{x}_2) = \underbrace{(\text{SO},\text{II})}_{\text{increasing-inflection sink}} \quad \text{if } \bar{x}_2 \in (a_{221}, a_{222}),\\
&(a_{111},\bar{x}_2) = \underbrace{(\text{SO},\text{DI})}_{\text{decreasing-inflection source}} \quad \text{if } \bar{x}_2 \in (a_{222}, \infty),\\
&\text{for } a_{110} > 0, a_{220} < 0;
\end{aligned}
\tag{5.128}
$$

$$
\begin{aligned}
&(a_{111},\bar{x}_2) = \underbrace{(\text{SI},\text{II})}_{\text{decreasing-inflection sink}} \quad \text{if } \bar{x}_2 \in (-\infty, a_{121}),\\
&(a_{111},\bar{x}_2) = \underbrace{(\text{SO},\text{II})}_{\text{increasing-inflection source}} \quad \text{if } \bar{x}_2 \in (a_{121}, a_{221}),\\
&(a_{111},\bar{x}_2) = \underbrace{(\text{SI},\text{DI})}_{\text{decreasing-inflection sink}} \quad \text{if } \bar{x}_2 \in (a_{221}, a_{222}),\\
&(a_{111},\bar{x}_2) = \underbrace{(\text{SI},\text{II})}_{\text{increasing-inflection sink}} \quad \text{if } \bar{x}_2 \in (a_{222}, \infty),\\
&\text{for } a_{110} < 0, a_{220} < 0.
\end{aligned}
\tag{5.129}
$$

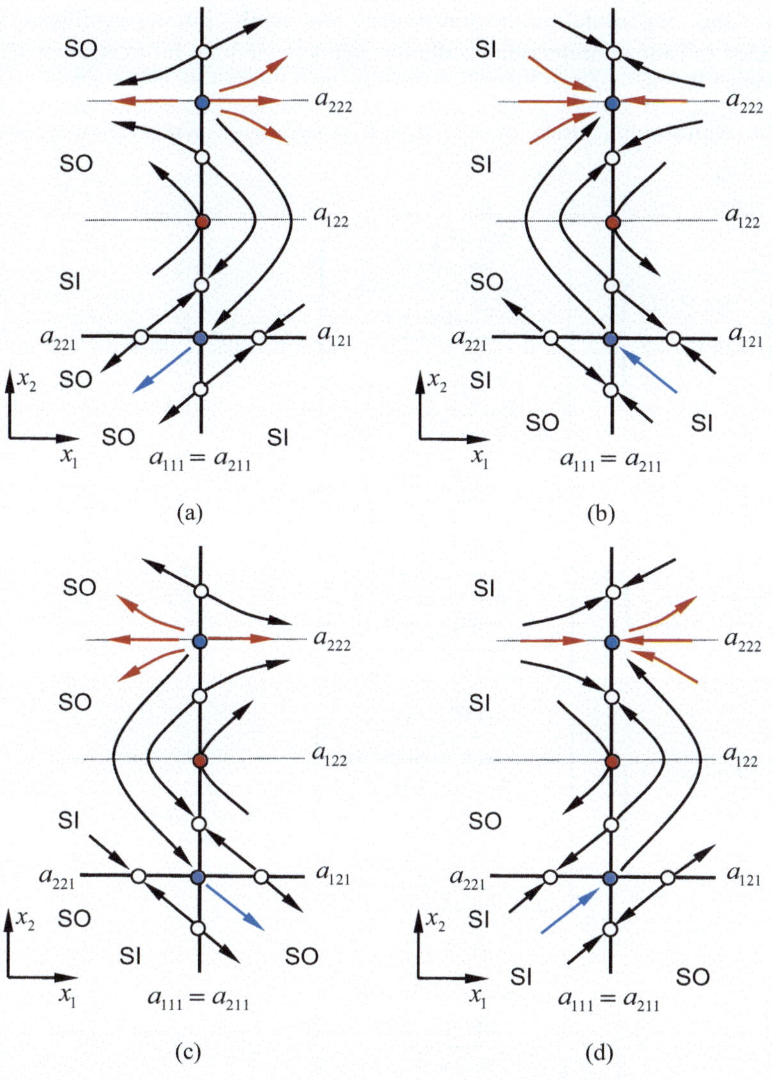

Fig. 5.18 Phase portraits ($a_{111} = a_{211}$ and $a_{121} = a_{221}$) for two-dimensional systems on the x_1-direction with $x_1^* = a_{111} = a_{211}$ and on the x_2-direction with $x_2^* = a_{121} = a_{221}, a_{122}, a_{222}$. The equilibrium flows: (**a**) ($a_{110} > 0$, $a_{220} > 0$), (**b**) ($a_{110} < 0$, $a_{220} > 0$), (**c**) ($a_{110} > 0$, $a_{220} < 0$), and (**d**) ($a_{110} < 0$, $a_{220} < 0$)

The inflection-source and sink infinite-equilibriums of $x_2^* = a_{121} = a_{221}$ and $x_1^* = a_{111} = a_{211}$ with $a_{122} \neq a_{222}$ are the switching bifurcations of sink, source, and saddles with hyperbolic flows for $a_{221} \in (a_{121}, a_{122})$ and $a_{221} < a_{121}$ with $a_{122} < a_{222}$ and for $a_{111} > a_{211}$ with $a_{111} < a_{211}$, as shown in Fig. 5.18. The horizontal inflection-sink and source infinite-equilibriums are at the bottom of the equilibrium series. The vertical and horizontal inflection-source and sink infinite-equilibriums are intersected to form an inflection-saddle for double-switching bifurcation in different directions.

(ii) The equilibrium series of $x_2^* = a_{121}, a_{122}, a_{221}, a_{222}$ with $x_1^* = a_{111}, a_{211}$ for $a_{221} = a_{121}$ with $a_{111} > a_{211}$ are

$$\left\{\begin{array}{c}(a_{122}, a_{211})\\(a_{111}, a_{222})\\(a_{111}, a_{221})\end{array}\right\} = \left\{\begin{array}{c}\underbrace{(\text{UP}, \text{US})}_{\text{up-parabola upper-saddle}}\\\underbrace{(\text{SI}, \text{II} : \text{DI})}_{\text{inflection-sink}}\\\underbrace{(_{\text{SO:SI}}\text{II}, {}_{\text{SO:SI}}\text{II})}_{\text{increasing-inflection saddle}}\end{array}\right\} \text{ for } a_{110} > 0, a_{220} > 0, \tag{5.130}$$

$$\left\{\begin{array}{c}(a_{122}, a_{211})\\(a_{111}, a_{222})\\(a_{111}, a_{221})\end{array}\right\} = \left\{\begin{array}{c}\underbrace{(\text{DP}, \text{LS})}_{\text{down-parabola lower-saddle}}\\\underbrace{(\text{SO}, \text{DI} : \text{II})}_{\text{inflection-source}}\\\underbrace{(_{\text{SO:SI}}\text{DI}, {}_{\text{SI:SO}}\text{DI})}_{\text{decreasing-inflection saddle}}\end{array}\right\} \text{ for } a_{110} < 0, a_{220} > 0, \tag{5.131}$$

$$\left\{\begin{array}{c}(a_{122}, a_{211})\\(a_{111}, a_{222})\\(a_{111}, a_{221})\end{array}\right\} = \left\{\begin{array}{c}\underbrace{(\text{DP}, \text{US})}_{\text{down-parabola upper-saddle}}\\\underbrace{(\text{SI}, \text{DI} : \text{II})}_{\text{inflection-sink}}\\\underbrace{(_{\text{SI:SO}}\text{DI}, {}_{\text{SO:SI}}\text{DI})}_{\text{decreasing-inflection saddle}}\end{array}\right\} \text{ for } a_{110} > 0, a_{220} < 0, \tag{5.132}$$

$$\left\{\begin{array}{c}(a_{122}, a_{211})\\(a_{111}, a_{222})\\(a_{111}, a_{221})\end{array}\right\} = \left\{\begin{array}{c}\underbrace{(\text{UP}, \text{LS})}_{\text{up-parabola lower-saddle}}\\\underbrace{(\text{SO}, \text{II} : \text{DI})}_{\text{inflection-source}}\\\underbrace{(_{\text{SI:SO}}\text{II}, {}_{\text{SI:SO}}\text{II})}_{\text{increasing-inflection saddle}}\end{array}\right\} \text{ for } a_{110} < 0, a_{220} < 0. \tag{5.133}$$

The infinite-equilibrium of $x_2^* = a_{121} = a_{221}$ is the same as in Eqs. (5.122)–(5.125). The infinite-equilibrium of $x_1^* = a_{111} = a_{211}$ is summarized as follows:

$$
\begin{aligned}
(a_{111}, \bar{x}_2) &= \underbrace{(\text{SO}, \text{II})}_{\text{increasing-inflection source}} && \text{if } \bar{x}_2 \in (-\infty, a_{121}),\\
(a_{111}, \bar{x}_2) &= \underbrace{(\text{SI}, \text{II})}_{\text{decreasing-inflection sink}} && \text{if } \bar{x}_2 \in (a_{121}, a_{222}),\\
(a_{111}, \bar{x}_2) &= \underbrace{(\text{SI}, \text{DI})}_{\text{decreasing-inflection sink}} && \text{if } \bar{x}_2 \in (a_{222}, a_{122}),\\
(\bar{x}_1, a_{221}) &= \underbrace{(\text{SO}, \text{II})}_{\text{increasing-inflection source}} && \text{if } \bar{x}_2 \in (a_{122}, \infty),\\
&\text{for } a_{110} > 0, a_{220} > 0;
\end{aligned} \tag{5.134}
$$

$$
\begin{aligned}
(a_{111}, \bar{x}_2) &= \underbrace{(\text{SI}, \text{DI})}_{\text{decreasing-inflection sink}} && \text{if } \bar{x}_2 \in (-\infty, a_{121}),\\
(a_{111}, \bar{x}_2) &= \underbrace{(\text{SO}, \text{DI})}_{\text{decreasing-inflection source}} && \text{if } \bar{x}_2 \in (a_{121}, a_{221}),\\
(a_{111}, \bar{x}_2) &= \underbrace{(\text{SO}, \text{II})}_{\text{increasing-inflection source}} && \text{if } \bar{x}_2 \in (a_{221}, a_{222}),\\
(\bar{x}_1, a_{221}) &= \underbrace{(\text{SI}, \text{DI})}_{\text{decreasing-inflection sink}} && \text{if } \bar{x}_2 \in (a_{222}, \infty),\\
&\text{for } a_{110} < 0, a_{220} > 0;
\end{aligned} \tag{5.135}
$$

$$
\begin{aligned}
(a_{111}, \bar{x}_2) &= \underbrace{(\text{SO}, \text{DI})}_{\text{decreasing-inflection source}} && \text{if } \bar{x}_2 \in (-\infty, a_{121}),\\
(a_{111}, \bar{x}_2) &= \underbrace{(\text{SI}, \text{DI})}_{\text{decreasing-inflection sink}} && \text{if } \bar{x}_2 \in (a_{121}, a_{221}),\\
(a_{111}, \bar{x}_2) &= \underbrace{(\text{SI}, \text{II})}_{\text{increasing-inflection sink}} && \text{if } \bar{x}_2 \in (a_{221}, a_{222}),\\
(\bar{x}_1, a_{221}) &= \underbrace{(\text{SO}, \text{DI})}_{\text{decreasing-inflection source}} && \text{if } \bar{x}_2 \in (a_{222}, \infty),\\
&\text{for } a_{110} > 0, a_{220} < 0;
\end{aligned} \tag{5.136}
$$

$$
\begin{aligned}
(a_{111}, \bar{x}_2) &= \underbrace{(\text{SI}, \text{II})}_{\text{decreasing-inflection sink}} && \text{if } \bar{x}_2 \in (-\infty, a_{121}),\\
(a_{111}, \bar{x}_2) &= \underbrace{(\text{SO}, \text{II})}_{\text{increasing-inflection source}} && \text{if } \bar{x}_2 \in (a_{121}, a_{221}),\\
(a_{111}, \bar{x}_2) &= \underbrace{(\text{SO}, \text{DI})}_{\text{decreasing-inflection source}} && \text{if } \bar{x}_2 \in (a_{221}, a_{222}),\\
(\bar{x}_1, a_{221}) &= \underbrace{(\text{SI}, \text{II})}_{\text{increasing-inflection sink}} && \text{if } \bar{x}_2 \in (a_{222}, \infty),\\
&\text{for } a_{110} < 0, a_{220} < 0.
\end{aligned} \tag{5.137}
$$

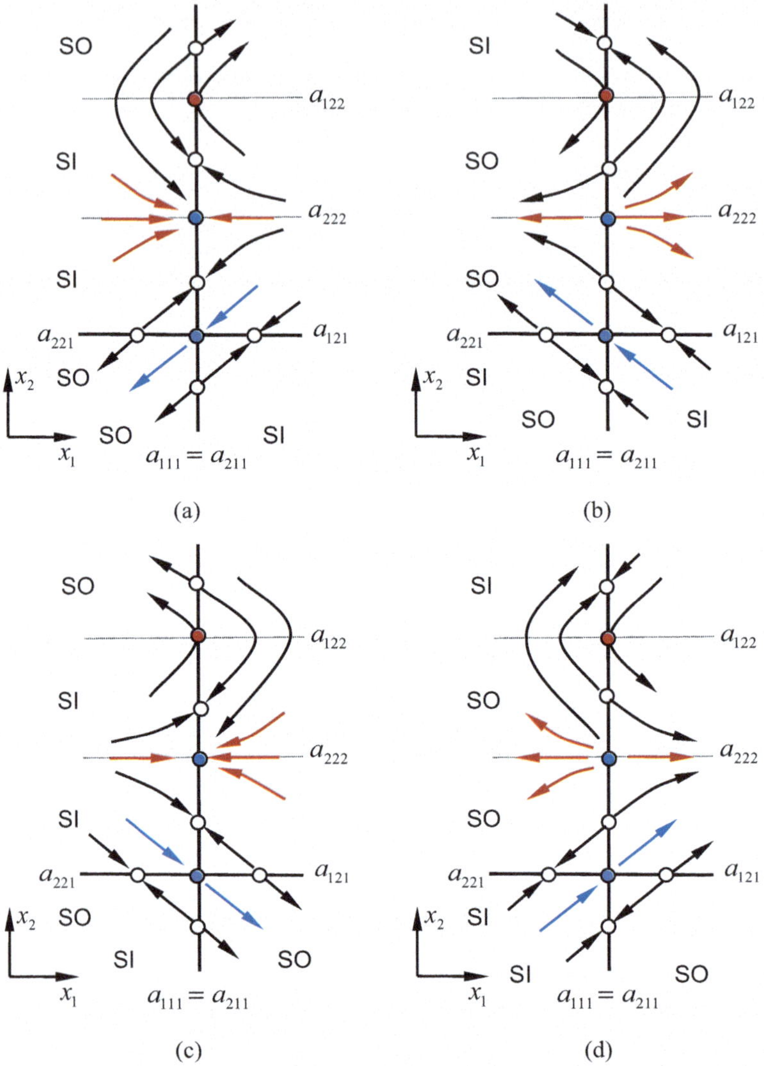

Fig. 5.19 Phase portraits ($a_{111} = a_{211}$ and $a_{121} = a_{221}$) for two-dimensional systems on the x_1-direction with $x_1^* = a_{111} = a_{211}$ and on the x_2-direction with $x_2^* = a_{121} = a_{221}, a_{122}, a_{222}$. The equilibrium flows: (**a**) ($a_{110} > 0$, $a_{220} > 0$), (**b**) ($a_{110} < 0$, $a_{220} > 0$), (**c**) ($a_{110} > 0$, $a_{220} < 0$), and (**d**) ($a_{110} < 0$, $a_{220} < 0$)

The inflection-source and sink infinite-equilibriums of $x_2^* = a_{121} = a_{221}$ and $x_1^* = a_{111} = a_{211}$ with $a_{122} \neq a_{222}$ are the switching bifurcations of sink, source, and saddles with hyperbolic flows for $a_{221} \in (a_{121}, a_{122})$ and $a_{221} < a_{121}$ with $a_{222} < a_{122}$ and for $a_{111} > a_{211}$ with $a_{111} < a_{211}$, as shown in Fig. 5.19. The horizontal inflection-sink and source infinite-equilibriums are at the bottom of the equilibrium series. The vertical and horizontal inflection-source and sink infinite-equilibriums are intersected to form an inflection-saddle for double-switching bifurcation in two different directions.

(iii) The equilibrium series of $x_2^* = a_{121}, a_{122}, a_{221}, a_{222}$ with $x_1^* = a_{111}, a_{211}$ for $a_{221} = a_{122}$ with $a_{111} = a_{211}$ are

$$\left\{\begin{array}{c}(a_{111}, a_{222})\\(a_{111}, a_{221})\\(a_{121}, a_{211})\end{array}\right\} = \left\{\begin{array}{c}\underbrace{(\mathrm{SO}, \mathrm{DI}:\mathrm{II})}_{\text{inflection-source}}\\ \underbrace{({}_{\mathrm{SO:SI}}\mathrm{DI}, {}_{\mathrm{SI:SO}}\mathrm{DI})}_{\text{decreasing-inflection saddle}}\\ \underbrace{(\mathrm{DP}, \mathrm{LS})}_{\text{down-parabola lower-saddle}}\end{array}\right\} \text{ for } a_{110} > 0, a_{220} > 0, \quad (5.138)$$

$$\left\{\begin{array}{c}(a_{111}, a_{222})\\(a_{111}, a_{221})\\(a_{121}, a_{211})\end{array}\right\} = \left\{\begin{array}{c}\underbrace{(\mathrm{SI}, \mathrm{II}:\mathrm{DI})}_{\text{inflection-sink}}\\ \underbrace{({}_{\mathrm{SO:SI}}\mathrm{II}, {}_{\mathrm{SO:SI}}\mathrm{II})}_{\text{increasing-inflection saddle}}\\ \underbrace{(\mathrm{UP}, \mathrm{US})}_{\text{up-parabola upper-saddle}}\end{array}\right\} \text{ for } a_{110} < 0, a_{220} > 0, \quad (5.139)$$

$$\left\{\begin{array}{c}(a_{111}, a_{222})\\(a_{111}, a_{221})\\(a_{121}, a_{211})\end{array}\right\} = \left\{\begin{array}{c}\underbrace{(\mathrm{SO}, \mathrm{II}:\mathrm{DI})}_{\text{inflection-source}}\\ \underbrace{({}_{\mathrm{SI:SO}}\mathrm{II}, {}_{\mathrm{SI:SO}}\mathrm{II})}_{\text{increasing-inflection saddle}}\\ \underbrace{(\mathrm{UP}, \mathrm{LS})}_{\text{up-parabola lower-saddle}}\end{array}\right\} \text{ for } a_{110} > 0, a_{220} < 0, \quad (5.140)$$

$$\left\{\begin{array}{c}(a_{111}, a_{222})\\(a_{111}, a_{221})\\(a_{121}, a_{211})\end{array}\right\} = \left\{\begin{array}{c}\underbrace{(\mathrm{SI}, \mathrm{DI}:\mathrm{II})}_{\text{inflection-sink}}\\ \underbrace{({}_{\mathrm{SI:SO}}\mathrm{DI}, {}_{\mathrm{SO:SI}}\mathrm{DI})}_{\text{decreasing-inflection saddle}}\\ \underbrace{(\mathrm{DP}, \mathrm{US})}_{\text{down-parabola upper-saddle}}\end{array}\right\} \text{ for } a_{110} < 0, a_{220} < 0. \quad (5.141)$$

The infinite-equilibrium of $x_2^* = a_{121} = a_{221}$ is summarized as follows:

$$\begin{aligned}(\bar{x}_1, a_{221}) &= \underbrace{(\mathrm{DI}, \mathrm{SO})}_{\text{decreasing-inflection source}} \quad \text{if } \bar{x}_1 \in (-\infty, a_{111}),\\ (\bar{x}_1, a_{221}) &= \underbrace{(\mathrm{DI}, \mathrm{SI})}_{\text{decreasing-inflection sink}} \quad \text{if } \bar{x}_1 \in (a_{111}, \infty),\end{aligned} \quad (5.142)$$

for $a_{110} > 0, a_{220} > 0$;

$$(\overline{x}_1, a_{221}) = \underbrace{(\text{II}, \text{SO})}_{\text{increasing-inflection source}} \quad \text{if } \overline{x}_1 \in (-\infty, a_{111}),$$
$$(\overline{x}_1, a_{221}) = \underbrace{(\text{II}, \text{SI})}_{\text{increasing-inflection sink}} \quad \text{if } \overline{x}_1 \in (a_{111}, \infty), \tag{5.143}$$
for $a_{110} < 0, a_{220} > 0$;

$$(\overline{x}_1, a_{221}) = \underbrace{(\text{II}, \text{SI})}_{\text{increasing-inflection sink}} \quad \text{if } \overline{x}_1 \in (-\infty, a_{111}),$$
$$(\overline{x}_1, a_{221}) = \underbrace{(\text{II}, \text{SO})}_{\text{increasing-inflection source}} \quad \text{if } \overline{x}_1 \in (a_{111}, \infty), \tag{5.144}$$
for $a_{110} > 0, a_{220} < 0$;

$$(\overline{x}_1, a_{221}) = \underbrace{(\text{DI}, \text{SI})}_{\text{decreasing-inflection sink}} \quad \text{if } \overline{x}_1 \in (-\infty, a_{111}),$$
$$(\overline{x}_1, a_{221}) = \underbrace{(\text{DI}, \text{SO})}_{\text{decreasing-inflection source}} \quad \text{if } \overline{x}_1 \in (a_{111}, \infty), \tag{5.145}$$
for $a_{110} < 0, a_{220} < 0$.

The infinite-equilibrium of $x_1^* = a_{111} = a_{211}$ is summarized as follows:

$$(a_{111}, \overline{x}_2) = \underbrace{(\text{SO}, \text{II})}_{\text{increasing-inflection source}} \quad \text{if } \overline{x}_2 \in (-\infty, a_{121}),$$
$$(a_{111}, \overline{x}_2) = \underbrace{(\text{SI}, \text{DI})}_{\text{decreasing-inflection sink}} \quad \text{if } \overline{x}_2 \in (a_{121}, a_{221}),$$
$$(a_{111}, \overline{x}_2) = \underbrace{(\text{SO}, \text{DI})}_{\text{decreasing-inflection source}} \quad \text{if } \overline{x}_2 \in (a_{221}, a_{222}), \tag{5.146}$$
$$(\overline{x}_1, a_{221}) = \underbrace{(\text{SO}, \text{II})}_{\text{increasing-inflection source}} \quad \text{if } \overline{x}_2 \in (a_{222}, \infty),$$
for $a_{110} > 0, a_{220} > 0$;

$$(a_{111}, \overline{x}_2) = \underbrace{(\text{SI}, \text{DI})}_{\text{decreasing-inflection sink}} \quad \text{if } \overline{x}_2 \in (-\infty, a_{121}),$$
$$(a_{111}, \overline{x}_2) = \underbrace{(\text{SO}, \text{II})}_{\text{increasing-inflection source}} \quad \text{if } \overline{x}_2 \in (a_{121}, a_{221}),$$
$$(a_{111}, \overline{x}_2) = \underbrace{(\text{SI}, \text{II})}_{\text{increasing-inflection sink}} \quad \text{if } \overline{x}_2 \in (a_{221}, a_{222}), \tag{5.147}$$
$$(\overline{x}_1, a_{221}) = \underbrace{(\text{SI}, \text{DI})}_{\text{decreasing-inflection sink}} \quad \text{if } \overline{x}_2 \in (a_{222}, \infty),$$
for $a_{110} < 0, a_{220} > 0$;

$$\begin{aligned}
(a_{111}, \bar{x}_2) &= \underbrace{(\mathrm{SO}, \mathrm{DI})}_{\text{decreasing-inflection source}} && \text{if } \bar{x}_2 \in (-\infty, a_{121}),\\
(a_{111}, \bar{x}_2) &= \underbrace{(\mathrm{SI}, \mathrm{II})}_{\text{increasing-inflection sink}} && \text{if } \bar{x}_2 \in (a_{121}, a_{221}),\\
(a_{111}, \bar{x}_2) &= \underbrace{(\mathrm{SO}, \mathrm{II})}_{\text{increasing-inflection source}} && \text{if } \bar{x}_2 \in (a_{221}, a_{222}),\\
(\bar{x}_1, a_{221}) &= \underbrace{(\mathrm{SO}, \mathrm{DI})}_{\text{decreasing-inflection source}} && \text{if } \bar{x}_2 \in (a_{222}, \infty),\\
&\text{for } a_{110} > 0, a_{220} < 0;
\end{aligned} \tag{5.148}$$

$$\begin{aligned}
(a_{111}, \bar{x}_2) &= \underbrace{(\mathrm{SI}, \mathrm{II})}_{\text{decreasing-inflection sink}} && \text{if } \bar{x}_2 \in (-\infty, a_{121}),\\
(a_{111}, \bar{x}_2) &= \underbrace{(\mathrm{SO}, \mathrm{DI})}_{\text{decreasing-inflection source}} && \text{if } \bar{x}_2 \in (a_{121}, a_{221}),\\
(a_{111}, \bar{x}_2) &= \underbrace{(\mathrm{SI}, \mathrm{DI})}_{\text{decreasing-inflection sink}} && \text{if } \bar{x}_2 \in (a_{221}, a_{222}),\\
(\bar{x}_1, a_{221}) &= \underbrace{(\mathrm{SI}, \mathrm{II})}_{\text{increasing-inflection sink}} && \text{if } \bar{x}_2 \in (a_{222}, \infty),\\
&\text{for } a_{110} < 0, a_{220} < 0.
\end{aligned} \tag{5.149}$$

The inflection-source and sink infinite-equilibriums of $x_2^* = a_{122} = a_{221}$ and $x_1^* = a_{111} = a_{211}$ are the switching bifurcations of sink, source, and saddles with hyperbolic flows for $a_{221} \in (a_{121}, a_{122})$ with $a_{122} < a_{221}$ and for $a_{111} > a_{211}$ with $a_{111} < a_{211}$, as shown in Fig. 5.20. The horizontal inflection-sink and source infinite-equilibriums are on the middle of the equilibrium series. The vertical and horizontal inflection-source and sink infinite-equilibriums are intersected to form an inflection-saddle for double-switching bifurcation in different directions.

(iv) The equilibrium series of $x_2^* = a_{121}, a_{122}, a_{221}, a_{222}$ with $x_1^* = a_{111}, a_{211}$ for $a_{121} = a_{222}$ with $a_{111} = a_{211}$ are

$$\left\{\begin{matrix}(a_{122}, a_{211})\\(a_{111}, a_{222})\\(a_{111}, a_{221})\end{matrix}\right\} = \left\{\begin{matrix}\underbrace{(\mathrm{UP}, \mathrm{US})}_{\text{up-parabola upper-saddle}}\\ \underbrace{({}_{\mathrm{SI:SO}}\mathrm{DI}, {}_{\mathrm{SO:SI}}\mathrm{DI})}_{\text{decreasing-inflection saddle}}\\ \underbrace{(\mathrm{SO}, \mathrm{II}:\mathrm{DI})}_{\text{inflection-source}}\end{matrix}\right\} \text{ for } a_{110} > 0, a_{220} > 0, \tag{5.150}$$

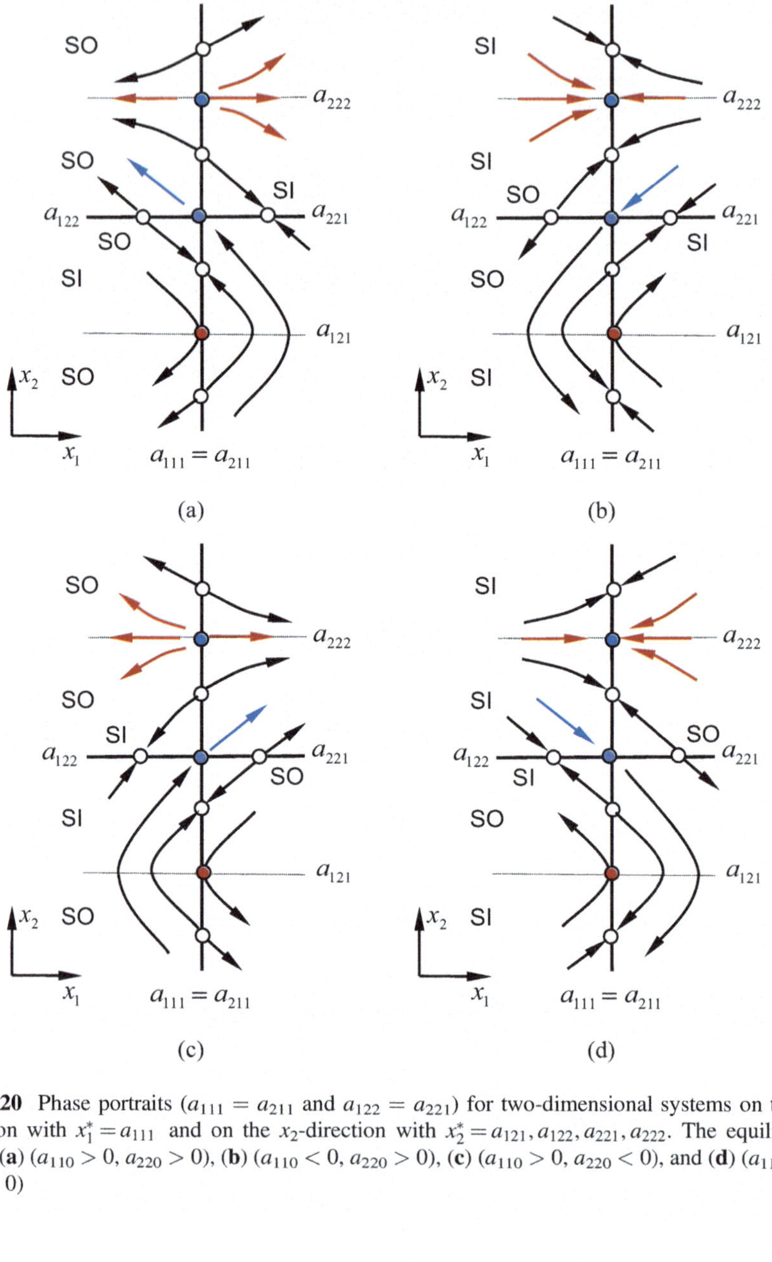

Fig. 5.20 Phase portraits ($a_{111} = a_{211}$ and $a_{122} = a_{221}$) for two-dimensional systems on the x_1-direction with $x_1^* = a_{111}$ and on the x_2-direction with $x_2^* = a_{121}, a_{122}, a_{221}, a_{222}$. The equilibrium flows: **(a)** ($a_{110} > 0$, $a_{220} > 0$), **(b)** ($a_{110} < 0$, $a_{220} > 0$), **(c)** ($a_{110} > 0$, $a_{220} < 0$), and **(d)** ($a_{110} < 0$, $a_{220} < 0$)

$$\left\{\begin{array}{c}(a_{122},a_{211})\\(a_{111},a_{222})\\(a_{111},a_{221})\end{array}\right\}=\left\{\begin{array}{c}\underbrace{(\text{DP},\text{LS})}_{\text{down-parabola lower-saddle}}\\\underbrace{({}_{\text{SI:SO}}\text{II},{}_{\text{SI:SO}}\text{II})}_{\text{increasing-inflection saddle}}\\\underbrace{(\text{SI},\text{DI}:\text{II})}_{\text{inflection-sink}}\end{array}\right\}\text{ for }a_{110}<0,a_{220}>0, \tag{5.151}$$

$$\left\{\begin{array}{c}(a_{122},a_{211})\\(a_{111},a_{222})\\(a_{111},a_{221})\end{array}\right\}=\left\{\begin{array}{c}\underbrace{(\text{DP},\text{US})}_{\text{down-parabola upper-saddle}}\\\underbrace{({}_{\text{SO:SI}}\text{II},{}_{\text{SO:SI}}\text{II})}_{\text{increasing-inflection saddle}}\\\underbrace{(\text{SO},\text{DI}:\text{II})}_{\text{inflection-source}}\end{array}\right\}\text{ for }a_{110}>0,a_{220}<0, \tag{5.152}$$

$$\left\{\begin{array}{c}(a_{122},a_{211})\\(a_{111},a_{222})\\(a_{111},a_{221})\end{array}\right\}=\left\{\begin{array}{c}\underbrace{(\text{UP},\text{LS})}_{\text{up-parabola lower-saddle}}\\\underbrace{({}_{\text{SO:SI}}\text{DI},{}_{\text{SI:SO}}\text{DI})}_{\text{decreasing-inflection saddle}}\\\underbrace{(\text{SI},\text{II}:\text{DI})}_{\text{inflection-sink}}\end{array}\right\}\text{ for }a_{110}<0,a_{220}<0. \tag{5.153}$$

The infinite-equilibrium of $x_2^* = a_{121} = a_{221}$ is summarized as follows:

$$\begin{array}{l}(\bar{x}_1,a_{221})=\underbrace{(\text{DI},\text{SI})}_{\text{decreasing-inflection sink}}\quad\text{if }\bar{x}_1\in(-\infty,a_{111}),\\(\bar{x}_1,a_{221})=\underbrace{(\text{DI},\text{SO})}_{\text{decreasing-inflection source}}\quad\text{if }\bar{x}_1\in(a_{111},\infty),\\\text{for }a_{110}>0,a_{220}>0;\end{array} \tag{5.154}$$

$$\begin{array}{l}(\bar{x}_1,a_{221})=\underbrace{(\text{II},\text{SI})}_{\text{increasing-inflection sink}}\quad\text{if }\bar{x}_1\in(-\infty,a_{111}),\\(\bar{x}_1,a_{221})=\underbrace{(\text{II},\text{SO})}_{\text{increasing-inflection source}}\quad\text{if }\bar{x}_1\in(a_{111},\infty),\\\text{for }a_{110}<0,a_{220}>0;\end{array} \tag{5.155}$$

$$\begin{array}{l}(\bar{x}_1,a_{221})=\underbrace{(\text{II},\text{SO})}_{\text{increasing-inflection source}}\quad\text{if }\bar{x}_1\in(-\infty,a_{111}),\\(\bar{x}_1,a_{221})=\underbrace{(\text{II},\text{SI})}_{\text{increasing-inflection sink}}\quad\text{if }\bar{x}_1\in(a_{111},\infty),\\\text{for }a_{110}>0,a_{220}<0;\end{array} \tag{5.156}$$

$$
\begin{aligned}
&(\bar{x}_1, a_{221}) = \underbrace{(\mathrm{DI}, \mathrm{SO})}_{\text{decreasing-inflection source}} \quad \text{if } \bar{x}_1 \in (-\infty, a_{111}),\\
&(\bar{x}_1, a_{221}) = \underbrace{(\mathrm{DI}, \mathrm{SI})}_{\text{decreasing-inflection sink}} \quad \text{if } \bar{x}_1 \in (a_{111}, \infty),\\
&\text{for } a_{110} < 0, a_{220} < 0.
\end{aligned}
\tag{5.157}
$$

The infinite-equilibrium of $x_1^* = a_{111} = a_{211}$ is summarized as follows:

$$
\begin{aligned}
&(a_{111}, \bar{x}_2) = \underbrace{(\mathrm{SO}, \mathrm{II})}_{\text{increasing-inflection source}} \quad \text{if } \bar{x}_2 \in (-\infty, a_{121}),\\
&(a_{111}, \bar{x}_2) = \underbrace{(\mathrm{SO}, \mathrm{DI})}_{\text{decreasing-inflection source}} \quad \text{if } \bar{x}_2 \in (a_{121}, a_{221}),\\
&(a_{111}, \bar{x}_2) = \underbrace{(\mathrm{SI}, \mathrm{DI})}_{\text{decreasing-inflection sink}} \quad \text{if } \bar{x}_2 \in (a_{221}, a_{222}),\\
&(\bar{x}_1, a_{221}) = \underbrace{(\mathrm{SO}, \mathrm{II})}_{\text{increasing-inflection source}} \quad \text{if } \bar{x}_2 \in (a_{222}, \infty),\\
&\text{for } a_{110} > 0, a_{220} > 0;
\end{aligned}
\tag{5.158}
$$

$$
\begin{aligned}
&(a_{111}, \bar{x}_2) = \underbrace{(\mathrm{SI}, \mathrm{DI})}_{\text{decreasing-inflection sink}} \quad \text{if } \bar{x}_2 \in (-\infty, a_{121}),\\
&(a_{111}, \bar{x}_2) = \underbrace{(\mathrm{SI}, \mathrm{II})}_{\text{increasing-inflection sink}} \quad \text{if } \bar{x}_2 \in (a_{121}, a_{221}),\\
&(a_{111}, \bar{x}_2) = \underbrace{(\mathrm{SO}, \mathrm{II})}_{\text{increasing-inflection source}} \quad \text{if } \bar{x}_2 \in (a_{221}, a_{222}),\\
&(\bar{x}_1, a_{221}) = \underbrace{(\mathrm{SI}, \mathrm{DI})}_{\text{decreasing-inflection sink}} \quad \text{if } \bar{x}_2 \in (a_{222}, \infty),\\
&\text{for } a_{110} < 0, a_{220} > 0;
\end{aligned}
\tag{5.159}
$$

$$
\begin{aligned}
&(a_{111}, \bar{x}_2) = \underbrace{(\mathrm{SO}, \mathrm{DI})}_{\text{decreasing-inflection source}} \quad \text{if } \bar{x}_2 \in (-\infty, a_{121}),\\
&(a_{111}, \bar{x}_2) = \underbrace{(\mathrm{SO}, \mathrm{II})}_{\text{increasing-inflection source}} \quad \text{if } \bar{x}_2 \in (a_{121}, a_{221}),\\
&(a_{111}, \bar{x}_2) = \underbrace{(\mathrm{SI}, \mathrm{II})}_{\text{increasing-inflection sink}} \quad \text{if } \bar{x}_2 \in (a_{221}, a_{222}),\\
&(\bar{x}_1, a_{221}) = \underbrace{(\mathrm{SO}, \mathrm{DI})}_{\text{decreasing-inflection source}} \quad \text{if } \bar{x}_2 \in (a_{222}, \infty),\\
&\text{for } a_{110} > 0, a_{220} < 0;
\end{aligned}
\tag{5.160}
$$

$$
\begin{aligned}
(a_{111}, \bar{x}_2) &= \underbrace{(\text{SI}, \text{II})}_{\text{decreasing-inflection sink}} && \text{if } \bar{x}_2 \in (-\infty, a_{121}),\\
(a_{111}, \bar{x}_2) &= \underbrace{(\text{SI}, \text{DI})}_{\text{decreasing-inflection sink}} && \text{if } \bar{x}_2 \in (a_{121}, a_{221}),\\
(a_{111}, \bar{x}_2) &= \underbrace{(\text{SO}, \text{DI})}_{\text{decreasing-inflection source}} && \text{if } \bar{x}_2 \in (a_{221}, a_{222}),\\
(\bar{x}_1, a_{221}) &= \underbrace{(\text{SI}, \text{II})}_{\text{increasing-inflection sink}} && \text{if } \bar{x}_2 \in (a_{222}, \infty),\\
&\text{for } a_{110} < 0, a_{220} < 0.
\end{aligned}
\tag{5.161}
$$

The inflection-source and sink infinite-equilibriums of $x_2^* = a_{222} = a_{121}$ and $x_1^* = a_{111} = a_{211}$ are the switching bifurcations of sink, source, and saddles with hyperbolic flows for $a_{121} \in (a_{221}, a_{222})$ with $a_{222} < a_{121}$ and for $a_{111} > a_{211}$ with $a_{111} < a_{211}$, as shown in Fig. 5.21. The horizontal inflection-sink and source infinite-equilibriums are on the middle of the equilibrium series. The vertical and horizontal inflection-source and sink infinite-equilibriums are intersected to form an inflection-saddle for double-switching bifurcation in different directions.

(v) The equilibrium series of $x_2^* = a_{121}, a_{122}, a_{221}, a_{222}$ with $x_1^* = a_{111}, a_{211}$ for $a_{122} = a_{222}$ and $a_{111} = a_{211}$ are

$$
\begin{Bmatrix} (a_{111}, a_{222}) \\ (a_{121}, a_{211}) \\ (a_{111}, a_{221}) \end{Bmatrix} = \begin{Bmatrix} \underbrace{({}_{\text{SI:SO}}\text{II}, {}_{\text{SI:SO}}\text{II})}_{\text{increasing-inflection saddle}} \\ \underbrace{(\text{UP}, \text{LS})}_{\text{up-parabola lower-saddle}} \\ \underbrace{(\text{SO}, \text{II}:\text{DI})}_{\text{inflection-source}} \end{Bmatrix} \text{ for } a_{110} > 0, a_{220} > 0, \tag{5.162}
$$

$$
\begin{Bmatrix} (a_{111}, a_{222}) \\ (a_{121}, a_{211}) \\ (a_{111}, a_{221}) \end{Bmatrix} = \begin{Bmatrix} \underbrace{({}_{\text{SI:SO}}\text{DI}, {}_{\text{SO:SI}}\text{DI})}_{\text{decreasing-inflection saddle}} \\ \underbrace{(\text{DP}, \text{US})}_{\text{down-parabola upper-saddle}} \\ \underbrace{(\text{SI}, \text{DI}:\text{II})}_{\text{inflection-sink}} \end{Bmatrix} \text{ for } a_{110} < 0, a_{220} > 0, \tag{5.163}
$$

$$
\begin{Bmatrix} (a_{111}, a_{222}) \\ (a_{121}, a_{211}) \\ (a_{111}, a_{221}) \end{Bmatrix} = \begin{Bmatrix} \underbrace{({}_{\text{SO:SI}}\text{DI}, {}_{\text{SI:SO}}\text{DI})}_{\text{decreasing-inflection saddle}} \\ \underbrace{(\text{DP}, \text{LS})}_{\text{down-parabola lower-saddle}} \\ \underbrace{(\text{SO}, \text{DI}:\text{II})}_{\text{inflection-source}} \end{Bmatrix} \text{ for } a_{110} > 0, a_{220} < 0, \tag{5.164}
$$

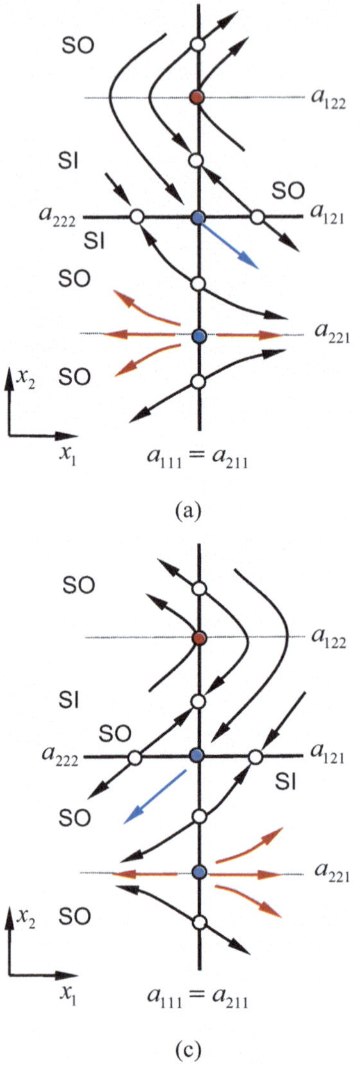

Fig. 5.21 Phase portraits ($a_{111} = a_{211}$ and $a_{121} = a_{222}$) for two-dimensional systems on the x_1-direction with $x_1^* = a_{111}$ and on the x_2-direction with $x_2^* = a_{121}, a_{221}, a_{222}$. The equilibrium matrix with infinite-equilibrium: **(a)** ($a_{110} > 0$, $a_{220} > 0$), **(b)** ($a_{110} < 0$, $a_{220} > 0$), **(c)** ($a_{110} > 0$, $a_{220} < 0$), and **(d)** ($a_{110} < 0$, $a_{220} < 0$)

$$\left\{\begin{array}{c}(a_{111},a_{222})\\(a_{121},a_{211})\\(a_{111},a_{221})\end{array}\right\}=\left\{\begin{array}{c}\underbrace{({}_{\text{SO:SI}}\text{II},{}_{\text{SO:SI}}\text{II})}_{\text{increasing-inflection saddle}}\\\underbrace{(\text{UP},\text{US})}_{\text{up-parabola upper-saddle}}\\\underbrace{(\text{SI},\text{II}:\text{DI})}_{\text{inflection-sink}}\end{array}\right\}\text{ for } a_{110}<0, a_{220}<0. \tag{5.165}$$

The infinite-equilibrium of $x_2^*=a_{121}=a_{221}$ is summarized as follows:

$$\begin{aligned}&(\bar{x}_1,a_{221})=\underbrace{(\text{II},\text{SI})}_{\text{increasing-inflection sink}}\quad \text{if } \bar{x}_1\in(-\infty,a_{111}),\\&(\bar{x}_1,a_{221})=\underbrace{(\text{II},\text{SO})}_{\text{increasing-inflection source}}\quad \text{if } \bar{x}_1\in(a_{111},\infty),\\&\text{for } a_{110}>0, a_{220}>0;\end{aligned}\tag{5.166}$$

$$\begin{aligned}&(\bar{x}_1,a_{221})=\underbrace{(\text{DI},\text{SI})}_{\text{decreasing-inflection sink}}\quad \text{if } \bar{x}_1\in(-\infty,a_{111}),\\&(\bar{x}_1,a_{221})=\underbrace{(\text{DI},\text{SO})}_{\text{decreasing-inflection source}}\quad \text{if } \bar{x}_1\in(a_{111},\infty),\\&\text{for } a_{110}<0, a_{220}>0;\end{aligned}\tag{5.167}$$

$$\begin{aligned}&(\bar{x}_1,a_{221})=\underbrace{(\text{DI},\text{SO})}_{\text{decreasing-inflection source}}\quad \text{if } \bar{x}_1\in(-\infty,a_{111}),\\&(\bar{x}_1,a_{221})=\underbrace{(\text{DI},\text{SI})}_{\text{decreasing-inflection sink}}\quad \text{if } \bar{x}_1\in(a_{111},\infty),\\&\text{for } a_{110}>0, a_{220}<0;\end{aligned}\tag{5.168}$$

$$\begin{aligned}&(\bar{x}_1,a_{221})=\underbrace{(\text{II},\text{SO})}_{\text{increasing-inflection source}}\quad \text{if } \bar{x}_1\in(-\infty,a_{111}),\\&(\bar{x}_1,a_{221})=\underbrace{(\text{II},\text{SI})}_{\text{increasing-inflection sink}}\quad \text{if } \bar{x}_1\in(a_{111},\infty),\\&\text{for } a_{110}<0, a_{220}<0.\end{aligned}\tag{5.169}$$

The infinite-equilibrium of $x_1^* = a_{111} = a_{211}$ is summarized as follows:

$$
\begin{aligned}
&(a_{111}, \bar{x}_2) = \underbrace{(\text{SO}, \text{II})}_{\text{increasing-inflection source}} \quad \text{if } \bar{x}_2 \in (-\infty, a_{221}),\\
&(a_{111}, \bar{x}_2) = \underbrace{(\text{SO}, \text{DI})}_{\text{decreasing-inflection source}} \quad \text{if } \bar{x}_2 \in (a_{221}, a_{121}),\\
&(a_{111}, \bar{x}_2) = \underbrace{(\text{SI}, \text{II})}_{\text{increasing-inflection sink}} \quad \text{if } \bar{x}_2 \in (a_{121}, a_{222}),\\
&(\bar{x}_1, a_{221}) = \underbrace{(\text{SO}, \text{II})}_{\text{increasing-inflection source}} \quad \text{if } \bar{x}_2 \in (a_{222}, \infty),\\
&\text{for } a_{110} > 0, a_{220} > 0;
\end{aligned}
\tag{5.170}
$$

$$
\begin{aligned}
&(a_{111}, \bar{x}_2) = \underbrace{(\text{SI}, \text{DI})}_{\text{decreasing-inflection sink}} \quad \text{if } \bar{x}_2 \in (-\infty, a_{221}),\\
&(a_{111}, \bar{x}_2) = \underbrace{(\text{SI}, \text{II})}_{\text{increasing-inflection sink}} \quad \text{if } \bar{x}_2 \in (a_{221}, a_{121}),\\
&(a_{111}, \bar{x}_2) = \underbrace{(\text{SO}, \text{DI})}_{\text{decreasing-inflection source}} \quad \text{if } \bar{x}_2 \in (a_{121}, a_{222}),\\
&(\bar{x}_1, a_{221}) = \underbrace{(\text{SI}, \text{DI})}_{\text{decreasing-inflection sink}} \quad \text{if } \bar{x}_2 \in (a_{222}, \infty),\\
&\text{for } a_{110} < 0, a_{220} > 0;
\end{aligned}
\tag{5.171}
$$

$$
\begin{aligned}
&(a_{111}, \bar{x}_2) = \underbrace{(\text{SO}, \text{DI})}_{\text{decreasing-inflection source}} \quad \text{if } \bar{x}_2 \in (-\infty, a_{221}),\\
&(a_{111}, \bar{x}_2) = \underbrace{(\text{SO}, \text{II})}_{\text{increasing-inflection source}} \quad \text{if } \bar{x}_2 \in (a_{221}, a_{121}),\\
&(a_{111}, \bar{x}_2) = \underbrace{(\text{SI}, \text{DI})}_{\text{decreasing-inflection sink}} \quad \text{if } \bar{x}_2 \in (a_{121}, a_{222}),\\
&(\bar{x}_1, a_{221}) = \underbrace{(\text{SO}, \text{DI})}_{\text{decreasing-inflection source}} \quad \text{if } \bar{x}_2 \in (a_{222}, \infty),\\
&\text{for } a_{110} > 0, a_{220} < 0;
\end{aligned}
\tag{5.172}
$$

$$\begin{aligned}
&(a_{111}, \bar{x}_2) = \underbrace{(\mathrm{SI}, \mathrm{II})}_{\text{decreasing-inflection sink}} && \text{if } \bar{x}_2 \in (-\infty, a_{221}),\\
&(a_{111}, \bar{x}_2) = \underbrace{(\mathrm{SI}, \mathrm{DI})}_{\text{decreasing-inflection sink}} && \text{if } \bar{x}_2 \in (a_{221}, a_{121}),\\
&(a_{111}, \bar{x}_2) = \underbrace{(\mathrm{SO}, \mathrm{II})}_{\text{increasing-inflection source}} && \text{if } \bar{x}_2 \in (a_{121}, a_{222}),\\
&(\bar{x}_1, a_{221}) = \underbrace{(\mathrm{SI}, \mathrm{II})}_{\text{increasing-inflection sink}} && \text{if } \bar{x}_2 \in (a_{222}, \infty),\\
&\text{for } a_{110} < 0, a_{220} < 0.
\end{aligned} \tag{5.173}$$

The inflection-source and sink infinite-equilibriums of $x_2^* = a_{222} = a_{122}$ and $x_1^* = a_{111} = a_{211}$ are the switching bifurcations of sink, source, and saddles with hyperbolic flows for $a_{122} \in (a_{221}, a_{222})$ with $a_{222} < a_{122}$ and for $a_{111} > a_{211}$ with $a_{111} < a_{211}$, as shown in Fig. 5.22. The horizontal inflection-sink and source infinite-equilibriums are on the top of the equilibrium series. The vertical and horizontal inflection-source and sink infinite-equilibriums are intersected to form an inflection-saddle for double-switching bifurcation in different directions.

(vi) The equilibrium series of $x_2^* = a_{121}, a_{122}, a_{221}, a_{222}$ with $x_1^* = a_{111}, a_{211}$ for $a_{122} = a_{222}$ and $a_{111} = a_{211}$ are

$$\left\{\begin{matrix} (a_{111}, a_{222}) \\ (a_{111}, a_{221}) \\ (a_{121}, a_{211}) \end{matrix}\right\} = \left\{\begin{matrix} \underbrace{({}_{\mathrm{SI:SO}}\mathrm{II}, {}_{\mathrm{SI:SO}}\mathrm{II})}_{\text{increasing-inflection saddle}} \\ \underbrace{(\mathrm{SI}, \mathrm{DI} : \mathrm{II})}_{\text{inflection sink}} \\ \underbrace{(\mathrm{DP}, \mathrm{LS})}_{\text{down-parabola lower-saddle}} \end{matrix}\right\} \text{for } a_{110} > 0, a_{220} > 0, \tag{5.174}$$

$$\left\{\begin{matrix} (a_{111}, a_{222}) \\ (a_{111}, a_{221}) \\ (a_{121}, a_{211}) \end{matrix}\right\} = \left\{\begin{matrix} \underbrace{({}_{\mathrm{SI:SO}}\mathrm{DI}, {}_{\mathrm{SO:SI}}\mathrm{DI})}_{\text{decreasing-inflection saddle}} \\ \underbrace{(\mathrm{SO}, \mathrm{II} : \mathrm{DI})}_{\text{inflection-source}} \\ \underbrace{(\mathrm{UP}, \mathrm{US})}_{\text{up-parabola upper-saddle}} \end{matrix}\right\} \text{for } a_{110} < 0, a_{220} > 0, \tag{5.175}$$

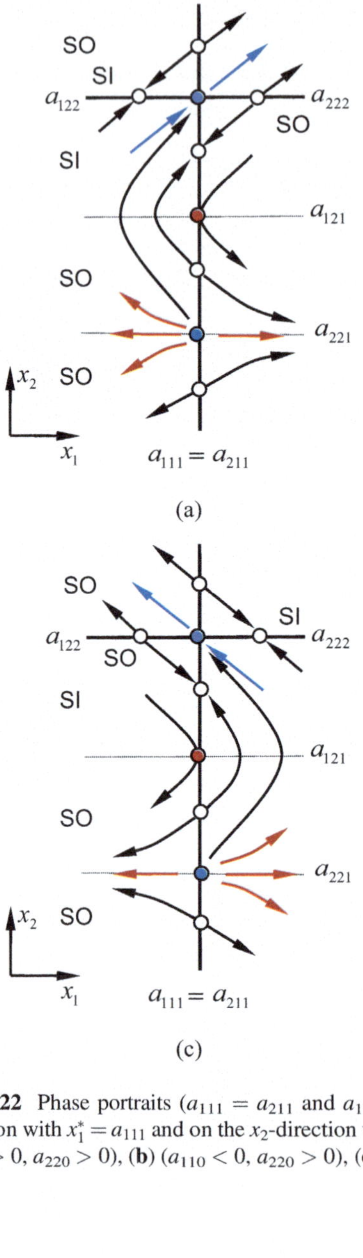

Fig. 5.22 Phase portraits ($a_{111} = a_{211}$ and $a_{122} = a_{222}$) for two-dimensional systems on the x_1-direction with $x_1^* = a_{111}$ and on the x_2-direction with $x_2^* = a_{121}, a_{221}, a_{222}$. The equilibrium flows: (**a**) ($a_{110} > 0$, $a_{220} > 0$), (**b**) ($a_{110} < 0$, $a_{220} > 0$), (**c**) ($a_{110} > 0$, $a_{220} < 0$), and (**d**) ($a_{110} < 0$, $a_{220} < 0$)

$$\left\{\begin{array}{l}(a_{111},a_{222})\\(a_{111},a_{221})\\(a_{121},a_{211})\end{array}\right\}=\left\{\begin{array}{c}\underbrace{({}_{SO:SI}DI,{}_{SI:SO}DI)}_{\text{decreasing-inflection saddle}}\\\underbrace{(SI,II:DI)}_{\text{inflection-sink}}\\\underbrace{(UP,LS)}_{\text{up-parabola lower-saddle}}\end{array}\right\}\text{ for }a_{110}>0,a_{220}<0,\tag{5.176}$$

$$\left\{\begin{array}{l}(a_{111},a_{222})\\(a_{111},a_{221})\\(a_{121},a_{211})\end{array}\right\}=\left\{\begin{array}{c}\underbrace{({}_{SO:SI}II,{}_{SO:SI}II)}_{\text{increasing-inflection saddle}}\\\underbrace{(SO,DI:II)}_{\text{inflection-source}}\\\underbrace{(DP,US)}_{\text{down-parabola upper-saddle}}\end{array}\right\}\text{ for }a_{110}<0,a_{220}<0.\tag{5.177}$$

The infinite-equilibrium of $x_2^* = a_{121} = a_{221}$ is the same as in Eqs. (5.166)–(5.169). The infinite-equilibrium of $x_1^* = a_{111} = a_{211}$ is summarized as follows:

$$\begin{array}{lll}(a_{111},\bar{x}_2)= & \underbrace{(SO,II)}_{\text{increasing-inflection source}} & \text{if }\bar{x}_2\in(-\infty,a_{121}),\\(a_{111},\bar{x}_2)= & \underbrace{(SI,DI)}_{\text{decreasing-inflection sink}} & \text{if }\bar{x}_2\in(a_{121},a_{221}),\\(a_{111},\bar{x}_2)= & \underbrace{(SI,II)}_{\text{increasing-inflection sink}} & \text{if }\bar{x}_2\in(a_{221},a_{222}),\\(\bar{x}_1,a_{221})= & \underbrace{(SO,II)}_{\text{increasing-inflection source}} & \text{if }\bar{x}_2\in(a_{222},\infty),\\\text{for }a_{110}>0,a_{220}>0; & & \end{array}\tag{5.178}$$

$$\begin{array}{lll}(a_{111},\bar{x}_2)= & \underbrace{(SI,DI)}_{\text{decreasing-inflection sink}} & \text{if }\bar{x}_2\in(-\infty,a_{121}),\\(a_{111},\bar{x}_2)= & \underbrace{(SO,II)}_{\text{increasing-inflection source}} & \text{if }\bar{x}_2\in(a_{121},a_{221}),\\(a_{111},\bar{x}_2)= & \underbrace{(SO,DI)}_{\text{decreasing-inflection source}} & \text{if }\bar{x}_2\in(a_{221},a_{222}),\\(\bar{x}_1,a_{221})= & \underbrace{(SI,DI)}_{\text{decreasing-inflection sink}} & \text{if }\bar{x}_2\in(a_{222},\infty),\\\text{for }a_{110}<0,a_{220}>0; & & \end{array}\tag{5.179}$$

$$\begin{aligned}
(a_{111},\bar{x}_2) &= \underbrace{(\mathrm{SO},\mathrm{DI})}_{\text{decreasing-inflection source}} && \text{if } \bar{x}_2 \in (-\infty, a_{121}),\\
(a_{111},\bar{x}_2) &= \underbrace{(\mathrm{SI},\mathrm{II})}_{\text{increasing-inflection sink}} && \text{if } \bar{x}_2 \in (a_{121}, a_{221}),\\
(a_{111},\bar{x}_2) &= \underbrace{(\mathrm{SI},\mathrm{DI})}_{\text{decreasing-inflection sink}} && \text{if } \bar{x}_2 \in (a_{221}, a_{222}),\\
(\bar{x}_1,a_{221}) &= \underbrace{(\mathrm{SO},\mathrm{DI})}_{\text{decreasing-inflection source}} && \text{if } \bar{x}_2 \in (a_{222}, \infty),\\
&\text{for } a_{110} > 0, a_{220} < 0;
\end{aligned} \tag{5.180}$$

$$\begin{aligned}
(a_{111},\bar{x}_2) &= \underbrace{(\mathrm{SI},\mathrm{II})}_{\text{decreasing-inflection sink}} && \text{if } \bar{x}_2 \in (-\infty, a_{121}),\\
(a_{111},\bar{x}_2) &= \underbrace{(\mathrm{SO},\mathrm{DI})}_{\text{decreasing-inflection source}} && \text{if } \bar{x}_2 \in (a_{121}, a_{221}),\\
(a_{111},\bar{x}_2) &= \underbrace{(\mathrm{SO},\mathrm{II})}_{\text{increasing-inflection source}} && \text{if } \bar{x}_2 \in (a_{221}, a_{222}),\\
(\bar{x}_1,a_{221}) &= \underbrace{(\mathrm{SI},\mathrm{II})}_{\text{increasing-inflection sink}} && \text{if } \bar{x}_2 \in (a_{222}, \infty),\\
&\text{for } a_{110} < 0, a_{220} < 0.
\end{aligned} \tag{5.181}$$

The inflection-source and sink infinite-equilibriums of $x_2^* = a_{222} = a_{122}$ and $x_1^* = a_{111} = a_{211}$ are the switching bifurcations of sink, source, and saddles with hyperbolic flows for $a_{122} \in (a_{221}, a_{222})$ with $a_{222} < a_{122}$ and for $a_{111} > a_{211}$ with $a_{111} < a_{211}$, as shown in Fig. 5.23. The horizontal inflection-sink and source infinite-equilibriums are on the top of the equilibrium series. The vertical and horizontal inflection-source and sink infinite-equilibriums are intersected to form an inflection-saddle for double-switching bifurcation in different directions.

B. Consider a dynamical system as

$$\begin{aligned}
\dot{x}_1 &= a_{110}(x_1 - a_{111})(x_2 - a_{121})(x_2 - a_{122}),\\
\dot{x}_2 &= a_{220}(x_1 - a_{211})(x_2 - a_{221})(x_2 - a_{222}),
\end{aligned} \tag{5.182}$$

and the corresponding first integral manifold for $a_{121} = a_{221}$ and $a_{122} = a_{222}$ with $a_{111} = a_{211}$ is

$$(x_1 - x_{10}) = \frac{a_{110}}{a_{220}}(x_2 - x_{20}). \tag{5.183}$$

The equilibrium matrices of $x_2^* = a_{121}, a_{122}, a_{221}, a_{222}$ with $x_1^* = a_{111}, a_{211}$ for $a_{121} = a_{221}$ and $a_{122} = a_{222}$ with $a_{111} = a_{211}$ are

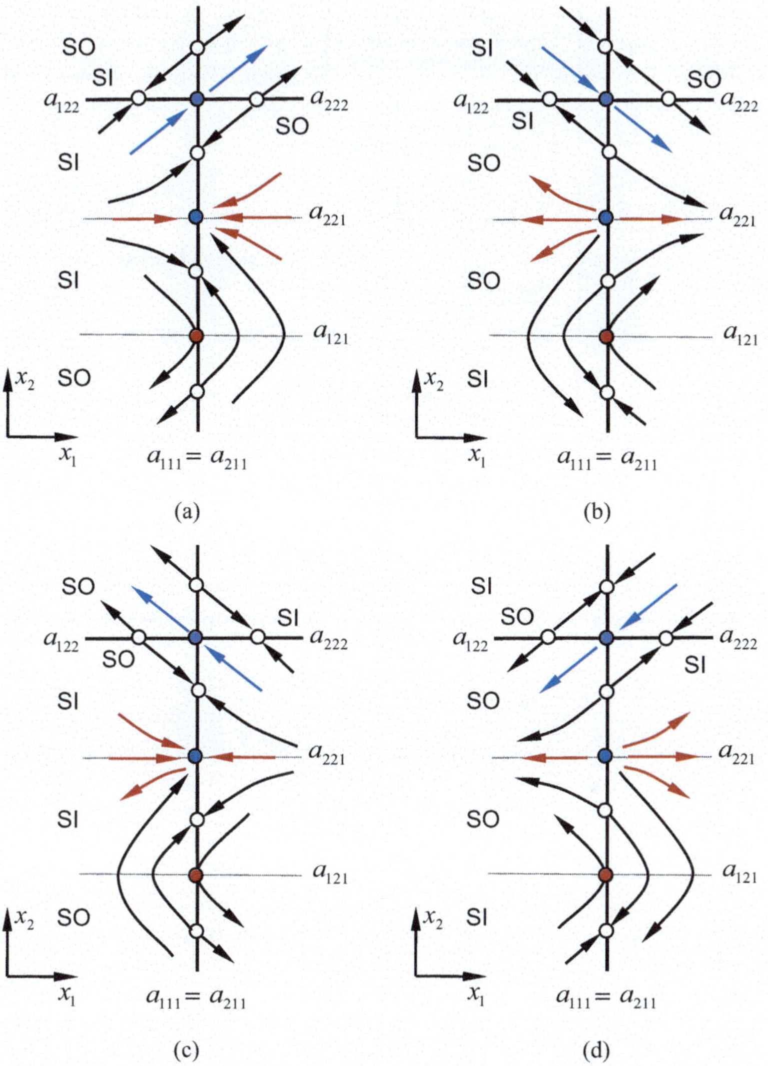

Fig. 5.23 Phase portraits ($a_{111} = a_{211}$ and $a_{122} = a_{222}$) for two-dimensional systems on the x_1-direction with $x_1^* = a_{111}$ and on the x_2-direction with $x_2^* = a_{121}, a_{221}, a_{222}$. The equilibrium flows: (**a**) ($a_{110} > 0$, $a_{220} > 0$), (**b**) ($a_{110} < 0$, $a_{220} > 0$), (**c**) ($a_{110} > 0$, $a_{220} < 0$), and (**d**) ($a_{110} < 0$, $a_{220} < 0$)

$$\left\{\begin{matrix}(a_{211}, a_{222}) \\ (a_{211}, a_{221})\end{matrix}\right\} = \left\{\begin{matrix}\underbrace{({}_{\text{SI:SO}}\text{II}, {}_{\text{SI:SO}}\text{II})}_{\text{increasing-inflection saddle}} \\ \underbrace{({}_{\text{SO:SI}}\text{II}, {}_{\text{SO:SI}}\text{II})}_{\text{increasing-inflection saddle}}\end{matrix}\right\}, \text{ for } a_{110} > 0, a_{220} > 0, \tag{5.184}$$

$$\left\{\begin{matrix}(a_{211},a_{222})\\(a_{211},a_{221})\end{matrix}\right\}=\left\{\begin{matrix}\underbrace{({}_{\text{SI:SO}}\text{DI},{}_{\text{SO:SI}}\text{DI})}_{\text{decreasing-inflection saddle}}\\\underbrace{({}_{\text{SO:SI}}\text{DI},{}_{\text{SI:SO}}\text{DI})}_{\text{decreasing-inflection saddle}}\end{matrix}\right\},\text{ for }a_{110}<0,a_{220}>0,\tag{5.185}$$

$$\left\{\begin{matrix}(a_{211},a_{222})\\(a_{211},a_{221})\end{matrix}\right\}=\left\{\begin{matrix}\underbrace{({}_{\text{SO:SI}}\text{DI},{}_{\text{SI:SO}}\text{DI})}_{\text{decreasing-inflection saddle}}\\\underbrace{({}_{\text{SI:SO}}\text{DI},{}_{\text{SO:SI}}\text{DI})}_{\text{decreasing-inflection saddle}}\end{matrix}\right\},\text{ for }a_{110}>0,a_{220}<0,\tag{5.186}$$

$$\left\{\begin{matrix}(a_{211},a_{222})\\(a_{211},a_{221})\end{matrix}\right\}=\left\{\begin{matrix}\underbrace{({}_{\text{SO:SI}}\text{II},{}_{\text{SO:SI}}\text{II})}_{\text{increasing-inflection saddle}}\\\underbrace{({}_{\text{SI:SO}}\text{II},{}_{\text{SI:SO}}\text{II})}_{\text{increasing-inflection saddle}}\end{matrix}\right\},\text{ for }a_{110}<0,a_{220}<0,\tag{5.187}$$

and the infinite-equilibrium of $x_2^* = a_{122} = a_{222}$ with $a_{111} = a_{211}$ is summarized as follows:

$$\begin{aligned}&(\bar{x}_1,a_{222})=\underbrace{(\text{II},\text{SI})}_{\text{increasing-inflection sink}}\quad\text{if }\bar{x}_1\in(-\infty,a_{111}),\\&(\bar{x}_1,a_{222})=\underbrace{(\text{II},\text{SO})}_{\text{increasing-inflection source}}\quad\text{if }\bar{x}_1\in(a_{111},\infty),\\&\text{for }a_{110}>0,a_{220}>0;\end{aligned}\tag{5.188}$$

$$\begin{aligned}&(\bar{x}_1,a_{222})=\underbrace{(\text{DI},\text{SI})}_{\text{decreasing-inflection sink}}\quad\text{if }\bar{x}_1\in(-\infty,a_{111}),\\&(\bar{x}_1,a_{222})=\underbrace{(\text{DI},\text{SO})}_{\text{decreasing-inflection source}}\quad\text{if }\bar{x}_1\in(a_{111},\infty),\\&\text{for }a_{110}<0,a_{220}>0.\end{aligned}\tag{5.189}$$

$$\begin{aligned}&(\bar{x}_1,a_{222})=\underbrace{(\text{DI},\text{SO})}_{\text{decreasing-inflection source}}\quad\text{if }\bar{x}_1\in(-\infty,a_{111})\\&(\bar{x}_1,a_{222})=\underbrace{(\text{DI},\text{SI})}_{\text{decreasing-inflection sink}}\quad\text{if }\bar{x}_1\in(a_{111},\infty),\\&\text{for }a_{110}>0,a_{220}<0;\end{aligned}\tag{5.190}$$

$$
\begin{aligned}
&(\bar{x}_1, a_{222}) = \underbrace{(\mathrm{II}, \mathrm{SO})}_{\text{increasing-inflection source}} &&\text{if } \bar{x}_1 \in (-\infty, a_{111})\\
&(\bar{x}_1, a_{222}) = \underbrace{(\mathrm{II}, \mathrm{SI})}_{\text{increasing-inflection sink}} &&\text{if } \bar{x}_1 \in (a_{111}, \infty),\\
&\text{for } a_{110} < 0, a_{220} < 0.
\end{aligned}
\tag{5.191}
$$

The infinite-equilibrium of $x_2^* = a_{121} = a_{221}$ with $a_{111} = a_{211}$ is summarized as follows:

$$
\begin{aligned}
&(\bar{x}_1, a_{221}) = \underbrace{(\mathrm{II}, \mathrm{SO})}_{\text{increasing-inflection source}} &&\text{if } \bar{x}_1 \in (-\infty, a_{111}),\\
&(\bar{x}_1, a_{221}) = \underbrace{(\mathrm{II}, \mathrm{SI})}_{\text{increasing-inflection sink}} &&\text{if } \bar{x}_1 \in (a_{111}, \infty),\\
&\text{for } a_{110} > 0, a_{220} > 0,
\end{aligned}
\tag{5.192}
$$

$$
\begin{aligned}
&(\bar{x}_1, a_{221}) = \underbrace{(\mathrm{DI}, \mathrm{SI})}_{\text{decreasing-inflection sink}} &&\text{if } \bar{x}_1 \in (-\infty, a_{111}),\\
&(\bar{x}_1, a_{221}) = \underbrace{(\mathrm{DI}, \mathrm{SO})}_{\text{decreasing-inflection source}} &&\text{if } \bar{x}_1 \in (a_{111}, \infty),\\
&\text{for } a_{110} < 0, a_{220} > 0;
\end{aligned}
\tag{5.193}
$$

$$
\begin{aligned}
&(\bar{x}_1, a_{221}) = \underbrace{(\mathrm{DI}, \mathrm{SO})}_{\text{decreasing-inflection source}} &&\text{if } \bar{x}_1 \in (-\infty, a_{111}),\\
&(\bar{x}_1, a_{221}) = \underbrace{(\mathrm{DI}, \mathrm{SI})}_{\text{decreasing-inflection sink}} &&\text{if } \bar{x}_1 \in (a_{111}, \infty),\\
&\text{for } a_{110} > 0, a_{220} < 0,
\end{aligned}
\tag{5.194}
$$

$$
\begin{aligned}
&(\bar{x}_1, a_{221}) = \underbrace{(\mathrm{II}, \mathrm{SI})}_{\text{increasing-inflection sink}} &&\text{if } \bar{x}_1 \in (-\infty, a_{111}),\\
&(\bar{x}_1, a_{221}) = \underbrace{(\mathrm{II}, \mathrm{SO})}_{\text{increasing-inflection source}} &&\text{if } \bar{x}_1 \in (a_{111}, \infty),\\
&\text{for } a_{110} < 0, a_{220} < 0.
\end{aligned}
\tag{5.195}
$$

The infinite-equilibrium of $x_1^* = a_{111} = a_{211}$ with $a_{121} = a_{221}$ and $a_{122} = a_{222}$ is summarized as follows:

$$
\begin{aligned}
&(a_{111}, \bar{x}_2) = \underbrace{(\mathrm{SO}, \mathrm{II})}_{\text{increasing-inflection source}} &&\text{if } \bar{x}_2 \in (-\infty, a_{121}),\\
&(a_{111}, \bar{x}_2) = \underbrace{(\mathrm{SI}, \mathrm{II})}_{\text{increasing-inflection sink}} &&\text{if } \bar{x}_2 \in (a_{121}, a_{122}),\\
&(a_{111}, \bar{x}_2) = \underbrace{(\mathrm{SO}, \mathrm{II})}_{\text{increasing-inflection source}} &&\text{if } \bar{x}_2 \in (a_{122}, \infty),\\
&\text{for } a_{110} > 0, a_{220} > 0,
\end{aligned}
\tag{5.196}
$$

$$
\begin{aligned}
&(a_{111}, \bar{x}_2) = \underbrace{(\mathrm{SI}, \mathrm{DI})}_{\text{decreasing-inflection sink}} \quad \text{if } \bar{x}_2 \in (-\infty, a_{121}),\\
&(a_{111}, \bar{x}_2) = \underbrace{(\mathrm{SO}, \mathrm{DI})}_{\text{decreasing-inflection source}} \quad \text{if } \bar{x}_2 \in (a_{121}, a_{122}),\\
&(a_{111}, \bar{x}_2) = \underbrace{(\mathrm{SI}, \mathrm{DI})}_{\text{decreasing-inflection sink}} \quad \text{if } \bar{x}_2 \in (a_{122}, \infty),\\
&\text{for } a_{110} < 0, a_{220} > 0,
\end{aligned}
\tag{5.197}
$$

$$
\begin{aligned}
&(a_{111}, \bar{x}_2) = \underbrace{(\mathrm{SO}, \mathrm{DI})}_{\text{decreasing-inflection source}} \quad \text{if } \bar{x}_2 \in (-\infty, a_{121}),\\
&(a_{111}, \bar{x}_2) = \underbrace{(\mathrm{SI}, \mathrm{DI})}_{\text{decreasing-inflection sink}} \quad \text{if } \bar{x}_2 \in (a_{121}, a_{122}),\\
&(a_{111}, \bar{x}_2) = \underbrace{(\mathrm{SO}, \mathrm{DI})}_{\text{decreasing-inflection source}} \quad \text{if } \bar{x}_2 \in (a_{122}, \infty),\\
&\text{for } a_{110} > 0, a_{220} < 0;
\end{aligned}
\tag{5.198}
$$

$$
\begin{aligned}
&(a_{111}, \bar{x}_2) = \underbrace{(\mathrm{SI}, \mathrm{II})}_{\text{increasing-inflection sink}} \quad \text{if } \bar{x}_2 \in (-\infty, a_{121}),\\
&(a_{111}, \bar{x}_2) = \underbrace{(\mathrm{SO}, \mathrm{II})}_{\text{increasing-inflection source}} \quad \text{if } \bar{x}_2 \in (a_{121}, a_{122}),\\
&(a_{111}, \bar{x}_2) = \underbrace{(\mathrm{SI}, \mathrm{II})}_{\text{increasing-inflection sink}} \quad \text{if } \bar{x}_2 \in (a_{122}, \infty),\\
&\text{for } a_{110} < 0, a_{220} < 0.
\end{aligned}
\tag{5.199}
$$

The three inflection-sink and source infinite-equilibriums of $x_2^* = a_{121} = a_{221}$ and $x_2^* = a_{122} = a_{222}$ are presented in Fig. 5.24.

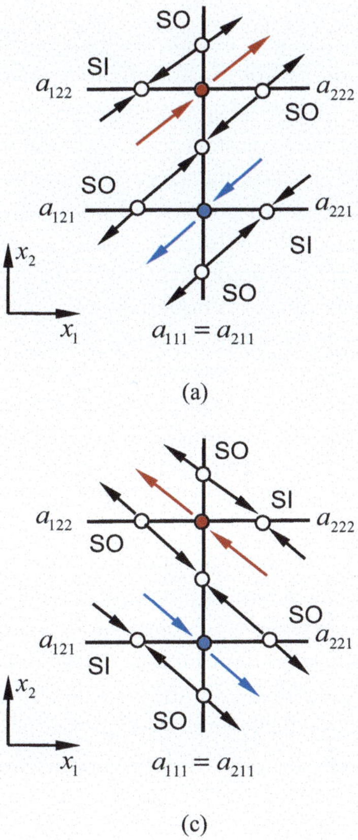

Fig. 5.24 Phase portraits ($a_{111} = a_{211}$, $a_{121} = a_{221}$, $a_{122} = a_{222}$) for two-dimensional systems on the x_1-direction with $x_1^* = a_{111} = a_{211}$ and on the x_2-direction with $x_2^* = a_{121} = a_{221}$, $x_2^* = a_{121} = a_{222}$. The equilibrium flows: (**a**) ($a_{110} > 0$, $a_{220} > 0$), (**b**) ($a_{110} < 0$, $a_{220} > 0$), (**c**) ($a_{110} > 0$, $a_{220} < 0$), and (**d**) ($a_{110} < 0$, $a_{220} < 0$)

Index

A. C. J. Luo, *Two-dimensional Two-product Cubic Systems, Vol. I*,
https://doi.org/10.1007/978-3-031-48487-2

SPRINGER NATURE

GPSR Compliance

The European Union's (EU) General Product Safety Regulation (GPSR) is a set of rules that requires consumer products to be safe and our obligations to ensure this.

If you have any concerns about our products, you can contact us on ProductSafety@springernature.com

In case Publisher is established outside the EU, the EU authorized representative is:

Springer Nature Customer Service Center GmbH
Europaplatz 3
69115 Heidelberg, Germany

The manufacturer's authorised representative in the EU is Springer Nature Customer Service Centre GmbH, Europaplatz 3, 69115 Heidelberg, Germany. If you have any concerns regarding our products, please contact ProductSafety@springernature.com

Printed and bound by CPI Group (UK) Ltd, Croydon, CR0 4YY
01/10/2026
02225965-0011